Embracing Analytics in the Drinking Water Industry

Embracing Analytics in the Drinking Water Industry

Edited by
Juneseok Lee and Jonathan Keck

Published by   **IWA Publishing**
Unit 104–105, Export Building
1 Clove Crescent
London E14 2BA, UK
Telephone: +44 (0)20 7654 5500
Fax: +44 (0)20 7654 5555
Email: publications@iwap.co.uk
Web: www.iwapublishing.com

First published 2022
© 2022 IWA Publishing

British Library Cataloguing in Publication Data
A CIP catalogue record for this book is available from the British Library

ISBN: 9781789062373 (Paperback)
ISBN: 9781789062380 (eBook)
ISBN: 9781789062397 (ePub)

This eBook was made Open Access in June 2022

Contents

Chapter 4

Water demand forecasting | time series data 75
Jarai Sanneh, A. Di Mauro, S. Venticinque, G. F. Santonastaso, A. Di Nardo,
Yi Wang and Juneseok Lee

Chapter 5

Use of cost-benefit analysis (CBA) in water infrastructure 123
Anita M. Chaudhry

Chapter 6
Water quality modeling and analysis *135*
Maria A. Palmegiani and Juneseok Lee

Chapter 7
Calibration and uncertainty analysis of hydraulic models *159*
Adell Moradi Sabzkouhi, Juneseok Lee and Jonathan Keck

Chapter 8
Optimal pump operation **187**
Adell Moradi Sabzkouhi, Juneseok Lee and Jonathan Keck

Chapter 9
Hydraulic transients in pipe systems
Juneseok Lee, Lu Xing and Lina Sela

Chapter 10
Innovative methods for optimal design of water network partitioning
Armando Di Nardo and Giovanni Francesco Santonastaso

Chapter 11
Reliability analysis using optimization *255*
Sangamreddi Chandramouli

Chapter 14
Water mains replacement decision using GIS analytics **335**
Diego Martinez Garcia

Chapter 15
Decision Analysis **381**
Eftila Tanellari and Juneseok Lee

Chapter 16
Non-revenue water, what are their determinants? ***399***
Gamze Güngör-Demirci and Juneseok Lee

Chapter 17
Water utility performance measurements using data envelopment analysis ***409***
Gamze Güngör-Demirci and Juneseok Lee

doi: 10.2166/9781789062380_xv

Preface

The purpose of this book is to introduce '*analytics*' to practicing water engineers so that they can incorporate the covered subjects, approaches, and detailed techniques within their daily operations, management, and decision-making processes. Also, undergraduate students as well as early graduate students who are in water and environmental systems concentration areas will be exposed to established analytical techniques, along with many methods that are currently considered to be new or emerging and maturing.

This book covers a broad spectrum of water industry analytics topics in an easy-to-follow manner. The overall background and context are motivated by (and directly drawn from) actual water utility projects that we have worked on over numerous recent years. Many chapter authors are the editor's previous students and collaborators that have worked together. We strongly believe that the water industry should embrace and integrate data-driven fundamentals and methods into their daily operations and decision-making process(es) in an effort to replace more traditional and established 'rule-of-thumb' and (arguably) weaker heuristic approaches – and an analytics viewpoint, approach, and culture is key to this industry transformation. Analytics can support numerous aspects of water utility planning, operations, and management, and the organization of this book naturally follows pace by including three principal sections – planning, operations, and management.

Water is essential for human well-being and survival, and throughout the water industry, it is becoming increasingly imperative that in-house analytics capability and championship be developed and integrated to address the current and transitional challenges we face. Again, one of our main contentions is that analytics will contribute substantially to future efforts aimed at providing innovative solutions that make the water industry more sustainable and resilient. We sincerely hope that this book provides a range of learning experiences that help to share and expand this view.

Juneseok Lee, Editor
Manhattan College
Jonathan Keck, Editor
Water First, LLC

doi: 10.2166/9781789062380_0001

Chapter 1

Introduction

*Jonathan Keck[1] and Juneseok Lee[2]**

[1]*Founder/Principal, Water First, LLC, Naperville, IL*
[2]*Department of Civil and Environmental Engineering, Manhattan College, Riverdale, NY 10471*
**Corresponding author: juneseok.lee@manhattan.edu*

Two decades into the 21st century, the water industry landscape is going through a major transformation brought about by the confluence of a number of powerful forces, including: (1) exposure to an increasingly complex and interdependent set of regulations and standards; (2) challenges in climate, environmental, and socio-economic patterns and processes (including citizen expectations); and (3) growing computational capacities paired with the accumulation of large amounts of performance data (from cheaper and more distributed sensors) coinciding with the fourth industrial revolution (IR4) of the internet of things (IoT), and data analytics. **We strongly believe that water industry needs a paradigm shift that is commensurate with these rapid transformations**.

Recent advances in **analytics** have the potential to fundamentally impact water industry planning, operations, and maintenance processes, particularly in complex interdependent infrastructure systems. Advanced analytics can be used to holistically identify and address problems at the system(s) level. This approach is particularly desirable in the case of complex infrastructure projects with multiple interdependent and interacting components. Successful system identification relies on the availability of abundant data for training algorithms such as artificial neural networks. Understanding data structures and the systematic storage and classification of data, particularly in the context of advanced data analytics/science methods such as machine learning (ML) and artificial intelligence (AI), are crucial skillsets that will be in high demand.

1.1 WHAT IS ANALYTICS?

Analytics is the process by which meaningful insights are extracted from available data. While *analysis* refers to the process itself, *analytics* includes the science behind the analysis and all the steps that precede (data needs, data collection, etc.) and follow (recommendations, implications, etc.) the analysis. The deep insights gained through analytics are primarily used for decision support, that is, recommending specific policies or actions. Analytics has evolved over the years from *descriptive* (What has happened?) to *diagnostic* (Why did it happen?) to *predictive* (What could happen?) to *prescriptive* (What action could be taken to promote/preempt a particular outcome?) (Keck & Lee, 2021). As many researchers and industry leaders have noted (see, e.g., Chastain-Howley, 2018; Karl and Wyatt, 2018; Lunani, 2018), the next significant paradigm shift will be towards *cognitive* analytics, which will exploit recent advances in high-performance computing (HPC) by combining AI and ML techniques. In particular, Karl and Wyatt (2018) pointed out that industries are reviewing or using less than 10% of their data, often overlooking key insights and opportunities to become more efficient in terms of operations and management. They concluded that society would benefit from the greater use of analytics to transform data into systems-level and actionable intelligence.

To cope with existing and emerging problems more effectively, our 21st-century infrastructure and quality of life goals and challenges demand a paradigm shift towards innovative approaches. According to the Engineer's Creed (first adopted by the National Society of Professional Engineers in June 1954), professional engineers should dedicate their professional knowledge and skill to the advancement and betterment of human welfare. This is, of course, especially true for water engineers who deal with our fundamental infrastructure, as these systems have a direct and significant impact on public safety, health, and welfare.

1.2 HOW CAN ANALYTICS HELP THE WATER INDUSTRY?

With sensors becoming less expensive and ubiquitous, many of the nation's water infrastructure elements are now being monitored in real-time, with vast amounts of data being collected. To augment this data, end-to-end simulations are being developed (e.g., digital twins) that have the predictive power to characterize region-wide performance of various systems under rare events for which observational data does not exist. These extensive datasets are waiting to be mined by system condition diagnosis tools that can be used to prioritize, plan, and carry out mitigative actions, including repairs and replacements, with sustainability and resilience becoming core objectives.

Drinking water industries protect public health and improve social wellbeing by operating and maintaining water infrastructure to provide safe and reliable water to customers. Having a better understanding of causality in drinking water infrastructure systems can help utilities and the entire water industry address gaps in the knowledge base and identify research needs. We strongly believe that analytics can support many aspects of drinking water industry planning, operations, and management. We also believe it is imperative that water utilities have in-house analytics championship as well as capacity to be integrated into their daily work to face the emerging challenges in the drinking water industry. In this vein, analytics will contribute significantly to providing innovative solutions toward more sustainable and resilient water industries. Therefore, it is critical that our drinking water industry adopt and integrate water-centered analytics practices, culture, and perceptions in-house. And finally, we strongly believe that the opportunity cost of *not keeping up* with these new industry trends will be extremely high in terms of missed opportunities for better systems management and improved public health and safety.

1.3 EFFECTIVE UTILITY MANAGEMENT

In May of 2006, the Association of Metropolitan Water Agencies (AMWA), the American Public Works Association (APWA), the American Water Works Association (AWWA), the National Association of Clean Water Agencies (NACWA), the National Association of Water Companies (NAWC), the United States Environmental Protection Agency (USEPA), and the Water Environment Federation (WEF) all entered into a *Statement of Intent* to 'formalize a collaborative effort among the signatory organizations in order to promote effective utility management'. These 'Collaborating Organizations' chartered the Effective Utility Management Steering Committee (Committee) to advise them on a future joint water utility sector management strategy applicable to water sector utilities across the country. The Committee found that water sector utilities across the country face numerous common challenges, such as rising costs and workforce complexities, and need to focus attention on these areas to deliver quality products and services and sustain community support. Within this context, the Committee identified four primary building blocks of effective water utility management, which would later become the basis of a future water utility sector management strategy. These foundational elements are listed next, and also described in more detail below: (1) Attributes of Effectively Managed Water Sector Utilities; (2) Keys to Management Success; (3) Water Utility Measures, and; (4) Water Utility Management Resources (USEPA, 2007).

1.3.1 Foundational element #1 – attributes of effectively managed water sector utilities

The Committee identified 'Ten Attributes of Effectively Managed Water Sector Utilities' (Attributes) that provide a focused overview of where effectively managed utilities should be active, and what they should strive to achieve. Further, the Committee recommended that the water utility sector adopt and utilize these Attributes as a basis for promoting improved management within the sector. The Ten Attributes further detailed in Table 1.1 are as follows: (1) Product Quality; (2) Customer Satisfaction; (3) Employee Leadership and Development; (4) Operational Optimization; (5) Financial Viability; (6) Operational Resilience; (7) Community Sustainability; (8) Infrastructure Stability; (9) Stakeholder Understanding and Support, and; (10) Water Resource Adequacy. The Ten Attributes can be viewed as a continuum of management improvement opportunities, and are not listed in any particular order, since utility managers will determine their relative and weighted importance and applicability based on individual utility circumstances (USEPA, 2017).

1.3.2 Foundational element #2 – keys to management success

As a complement to the Ten Attributes, the Committee also identified five 'Keys to Management Success', which are considered to be approaches and systems that foster and continually support utility management success. The Committee recommended that the Keys to Management Success be referenced and promoted with the Attributes to enable more effective utility management within the sector.

1.3.2.1 Leadership

Leadership plays a critical role in effective utility management, particularly within the context of driving and inspiring change within an organization. In this context, the term 'leaders' refers to both individuals who champion improvement, and also to leadership teams that provide resilient, day-to-day oversight, management continuity, and direction. Effective leadership ensures that the utility's direction is understood, embraced, and followed on an ongoing basis throughout the management cycle.

1.3.2.2 Strategic business planning

Strategic business planning helps utilities balance and drive integration and cohesion across the Attributes. It involves taking a long-term view of utility goals and operations and establishing an explicit vision and mission that guide utility objectives, measurement efforts, investments, and operations.

1.3.2.3 Organizational approaches

A variety of organizational approaches can be critical to management improvement. These approaches include establishing a 'participatory organizational culture', which seeks to actively engage employees in improvement efforts, deploys an explicit change management process, and uses implementation strategies that seek early, stepwise victories to build momentum and motivation.

1.3.2.4 Measurement

A focus and emphasis on measurement is the backbone of successful continual improvement in management and strategic business planning. Successful measurement efforts are reasonably viewed on a continuum, starting with basic internal tracking.

Table 1.1 Ten attributes of effectively managed water sector utilities.

Product quality	Customer satisfaction
Produces potable water, treated effluent, and process residuals in full compliance with regulatory and reliability requirements and consistent with customer, public health, and ecological needs	Provides reliable, responsive, and affordable services in line with explicit, customer-accepted service levels. Receives timely customer feedback to maintain responsiveness to customer needs and emergencies
Employee and leadership development	**Operational optimization**
Recruits and retains a workforce that is competent, motivated, adaptive, and safe-working. Establishes a participatory, collaborative organization dedicated to continual learning and improvement. Ensures employee institutional knowledge is retained and improved upon over time. Provides a focus on and emphasizes opportunities for professional and leadership development and strives to create an integrated and well-coordinated senior leadership team	Ensures ongoing, timely, cost-effective, reliable, and sustainable performance improvements in all facets of its operations. Minimizes resource use, loss, and impacts from day-to-day operations. Maintains awareness of information and operational technology developments to anticipate and support timely adoption of improvements
Financial viability	**Operational resiliency**
Understands the full life-cycle cost of the utility and establishes and maintains an effective balance between long-term debt, asset values, operations and maintenance expenditures, and operating revenues. Establishes predictable rates that are consistent with community expectations and acceptability, and are adequate to recover costs, provide for reserves, maintain support from bond rating agencies, and plan and invest for future needs.	Ensures utility leadership and staff work together to anticipate and avoid problems. Proactively identifies, assesses, establishes tolerance levels for, and effectively manages, a full range of business risks (including legal, regulatory, financial, environmental, safety, security, and natural disaster-related) in a proactive way consistent with industry trends and system reliability goals
Community sustainability	**Infrastructure stability**
Is explicitly cognizant of and attentive to the impacts its decisions have on current and long-term future community and watershed health and welfare. Manages operations, infrastructure, and investments to protect, restore, and enhance the natural environment; efficiently use water and energy resources; promote economic vitality; and engender overall community improvement. Explicitly considers a variety of pollution prevention, watershed, and source water protection approaches as part of an overall strategy to maintain and enhance ecological and community sustainability	Understands the condition of and costs associated with critical infrastructure assets. Maintains and enhances the condition of all assets over the long-term at the lowest possible life-cycle cost and acceptable risk consistent with customer, com- munity, and regulator-supported service levels, and consistent with anticipated growth and system reliability goals. Assures asset repair, rehabilitation, and replacement efforts are coordinated within the community to minimize disruptions and other negative consequences
Stakeholder understanding and support	**Water resource adequacy**
Engenders understanding and support from over- sight bodies, community and watershed interests, and regulatory bodies for service levels, rate structures, operating budgets, capital improvement programs, and risk management decisions. Actively involves stakeholders in the decisions that will affect them	Ensures water availability consistent with cur- rent and future customer needs through long-term resource supply and demand analysis, conservation, and public education. Explicitly considers its role in water availability and manages operations to provide for long-term aquifer and surface water sustainability and replenishment

1.3.2.5 *Continual improvement management framework*

A 'plan, do, check, act' (PDCA) continual improvement management framework typically includes several components, such as conducting honest and comprehensive self-assessments, establishing explicit performance objectives and targets, implementing measurement activities, and responding to evaluations through the use of an explicit change management process (Figure 1.1).

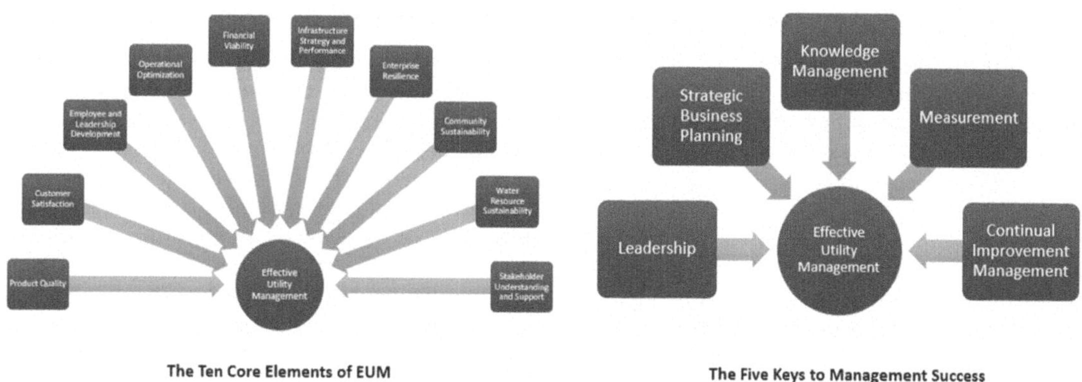

The Ten Core Elements of EUM   **The Five Keys to Management Success**

Source: Adapted from EUM Workgroup, 2017

Figure 1.1 Ten attributes and five management keys of effectively managed water sector utilities.

1.3.3 Foundational element #3 – water utility measures

The Committee strongly affirmed measurement as a critical element of effective utility management. The Committee also noted that utility measurement is complicated and needs to be done carefully in order to be useful. The challenges presented by performance measurement include deciding what to measure, identifying meaningful measures, and making sure that data is collected in such a way as to support meaningful analyses and comparisons. Consideration of these factors is important if the data are to be used to make real improvements and to communicate accurate information. Careful scrutiny here also helps to ensure that the resulting information is interpreted correctly.

Within this context, the Committee identified a set of high-level, illustrative example water utility measures related to the Ten Attributes, and recommended that, to get started on simple terms, these or similar utility measures become part of a first-level assessment. These preliminary example measures included, for instance, under *Operational Optimization*, the amount of distribution system water loss, while under *Operational Resiliency*, whether the utility has in place a current all-hazards disaster readiness response plan (yes/no?). A further example under *Stakeholder Understanding and Support*, includes whether the utility regularly consults with stakeholders (yes/no?). The Committee also recommended a longer-term initiative to identify a cohesive set of targeted, generally applicable, individual water sector utility measures. The goal would be to provide robust measures for individual utilities to use in gauging and improving operational and managerial practices and for communicating with external audiences such as boards, rate payers, and community leaders.

1.3.4 Foundational element #4 – water utility management resources

Based on the overall findings of the Statement of Intent Workshop, the Committee believed that water utilities are interested in tools that can support management progress, and that many utilities would benefit from a 'helping hand' that can guide them to useful management resources, particularly in the context of the Attributes. Therefore, the Committee recommended that the future sector strategy include a 'resource toolbox' linked to the Attributes and submitted a preliminary list of management resources that could be used as a starting point. One of the key deliverables in this regard was to develop a 'primer' to help utility managers understand the background and objectives of the initiative and help them use the Attributes and apply the Keys to Management Success.

1.4 EFFECTIVE UTILITY MANAGEMENT (EUM) AND WATER ANALYTICS

Water utilities protect public health and improve social well-being by operating and maintaining drinking water infrastructure to provide safe and reliable water to customers. Having a better understanding of causality in drinking water infrastructure systems can help utilities and the entire water industry address gaps in the knowledge base and identify research needs. Williams (2013) introduced the term 'information engineering' in water management – that is, the holistic application of information technology (IT) to the water industry via integration of data and optimization. Neemann *et al.* (2013) emphasized the importance of transforming data into information, then into knowledge and wisdom, which will have a large strategic impact on the utility as well as customers. The authors also recommended that utilities start by identifying business domains that increase insights that can yield high value and return on investment. A strong EUM viewpoint and orientation, combined with knowledge and appreciation of the power of water analytics, clearly shows that analytics has the potential to enhance *all* of the important aspects of EUM. Having stated this, a handful of domain areas are highlighted below in order to provide examples and illustrative detail.

1.4.1 Supply and demand management

When applying analytics to automated metering infrastructure to establish demand characterization and management strategies, the basic objective has been to understand the factors driving water demand in conjunction with conservation and sustainability goals (e.g., incentive programs), along with making reliable forecasts. However, this barely scratches the surface of what is possible – internal information about customer demand as well as data from utility commissions, state and local data repositories, local boards, and other stakeholders can also be used (added) to develop more robust local and regional models that can better predict future service levels over wider scales, thus providing greater insight into the hydrologic, socio-economic, and infrastructure performance dependencies naturally present in many of our more developed cities and regions. Relative to these regional – and even national or world-wide water supply questions – *block-chain technology* has the ability to support a far-reaching and secure transactional ecosystem around water rights, allocations, and transfers, and can even help to better illustrate 'true' resource quality and availability by virtue of its underlying distributed design and ledger transparency (The Water Network, 2020; Zuckerman, 2018). Analytics can also be used to shed new light on a broad spectrum of nonrevenue water issues in conjunction with a number of asset management and modeling applications that are explored in the following sections.

1.4.2 Enterprise asset management

According to the 2021 *State of the Water Industry Report* prepared by AWWA, aging infrastructure is the most critical challenge facing the water industry, followed by financing for capital improvements, long-term water supply availability, emergency preparedness, and a host of other concerns related to utility/system integrity as well as public views and outreach. Analytics can be used to improve the understanding of key physical processes related to water utility system integrity, including performance-driven screening and assessment (e.g., capacity, efficiency, and level of service), failure modes and effects (e.g., mortality and outage consequence), operations and maintenance, risk identification and characterization, and capital investment allocation and prioritization. Performance management is particularly crucial because it encompasses every aspect of a utility's asset management program, typically defined by the quantity, quality, and reliability levels achieved, along with short- and long-term environmental standards. A strong analytics-based understanding in these areas will lead to better life-cycle planning, analysis, design, and operational decision-making because of improved business/enterprise intelligence. Given that asset management activities generally entail sizable amounts of transactional data (travel, works orders and repair activities, invoices, etc.), here again a future move to block-chain technology can (conceptually) yield many of the same data architectural benefits noted above for water management (though in this case, through asset-activity tracking and linking, in addition to ledger transparency). This overall tracking and linking construct will also support improved life-cycle cost accounting, auditing, and other forms of corporate/organizational governance.

1.4.3 Distribution system modeling

Analytics can also support hydraulic, energy, and water quality modeling in a multitude of ways. Many of these ultimately link to a powerful and granular data ecosystem built upon pressure and water quality surveys, surface and groundwater reservoir profiling, pump tests and energy audits, district metering areas (DMA) and other forms of subzone monitoring, SCADA, and advanced metering infrastructure (AMI), and so on. with the following benefits:

- A greater ability to develop systems-level integrated views of environmental boundary conditions, control inputs, dynamic stresses and loading, and resulting system behavior;
- More effective planning, deployment, and implementation of pressure management, leak detection, and water quality monitoring programs – say through sensor placement and central event management (CEM) platforms;
- Improved capacity to more effectively manage system-wide energy consumption and efficiency (intensity), as well as water quality. Advanced analytics, when lock-stepped with robust modeling and optimization processes, can support 'a new era' relative to distribution system energy and water quality management systems (EWQMS);
- Improved emergency planning, response, and recovery – say through extended period simulation (EPS) of flow and pressure, along with source tracing and other forms of water age and quality forecasting;
- Better business risk assessments linked to improved estimations of likelihood of failure (LOF) and consequence of failure (COF). More specifically, well-calibrated hydraulic models now enable rich assessments of network outages, thus adding a much-needed layer of dynamic and operational insight to risk characterizations that have (to date) not considered the full hydraulic and water quality impacts of network failure;
- More robust, streamlined, and accurate processes to create, calibrate, validate, and maintain system models, which ultimately lead to wider application and higher confidence in modeling program outcomes.

In addition, real-time modeling provides a continuous baseline to facilitate operational optimization decisions as well as troubleshoot and reconcile problems, while SCADA data can support a more-or-less continuous form of model calibration/validation. Juxtaposing these two considerations leads to the now well-known 'digital twin'. In the short term, this can simply help utilities better characterize and observe assets and their performance (through formalized and programmatic linkages to asset management), while in the long term, the digital twin framework can be used to optimize broad and high-impact enterprise programs like energy and water quality management, water loss, and capital investment, renewal, and prioritization. Such an approach will also make it possible for decision makers to account for a broader set of value-engineering factors when considering topics such as long-term capital expenditures, emergency response planning, and level-of-service definitions and metrics.

1.4.4 Long-range planning

It is beneficial to establish a formal system to analyze and optimize the underlying decision space of a project – the span of options that go into a utility's long-range and enterprise-level planning portfolios and submittals. Doing so will increase opportunities to rationally plan, while also making the best use of capital and operational projects and programs. Successful long-range planning programs generally encompass the following:

- Holistic knowledge and vision of resource availability, customer demands, water and energy supply portfolio attributes, product quality and quality control levers, operational characterizations, energy use, and carbon footprint considerations;
- Frameworks and programs for project planning, justifications, approval, design, and delivery;
- Asset management information that supports life-cycle cost–benefit analyses, including programmatic repair and replacement programs, as well as risk control;
- Financial considerations such as rate design and advanced budgets;
- Considerations of customer service and industry reputation.

Other issues to consider in long-range planning are formal regulatory criteria (including emerging regulations and legislation), non-regulatory criteria (which still should consider best practice and technology), enterprise goals and mandates, triple-bottom-line considerations, customer confidence, affordability, environmental considerations (including climate variation), and infrastructure and utility level resilience. From this starting point, there are at least five dimensions where an analytics viewpoint and approach can both drive, and positively affect, long-range planning outcomes:

- A resulting need for rigorous problem formulation and structure;
- Formalized and standardized goals, objectives, constraints, and analytical processes;
- Improved articulation and transparency around governing assumptions, processes, and results;
- More powerful and efficient means of confronting large decision spaces, as well as solving the technical and computational challenges associated with them (i.e., creating and assessing options – lots of them);
- An enhanced ability to perform sensitivity analyses, which produces a deeper understanding of underlying or embedded trade-offs, as well as a greater appreciation of the range of outcomes and potential impacts that accompany current and future decisions and actions.

1.4.5 Systems optimization

Modeling as previously described can be enlarged and synthesized using an analytics perspective to include systems-level multi-objective problem definitions that balance the cost of investment against the net benefits gained to establish effective prioritization models. To do this, it is first necessary to clearly define level of service goals, assumptions, and key performance indicators, all of which necessarily include a careful consideration of reliability, customer satisfaction, and other strategic variables. A vastly improved organizational arrangement of water utility IT systems, which can often be highly fragmented, can help to streamline the many disparate databases, systems, and processes involved in operating the water utility's system. The important step of establishing a data-driven objective and constraint model, the utility's common operating picture or framework, will first augment and then slowly replace various aspects of 'ad-hoc' and 'rule-of-thumb' engineering judgments that currently drive utility decision-making. Over time, this will allow water distribution systems to operate at greater levels of efficiency and with higher levels of confidence and transparency (Figure 1.2).

1.5 RECOMMENDATIONS

To create the conditions necessary for water utilities to fully implement analytics and maximize their associated benefits, actions in the following areas are recommended (Figure 1.3).

1.5.1 Analytics leadership

Exemplary enterprise-level analytics requires leadership, which should start at the highest levels of the organization, for example, board and council members, C-suite representatives, department heads, and directors. Analytics leadership should have, or take the form of, articulating and adopting a strong and explicit charter or mission statement that underscores the value of data, and the utility's long-term commitment to use data within the context of decision-making. In some water organizations, it

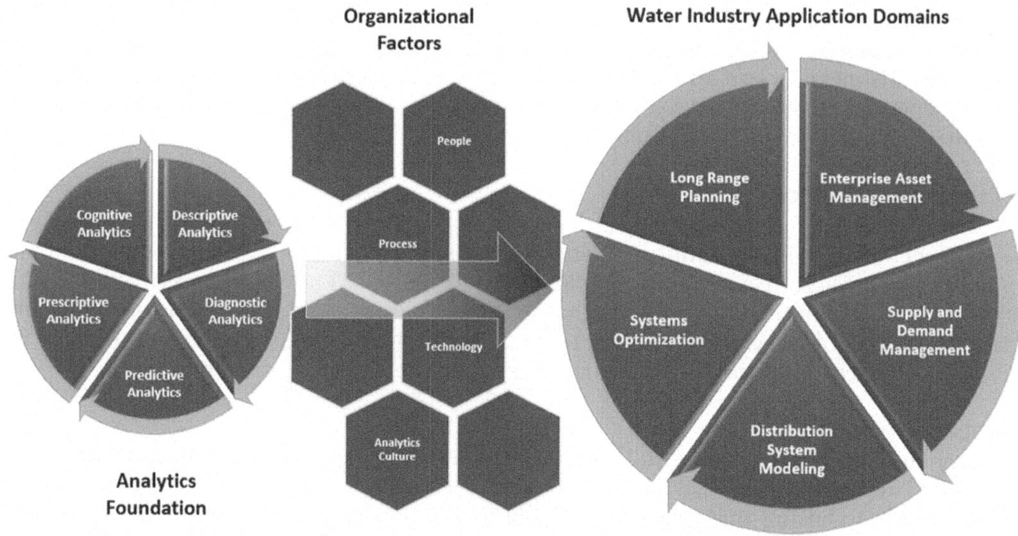

Figure 1.2 Water analytics and effective utility management.

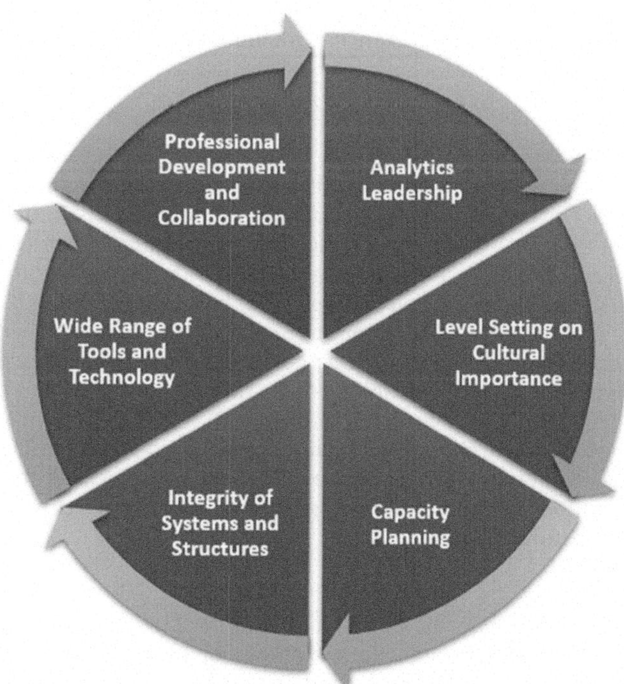

Figure 1.3 Utility planning and capacity areas for water analytics within EUM.

may even make sense to designate a chief data officer (CDO) or chief analytics officer (CAO), whose supervisory mandate spans engineering, information technology (IT), and operational technology (OT), to carry this message and array of tasks. Finally, a sustained managerial commitment to these charter elements and other day-to-day analytics principles should be fully evident and should permeate all divisions, departments, and groups within the utility.

1.5.2 Cultural importance

A second significant building block, which ultimately ties to analytics leadership, is helping people at all levels and functions within the utility understand the importance of data, data integrity, and the value/ability of being able to extract insights out of data – otherwise known as level setting on cultural importance. Once a cultural/organizational norm of this nature is set, other (downstream) efforts around capacity planning, system structure, tool and skill set choices, and so on. will become more congenial and efficient by virtue of this common viewpoint and frame of reference.

1.5.3 Capacity planning

A third key building block to more fully embrace an analytics culture within water utilities rests on planning. This view means that utilities should periodically review their 'people, process, and technology' chain to ensure that their overall suite/foundation of analytics architecture, processes, tools and technology, and skill sets are of sufficient bandwidth, and also properly link to mission-centric outcomes in both current and forecasted settings (goals identification, process mapping, and needs assessment). This effort will ultimately identify functional areas where a stronger analytics view can unlock additional value, while also helping to find duplicate processes and capacities that can be suitably consolidated to make them more efficient, and without loss of performance. The enterprise analytics planning effort is also an ideal place where analytics leadership tenants can be reinforced and deployed in both current and go-forward settings, while also (simultaneously) maintaining a consistent cultural message about the importance of an analytics orientation being an integral part of the utility's future.

1.5.4 Systems and structure

A fourth key building block to more fully embracing an analytics culture within water utilities rests on recordkeeping, appropriate systems analysis, and timely renewal of facilities. To instill confidence in methods used to assess risk and plan for sustainable programs, institutional structures should ensure data management integrity, that is, data collection, processing, interpretation, and integration, that establishes a coherent database. Data management standards and protocols must be set and maintained at all levels, including in the field, office, and laboratory, along with appropriate-cost data acquisition procedures. This requires regular communications across departments to improve overall data flow and maintain a consistent data structure and architecture. With suitable analytics protocols applied, accumulated data should yield valuable insights that facilitate better predictions and support logical decisions. Also, technical as well as non-technical staff will benefit from a better understanding of the overall data ecosystem and architecture, including any downstream and case-specific decision-modeling sensitivity. Finally, network and database cyber security concerns and factors should figure prominently here, and right-sized mitigation responses should be thoroughly woven into any and all subsequent systems architecture efforts.

1.5.5 Tools and technology

Tools and technology are a fifth major building block of an analytics culture and orientation within water utilities. More specifically, through an analytics capacity planning and needs assessment exercise, utilities must determine which core tools it will be using so that it can align this array against current and future skill sets and training expectations, data systems and structures, hosting and dissemination architecture, computational power, as well as rights, permissions, owners, and gatekeepers. Considerations of day-to-day as well as long-term maintenance of this toolbox and software stack should also figure prominently in the selection and stand-up process.

1.5.6 Professional development and collaborative research

Finally, linking back to the norm of organizational importance, in order to establish in-house analytics capabilities and champions, it is vitally important to provide professional development opportunities with regard to analytics training. Having sufficiently trained staff will help utilities more effectively incorporate analytics elements into their culture and operations. In particular, collaboration with university and laboratory researchers, regulatory representatives, and other technical and professional organizations (both public and private) is often rewarding, and therefore strongly recommended. In addition, the outreach to (and inclusion of) young professionals (YP) within a utility-analytics culture is also vitally important, as YPs are often 'early adopters' and 'profound innovators' within the overall analytics and data science realm(s), and they also constitute the next generation of water industry practitioners. Collectively, collaborations, such as the ones outlined here, enable industry representatives across a range of backgrounds and experience levels to work together to explore issues facing water utilities, while also improving the means with which to develop tangible and deployable technology (Keck & Lee, 2015).

1.6 A CLEAR FUTURE FOR ANALYTICS

Analytics can support numerous aspects of water utility planning and operations. Throughout the water industry it is becoming increasingly imperative that in-house analytics capability and championship be developed and integrated to address the current and transitional challenges we face. Analytics will contribute substantially to future efforts aimed at providing innovative solutions that make the water industry more sustainable and resilient.

1.7 ROADMAP OF THE BOOK

This book is composed of 17 chapters categorized into three sections: Planning, Operations, and Management. The Planning section covers Chapters 2–5, the Operations section covers Chapters 6–12, and the Management section covers Chapters 13–17.

1.7.1 Planning section

The planning section covers the context of water demand management as well as cost-benefit analysis for water infrastructure. Specifically, in Chapter 2, 'Water Demand Analysis | Regression', Tanverakul discusses advanced regression analysis to explore the relationships between water demand and their influencing factors. Water supply and demand problems, and their solutions, are often rife with unique challenges involving many aspects of hydraulics, environmental science, socioeconomics, finance, laws and regulations, and politics. Because water is difficult and expensive to transport, available water sources are often relatively near their users and tied to local conditions such as local climate and level of treatment necessary. Modeling water demand is modeling human behavior by evaluating how water use is influenced by user characteristics and various external factors like weather, price, or other constraints. Also, future water demand estimates are key inputs in water resources planning and management. Ensuring a sufficient and reliable volume of water is available to meet demand is a core function of all water suppliers and distributors. Accurate future forecasts are critical since water supply availability is highly variable and water infrastructure projects, often large and expensive, are designed and constructed with long useful lives typically upwards of 50+ years. For these reasons, the ability to make accurate future water demand estimates has long-term consequences. Regression is a popular and well-demonstrated choice and has been chosen for this discussion because of its ability to produce valuable insights on water demand behavior and to provide practical results. The chapter notes the challenging aspect of regression as the set-up and interpretation of which requires knowledge and intuition of water use, and careful consideration of the theories behind regression analysis.

In Chapter 3, 'Water Demand Forecasting | Machine Learning,' Xenochristou discusses a basic machine learning (ML) pipeline for water demand forecasting. ML is a subfield of artificial intelligence (AI), where algorithms are recognizing and assimilating patterns from data. In this chapter, we focus on supervised learning, a field of ML where an algorithm learns how to map an input to an output, given a set of examples. Each training example constitutes a sample in our dataset and includes a set of features (predictors/independent variables/explanatory variables), as well as one or more target variables (i.e., dependent variables). In water demand forecasting problems, the target variable is often water demand at a given temporal (e.g., daily or monthly) and spatial (e.g., at the household or city level) scale, while the features are variables that are suspected to influence water demand, such as air temperature or day of the week. ML methods have recently dominated the water demand forecasting literature, due to their superior accuracy compared to traditional statistical methods. This chapter introduces basic ML concepts and describes a ML pipeline, from data collection to deployment.

In Chapter 4, 'Water Demand Forecasting | Time Series,' Sanneh *et al.* discuss the vital role of water demand forecasting in many aspects of Water Distribution Systems (WDS) because it helps minimize cost, optimize operations, and provide strategies for water conservation. Demand forecasting also plays a vital role in the planning, operations, and management of physical assets for water utilities such as pumping stations, treatment plants, tanks, and distribution networks, which rely on future consumption forecasts. In this chapter, traditional time series forecasting methods such as Auto-Regressive Moving Average (ARMA), Auto-Regressive Integrated Moving Average (ARIMA) and Seasonal Auto-Regressive Integrated Moving Average (SARIMA) are introduced to forecast water demand using time series historical data. In addition, various ML techniques are introduced to time series-based water demand forecasting problems. They have the advantage of being able to forecast nonlinear relationships between response variables and their predictors in time series models with the presence of noisy data. The increasing use of smart water metering in the water sector has made available a great amount of data which cannot be processed with traditional methods. Therefore, the need to identify new data analysis techniques able to extract valuable information from available data and support water utilities in their decision systems has proven to be paramount. Analytics in the Drinking Water Industry illustrates how to improve demand side management and water distribution network efficiencies, which can lead to significant water savings, promote sustainable customer behaviors, identify peak hours of use, and facilitate water forecast demand modelling.

In Chapter 5, 'Cost-Benefit Analysis for Water Infrastructure,' Chaudhry discusses Cost-Benefit Analysis (CBA) as one of the most prominent and widely used policy evaluation and decision-making tools in public policy. CBA has played a key role in water infrastructure project analysis, and at the same time, application of CBA tools and methods in water industry have also contributed to the development and refinement of tools and approaches now used in CBA. This chapter gives an overview of the methods within CBA, with a brief outline of the history and the regulatory requirements of using CBA in the water industry. CBA is an economic tool for helping decision-makers assess the economic efficiency of a policy or a project. As this chapter shows, CBA does this by quantifying all the benefits and costs of the project for the relevant population. Although it seems straightforward to fill in the empty cells and determine the benefits and costs, a CBA is more than just net present value (NPV) for several reasons: First, it can be quite hard to reduce all of the impacts (costs or benefits) of a project to a single metric. For practical reasons an NPV will not include all important project consequences. However, a well-done CBA includes determination and disclosure of all project impacts, not just those that can be readily quantified in dollar terms. Therefore, the researcher often must make decisions on which impacts to include in the calculation of NPV and which to leave aside. Also, the choice of the discount rate to convert future benefits and costs to present values is an important choice. These decisions can lead to substantial impacts on the calculated NPV. It is imperative that researchers and practitioners clearly disclose all assumptions and make modeling decisions transparent so that the audience understands the true scope of the analysis and results (including limitations).

1.7.2 Operations section

The Operations section covers diverse aspects of water utility operations. In Chapter 6, 'Water Quality Analysis | Modeling and Optimization,' Palmegiani and Lee discuss water quality modeling and calibration for water distribution systems. Water quality within water distribution systems is a highly complex, and rapidly changing issue that is driven by many factors and is difficult to intuitively predict. This is because it depends on many factors such as the pipe materials, system layout, incoming water to the system, water use patterns, corrosion levels, flowrates, and other hydraulic factors. Also, variations of water quality due to seasonal temperature have been previously observed. Many Opportunistic Premise Plumbing Pathogens (OPPPs) and complex chemical species can exist within a building water system, which can expose communities to waterborne diseases such as Legionnaire's disease and cause outbreaks. Issues often occur as the water ages in the plumbing system. Drinking water is often treated with a chlorine disinfectant to prevent growth of harmful chemical and microbial contaminants, as well as corrosion control inhibitors to prevent metal leaching from the pipes. However, as the water age increases, the system experiences decay of both the disinfectant and the corrosion control inhibitors, allowing for contaminants and pathogens to grow inside the system and biofilm. It is critical to perform in-depth water quality modeling to understand the complex dynamics of the system.

In Chapter 7, 'Hydraulic Analysis | Calibration and Uncertainty Analysis,' Moradi *et al.* discuss calibration and uncertainty issues in hydraulic modeling. Today, hydraulic models play an undeniable facilitating role in various stages of design/development, rehabilitation, operation and management of urban water distribution networks. Models represent an estimate of the behavior of Water Distribution Networks (WDNs), not their entire reality, and this is because hydraulic models are prone to different sources of uncertainty. Uncertainties due to incomplete understanding of the dynamics of phenomena, uncertainties in the structure of models and uncertainties in data and parameters are the most important types of uncertainty associated with modeling WDNs. In WDNs modeling, parameters are unknowns (constants or non-constants) that appear in the governing equations describing the system dynamics, mainly as coefficients or exponents that can be spatiotemporal variable. Roughness coefficients of pipes, nodal demand patterns, bulk and wall reaction rate coefficient of chemicals and so on., are examples of parameters in WDNs modeling. Parameters may be estimated by laboratory tests (e.g., new pipe roughness coefficients) or by analysis of field measurements (e.g., demand patterns or pipe roughness coefficients for systems under operation) or by a combination of them. Calibration of water distribution models is a process that adjusts network parameters to minimize the differences between simulation results in the model and real measurements in the network. Any parameter calibration is prone to inaccuracy since we just have to make an estimate of the parameters. Hence, parameter calibration is generally accompanied by an uncertainty analysis. Uncertainty analysis is performed to quantify to what extent the inaccuracies of parameter estimation make the model results imprecise (e.g., nodal heads, velocity in pipes, concentration of chemicals etc.). Such analysis is called parameter 'uncertainty quantification' or 'uncertainty analysis' (UA). An important function of UA for operators could be awareness of the expected range of fluctuations in model results. In this chapter we are going to review the concepts of WDNs calibration and UA, and represent how to apply these concepts on practical examples.

In Chapter 8, 'Optimal Pump Operations | Optimization,' Moradi *et al.* discuss pump operations within the WDN using optimization concepts. Specifically, this chapter presents the framework and requirements for a WDN modeling with optimal pump operations/scheduling. At the end of the chapter, an example of EWQMS is also provided. Pumps are the beating hearts of many civil and industrial projects around the world, and without these critical elements, proper performance of many civil infrastructures such as irrigation and drainage networks, water and wastewater treatment plants, sewer and storm water collection systems, urban and industrials water/oil/gas supply systems, and so on. could not be conceivable. The structural, geometric and mechanical features of pumps are designed considering a variety of hydraulic performance expected in operation. Although in the design stage of a pumping station taking variable demands would result in a more flexible system with more realistic insight into operational variation, designers classically consider the most conservative data to size system's components. Operators, however, are generally more interested in managing the systems in a way that they have an optimum operation condition to achieve the best system performance (e.g., minimum energy consumption, improving water quality etc.).

Optimum operation could have different meanings based on defined objectives. For an aged WDS that suffers from a high rate of leakage, optimum system operation may be defined as maintaining pressure of the network as low as possible to minimize water loss, while meeting the minimum pressure requirements. For a network having a substantially high rate energy tariff over the peak water demand hours of the day, optimum system operation relates to setting the pumps schedule to have the minimum energy cost. Moreover, a multi-purpose approach may consider the optimum operation of network to find the trade-off among different conflicting objectives such as energy consumption and/or energy cost, and water quality measure. Today, challenges with key resources including water shortage, limitations on energy and finance, environmental pollutions and other aspects of sustainable development have compelled decision-takers to inevitably adopt an integrated approach to make better informed decisions in practice. Hence, water organizations should invest in novel multi-objective approaches such as EWQMS to better understand and efficiently resolve problems, covering different concerns associated with available resources.

In Chapter 9, 'Hydraulic Transients | Numerical Analysis,' Lee *et al.* discuss hydraulic transients and a modeling framework in addition to phenomena within the systems. Many water utilities have in-house hydraulic modeling capacities to analyze their systems in terms of planning, design, operations, and management. However, many of the modeling efforts are geared toward or limited to steady state or extended period simulations, which assume that the water is completely incompressible, and that pipe materials are inelastic. Clearly, the mass continuity and energy equations neglect to explain rapid changes that should be described by momentum equations (i.e., transient pressure waves generated due to sudden changes in flow). As is well known, the resulting pressure can result in pipe bursts and structural damage to other critical appurtenances. In addition, low flow due to transients can induce contamination intrusion in the systems. This chapter introduces basic theories and TSNET, so readers can run and see the impacts of hydraulic transients in the system.

In Chapter 10, 'Network Partitioning,' Di Nardo *et al.* discuss one of the most effective ways to reduce WDN complexity within the context or paradigm of 'divide and conquer', which exploits the property that complex systems can be better analyzed if they can be split into many sub-parts. This technique was proposed in England in the early 1980s and is now implemented in many countries. It consists of defining smaller water districts or sectors, defined as district meter area (DMA), obtained with the permanent insertion of boundary valves and flow meters along properly selected pipes. This can significantly improve the management, the maintenance and, specifically, the water balance estimation for water leakage detection, along with supporting/enhancing potential pressure control and emergency response strategies to reduce water losses and water security from intentional contaminations. This technique provides a series of interventions on the WDN that require a careful economic planning by the managing authority; furthermore, it envisages the use of modern monitoring systems (remote control, etc.) which no longer have a prohibitive cost and which, to be implemented, only await a new management policy. It is evident that having a network divided into smaller sub-regions makes it easier to study and manage the system.

In Chapter 11, 'Pipe Network Reliability Analysis | Optimization,' Chandramouli discusses the linking of EPANET tool kit functions within the MATLAB Dynamic Link Library, use of a genetic algorithm tool in MATLAB, the concepts of fuzzy logic, as well as optimization and reliability. Reliability of water distribution networks is another aspect on which considerable research has been carried out. Reliability of water distribution systems is concerned with the ability of the network to provide an adequate supply to the consumers under both normal and abnormal operating conditions. The chapter develops a reliability-based optimization model for design of water supply pipe networks in MATLAB by combining EPANET toolkit functions and the readers will be able to appreciate the difference between binary logic and fuzzy logic in terms of reliability achievement for the water supply pipe networks by working with different types of networks of water supply for their design.

In Chapter 12, 'Resilience | WNTR,' Chu-Ketterer *et al.* discuss: (1) the challenges that disasters pose on WDN infrastructure and how WNTR can be used to assess these challenges; (2) steps to install WNTR; (3) types of disasters that can be currently modeled; (4) available resilience metrics; and (5) tutorials. WNTR is actively being used and extended within the Water Distribution Systems Analysis community for a variety of topic areas. Resilience has many different definitions, but it can be described as the capability of an object to recover or adjust after a source of strain or change. In the context of drinking WDN, resilience is the ability of the system to continue delivering water in a damaged state or how fast the system can return to service after damage. Predicting and measuring resilience in WDN is helpful to prioritize strategies to improve resilience, perform cost-benefit analyses, measure progress, and clarify what is meant by resilience. Tools that can quantify system resilience are important and help improve system security and general operations even when confronted with natural or other disruptions. Simulation analysis can be used to evaluate and potentially improve response actions through failure planning exercises and to develop more effective mitigation strategies for the future. WNTR can also be used to run more routine modeling exercises such as fire flow analysis to access WDN ability to respond to everyday incidents.

1.7.3 Management section

The management section covers critical aspects of effective utility management. In Chapter 13, 'Water Mains Optimal Replacement Time | Optimization,' Lee discusses optimal replacement analysis using historical failure data. Asset management (AM) is defined as 'maintaining a desired level of service for what you want your assets to provide at the lowest life-cycle cost. Lowest life-cycle cost refers to 'the best appropriate cost for rehabilitating, repairing or replacing an asset'. In a water distribution system, the repair/replacement cost and possible water damage cost must be balanced by the water utility when deciding at the time of a leak/break whether to repair or replace the system. Accelerated replacement refers to replacing the system well in advance of the optimal replacement time, while delaying replacement beyond the optimal replacement time will lead to consequences through neglecting repairs, which may effectively amount to the utility paying a penalty to compensate for the high replacement cost. To manage the integrity of water main infrastructure through its entire life-cycle, we introduce a replacement program for water utilities in this section. This program is expected to ensure affordability, manage risk, and support a high level of confidence in the decisions reached.

In Chapter 14, 'Water Mains Replacement Decision | GIS Analytics,' Martinez Garcia discusses water infrastructure asset management issues using GIS. Depending on the number of served customers, large water utilities can manage hundreds of miles of water mains made of different materials and diameters. When water mains fail, utilities are affected by the loss of treated and energized water. Additionally, rising failure rates in distribution systems increase the capital improvement and maintenance budgets which likely lead to higher bills to their customers and a negative public perception. Although an aggressive capital program to repair or replace all affected water mains will reduce the amount of revenue loss, economic and financial constraints make it impossible to replace all failed water mains at the same time. Therefore, supporting water utilities to make informed decisions about the time and location to perform water mains repairs or replacements has attracted attention from researchers in the water industry. The tools presented in this chapter can provide valuable information about the spatiotemporal trend of water main failures. By applying these techniques, water utilities can save economic resources in avoided failures, reduced water loss and energy savings. In addition, an asset management program (or water mains integrity program) can help select improved materials and sizing can provide other benefits to customers such as improvement in water supply reliability, overall system resilience, and overall levels of service.

In Chapter 15, 'Decision Analysis | CA, CV, and AHP,' Tanellari and Lee discuss critical decision analysis tools that can be used for water resources in general. First, nonmarket valuation is a method that is used to estimate the total willingness to pay for a good or a service that is not traded in the market. For goods that are traded in the market, the total willingness to pay can be easily estimated by the area under the demand curve. However, this is a more challenging task in the case of nonmarket goods. Because these goods and services are not sold in the market, the demand curve does not exist. Instead, the willingness to pay is either revealed through consumers' choices or directly elicited through surveys. There are two broad categories of valuation methods, revealed preference methods and stated preference methods. Revealed preference methods are based on actual choices that individuals make which in turn reveal the values that they may place on the good or service of interest. For example, by calculating how much households spend on bottled water, filters and water treatment devices in a given time period, a revealed preference method may infer the value that households place on clean water. The cost of such treatments and devices is directly incurred by households and is observable through the prices they pay. Stated preference methods elicit willingness to pay directly from consumers through surveys. Consumers are directly or indirectly asked to state their willingness to pay for a good or service. In this section, we will examine two widely used stated preference methods, contingent valuation and conjoint analysis. In addition, the chapter covers AHP, which determines the preference for a decision-making unit by pair-wise comparison of attributes. Assessing pair-wise preferences enables the decision maker to concentrate his/her judgment on two elements with regards to a single property. So, in this case, the decision maker does not need to think of other properties or elements while comparing and deriving the final decision. We will introduce all steps using spreadsheet.

In Chapter 16, 'Non-revenue water,' Gungor Demirci and Lee discuss one of the critical management issues for the water utilities, namely, non-revenue- water. Around the world, more than $14 billion per year is lost due to water loss, and these losses are covered by paying customers. Water loss is a huge challenge for water utilities, which require fundamental understanding of the influencing factors. The Organization for Economic Co-operation and Development (OECD) found that water loss can be as high as 65% for developing countries. It is a challenging task to reduce the water loss even in highly developed countries as well. For an effective water loss reduction program, it is critical to have a deep understanding of the causal factors as well as why its reduction is so challenging. Many literatures cited environmental, managerial, physical, sociological, and technical factors. The chapter examples include system age, pipe length/layouts of the systems, hydraulic conditions, external soil characteristics/topography, traffic loading, service connection densities. The problem is solved using R.

In Chapter 17, 'Performance Assessment of Water Industry | DEA,' Gungor Demirci and Lee discuss water utility performance and performance measurement methodologies. A water utility's efficient management practice has become more vital than ever because of the large gap between the available water supply and the rising demand, as well as unpredictable climate patterns due to changing climate. Not all water utilities are functioning at the same level of efficiency in their operations. In this chapter, we will develop a useful performance measurement tool and apply it to the individual water utility's operations. Measurement of performance assessments for each water utility will identify the opportunities to improve their management deficiencies and economic performances. Also, the performance measurements will provide in-depth insights toward a fully efficient water utility. Data Envelopment Analysis (DEA) is an optimization tool for measuring efficiencies of the units in any organization. In addition to conventional DEA methods, we will explore two additional stages to examine the exogenous variables' impacts on the individual water utility's performance: double bootstrap truncated regression and Tobit regression. This chapter is based on our previous publications.

All chapters are independent, so you can study based on your interests and needs. We hope you enjoy reading and practicing each chapter!

REFERENCES

Chastain-Howley A. (2018). How big is big data among water utilities? Water Online. Available at: www.wateronline.com/doc/how-big-is-big-data-among-water-utilities-0001 (last accessed 10 May 2022)

Karl M. and Wyatt G. (2018). Water Online. Available at: www.wateronline.com/doc/smart-utility-building-a-foundation-for-artificial-intelligence-0001

Keck J. C. and Lee J. (2015). A new model for industry–university partnerships. *Journal – American Water Works Association*, **107**(11), 84–90, https://doi.org/10.5942/jawwa.2015.107.0161

Keck J. and Lee J. (2021). Embracing analytics in the Water Industry. *ASCE Journal of Water Resources Planning and Management*, **147**(5), 02521002.

Lunani M. (2018). Artificial intelligence for water and wastewater: friend or foe? *Opflow*, **44**(6), 6–7, https://doi.org/10.1002/opfl.1017

Neemann J., Roberts D., Kenel P., Chastain-Howley A. and Stallard S. (2013). Will data analytics change the way we deliver water?. *Journal-American Water Works Association*, **105**(11), 25–27, https://doi.org/10.5942/jawwa.2013.105.0163

The Water Network. (2020). Australian Government in Water Ledger Blockchain for Trading Water Rights. Available at: https://thewaternetwork.com/_/rising-water-technologies/article-FfV/australian-government-in-water-ledger-blockchain-for-trading-water-rights-ZufchC63jUarM_Q8K5cfCw

US Environmental Protection Agency (USEPA). (2007). Water Sector Collaboration on Effective Utility Management Fact Sheet. Available at: http://www.epa.gov/water/infrastructure/utility-mgmt-joint-statement.pdf

US Environmental Protection Agency (USEPA). (2017). Effective Utility Management: A Primer for Water and Wastewater Utilities. Prepared by the EUM Utility Leadership Group. USEPA, Washington.

Williams P. (2013). Information engineering: an integrated approach to water system management. *Journal – American Water Works Association*, **105**(6), 61–66, https://doi.org/10.5942/jawwa.2013.105.0081

Zuckerman S. (2018). A New Internet – How Berkeley Haas is Accelerating Blockchain's Potential. Berkeley Haas Magazine, Summer 2018.

Part I
Planning

doi: 10.2166/9781789062380_0021

Chapter 2
Water demand analysis | regression

*Stephanie A. Tanverakul**

California Water Service, Project Engineer, San Jose, California, United States
**Corresponding author. E-mail: stanverakul@gmail.com*

LEARNING OBJECTIVES

(1) At the end of this chapter, you will be able to:
(2) Apply regression methods to forecast water demand.
(3) Discuss the practical aspects and implications of using ordinary least squares estimation in regression analysis.
(4) Build and run a regression model with panel data in R.
(5) Interpret linear regression results.

2.1 INTRODUCTION

Future water demand estimates are key inputs in water resources planning and management. Ensuring a sufficient and reliable volume of water is available to meet demand is a core function of all water suppliers and distributors. Meeting demand requires knowing how much water is needed now and will be needed in the future. Accurate future forecasts are critical since water supply availability is highly variable and water infrastructure projects, often large and expensive, are designed and constructed with long useful lives upwards of 20–50+ years. For these reasons, the ability to make accurate future water demand estimates has long-term consequences.

Water demand forecasts can be derived from various sources. Historical use data, where available, can be useful in projecting demand under certain circumstances. However, changes from differing housing and commercial development patterns, changing demographics, and shifting weather patterns will often alter water demand patterns reducing the confidence of projections based on historical use alone. **Understanding what factors influence demand can help project future demand with greater +accuracy**.

Water supply and demand problems, and their solutions, are often localized with unique challenges involving many aspects of hydraulics, environmental sciences, socioeconomics, finance, laws and regulations, and politics. Because water is difficult and expensive to transport (think of density/specific weight of water!), available water sources are most often near their users and tied to local conditions such as local climate and level of treatment necessary. The uniqueness of water use behavior by location is relevant, even critical, for forecasting water demand when determining the scope and application of the demand model. A demand model using residential water demand data from a city in California will likely not be appropriate to use for a city in New York. Also, models of regional demand for the agriculture region of Iowa would not be useful to use in a heavy industrial region. Modeling water demand must always consider how water use volume and behavior differs by user type and location.

Modeling water demand is modeling human behavior by evaluating how water use is influenced by user characteristics and various external factors like weather, price, or other constraints. Unfortunately, for building the models, behavior is often not straightforward or linear. There may be user-specific characteristics that determine water demand. For example, a factory may have a set volume requirement for their process water and other functioning needs, or a residential home with a minimum amount for essential needs and additional uses of lawn irrigation. Combined with those factors are other variables like weather or water price that may affect the amount of water needed or influence the amount of discretionary use. For example, residential customers with outdoor water needs tend to increase water use during dry months and decrease during wet months, but may choose to reduce irrigation water use if requested by their utility to do so during drought periods, or a factory or business may change their processes if water prices rise enough. Another example can be during COVID-19. Overall residential demand increased (due to lifestyle changes) while commercial demand decreased due to lockdown. So, identifying these types of factors that impact water use is a principal step in setting up water demand forecast models.

This chapter discusses regression analysis as a useful method to explore the relationships between water demand and influencing factors. Over the previous decades, numerous studies have been performed measuring and modeling water demand using many different techniques (Arbués *et al.*, 2003; Donkor *et al.*, 2014; Gracia-de-Rentería & Barberán, 2021). Regression is a popular and well-demonstrated choice and has been chosen for this discussion because of its relative simplicity to perform with (free) software programs (e.g. R, Python, etc.), and its ability to produce valuable insights on water demand behavior and to provide practical results. With that said, the challenging aspect of regression is the set-up and interpretation which require knowledge and intuition of water use, and careful consideration of the theories behind regression analysis. *The ease of running regression models can easily lead to misinterpretation!*

The basics of regression are presented here and are applied to water demand forecasting with the objective that you will be able to perform and understand their own analysis. The theories behind regression can get very complicated quickly and this chapter does not touch upon every aspect. You are encouraged to consult other econometric sources, particularly if deviating far from the examples discussed herein.

The structure of the chapter begins with an introduction to regression analysis with an example problem, followed by discussions on model specification, model estimation, and ends with model interpretation.

2.2 PRINCIPLES OF REGRESSION

2.2.1 What is regression?

Regression methods can help answer how different factors affect one variable of interest. In the case of estimating water demand, regression methods can be used to characterize relationships between demand and influencing factors such as weather, demographics, pricing, and other identified factors. Water demand is the variable of interest, taken as the *dependent variable*. All other factors used to characterize demand are the *explanatory, or independent variables*. A simple linear regression example using residential water demand and one explanatory variable is used in the next subsection to introduce the regression equation.

2.2.2 Basic regression equation – water demand and lot size example

Simple linear regression deals with a single explanatory variable, and its relationship with the dependent variable. When estimating residential water demand, one variable that may be useful to estimate demand is lot size. A larger lot size may be assumed to explain higher water use since a lot size is correlated with a large yard and larger yards may have increased use of irrigation water. Choosing appropriate variables to explain the dependent variable (i.e. water demand) is further discussed in the next section and is an important decision in performing a good regression analysis.

Plotting water demand data with lot size is a useful first step to check the assumption that lot size may assist in explaining water demand. Figure 2.1 plots all the data from a fictionalized data set containing household water demand (in liters per day (lpd) and household lot size (in square meters). It appears there is strong correlation between the demand and lot size, and on average, water demand is higher on larger lot sizes. Using only a visual assessment, a trend line could be drawn demonstrating the increasing trend.

The trend line follows the equation of a line: $y = b + mx$, where m represents the line slope and b is the y-intercept. Applying this to the example, the equation becomes:

$$\text{Water demand (lpd)} = \text{intercept} + m * [\text{lot size, sq meters}] \tag{2.1}$$

The slope, m, in Equation (2.1) represents how much water demand changes with a change in lot size. It can also be deduced that a steeper slope means a larger change in water demand from a smaller change in lot size. This concept is referred to as *elasticity*. The y-intercept has less direct meaning here since it would not be useful to know the water demand on lot sizes of zero.

Moving towards a more rigorous analysis to estimate a trend line is simple linear regression. The ordinary least squares (OLS) estimator is used to estimate the slope by minimizing the difference between each data point and the average of all points. Figure 2.2 illustrates this difference. This can be calculated by hand, but can also be done quickly with a spreadsheet like Microsoft Excel's trend line feature, which was done for this fictionalized example to produce the following:

$$\text{Water demand (lpd)} = 114.08 + 3.05 * [\text{lot size, sq meters}] \tag{2.2}$$

The interpretation of Equation (2.2) is that water demand will, on average, increase by a factor of 3.05 for every square meter increase in lot size. The equation is useful to determine average water demand patterns from house lot sizes, but there are several caveats to consider. The first being the

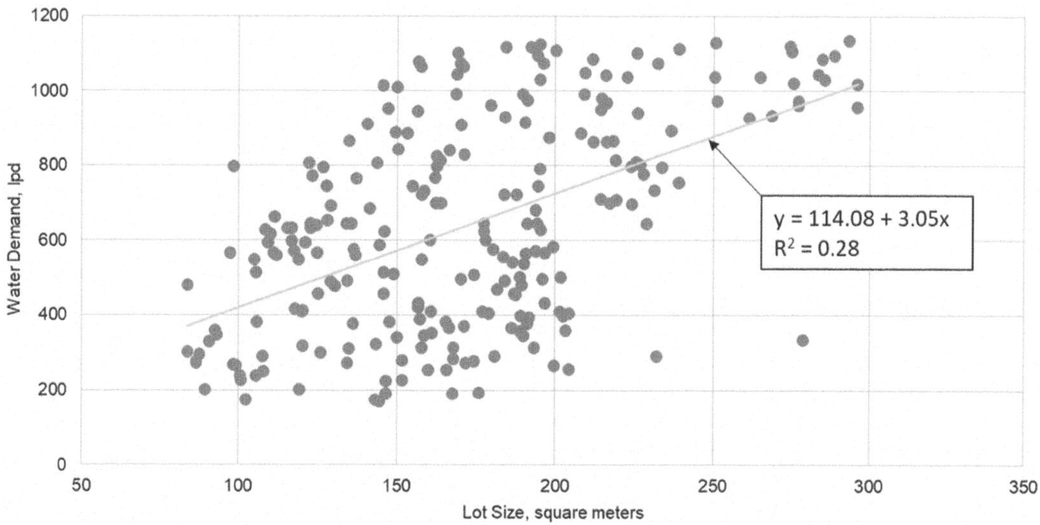

Figure 2.1 Water demand versus lot size, fictionalized data example.

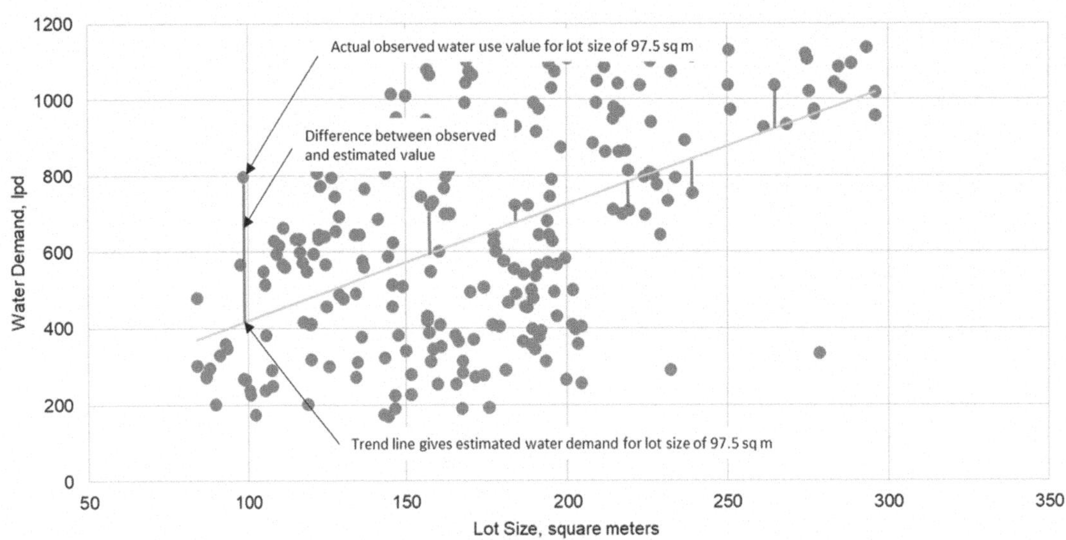

Figure 2.2 Water demand versus lot size, observed versus estimated difference.

equation is only adequate to determine water demand from the range of lot sizes that were used to develop the equation. In this case, the range of lot sizes were between 84 and 296 square meters. Estimating demand for a 500 square meter lot would not be appropriate. Another consideration is time. The data was from a single point in time. The data may be significantly different depending on the season or location. If this data came from a rural, dry region, it would not be appropriate for an urban city with high precipitation. The equation is only appropriate for locations at a certain time with other similar characteristics (e.g. socioeconomic status, temperature, etc.).

A serious consideration when evaluating the analysis is that lot size may not be the strongest single factor to estimate residential water demand. This puts the validity of the equation into question and should always be considered. The r-squared value is often estimated to measure the strength of the relationship between the two variables. For this equation, r-square (shown in Figure 2.1) was 0.28, meaning 28% of the variability in water demand could be explained by lot size. An r-squared value of 1.0 would signal a perfect linear relationship. This is never observed with collected data except for a perfectly controlled laboratory setting. The r-squared value here could be considered adequate for the given data type but the relationship could still be questioned. It could be reasoned that larger lot sizes would have larger homes with multiple stories, more water-intensive appliances, and more occupants. Temperature is another possible variable that could explain higher water use in place of lot size, since higher water use may be expected during summer months, assuming higher temperatures require more water used in irrigation. Is it larger lot sizes, or perhaps higher temperature that influences higher water use during summer months? Higher temperatures may have a stronger relationship to water use in locations with houses with large yards compared to highly dense urban neighborhoods. Considering all these additional factors, perhaps the number of people per house, the number of bathrooms, or a weather variable would produce a stronger correlation with water demand. *This process is a central challenge to the validity of regression equations.*

Before moving on, looking at the generalized simple regression equation may be helpful:

$$Y_i = \alpha + \beta_1 X_i + \varepsilon_i \tag{2.3}$$

where Y_i is the dependent variable, α_i is the intercept, β_1 is the regression coefficient, X_i is the independent variable, and ε_i is the residual, or error term. This holds for individual observation, $i = 1$, ..., n. The equation is the same as the line for an equation used above with the addition of ε to express the residuals, or the error term. The error term accounts for the differences between the predicted values of Y versus the actual observed values of Y. Shown in Figure 2.2, this difference is the distance between the predicted regression line and each observed individual data point. This difference partly arises because X (lot size) is not the single, perfect predictor of Y (residential water demand). Lot size alone cannot provide a perfect estimate of water demand. There are many other factors influencing demand. *In this way, the error term can be thought of as the amount of variability in water demand (Y) that cannot be explained by lot size (X).* The error term also absorbs other errors that may exist such as errors in how the data was measured. For the example of the lot size, questions to be asked would be how the data was collected; was it taken from an online repository, or was it self-reported by homeowners? Any of these options could have incurred mistakes/errors. invalidating some values. If there are significant outliers, the errors could have an impact on the regression model as well.

2.2.3 OLS assumptions

OLS has a vast decades-long precedence of being used across different disciplines. At the core of OLS is estimating parameters that minimize the sum of squares of distance between a predicted regression line and sample observations, while seemingly simple to correctly use OLS requires certain assumptions be met. These assumptions have a deeper theoretical and mathematical foundation, but the focus here will be on the practical implications of what the assumptions mean and how violating the assumptions can affect the model results.

2.2.3.1 Assuming linearity

The general multiple regression model, shown in Equation (2.3), has a linear form. The linear form is defined as each of the explanatory variables (the X's) multiplied by a parameter (β's) which are then added together with the addition of the constant term. In this form, the model is 'linear in parameters'.

Note this is a bit different than the assumption that the relationship between an explanatory variable and water demand is linear. If that relationship is not linear, the variables can be transformed. In this manner, the linear model can fit a non-linear relationship between variables. Logs, inverses, or squares can be used to satisfy the linear assumption, for example, the following Equations (2.4) and (2.5) use non-linear transformation, but the equation is still linear:

$$\text{Log}(y) = \alpha_i + \beta_1 X_{1t} + \beta_2 X_{2t} + \cdots + \beta_n X_n + \varepsilon_{it} \tag{2.4}$$

Or

$$\text{Log}(y) = \alpha_i + \beta_1 X_{1t} + \beta 2 \log(X_{2t}) + \cdots + \beta_n X_n + \varepsilon_{it} \tag{2.5}$$

If the data is not linear and OLS is used without first transforming the data to achieve linearity, the results will not be reliable. To check for linear relationships in the model once results are produced, a graph of observed data versus predicted values is helpful. If linearity is not observed in the plot (45° line should be clear) a non-linear (e.g. log) transformation can be performed on the independent/dependent variables. The model can then be re-estimated and checked for linearity once again.

Figure 2.3 shows a plot of the actual versus predicted values from the demand versus lot size example. A perfect predictive model would show all point along the 45° plotted line. Within the middle ranges of 150 and 200 (circled in Figure 2.3) there is good linearity. Both below and above this range, however, the predictions are higher and lower, respectively. Performing a transformation on the data and replotting can be performed to check if a better estimate of the relationship may be possible first, without changing other aspects of the model. Figure 2.4 shows the example data with a log transformation. The predicted values appear closer to the 45° line for values above 175. Below 175

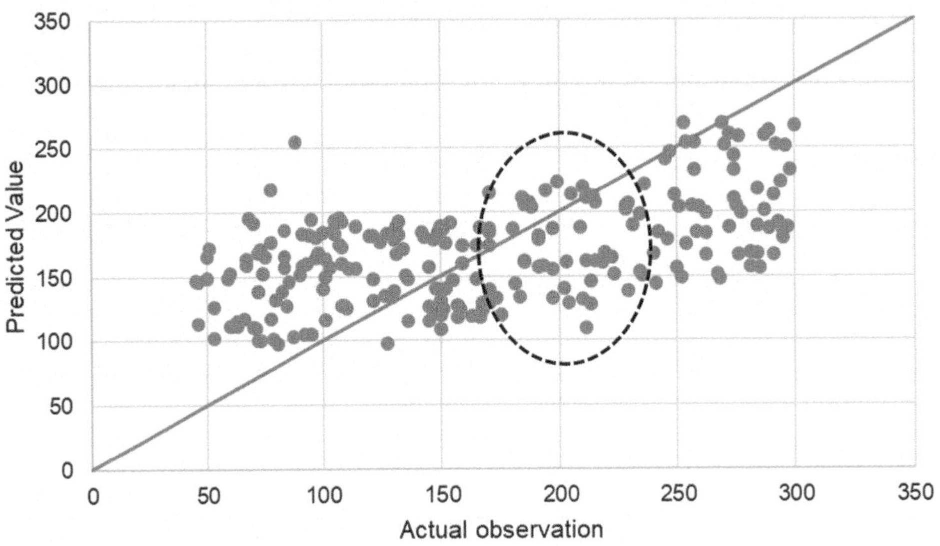

Figure 2.3 Predicted versus actual value plot.

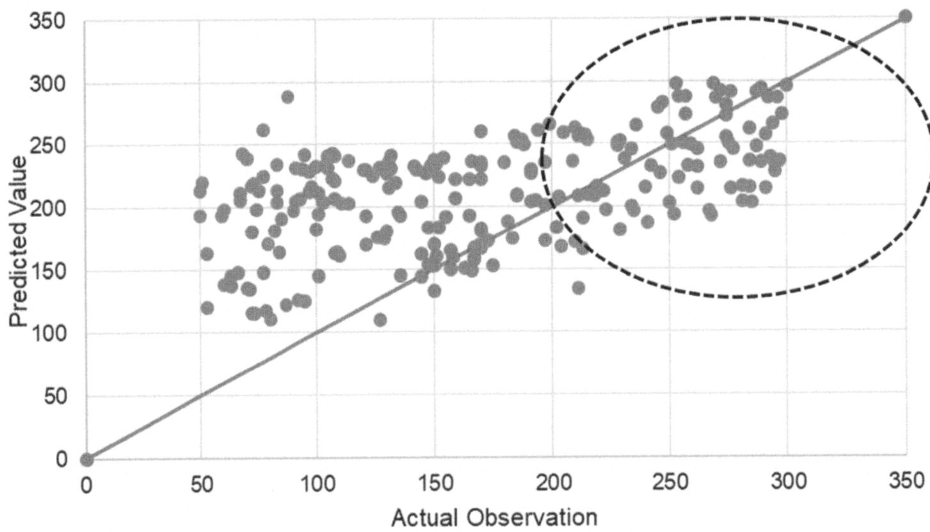

Figure 2.4 Log-transformation – predicted versus actual value plot.

the predicted values are all much higher than the actual observation. In this case, the transformation helps with the higher values but does not fully provide a solution.

2.2.3.2 Assuming independence between explanatory variables (multicollinearity)

In multiple regression, the intent is to estimate how individual variables (independent variables) help explain water demand changes (dependent variable). What is being estimated is the marginal one-unit change in an independent variable, holding all other variables constant. *For this to be most accurate, all independent variables must be independent of each other.* If the independent variables are correlated with each other, it can create an incorrect model! For example, rainfall and evapotranspiration (ET) are both variables that could be used to estimate water demand. However, rainfall is used to estimate ET. In this case, it would be impossible to discuss the marginal change in ET, holding all other variables constant since rainfall is a factor of ET and the two variables move together.

Correlation between independent variables is referred to as **multicollinearity**. Possible relationships between the explanatory variables should be explored. If any variables are strongly related, then they should not be used together. If multicollinearity does exist, it can decrease the reliability of the estimated parameters and lead to incorrect interpretation. Multicollinearity may be a suspected cause if the expected sign of a regression coefficient (β) is reversed in the regression results. For example, high temperatures are (generally) expected to increase water demand. If temperature was used an explanatory variable and its coefficient was negative, it would imply that high temperatures decrease water demand. Since this goes against intuition, it would be important to further investigate what else is happening with the equation. One item to check is whether another included explanatory variables, likely another weather variable correlated with temperature, was affecting the temperature coefficient.

Correlation matrices between variables are useful in checking for strong correlation. One type of correlation matrix is discussed in Section 2.4 and shown in Figure 2.7. While plotting water demand with each explanatory variable is helpful to check if that single explanatory variable should be added to the model, plotting the explanatory variables with one another can cause multicollinearity concerns.

Variable inflation factor (VIF) is a tool used to detect multicollinearity. VIF compares the amount of inflation to variance from the addition of a single explanatory variable compared with the total model with all explanatory variables included. VIF is estimated for each explanatory variable in a regression model. A high VIF would mean the variable could be highly correlated with another explanatory variable:

$$VIF_i = \frac{1}{1 - R_i^2} \tag{2.6}$$

If multicollinearity is suspected using one of the tools above, removing one of the explanatory variables from the model may help. Thinking through whether an explanatory variable is important may provide an argument for removing or keeping a variable. Combining the variables to create a new variable can also be a solution or there are other methods that can be used besides OLS. Key takeaways are to always explore the data and understand how variables are expected to impact water demand. For presenting and discussing regression results, it is often good practice to include all variables that were removed. This can be done by presenting more than one set of results with and without variables that were removed.

2.2.3.3 Independent observations
The coefficients in the regression model are only estimates of the actual sample parameters. In essence, data is collected as a random sample of a population. The sample is used to estimate/infer population properties. An objective is to minimize the difference between estimated and actual parameters. Random sampling helps to ensure the differences are not skewed in one direction (i.e. that could cause errors in one direction). We want to make sure that sample estimates/inferences are representing the whole population.

2.2.3.4 Several assumptions dealing with error term
The error term in the model accounts for the residual, or the difference between the actual observation and the predicted. It is the variability of Y that is not explained by the explanatory variables. There are several assumptions that deal with the error term that are all concerned with checking that the model

is correctly designed. The assumptions involving the error term are listed below. Again, each of these assumptions have a deeper mathematical or theoretical underpinning in regression modeling with OLS estimation. The objective in this chapter is to highlight the practical aspects to verify the model specification and interpret results.

(1) **No systematic errors**. The error term, on average, should equal zero. This will ensure that the error in the model is random and there are not systematic errors. If there are systematic errors, then it can be assumed that the residuals are predictable. If the residuals are predictable then that means there is predictable variation that could have been captured with the model.

(2) **Homoscedasticity**. Errors should have the same variance across all the observed values. Constant variance in the errors is referred to as *homoscedasticity*, or having no heteroscedasticity. A problem with *heteroscedasticity* can uncover that the model is putting too much importance to one range of observations. When interpreting regression results, heteroscedasticity can impact the test for variable significance and result in an explanatory variable appearing significant in influencing water demand, when in reality it has no impact (see Section 2.5.2). A plot like the one in Figure 2.5 showing residuals versus the predicted values can be used to check for heteroscedasticity. When heteroscedasticity is present, a discernable pattern can be seen, such as the diamond shape in Figure 2.5. Another easily spotted sign of heteroscedasticity is a cone shape with the residuals fanning out or fanning in.

If there was no heteroscedasticity, the expectation would be what is shown in Figure 2.6, where no discernable pattern is seen with the plotted dots, and they appear to be roughly even around the zero-residual line.

Heteroscedasticity is commonly seen with small data sets with large variation or when one explanatory variable has a wide range of input values. A possible method to reduce heteroscedasticity includes transforming a suspected explanatory variable by taking the log or square root, for example.

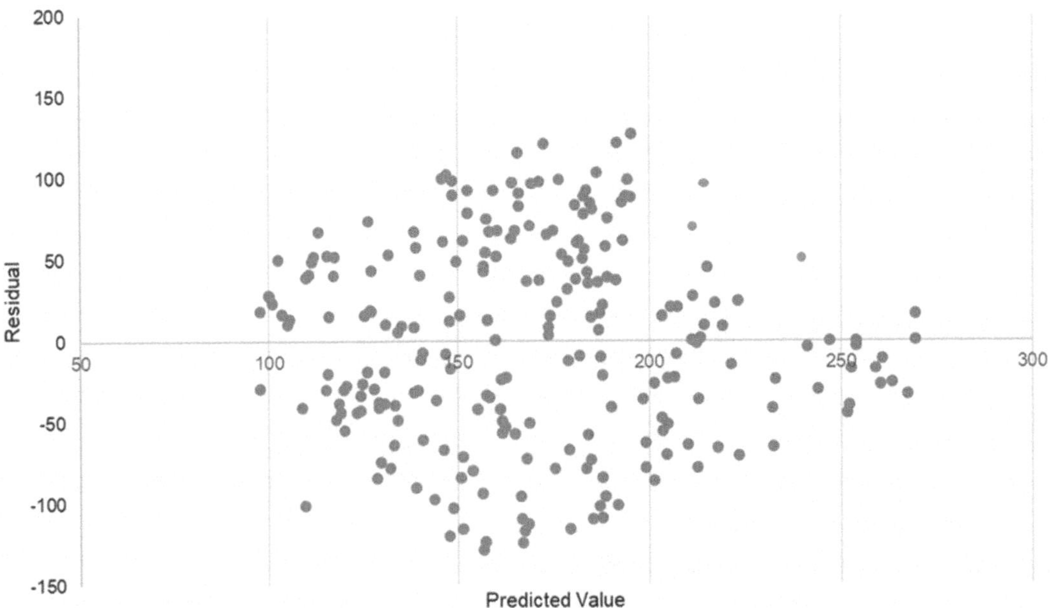

Figure 2.5 Predicted versus residual plot.

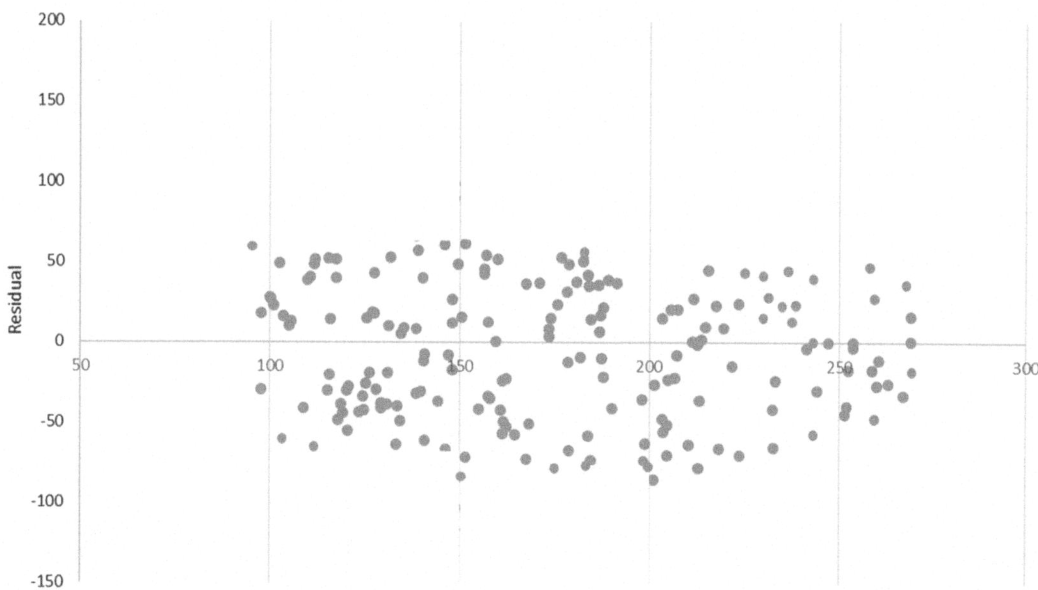

Figure 2.6 Predicted versus residual plot – no heteroscedasticity.

Changing a variable in this manner can often eliminate or reduce heteroscedasticity, and thereby also strengthen the model.

(3) **No autocorrelation.** Errors should be independent of each other, which is known as having no autocorrelation. Autocorrelation is often a problem with time series data, when each subsequent observation is correlated with the previous (see **Chapter 4's Time series analysis**). Seasonal correlation is an example that would be solved by adding seasonal dummy variables to the model. This is done by including the season as an explanatory value as a number. For example, summer would be 1, winter would be 2, and so forth. The idea is to add an additional variable that accounts for the seasonal pattern. Another solution is adding a time-lagged variable to the regression model. A time-lagged variable would be an additional variable added to the model representing a lag of one time period, for example.

(4) **Random error.** Errors should be uncorrelated with the explanatory variables. When there is correlation, this is called *endogeneity bias*. Endogeneity is a problem because it violates the random error assumption because the correlation implies it is possible to predict a part of the error term with that explanatory variable. The result is it biases the coefficients. The cause of endogeneity is often due to measurement errors in the explanatory variable or omitted variables. Omitted variables are important factors influencing water demand that were not included in the model. Also, error terms should follow a normal distribution. This can be checked with a normal probability plot, or q-q plot for the errors. If the linearity assumption is violated, then error terms may not follow a normal distribution. The consequence to the results is large confidence intervals that are too wide or too narrow which make interpretations less reliable.

2.2.4 Panel data regression

In this section, we would like to explore more real-world datasets. Observation data is often categorized as *time-series, cross-sectional, and panel*. Time-series data consist of one data point being measured over time. This could be one customer's water use measured monthly. Cross-sectional data refers

<table>
<tr><td colspan="2" align="center">Time-Series</td><td colspan="2" align="center">Cross-Sectional</td><td colspan="3" align="center">Panel</td></tr>
<tr><td>Time Period</td><td>Value</td><td>Individual ID</td><td>Value</td><td>Individual ID</td><td>Time Period</td><td>Value</td></tr>
<tr><td>1//2020</td><td>96</td><td>1</td><td>120</td><td>1</td><td>1/2020</td><td>143</td></tr>
<tr><td>2/2020</td><td>105</td><td>2</td><td>123</td><td>1</td><td>2/2020</td><td>141</td></tr>
<tr><td>3/2020</td><td>115</td><td>3</td><td>178</td><td>1</td><td>3/2020</td><td>150</td></tr>
<tr><td>4/2020</td><td>117</td><td>4</td><td>145</td><td>2</td><td>1//2020</td><td>210</td></tr>
<tr><td>5/2020</td><td>125</td><td>5</td><td>163</td><td>2</td><td>2/2020</td><td>243</td></tr>
<tr><td>.</td><td>.</td><td>.</td><td>.</td><td>2</td><td>3/2020</td><td>212</td></tr>
<tr><td>.</td><td>.</td><td>.</td><td>.</td><td>.</td><td>.</td><td>.</td></tr>
<tr><td>12/2020</td><td>92</td><td>1,900</td><td>124</td><td>.</td><td>.</td><td>.</td></tr>
<tr><td></td><td></td><td></td><td></td><td>.</td><td>.</td><td>.</td></tr>
<tr><td></td><td></td><td></td><td></td><td>n</td><td>t</td><td>X</td></tr>
<tr><td></td><td></td><td></td><td></td><td>n</td><td>t</td><td>X</td></tr>
</table>

Figure 2.7 Example of data type.

to data that represents a swatch of different measurements at a single point in time. This could be a single reading of average monthly water use for 20 000 customers, for example. Panel-data is the combination, where many readings are available over time for different entities. Figure 2.7 presents example data types for time-series, cross-section, and panel. Our experiences taught us that panel-data is the most useful for accurate water demand forecasting

The use of panel data expands the regression Equation (2.3) into:

$$Y_{it} = \alpha + \beta_1 X_{it} + \varepsilon_{it} \tag{2.7}$$

where Y_{it} is the dependent variable for individual i at time period t, α is the intercept, β_1 is the regression coefficient, X_{it} is the independent variable, and ε_{it} is the residual, or error term. This holds for time period, $t = 1, ..., t$ and individual, $i = 1, ..., n$.

Estimating panel data regression models can be done using different estimation methods. We will consider pooled, fixed, and random effects for panel data in the Estimation Section.

2.2.5 Multiple regression

Multiple regression expands on the case of one explanatory variable to include more than one variable to describe change in water demand. The general equation expands on Equation (2.7) and becomes:

$$Y_{it} = \alpha + \beta_1 X_{1t} + \beta_2 X_{2t} + \cdots + \beta_n X_n + \varepsilon_{it} \tag{2.8}$$

where Y_{it} is the dependent variable, α_i is the individual intercept, $\beta_1, \beta_2, \beta_n$ are the regression coefficients, X_{1t}, X_{2t}, X_{3t} are the independent variables, and ε_{it} is the residual, or error term. This holds for time period, $t = 1, ..., t$ and individual, $i = 1, ..., n$.

The estimation of the multiple regression equation quickly increases in complexity from the simple linear regression example. With multiple regression, the dependent variable of interest is being explained by more than one variable. Each of the added explanatory variables are assumed to be independent of each other and the dependent variable, so that the individual impact of each explanatory variable on the dependent variable can be estimated.

2.2.5.1 Problem 1

The provided file Regression Chapter – Ex1.xls contains monthly water demand and rainfall data for a period of six years. Using Excel spreadsheet, plot demand and rainfall, add a tread line (regression) in Excel. (Excel uses the least square estimator.) Answer the following questions:

(a) What type of data is this? Cross-sectional, time-series, panel? Are there limitations to using this data to estimate water demand? Explain.
(b) Is there visible correlation between the water data and weather data? Would you expect to see correlation between water demand and the weather data? Why or why not? What questions could be asked about the data to further investigate your assumptions?
(c) What other analysis could be done with these data to further evaluate the data trends?
(d) Interpret what the regression equation means. Does the weather variable help to explain water demand in the data?

2.2.5.2 Brief suggested solutions

(a) What type of data is this? Cross-sectional, time-series, panel? Are there limitations to using this data to estimate water demand? Explain.
Data is time-series, characterized by observations over time for one entity (labeled Customer_ Group). This data is aggregated to the level of only one entity and as such, cannot account for differences across entities; the data only provides the water demand trend across time.
(b) Is there visible correlation between the water data and weather data? Would you expect to see correlation between water demand and the weather data? Why or why not? What questions could be asked about the data to further investigate your assumptions?
Visually there does appear to be negative correlation between the average monthly water demand and total monthly rainfall. Plot is shown in the figure below. The negative correlation could be attributed to lower water use when there is precipitation, perhaps from reduced outdoor water use for plant and lawn irrigation. Further investigation into the water demand source may support or refute the irrigation assumption. Is the data from a rural or urban area? Do the houses have large lots? What are other weather conditions? Do the temperatures rise during the summer months?
See Section 2.3.4 for discussion on the zero precipitation values.

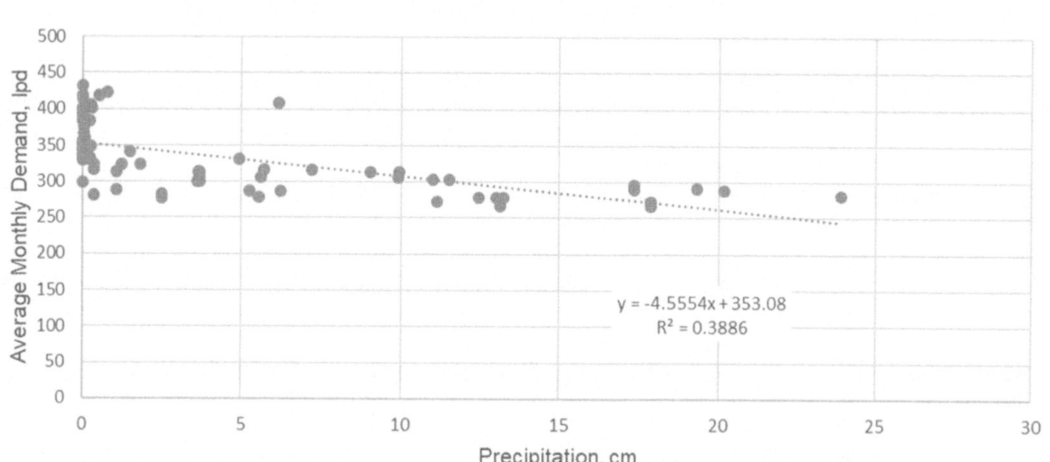

(c) What other analysis could be done with these data to further evaluate the data trends?

Plotting the water demand over time (figure below) can visually provide seasonal trend information. In this example, higher demand is observed annually between July and October. Although the annual trend appears steady over the entire time period (2015–2020), the peak does appear to slightly change between the years.

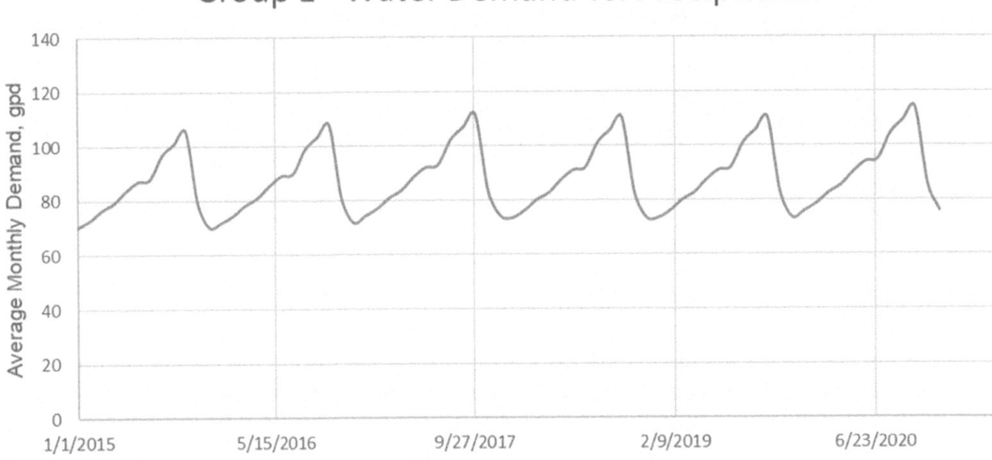

(d)   Interpret what the regression equation means. Does the weather variable help to explain water demand in the data?

Using excel, the regression line follows the equation: water demand (lpd) $=-4.54$ (precipitation, cm) $+353.08$. The r-squared value is 0.39. The negative value on the precipitation coefficient represents a negative impact on water demand. For every one unit increase in precipitation, a 4.54 decrease in liters per day is expected. The intercept for this simple regression can be interpreted as the average monthly water demand when there is no precipitation in the month. Unlike the water demand versus lot size example, the intercept value holds importance since the data has several demand observations with zero precipitation.

2.3 MODEL SPECIFICATION

Model specification involves deciding what explanatory variables (e.g. X_{1t}, X_{2t}, X_{3t}) to include in the regression model. This is an iterative process and requires an understanding of what factors influence water use. However, specification or model structure depends on what data is readily accessible and of sufficient quality, time length, and number of observations.

Data availability continues to grow with new technologies making it easier and cheaper to invest, deploy, and collect large amounts of information. The deployment of more water meters (e.g. AMI – Advanced Metering Infrastructure) has provided the opportunity to measure and therefore, forecast use in more water sectors. Further, finer resolution data (e.g. time interval of seconds) has allowed for more detailed information on how water is used for specific end uses. For residential water demand this has translated to understanding water use by end use for particular appliances (e.g. kitchen sink, bath shower, etc.). More data also means more time spent on investigating the data quality and patterns.

In the next section, we will delve into choosing the best variables starting with fundamental theories of water use, method of exploring available data, and ending with common mistakes around misspecification/interpretation on regression models.

Table 2.1 Factors possibly influencing water demand.

Category	Factor
Social-demographic	Income
	Education level
	Number of adults and children in household
	Level of environmental concern (e.g. water conservation, recycling, energy saving)
Utility or supplier controlled	Water rates
	Rate structure (e.g. increasing tiered rate)
	Mandatory conservation measures
	Voluntary conservation measures
	Metering
	Detailed water use information available
Location	Population growth
	Population density
	Neighborhood characteristics and average demographics
Environmental	Temperature
	Precipitation
	Evapotranspiration
	Droughts
House/building	Lot size
	Building square meters
	Number of bathrooms
	Number of water intensive appliances/high efficient fixtures
	Age of house
User type	Mix of residential, commercial, industrial, agriculture

2.3.1 Water use relationships

The best starting point in identifying explanatory variables is to review the question that needs to be answered. The objective of analysis will help shape what should be included in the regression model. The form of the dependent water demand variable may also change based on the intended analysis. For water utilities, per capita daily information by customer type may be most useful; and for wholesale suppliers, monthly or yearly information may be more practical.

Previous literature review studies can provide useful information and support arguments for choosing explanatory variables. A few review studies that can be helpful are the following: Worthington and Hoffman (2008), Sebri (2014) and Tanverakul and Lee (2016). Table 2.1 provides a list of possible factors that have been explored as possibly influencing water demand. There may be many more factors that could potentially impact water demand and some of the listed factors may not be impactful. You should give careful consideration in determining what factors make sense for the given objective and region.

For every factor that may influence water demand, an explanation should be given as to how that factor influences demand. This is important when interpreting and using the regression results since the model itself is easy to run with software programs and it may be tempting to add in all variables that may possibly affect water demand. As discussed in the next section, not being selective with the explanatory variables can cause problems with the model results and violate key model assumptions. The challenge is constructing an appropriate model and making reasonable and fair interpretations.

Thinking through potential causal relationships can aid in narrowing down the important explanatory variables to include in the model and check for correlation between explanatory variables. Correlation between explanatory variables can obscure and invalidate the impact of each individual explanatory variable on water demand. One example is including house size and number of bathrooms. Both of these variables could reasonably be used to explain household water demand. However, house size could also be correlated with number of bathrooms since larger houses could be expected to have more bathrooms. Because of this relationship, the regression equation would not be able to accurately predict the impact of the number of bathrooms on water demand because some of that impact could be absorbed into the impact from lot size. Correlation between explanatory variables is referred to as multicollinearity (as mentioned earlier in this chapter) and is a violation of a key assumption of regression analysis.

2.3.2 Data exploration

In this section, we will look at ways to explore and choose available data. You should be careful to not pick data only to fit a model and vice versa. Many common issues with data can be prevented through utilizing the considerations and tools further discussed below.

2.3.2.1 Data collection

A major consideration of available data is how the data is collected. Measures to avoid bias and correlation in data collection is ensuring that the data is representative of the entire population being explored. If data on the entire population is not available and a sample of demand data must be used, the sampled data often must be randomly collected to be representative of the entire population.

Also, there are other issues that can affect the accuracy and precision of data. Some of these items are the source, unsuitable method of collection, instrument measurement errors, or mistakes in manual data inputting into databases. Certain methods of collection, such as self-reported use or beliefs, carry a level of uncertainty of whether accurate answers were given, intentionally or unintentionally. Errors in measurement, as possible with metering for example, should be expected and investigated for obvious errors that can be further evaluated. Since it is practically impossible to accurately measure natural systems and collect flawless data on large samples, the importance is not to attempt to fully remove all errors, but to be aware and make appropriate interpretations by considering the involved uncertainties.

2.3.2.2 Data time series length

For water demand estimation, the length of available record is important to consider because of the longer cyclical nature of demand over monthly weather changes and annual patterns of higher and lower temperatures and weather event frequency changes. Other examples besides weather could be development growth and density patterns, or long stretches of mandatory conservation measures during drought periods. Having a long enough period of record will determine whether the model can pick up on these changes and offer predictions that will include these variations. If not possible then any significant events that could have impacted the analysis should be noted so any use of the results will be able to consider and use caution when necessary.

2.3.2.3 Data management and cleaning

A decent assumption is that raw data will always require some sort of cleaning. Documenting any changes to raw data is critical for model accountability. Being able to clearly describe any changes to model and the reasoning for doing so is necessary for a full understanding of the model results. If the model is ever to be reproduced or applied to different situations, these notes will be required. Note that many of the academic journal articles strongly recommend open access and data transparency, which will help increase the accuracy/transparency of analytical processes and research outcomes.

Looking through time series water data may have zeros or missed readings. This is not uncommon with metered data. Whether to include or exclude these readings will have implications for the model and interpretation. Questions to consider are whether the zeros are accurate and are representative of shutoffs or a missed reading (e.g. electrical/mechanical failures).

Demographic data can have errors or missing information based on collection methods. Self-reported data has an added layer of inaccurate information that cannot be checked often. In large data sets, the data input process may have added errors. Some of these mistakes can be spotted easily through data exploration methods but they can also go unnoticed or may sometimes be a true outlier. Noting these points is a good practice and deciding how to handle them can be done in later steps, if the outliers are making significant impacts to the data set and the results. Depending on the model objective, arguments to remove these outliers may be justified, but again, always should be noted as exceptions.

2.3.2.4 Descriptive statistics and visualizations

Various methods should be employed to explore the available data. Initial exploration assists in understanding the data patterns and helps with model estimation and interpretation. Being able to describe the collected data (i.e. what is the story from the data?) provides context for the model results and helps make important choices such as what explanatory variables should be included in the model.

Combining basic statistical and visual tools can present an overall, summary view of all information. These tools provide a benchmark, or gut check, to interpret model results and can provide valuable insight on their own. Often, interesting, and important information can be seen by initial data plots, basic statistics, and mapping if geographical is available (see Chapter 15 for use of GIS).

Water demand data can quickly be plotted against suspected influencing factors to determine if there is an observable relationship and the strength of that relationship. A simple *correlation graph* can explain a lot without much expended effort. These plots are also useful in inspecting the data for possible errors or outliers. One type of correlation plot is discussed in Section 2.4 and shown in Figure 2.7. Plots should be done for each considered explanatory variable against the water demand dependent variable.

Preparing *time series plots* of water demand unveils patterns and cycles that may need to be included in the model specification (Chapter 4 will discuss the water demand in time series and their forecasting). Figure 2.8 presents an example of average monthly water demand plotted over time. Any

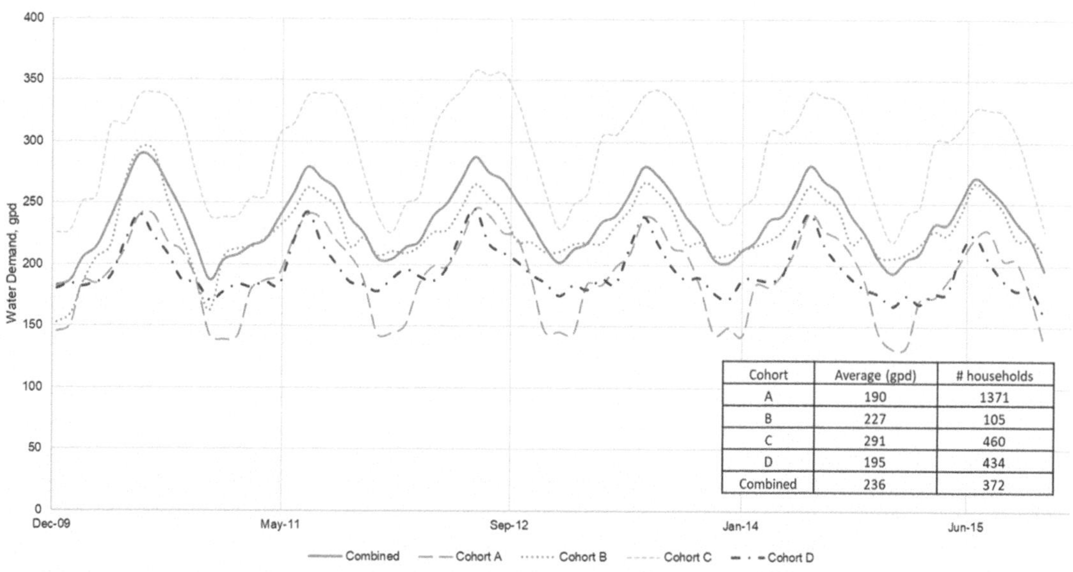

Cohort	Average (gpd)	# households
A	190	1371
B	227	105
C	291	460
D	195	434
Combined	236	372

Figure 2.8 Average monthly water demand by group.

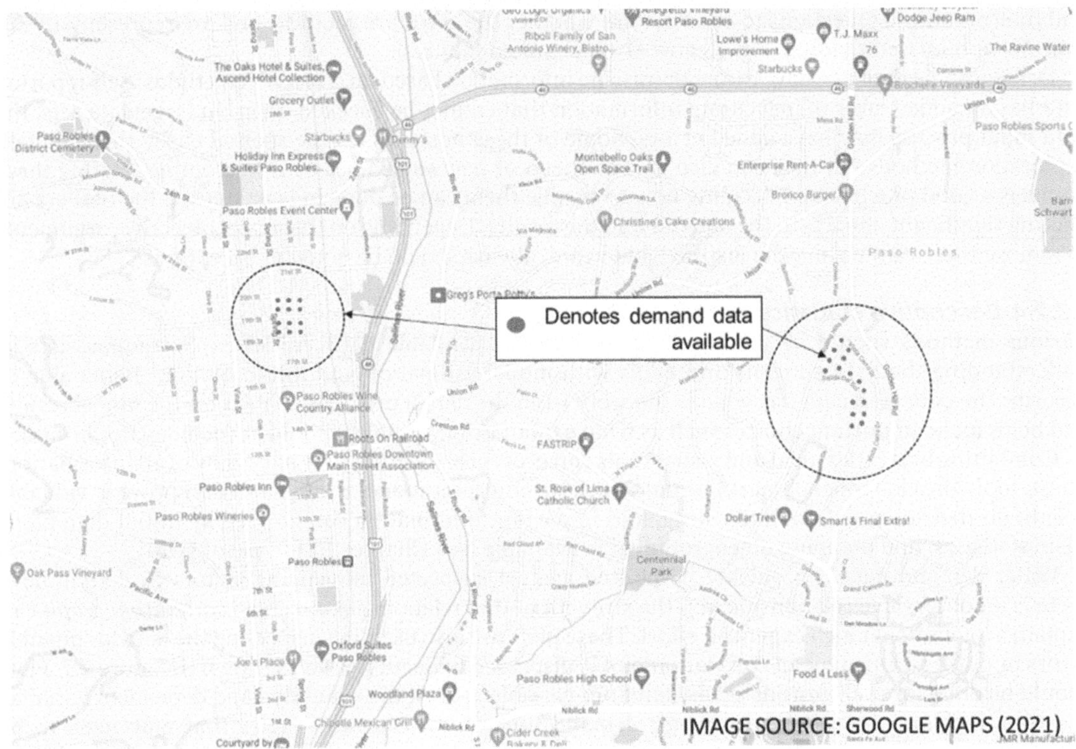

Figure 2.9 Cohort example.

large changes may require further investigation as to the cause and whether it can be captured in the model. Plotting multiple variables across time can also show correlation through time. A seasonal peak during summer months is discernable in the time series. There does not appear to be much variation across the years, except for a slightly noticeable decrease in the final year.

Figure 2.8 also plots water demand data from four separated neighborhood areas, identified in the graph as cohorts A, B, C, and D. By separating out the demand data in this manner, different use is observed. Cohort C appears to have significantly higher average use than Cohort A, for example. Looking at only the combined average line erases the differences between neighborhoods.

Since water use is often localized and may vary greatly between cities or regions, water use data should be explored spatially when possible. Mapping the water demand points can be useful for specific characteristics about location. This type of spatial clustering is a specific occurrence that should be included in the model. For example, if demand data is heavily concentrated in clusters in different neighborhoods, it may be necessary to include neighborhood indicators in the regression model. Figure 2.9 presents a fictionalized example of how useful information can be revealed through mapping. The available water demand information is concentrated in two areas on the map. One area appears to be in a dense, downtown location and the other in a residential area. Since these two types of locations often have different house characteristics, the water demand uses may be different as well.

Descriptive statistics include averages, quartiles, medians, ranges, standard deviations and any other statistic that may be of interest. These calculations can create a picture of the entire data set and can be useful in further investigating data features such as the possible neighborhood specific demand as identified with the time series plotting. Separating the data and running basic statistics helps to quantify the use variation between the neighborhoods. The table within Figure 2.8 shows the variation in average and number of observations between the four cohorts.

Spatial clustering or significant difference between groups in the data can be included in the regression model in different ways. Due to these differences, it may be useful to separate the data into separate models or include a grouping (or cohort) indicator in a single model as an explanatory variable. One way is to run separate models for each group. Another method is to add an indicator variable, sometimes referred to as dummies, for each cohort, or localized area. Since different locations or groups may have unobservable or unquantifiable characteristics affecting use, dummy variables work to capture the expected mean of water demand for that group relative to one group, holding all other variables constant. More details will be explained in Section 2.5.

2.3.3 Level of aggregation

The level of data aggregation may shape what information can be input and what we can extract from the model. It may be necessary to separate data and run separate models for different regions or it may be best to aggregate available data to use for regional or state models. Depending on the objective, it may be necessary to distinguish between different sectors (e.g. agriculture, residential, industrial, or environmental) or scale (e.g. individual household, census block, city, or state), as was evident in the above example of water demand by neighborhood cohorts.

2.3.4 Data range and variation

Regression methods estimate the change in one variable based on changes on other chosen variables. To quantify this change accurately, there must be enough change in the data set. Deciding if data is sufficient and appropriate can be very subjective at times and judgment and experience must be used.

Using the data from Example Problem 1 can help illustrate problems that can arise from lack of data variation. Average monthly water demand was provided along with total monthly precipitation. The precipitation data contained many zeros and many small values. The range of precipitation was zero to 23.9 cm but with an average of 4.57. Out of 72 observations, 15 (20%) were zero. Depending on location, zero precipitation values would be expected so they should arguably not be removed from the data set. If precipitation is the only explanatory variable being used, there will likely be a lot of variation in water demand values associated with zero precipitation. Since all the precipitation values are zero, the variation in those water demand values cannot be explained with a change in precipitation, diminishing the strength of predicative power in the model.

In the case of Example Problem 1, there was enough variation in precipitation to get a regression model with a decent r-squared value. The variation of water demand observations in zero precipitation months was low and there was sufficient variation and correlation in the other values. This may not always be the case and should be considered if the available data has many expected zeros or a small value range. Possible mitigations are adding additional or different explanatory variables, if possible. Transforming the data, such as taking the log of the variable, may also help if the range is small.

2.3.5 Misspecification

When important factors are left out of a regression model then the model is not clearly a 'good or reliable' model (i.e. mis-specified model). Natural systems and human behavior are both challenging to accurately predict. Since a model is only ever an approximation, the objective should be to get as close to the actual phenomena as possible. It may be helpful to remember that it will likely never be possible to precisely explain water demand patterns even if data for all identified influencing factors were available.

If we accept that most models are mis-specified in some manner, we must consider what that means for model interpretation and application. A thorough understanding of the system being modeled helps to appropriately assess and consider the limitations of the model results. The growing availability of large data sets (e.g. 'big data') is a good example of how disciplinary expertise is critical for drawing appropriate conclusions. With big data it can be easy to find correlation between variables that have zero causality. For example, residential water demand tends to peak during the summer months, but so do ice-cream sales. Of course, it would not stand-up to reasoning that to reduce summer water demand, we should restrict ice cream sales.

It can be tempting to add in as many explanatory variables as likely to explain water demand accurately. However, more variables are not always better. Including all possible variables could have an effect that would do more harm than good. Using the above example of water demand and ice cream sales, we should all remember that the common refrain correlation does not mean causation. You should always have a reasonable argument for how each variable influences demand. There are several problems that occur if too many explanatory variables are included in the model without reason. One problem is it makes the model appear to have a stronger explanatory power than it actually does. Another is the increased chance of including explanatory variables that interact with other. When this occurs, the impact of each individual variable on the dependent variable is no longer straightforward. The model may over- or underestimate the impact of the related explanatory variables. On the other hand, omitting important variables is another problem with serious consequences. We will go over this important topic in the following section.

Specifying a model is ***an iterative process***. As discussed in the next section on estimating parameters, running the model and testing the model may lead to further investigation of the data, the model set-up, and may even require reframing of the initial research question.

2.3.5.1 Problem 2

Using water demand as the dependent variable, discuss the reasoning why it was chosen (e.g. would like to project future water supply under changing weather patterns, or evaluate residential water use under drought conservation measures). What are 3–5 explanatory variables that could influence the chosen dependent variable? Find previous literature to support the choice of explanatory variables. Are there factors that could be a strong influencer on the dependent variable but would be difficult to find good data? For the chosen explanatory variables, are there any mechanisms or relationships, showing correlation, between the individual explanatory variables chosen? For the objective, what time period of data would be ideal? Explain your reasoning.

2.4 ESTIMATING PARAMETERS

For regression models considering only one explanatory variable, a simple line could be drawn to estimate the regression line. As mentioned earlier, this type of initial visual estimation can provide a quick snapshot of a linear relationship between two variables. However simple, this method is highly subjective and tends to ignore outliers. Therefore, we need a systematic method to estimate parameters. When multiple explanatory variables are considered, there is no simple graphical method. Ordinary least squared method (OLS) is a widely used for linear models that we will discuss herein. It can be computationally quick and simple to execute with various software programs (e.g. Excel, R, Python, etc.). It is great that we can easily access them, but care should be taken to understand the **assumptions behind the method to ensure reliable results/interpretations**.

2.4.1 Panel regression – pooled, fixed effects, and random effects

When working with panel data, there are three types of regression: **pooled, fixed, or random effects**. A summary of each is given in Figure 2.10. The **pooled OLS** estimator does not consider the panel nature of the data and is what was described in the first example estimating water demand using lot size. Also, the data used in that example is considered *cross-sectional* since there was only a single time period. If panel data is used with the pooled OLS estimator, all the data is pooled together and there would not be any way to track how an individual household water demand changed over time. The intercept

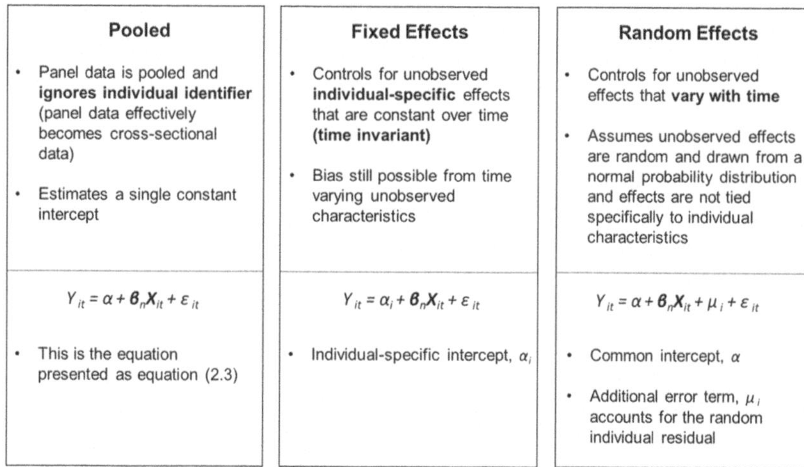

Pooled	Fixed Effects	Random Effects
• Panel data is pooled and **ignores individual identifier** (panel data effectively becomes cross-sectional data) • Estimates a single constant intercept	• Controls for unobserved **individual-specific** effects that are constant over time **(time invariant)** • Bias still possible from time varying unobserved characteristics	• Controls for unobserved effects that **vary with time** • Assumes unobserved effects are random and drawn from a normal probability distribution and effects are not tied specifically to individual characteristics
$Y_{it} = \alpha + \boldsymbol{B}_n \boldsymbol{X}_{it} + \varepsilon_{it}$ • This is the equation presented as equation (2.3)	$Y_{it} = \alpha_i + \boldsymbol{B}_n \boldsymbol{X}_{it} + \varepsilon_{it}$ • Individual-specific intercept, α_i	$Y_{it} = \alpha + \boldsymbol{B}_n \boldsymbol{X}_{it} + \mu_i + \varepsilon_{it}$ • Common intercept, α • Additional error term, μ_i accounts for the random individual residual

Figure 2.10 Panel data regression method summary.

would be a constant value for all entities. The regression intercept in this case would be the average of all water demand for every individual over time. If one individual had a significant different demand pattern, pooling all the data together would ignore the variation within that one individual. Using the neighborhood cohort example from above, this pooled method would ignore specifics about each cohort.

In a **fixed effects model**, individual intercepts are estimated for each individual. In this manner, all unobserved characteristics about a single customer (that would not change with time) is absorbed into an individual-specific intercept. Fixed effects attempt to control for unmeasurable variables that are constant over time, but may vary between individuals. An assumption is that there are characteristics of each household, or group, that effect the amount of water used and for which these characteristics cannot be observed and added to the model as an explanatory variable. A household specific example could be that some houses have older service lines which may be prone to leaks leading to higher water use recordings. This is not something easily known so cannot be included in the model as an explanatory variable. Another example could be household-specific behaviors and attitudes such as frequency of clothes washing or bathing. These behaviors are difficult to accurately model but do account for household specific water use patterns. For the neighborhood cohort example this would be the assumption that there are specific aspects of the neighborhood that cannot be measured or added as an explanatory variable, but there are features, perhaps a conservation culture or a shared love of green lawns, that is not easily measurable or observed.

Lastly, a **random effects** model assumes unobserved individual-specific variables are random, or follow a certain probability distribution, rather than assuming there is some individual-specific characteristics that are correlated with the explanatory variables. Using random effects assumes there is no related individual specific effects. Because of this assumption and the difficulty of proving it, a fixed effects model is most often proposed and will be discussed herein.

2.4.2 Estimation example walk-through problem in R

In this section, a water demand regression problem will be estimated and evaluated using the R program. We would like to estimate a forecasting equation given household-level water utility data consisting of monthly residential water demand over a period of five years. The resulting regression equation can be used to forecast demand for short-term planning and operations of the water distributer. Data for this example is provided in the file: 'Demand_Data_Ex.csv'.

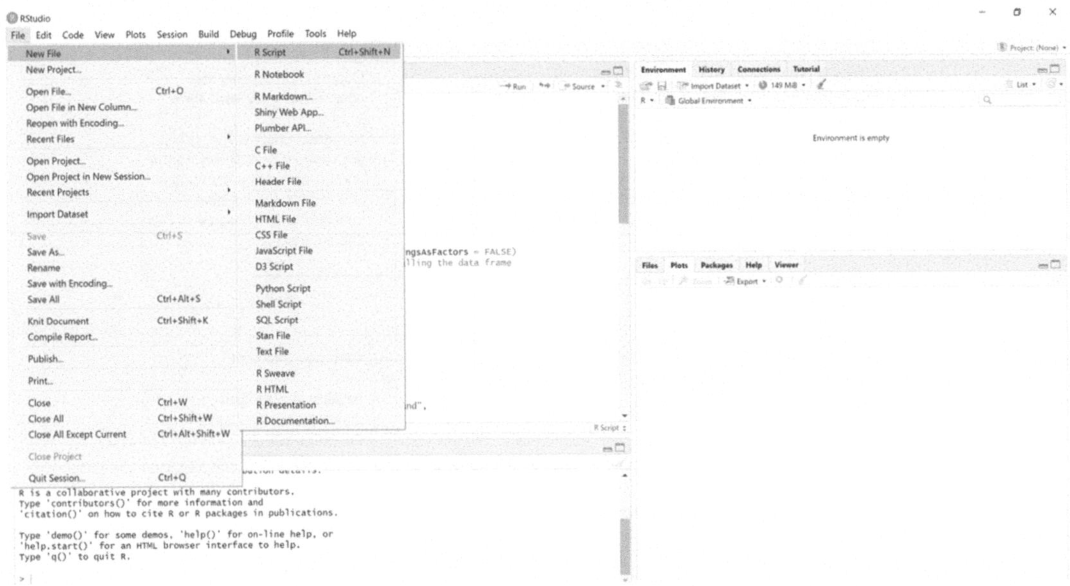

Figure 2.11 RStudio environment – create R scripts.

To begin, let us explore the provided data. The file has already been structured in a format that is ready to use with popular regression packages in the software program R. R can be downloaded freely from *The R Project for Statistical Computing* website (www.r-project.org). R Studio is an additional product that provides a useful editor and tools for R.

Once R Studio is downloaded, there are a few quick steps recommended for set-up. R Studio default layout includes the console where code can be directly run, or code can be written and saved in scripts. Scripts are useful to save, share, and keep a neat record of what is being done. One way a new R script can be opened is through File > New File > R Script (Figure 2.11). Figure 2.12 is a screenshot of the first lines of commands for set-up as written in an R script. The first line has been added to ensure a clean directory and removes data from previous sessions. This is helpful to ensure previous data and objects do not interfere with the current session. The second line sets a working directory so that all files later can be called in reference to that default location. R is case-sensitive so take note of command capitalization and when setting object names. The hashtag on the lines shown in Figure 2.12 represent notes that can be added for reference and will not be executed. Each of the lines in the R script can be run individually with ctrl + enter.

R has default base commands but has many packages that can be installed and loaded. For this problem, we will load several packages. The next few command lines shown in Figure 2.13 show which programs to install and load for this example. Documentation on each of these packages is available and recommended to learn their full capabilities (e.g. Croissant *et al.* 2021). R programmers are constantly improving and writing new packages. The ones shown here are suggestions to use but other packages, including writing your own packages, can be used to achieve the same results presented in this example.

```
rm(list = ls())                        #resets environment of objects
setwd("C:/Users/Steph/Desktop")        #set working directly
```

Figure 2.12 R example set-up commands.

```
#install packages
install.packages("plm")
install.packages("tidyverse")
install.packages("corrplot")

#load packages
library(plm)
library(tidyverse)
library(corrplot)
```

Figure 2.13 R package install and load.

Importing the data file is shown in Figure 2.14. The file is being read into R and is named *data*. The lines below show several ways to explore the data file and its structure. The file has been structured to import as a *data.frame* in R as noted with the str() command. There are eight variables and 57 060 variables. With the head() command, the column names and first few rows are shown. The Summary() command provides basic statistics on each of the variables. Combined, these commands present a quick view of the provided data. In summary, there are five years (60 months) of monthly water

```
#data plots
hist(Demand)                        #histogram - check for normal distribution

#more plots
demandbygroup <- ggplot(data=data, aes(x=Time,y=Demand,group=Group))+
   stat_summary(aes(color=Group), geom="line", fun=mean, size=1)
plot(demandbygroup)

#plot average demand across all observations
matrix <- as.matrix(tapply(Demand,Time,mean))
plot(row.names(matrix),matrix, type="l", main="Average Monthly Demand",
     xlab="Time", ylab="Monthly Water Demand, gpd", col="blue")
```

```
> summary(data)        #basic statistical summary
      ID              Time            Demand           Group            Temp           Rainfall             ET              Bath
 Min.   :  1    Min.   : 1.00   Min.   : 84.0   Min.   :1.000   Min.   :48.70   Min.   : 0.000   Min.   :1.520   Min.   :1.000
 1st Qu.:238    1st Qu.:15.75   1st Qu.:191.0   1st Qu.:2.000   1st Qu.:56.60   1st Qu.: 0.000   1st Qu.:2.822   1st Qu.:1.000
 Median :476    Median :30.50   Median :239.0   Median :3.000   Median :61.40   Median : 0.075   Median :4.535   Median :2.000
 Mean   :476    Mean   :30.50   Mean   :239.7   Mean   :2.874   Mean   :62.96   Mean   : 1.913   Mean   :4.438   Mean   :2.478
 3rd Qu.:714    3rd Qu.:45.25   3rd Qu.:281.0   3rd Qu.:4.000   3rd Qu.:69.12   3rd Qu.: 1.000   3rd Qu.:5.822   3rd Qu.:3.000
 Max.   :951    Max.   :60.00   Max.   :449.0   Max.   :4.000   Max.   :78.30   Max.   :74.000   Max.   :7.830   Max.   :4.000
> colnames(data)       #view column names
[1] "ID"        "Time"      "Demand"    "Group"     "Temp"      "Rainfall" "ET"       "Bath"
> head(data)           #view first few rows of data
  ID Time Demand Group Temp Rainfall   ET Bath
1  1    1    146     1 50.7     0.03 2.44    4
2  2    1    122     1 50.7     0.03 2.44    1
3  3    1    189     1 50.7     0.03 2.44    1
4  4    1    124     1 50.7     0.03 2.44    2
5  5    1    104     1 50.7     0.03 2.44    4
6  6    1    166     1 50.7     0.03 2.44    4
> str(data)            #view data structure
'data.frame':   57060 obs. of  8 variables:
 $ ID      : int  1 2 3 4 5 6 7 8 9 10 ...
 $ Time    : int  1 1 1 1 1 1 1 1 1 1 ...
 $ Demand  : int  146 122 189 124 104 166 160 180 94 184 ...
 $ Group   : int  1 1 1 1 1 1 1 1 1 1 ...
 $ Temp    : num  50.7 50.7 50.7 50.7 50.7 50.7 50.7 50.7 50.7 50.7 ...
 $ Rainfall: num  0.03 0.03 0.03 0.03 0.03 0.03 0.03 0.03 0.03 0.03 ...
 $ ET      : num  2.44 2.44 2.44 2.44 2.44 2.44 2.44 2.44 2.44 2.44 ...
 $ Bath    : int  4 1 1 2 4 4 2 2 3 1 ...
```

Figure 2.14 R import.csv file.

```
#data plots
hist(Demand)                    #histogram - check for normal distribution

#more plots
demandbygroup <- ggplot(data=data, aes(x=Time,y=Demand,group=Group))+
  stat_summary(aes(color=Group), geom="line", fun=mean, size=1)
plot(demandbygroup)

#plot average demand across all observations
matrix <- as.matrix(tapply(Demand,Time,mean))
plot(row.names(matrix),matrix, type="l", main="Average Monthly Demand",
     xlab="Time", ylab="Monthly Water Demand, gpd", col="blue")
```

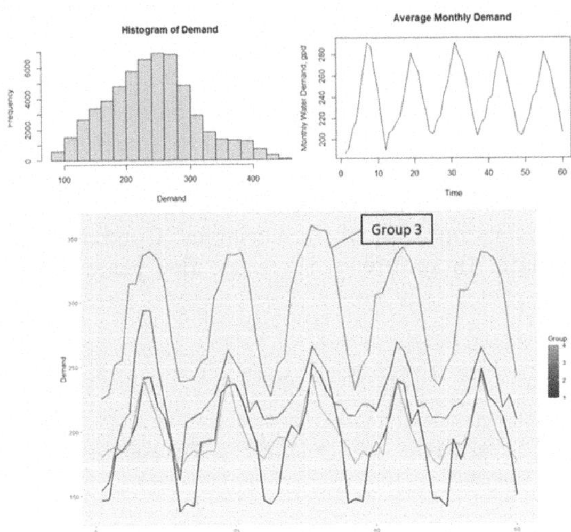

Figure 2.15 R select commands and plots.

demand (lpd) for 951 individuals. For each individual, the number of household bathrooms is provided. Accompanying weather data includes monthly average temperature (degrees Celsius), monthly average rainfall (cm), and monthly average adjusted evapotranspiration (cm). The remaining column, Group, is an identifier categorizing the individual household as being in one of four geographic groups.

Graphing is another way to explore the data as shown in a few selected commands in Figure 2.14. The first is a histogram of the demand variable to check for normal distribution and to view the range of demand data. The next command plots all the demand data for all individuals over time.

In the next plot, only the average monthly average is plotted and is divided into the four groups. The next command lines show a method to check for correlation among all the variables as well as a method to individual check correlation between two variables (Figure 2.16).

As shown in the bottom plot in Figure 2.15, Group 3 demand is significantly higher than the other groups. Because of this notable difference, we will run regression models separately for the groups to capture this difference. For this example, we will show the regression analysis for Group 3 which can be replicated for the other three groups.

A pooled regression is performed first. The plm function is used in this example (Figure 2.17). Within this function, pooling is denoted with specifying the *model* and the data being called is a subset of the larger data file. In the first regression model, called *Pooled_all*, all weather variables

```
#check for correlation between all variables
cor = cor(data)
view(cor)
#check correlation between individual variables
cor(Temp,ET)
cor(ET,Rainfall)
cor(Rainfall,Temp)
```

	ID	Time	Demand	Group	Temp	Rainfall	ET	Bath
ID	1.000000e+00	0.000000e+00	-0.06410828	9.455899e-01	-5.236192e-19	0.0002351120	1.775682e-19	3.670335e-02
Time	0.000000e+00	1.000000e+00	0.05211628	2.502309e-21	1.824468e-01	0.1391279794	1.146822e-01	0.000000e+00
Demand	-6.410828e-02	5.211628e-02	1.00000000	-4.326933e-02	3.731991e-01	-0.0175914152	3.374287e-01	-1.455826e-02
Group	9.455899e-01	2.502309e-21	-0.04326933	1.000000e+00	-2.379971e-19	0.0002835507	5.462671e-19	2.863275e-02
Temp	-5.236192e-19	1.824468e-01	0.37319913	-2.379971e-19	1.000000e+00	-0.0040395614	8.731471e-01	3.150541e-21
Rainfall	2.351120e-04	1.391280e-01	-0.01759142	2.835507e-04	-4.039561e-03	1.0000000000	-9.482133e-02	-1.844205e-04
ET	1.775682e-19	1.146822e-01	0.33742867	5.462671e-19	8.731471e-01	-0.0948213278	1.000000e+00	-2.671779e-20
Bath	3.670335e-02	0.000000e+00	-0.01455826	2.863275e-02	3.150541e-21	-0.0001844205	-2.671779e-20	1.000000e+00

Figure 2.16 R correlation plots.

and the number of bathrooms is used. From the results, bathroom is not significant (p-value greater than 0.05) so the next model, named *Pooled2*, is run without the bathroom variable. Temperature has also been removed, recalling that there was strong correlation between temperature and ET in the correlation matrix which violates one of the basic OLS assumptions. Results in *Pooled2* show rainfall is not significant so another model is run with the remaining explanatory variable, ET. Regression results for the final pooled model are shown in Figure 2.18.

Next, a fixed effects (FE) model is estimated to account for individual-specific effects that do not change over time. Since bathroom is a time invariant individual specific characteristic, it would not be included as an explanatory variable in an FE model. If it was added to the equation shown in Figure 2.19, a coefficient could not be estimated.

A few tests are shown next in Figure 2.20. The first tests for time-fixed effects to check if the pooled or fixed effects model would be most appropriate. For this example, time-fixed effects were observed (p-value less than 0.05 for this test), making an argument that the fixed-effects model should be used. The next lines are selected commands to test the model based on the OLS assumptions. The errors appear relatively normally distributed, and the residual variance appears mostly random.

Further interpretation of these results are discussed in the next section on interpretation.

```
#Pooling Model (Group 3)
Pooled_all<- plm(Demand~ET+Rainfall+Temp+Bath,data=subset(data,Group==3), model="pooling")
summary(Pooled_all)

Pooled2<- plm(Demand~ET+Rainfall,data=subset(data,Group==3), model="pooling")
summary(Pooled2)

Pooled3<- plm(Demand~ET,data=subset(data,Group==3), model="pooling")
summary(Pooled3)
```

Figure 2.17 R pooled regression.

```
> Pooled3<- plm(Demand~ET,data=subset(data,Group==3), model="pooling")
> summary(Pooled3)
Pooling Model

Call:
plm(formula = Demand ~ ET, data = subset(data, Group == 3), model = "pooling")

Balanced Panel: n = 340, T = 60, N = 20400

Residuals:
      Min.    1st Qu.    Median    3rd Qu.       Max.
-164.9757   -38.3587   -5.7707    34.5033   155.4185

Coefficients:
               Estimate Std. Error t-value  Pr(>|t|)
(Intercept) 209.10309     0.97800 213.807 < 2.2e-16 ***
ET           18.85678     0.20533  91.835 < 2.2e-16 ***
---
Signif. codes:  0 '***' 0.001 '**' 0.01 '*' 0.05 '.' 0.1 ' ' 1

Total Sum of Squares:    74261000
Residual Sum of Squares: 52539000
R-Squared:        0.29251
Adj. R-Squared: 0.29248
F-statistic: 8433.68 on 1 and 20398 DF, p-value: < 2.22e-16
```

Figure 2.18 R pooled regression results.

```
#Fixed Effects Model
FE<- plm(Demand~ET, data=subset(data,Group==3), model="within")
summary(FE)

summary(fixef(FE,type='dmean')) #Individual effects, deviating from overall intercept
> FE<- plm(Demand~ET, data=subset(data,Group==3), model="within")
> summary(FE)
Oneway (individual) effect Within Model

Call:
plm(formula = Demand ~ ET, data = subset(data, Group == 3), model = "within")

Balanced Panel: n = 340, T = 60, N = 20400

Residuals:
      Min.    1st Qu.    Median    3rd Qu.       Max.
-113.1550   -38.0036   -5.5821    34.1413   159.2319

Coefficients:
   Estimate Std. Error t-value  Pr(>|t|)
ET  18.8568     0.2054  91.806 < 2.2e-16 ***
---
Signif. codes:  0 '***' 0.001 '**' 0.01 '*' 0.05 '.' 0.1 ' ' 1

Total Sum of Squares:    73421000
Residual Sum of Squares: 51699000
R-Squared:        0.29586
Adj. R-Squared: 0.28393
F-statistic: 8428.31 on 1 and 20059 DF, p-value: < 2.22e-16
```

Figure 2.19 R fixed effects regression results.

```
#Test for time-fixed effects
pFtest(FE,Pooled3) #test for individual effect; p-value < 0.05 then use fixed-effects

#normally distributed errors
hist(residuals(FE), xlab='Residuals')

#fitted values
fitted <- as.numeric(FE$model[[1]]-FE$residuals)
plot(fitted,residuals(FE))
```

Figure 2.20 R fixed effects regression results.

2.5 INTERPRETATION

Regression methods can be employed in various ways. The example in Section 2.4.2 was centered on creating a forecast model. Interpreting those results will be presented next, followed by the presentation of a real example problem that uses regression techniques to evaluate the impact of metering on residential water demand.

2.5.1 Regression example – forecasting

The results from the regression analysis in Section 2.4.2 is shown in Figure 2.14. From the fixed effects model, the regression equation that can be used to represent and forecast average monthly water demand for households within the geographic Group 3 is:

$$\text{Monthly Water demand, lpd} = 791 + 28.1 \times \text{adjusted ET (cm)}$$

Holding everything else constant, for a 1 cm increase in adjusted ET, monthly water demand is expected to increase by 28.1 lpd. This is a rather simple equation that can be quickly used to provide estimates of water demand as it changes on a monthly level. The caveat of its simplicity is the equation provides only an average and would not be useful to predict individual household use. Finer resolution data of end-use appliances as input would be needed to build a finer resolution model for individual households.

ET was estimated to have a significant relationship with water demand but there may be other variables that were not evaluated but could be more meaningful to predict water demand. An example of this could be the price of water. Omitting water price may be relevant if large changes in price occur, for example, since this equation essentially assumes no changes in water price will occur. If forecast equations such as these are consistently used, the models should be updated as more data is collected.

2.5.2 Regression example – metering impacts

A real example problem will be discussed and evaluated in this section to walk through how regression methods can be used to evaluate impacts to water demand over time with changes to particular variables. In this example, the research question involved whether residential water demand would be impacted by the installation of water meters and associated volumetric pricing on previously unmetered residential households. This is following Tanverakul and Lee (2015). Monthly data was collected over 10 years for 1572 residential customers; some of which underwent metering while others did not. The metered group was considered as a *treatment group* and the non-metered households were considered as a *control group*. The control group was utilized as a proxy to account for variation in water demand that would have occurred regardless of the meter installation. All data was collected within one California city with above average demand for the state. A fixed effects regression model was chosen to be able to account for individual household effects.

To deal with the question of pre- and post-metering time periods, three time periods were differentiated and added to the regression model as explanatory variables. A pre-metering period was distinguished, and post-metered time periods were divided into two periods, accounting for a first post-metered period of two billing cycles past metering and a second post-metered period including two later billing cycles. This was done to evaluate whether metering had a short- and longer-term impact to water demand. Of importance is that the time period of metering was not identical for all households as the metering installation program occurred over time. To account for the seasonal effects that could mask changes from metering, a weather variable was added to the model.

The specified regression equation was:

$$\text{Monthly water demand (gpd)}_{it} = \alpha_i + \beta_1(\text{Pre-metered})_{\text{treatment}}$$
$$+ \beta_2(\text{Post-metered Time Period 1})_{\text{treatment}} + \beta_3(\text{Post-metered Time Period 2})_{\text{treatment}}$$
$$+ \beta_4(\text{Pre-metered})_{\text{control}} + \beta_5(\text{Post-metered Time Period 1})_{\text{control}}$$
$$+ \beta_6(\text{Post-metered Time Period 2})_{\text{control}}$$
$$+ \beta_7(\text{Evapotranspiration in inches, ET})_{it} + \varepsilon_i$$

This example problem uses dummy variables to identify whether the observed data is from the control or treatment group and what time period matches the observed data. The dummy variable takes on a value of either zero or one. In the way this regression equation was built, a value of one represents a single time period and group (either treatment or control). For example, when pre-metered water demand in the treatment group is wanted, that variable becomes one in the above equation and all other variables representing time and group are zero. The evapotranspiration variable was used to account and control for monthly and seasonal weather fluctuations.

The equation is estimating monthly water demand based on if a household was metered, time length after being metered (if metered), and ET. The assumption is water demand can be predicted based on these influencing factors. Using fixed effects will allow individual household effects to be controlled. In the above equation, the fixed effects are represented by the intercept value. An individual intercept value will be estimated for each household. We also tested lot sizes, number of bathrooms, and house age for their explanatory strength, but found they were not significant. Significance was evaluated as further discussed below.

The results of the regression model are shown in Table 2.2.

The estimates shown for each explanatory variable represent the impact on water demand. For the ET coefficient estimate of 19.4, average monthly water demand can be expected to increase by a factor of 19.4 gpd (73.4 lpd) with a one-unit change in ET. The rest of the estimates show the average amount

Table 2.2 Regression results.

	Estimate	Standard Error	t-Value	$\Pr(>|t|)$
Pre-metered treatment	721.2	37.886	25.703	$<2.2 \times 10^{-16}$
Post-metered treatment	510.2	34.033	14.386	$<2.2 \times 10^{-16}$
Second-post-metered treatment	501.2	33.900	13.733	$<2.2 \times 10^{-16}$
Pre-metered control	592.9	35.55	17.068	$<2.2 \times 10^{-16}$
Post-metered control	498.7	34.909	9.193	$<2.2 \times 10^{-16}$
Second-post-metered control	465.2	34.909	9.242	$<2.2 \times 10^{-16}$
Adjusted ET (inches)	19.4	5.847	2.155	0.03127
Total sum of squares	1 615 700 000			
Residual sum of squares	1 408 900 000			
R-squared	0.128			
Adjusted R-squared	0.127			
F-statistic	81.25			
p-value	2.2×10^{-16}			
DF	3773			

of water demand for the given group (metered or unmetered) and in what time period (relative to the time of metering).

From the estimates, the difference in demand between the control and treatment groups was 128 gpd (=721.2–592.9) (484.5 lpd), showing that the treatment group used more water on average than the control group. After having a meter installed and moving to volumetric pricing, the treatment (metered) households decreased use by 211 gpd (=721.2–510.2) (798.7 lpd) in the first post-metered time period and 220 gpd (721.2–501.2) (832.8 lpd) by the second time period. Accounting for the decrease in demand that also occurred in the control group, the decrease in demand from metering after six months had a 13% decrease (=((721.2–501.2) – (592.9–465.2))/721.2).

The rest of the information in the table can be used to verify the model. The final column lists two-tail p-values that tests whether each coefficient is different from zero. A zero coefficient would indicate no significant influence of the explanatory variable on water demand. It is common to set the significance level at less than 0.05, so if it is less than 0.05 then the explanatory variable has a statistically significant influence on the dependent variable. The F-statistic does something similar but for the entire model. If the p-value for the F-statistic is less than 0.05 then all regression coefficients on the explanatory variables are significant. Significance here can be thought as the values for all coefficients are different than zero, representing some effect.

2.5.3 Presentation of results

Presentation of the results depends much on the objective of the analysis. At a minimum, basic statistics, regression results, and any statistical tests to validate the regression model should be included for a complete picture of the regression equation and results.

Since all models are approximations, they are riddled with limitations. Including the known limitations as discussed through this chapter is good practice. For most water demand models, because water can be a local affair, acknowledging the demographics and other regional uniqueness is helpful to know where the results and model predictions would have the most appropriate and accurate application.

After the model and results have been presented, critical remaining questions are: What could be done in the next model? What could be improved? Are more or better quality data observations available? Is there a way to improve modeling or understanding of weather patterns?

2.5.3.1 Problem 3

Describe the following results from a fixed effects regression model and write the general regression equation. The dependent variable is average monthly demand given in liters per day. What do the estimates represent? How can you test if each explanatory variable is significant and are there recommendations for deciding to rerun the model with less or different variables? What other information would be helpful to determine if these results were from a properly specified model? How could these results be useful for policy related decisions?

	Estimate	**Standard Error**	**t-Value**	**Pr($>|t|$)**
Number of bathrooms	25.1	55.887	19.901	0.071
House age	0.003	102.03	11.511	0.111
Total bill price	7.59	33.900	14.444	$<2.2 \times 10^{-10}$
Adjusted ET (cm)	16.12	4.899	1.015	0.025
R-squared	0.09			
Adjusted R-squared	0.011			
F-statistic	101.25			
p-value	2.2×10^{-16}			

2.5.3.2 *Brief suggested answer*

The general regression equation is as follows:

$$\text{Water demand (lpd)} = 9.1(\text{number of bathrooms})$$
$$+ 0.003(\text{house age, year}) + 3.39(\text{total bill price, \$}) + 4.89(\text{ET, cm})$$

The average monthly demand is positively influenced with the number of household bathrooms, the age of the house, the total household water bill, and ET values. A greater number of bathrooms, older houses, higher water bills, and higher ET values are expected to increase average monthly water demand. An increase in any of these variables will produce an expected increase in monthly average water demand.

Number of bathrooms, total bill price, and ET values are all significant. House age is not significant.

Holding everything else constant, for every additional bathroom, water demand in lpd is expected to increase, on average, by 25.1 lpd. For every dollar increase in total monthly bill price, expected monthly water demand will increase by 7.59 lpd. An increase in ET of 1 cm is expected to increase monthly water demand by 16.12 lpd.

2.6 CONCLUSION

Water resource management requires a thorough understanding of the significant factors that influence demand. How much water is needed by different sectors and regions is necessary for planning water sources supply, future capital infrastructure programs, water agreements, and alternative and emergency planning. Knowledge of what factors can influence demand, and for what sectors, can be helpful for strategizing conservation programs and other management policies. Regression techniques have a well-demonstrated history of being useful in estimating water demand. This chapter focused on some of the significant aspects of specifying a regression model, estimating, and interpretation. Emphasis throughout the chapter focuses on the importance of understanding how factors influence demand and key things to consider during model estimation and caveats during interpretation.

The multiple linear regression models estimated with ordinary least squares can be simply performed with software programs, making it an ideal choice to perform analysis. The greater challenge is building the regression model and appropriately interpreting results. The mathematical underpinnings of the models should be understood, but the OLS method and fixed effects panel regression was specifically reviewed here to highlight the practical use and effectiveness of these model in providing powerful predictions to manage critical water resources now and for future generations.

REFERENCES

Arbués F., García-Valiñas M. Á. and Martínez-Espiñeira R. (2003). Estimation of residential water demand: a state-of-the-art review. *The Journal of Socio-Economics*, **32**(1), 81–102, https://doi.org/10.1016/S1053-5357(03)00005-2

Croissant Y., Millo G., Tappe K., Toomet O., Kleiber C., Zeileis A., Henningsen A., Andronic L. and Schoenfelder N. (2021). PLM: Linear Models for Panel Data. Available at: https://cran.r-project.org/package=plm; https://r-forge.r-project.org/projects/plm/ (last accessed 5 March 2022).

Donkor E. A., Mazzuchi T. A., Soyer R. and Roberson J. A. (2014). Urban water demand forecasting: review of methods and models. *Journal of Water Resources Planning and Management*, **140**(2), 146–159, https://doi.org/10.1061/(ASCE)WR.1943-5452.0000314

Gracia-de-Rentería P. and Barberán R. (2021). Economic determinants of industrial water demand: a review of the applied research literature. *Water*, **13**(12), 1684, https://doi.org/10.3390/w13121684

Sebri M. (2014). A meta-analysis of residential water demand studies. *Environment, Development and Sustainability*, **16**, 499–520, https://doi.org/10.1007/s10668-013-9490-9

Tanverakul S. and Lee J. (2015). Impacts of metering on residential water demand in California. *Journal of Water Supply: Research and Technology – AQUA*, **64**(2), 211–218, https://doi.org/10.2166/aqua.2014.082

Tanverakul S. and Lee J. (2016). Decadal review of residential water demand analysis from a practical perspective. *Water Practice and Technology*, **11**(2), 433–447, https://doi.org/10.2166/wpt.2016.050

Worthington A. C. and Hoffman M. (2008). An empirical survey of residential water demand modelling. *Journal of Economic Surveys*, **22**(5), 842–871, https://doi.org/10.1111/j.1467-6419.2008.00551.x

doi: 10.2166/9781789062380_0051

Chapter 3

Water demand forecasting – machine learning

*Maria Xenochristou**

Department of Biomedical Data Sciences, Stanford University, Stanford, CA, USA
**Corresponding author: mxenoc@stanford.edu*

LEARNING OBJECTIVES

At the end of this chapter, you will be able to:

(1) Install and run the necessary software.
(2) Perform data preprocessing.
(3) Run a basic ML model.
(4) Assess and interpret the model.
(5) Visualize findings.

3.1 INTRODUCTION

Machine learning (ML) is a subfield of artificial intelligence (AI), where algorithms are learning patterns from data, rather than being rigidly programmed (Radakovich *et al.*, 2020). In this chapter, we focus on *supervised learning*, a field of ML where an algorithm learns how to *map an input to an output*, given a set of examples. Each training example constitutes a sample in our dataset and includes a set of features (predictors/explanatory variables), as well as one or more target variables. In water demand forecasting problems, the target variable is often water demand at a given temporal (e.g., daily or monthly) and spatial (e.g., at the household or city level) scale, while the features are variables that are suspected to influence water demand, such as air temperature or day of the week. ML methods have dominated the water demand forecasting literature (Anele *et al.*, 2017; Antunes *et al.*, 2018; Fiorillo *et al.*, 2021; Menapace *et al.*, 2021; Pesantez *et al.*, 2020; Romano & Kapelan, 2014; Xenochristou, 2019; Xenochristou & Kapelan, 2020), due to their superior accuracy compared to statistical methods. In this chapter, we will introduce basic ML concepts and describe a ML pipeline, from data collection to deployment.

In the following, we outline a basic ML pipeline for water demand forecasting (Figure 3.1) based on tabular data. The first step is understanding the drivers of water demand and defining the types and sources of data we need to collect. Next, we need to follow the necessary preprocessing steps to prepare the data for modeling. The specific methods may vary depending on the project goal, modeling strategy, and data characteristics, but a form of data cleaning, feature engineering, feature selection, and data transformation is often required. Next, we choose a model for our application and determine the optimal set of hyperparameter values, that is model parameters that need to be

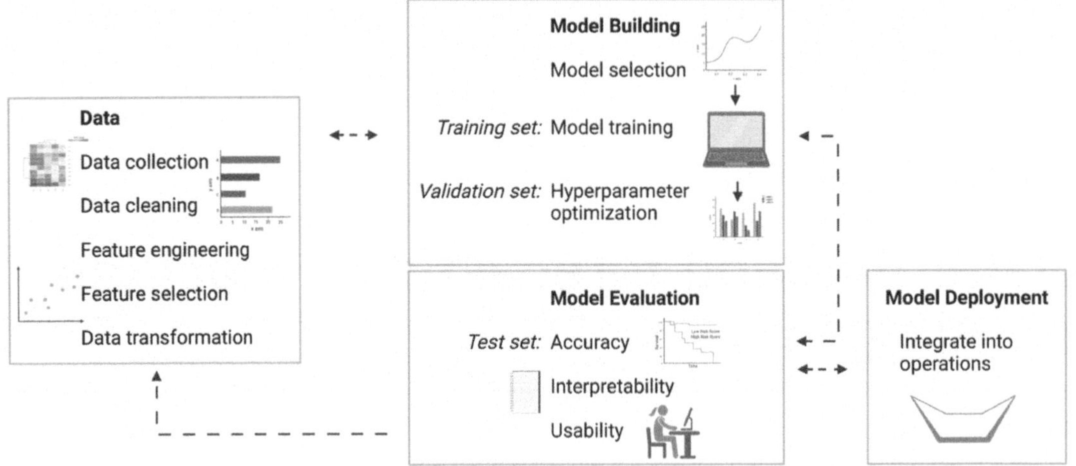

Figure 3.1 The ML pipeline. A simple ML pipeline consists of four interconnected parts, data collection and preprocessing, model building, model evaluation, and model deployment.

determined through trial and error and are not learnt during training. There are several ways to assess the success of the modeling strategy, including model prediction accuracy, interpretability, and usability. Accuracy refers to how well the model predictions match the ground truth. Model interpretability reflects how well we understand how the model makes decisions, while the usability metric incorporates all other constraints that we may need to consider, such as memory and time resources as well as human expertise.

The above process is not linear, as results from each part can be used to update a different step of the pipeline. Insights from model interpretability metrics can inform the data collection process by assessing which features improve model predictions, while a low accuracy may indicate that the model building phase and/or data inputs need to be updated. Reaching the desired outcomes will likely require several iterations of the above process. The final step is model deployment, which loosely refers to integrating the model into operations. After deployment, we need to continuously monitor performance and adjust all parts of the ML pipeline as needed.

In the following, we describe in detail all parts of the above process and list useful software tools. Finally, we present a set of practice problems that will help you understand the fundamental theory and build your first ML model!

3.2 DATA

3.2.1 Data collection

The most important predictor of future water demand is past water demand (Xenochristou *et al.*, 2021), which in most cases is available by the water utility/company. Researchers and practitioners often use additional predictors, that is variables that influence water demand, available from different sources. There are four categories of predictors that are most frequently used in the water demand forecasting literature:

(1) Household and socioeconomic characteristics, such as income, occupancy rate, water price/ rate/rate structure, floor space, property type, and the presence/size of garden. Higher income is linked to a larger number of water-using appliances and higher outdoor consumption (Butler & Memon, 2006; Chang *et al.*, 2010; Domene & Sauri, 2006). Detached houses with larger

floor space are also linked to higher consumption (Butler & Memon, 2006), while in a study by Xenochristou *et al.* (2021), single-occupancy households consumed almost double the daily amount per capita compared to properties with three or more occupants.

(2) Temporal characteristics, such as the day of the week, the season, as well as the time and the type of day (working day or weekend/holiday). Changes in water demand follow seasonal, weekly, and daily patterns. Typically, water demand is higher during the summer months, when water is used for outdoor activities (Cole & Stewart, 2013), as well as weekends (Parker, 2014), when people tend to spend more time at home. In addition, water use follows a diurnal pattern during the day, with peak consumption during the morning (7–8 am) and evening (6–8 pm) hours (Kowalski & Marshallsay, 2005), when most people wake up and come back from work, respectively.

(3) Weather characteristics, such as air temperature, humidity, soil moisture, irradiation, sunshine hours, rainfall, evapotranspiration, and days without rain (Bakker *et al.*, 2014; Dos Santos & Pereira, 2014; Xenochristou *et al.*, 2020a). Out of the weather variables appearing in the literature, air temperature is most strongly linked to water use (Beal & Stewart, 2014; Fiorillo *et al.*, 2021; Willis *et al.*, 2013; Xenochristou *et al.*, 2020b), while there is a much weaker association between water use and rainfall (Beal & Stewart, 2014; Cole & Stewart, 2013; Xenochristou *et al.*, 2020a).

(4) Past water demand incorporates a lot of the above information related to weather, temporal, and household characteristics, as well as water use habits, which make it a valuable source of information. In a study by Xenochristou *et al.* (2020b), the authors found that the importance of additional predictors becomes significantly stronger when past consumption is not included as an explanatory factor.

There are several issues we should consider when drafting the data collection process. The effect of household, socio-economic, temporal, and weather predictors is often considered univariate across different types of customers, properties, and times of the day, the week, or the year. This means that the same increase of 5°C in temperature is assumed to have the same impact in properties with different garden sizes. In reality, the effect of that same increase in temperature on water demand can vary significantly among different types of properties or times of the year (Xenochristou *et al.*, 2021). Therefore, it is important to consider the interactions between these variables (e.g., temperature and garden size) and use forecasting strategies that can capture the complicated relationships among those predictors.

Finally, we need to account for the cost and time required for data collection, data storage and transfer, and ensure the privacy of the related approaches. While the cost of collecting additional data may be justified in a water scarce area where high forecasting accuracy is necessary to ensure water availability, the same cost may not be justified in a different area with higher water availability. In both cases, the data collection strategy should be continuously updated based on the evaluation of the modeling results.

3.2.2 Data cleaning

The data cleaning step aims to reduce the number of errors, gaps, and inconsistencies in the data, as well as remove redundant information. Common data cleaning steps consist of addressing missing and erroneous measurements, identifying outliers, and removing duplicate features and samples. Incorrect or missing measurements can occur due to faults in data recordings (e.g., faulty water meters) or transmission. Pipe bursts can result in large, short-lived spikes in consumption that are relatively easy to identify and remove. However, smaller, ongoing leakages are likely to go undetected by water utilities, customers, and ML practitioners. Using nighttime demand is often a good metric of such leakages as water consumption over the night and early morning hours is expected to be near zero.

On the other hand, while a pipe burst should be excluded from the dataset, days with abnormally high consumption due to other reasons, such as high temperatures overlapping with a weekend or holiday, can provide valuable information to the model. Thus, excluding outliers from the dataset should be handled with care.

Depending on the extent of missing or erroneous values for a certain feature or sample, we can choose to remove it from the dataset or impute the missing values. Simple and commonly used data imputation methods vary depending on the type of data. For time series of water demand, we can impute missing values by linear interpolation. This means that if we draw a straight line between the data point immediately before and immediately after the missing value, we assume that the missing value will fall on that straight line. Alternatively, we can impute missing values with the mean or median across all samples for numeric variables or with the mode for both categorical and numeric variables. Finally, there are specific methods and packages dedicated to missing data imputation, such as the *missForest* package in *R* that can be used to impute continuous and categorical data (Stekhoven & Buehlmann, 2012). The appropriate method for each scenario will depend on the dataset characteristics and level of accuracy required.

3.2.3 Feature engineering

At this step, we need to decide the level of data granularity required for our problem, as both prediction accuracy and feature importance are dependent on the level of temporal and spatial scale (Xenochristou *et al.*, 2020a). This decision will depend on the problem objective and data availability. While understanding water consumption to influence customer attitudes requires water demand modeling at the household or micro-component level, city-level forecasts may be sufficient for planning infrastructure investments. Aggregating data spatially or temporally ultimately results in new features (e.g., from daily to weekly air temperature).

High data granularity is associated with high variability in the consumption signal, partly due to the inherent randomness of water use. Averaging over a longer time period and number of users results in a smoother signal as it averages out individual differences and random effects. Since these are hard to predict, prediction accuracy drops at lower aggregation levels. In a study by Xenochristou *et al.* (2020b), the mean absolute percentage error (MAPE) of daily predictions of water demand increased exponentially from ~5%, for a household group size of 200 households, to ~17% for a group of five households.

Another way of forming new features is by binning categorical or numerical feature values into categories. For example, instead of using the exact size of the garden for each property, we may create groups that contain certain ranges of garden sizes (e.g., 0–10, 10–30 and >30 m^2). This strategy can help reduce the number of classes, balance out class imbalances and increase the number of examples within a certain class. Another scenario where this strategy can be particularly useful is when we know or suspect that a feature has an effect after its value exceeds a certain threshold. An example would be creating a binary variable (a variable that can only take one of two values), indicating if the maximum air temperature exceeded 35°C, or using the daily amount of rainfall to create a new feature that corresponds to the number of consecutive days without rain. This would be particularly useful if we think that the presence of an event (e.g., whether it rained or the temperature exceeded a certain threshold) is what drives water demand. Finally, we can create new features using dates, as water use follows a seasonal, weekly, and daily pattern, thus we can use the season, month, day of the week, and time of day as predictors of water demand.

3.2.4 Feature selection

One caveat of ML models is that since they do not make any underlying assumptions about the relationship between inputs and target, but learn based on a set of examples, they are prone to *overfitting* on the training data. This means that they learn to fit the training set too well, and thus fail to generalize on new, unseen data (Figure 3.2).

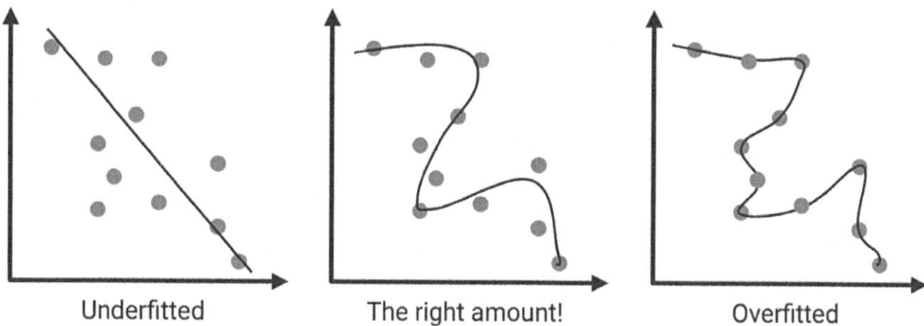

Figure 3.2 Different model fitting scenarios. The dots represent the data points while the line represents the model fit. A model that is underfitted has not learned meaningful relationships between the input and target variables, while an overfitted model has learned the training set too well and is not able to generalize on unseen data.

Feature selection aims to reduce noise by removing the features that are less likely to contain meaningful information. Using too many features as model predictors can increase the risk of overfitting, also known as the *curse of dimensionality* (Indyk & Motwani, 1998), particularly when the model does not have enough samples to learn from. For this reason, we want to remove uninformative or redundant features. If the number of both features and samples is too small on the other hand, the model may underfit on the training data. In other words, it may not have enough examples and/or features to learn meaningful relationships between predictors and target.

A simple feature selection approach is to filter out features that are strongly correlated and features with zero (or near-zero) variance. Including strongly correlated predictors that provide similar information can bias the model towards these predictors (e.g., house size and lawn size), while features that have the same value for all samples are unlikely to explain the variability in the target.

Another option is to filter features based on importance. The correlation between feature and target can provide a first indication of feature importance. However, this method does not account for feature interactions that can provide additional information to the model. For this reason, methods that use the model as part of the feature selection process are preferred. Sequential feature selection iteratively finds the best feature to add to the model to maximize performance, according to a scoring metric (e.g., by minimizing mean absolute error). Backward sequential feature selection applies the reverse of the above method; it iteratively removes the feature that causes the smallest reduction in model performance. Finally, linear models such as the Lasso algorithm (Tibshirani, 1996) that model linear relationships between a set of features and a target can be used for feature selection. The Lasso algorithm performs feature selection by applying an L1 sparsity penalty that forces many coefficients (the ones with the smallest effect on the cost function) to zero. By forcing a coefficient to zero, the corresponding feature is not used as a model predictor. Using a Lasso model as a preprocessing step for feature selection has the benefit of accounting for interactions between features and their influence on the target. However, this only applies to linear relationships between model features and target.

We can also reduce the number of predictors using *dimensionality reduction methods*. These refer to the transformation of a high dimensional space (in this case a set of features) to a lower dimensional space, while maintaining most of the qualities of the original feature set. Some techniques we use to achieve dimensionality reduction are *Principal Component Analysis (PCA), t-distributed stochastic neighbor embedding (t-SNE), and Autoencoders*. In most water demand forecasting studies, the number of predictors is relatively small and thus dimensionality reduction methods are not typically used.

3.2.5 Data transformations

Different types of algorithms require different transformation steps that depend on their structure and assumptions. Common data transformation methods include data normalization and standardization, and data encodings:

- Data normalization refers to scaling predictors, often between 0 and 1. Predictors can have vastly different scales, such as 1000–150 000 USD for income and −20 to 40°C for temperature, which can cause issues during model training. Data normalization is particularly important for distance algorithms such as *k-nearest neighbors* (k-NN), algorithms that use *regularization* (such as Ridge Regression and Lasso), and algorithms that use *gradient descent* (such as neural networks).
- Data standardization is the process of transforming data to have zero mean and unit variance N (0, 1). It is used when an algorithm assumes the data to be normally distributed.
- Categorical data encoding is the process of turning categorical labels into numerical values. This is often required as most models can only use numerical inputs. The type of encoding that is recommended depends on the nature of the categorical data. If the categorical values are ordinal (e.g., garden size bins), then ordinal encoding assigns a numerical value to each category. For categorical data that lack this structure, one-hot-encoding transforms each class into a feature with a binary value for each sample in the data. For example, the property type could be encoded as three different features (single-family home, townhouse, and condominium), where the value indicates if a property belongs to the corresponding property type (1) or not (0).

For a visual guide of the effect of different data transformation methods, see the scikit-learn package guide (https://scikit-learn.org/stable/auto_examples/preprocessing/plot_all_scaling.html) (Pedregosa *et al.*, 2011).

3.3 MODEL BUILDING

3.3.1 Model selection

There are many types of ML models with varying levels of complexity, requirements, and use cases. The choice of ML model should account for many factors, such as data availability, cost, project aim, and vulnerability of research area.

3.3.2 Hyperparameter optimization

Hyperparameters are model parameters that are defined prior to model training. They determine various model characteristics such as how quickly the model learns or how much randomness is induced in the training process and need to be tuned for each individual dataset. The selection of the right set of hyperparameters is called *hyperparameter optimization* or *hyperparameter tuning*.

There are four methods commonly used for hyperparameter tuning: *manual search, grid search, random search, and Bayesian optimization*. The simplest but also the most labor-intensive way to do hyperparameter optimization is by manually testing model performance for different combinations of hyperparameter values. In grid search, we automate the process by defining a search grid for each hyperparameter and iteratively testing all combinations within this multi-dimensional grid, where each hyperparameter is one dimension. In random search, values are selected randomly from within the search space. Finally, in Bayesian optimization, hyperparameter combinations that have higher probability of resulting in higher prediction accuracy are selected. Many R packages have methods for hyperparameter tuning already implemented and ready to use. The caret (Kuhn, 2008) and h2o (h2o.ai 2020) packages in R provide the capability for grid search and random search for a number of algorithms and hyperparameters.

3.3.3 Training, validation, and testing

Since ML models are prone to overfitting, we need to ensure that the trained model can generalize on new data. For this reason, we divide our data into three sets used for *model training, validation, and testing*. The training and validation sets are used for model development. Specifically, the training set is used to train the model, that is to learn on a set of examples, while the validation set is used for hyperparameter optimization (Figure 3.1). The test set provides an unbiased estimate of model performance on unseen data, that is data that was not used during the model development phase. When modeling time series, the training set should only include samples that are chronologically prior to the validation set. Similarly, the validation set should only include samples that are chronologically prior to the test set.

Cross-validation (Kohavi, 1995) is a sampling technique we can use to divide data into training and validation. It is used to provide a robust estimate of model performance and it is particularly useful when the number of samples is limited. A basic form of cross validation is based on dividing the dataset into k equal parts (*k-fold cross-validation*). At each iteration, one fold is used as the validation set, while the rest of the folds are used for training.

3.4 MODEL EVALUATION

3.4.1 Model accuracy

Assessing model performance depends on the problem definition, requirements, and constraints. In water scarce areas, where water utilities are at risk of being unable to cover demand, accurate predictions are essential to ensure water availability and inform decision making. In this case, sacrificing cost and interpretability to obtain extra accuracy is likely a worthy investment.

Accuracy metrics that are often used in the water demand forecasting literature are *Mean Absolute Error – MAE* (Antunes *et al.*, 2018; Dos Santos & Pereira, 2014; Herrera *et al.*, 2010; Kofinas *et al.*, 2014; Shabani *et al.*, 2016), *Mean Absolute Percentage Error – MAPE* (Bai *et al.*, 2014; Candelieri *et al.*, 2015; Kofinas *et al.*, 2014; Tiwari *et al.*, 2016), *Root Mean Square Error – RMSE* (Dos Santos & Pereira, 2014; Kofinas *et al.*, 2014; Shabani *et al.*, 2016; Tiwari *et al.*, 2016), and R^2 *coefficient of determination* (Babel *et al.*, 2007; Bakker *et al.*, 2014; Dos Santos & Pereira, 2014; Haque *et al.*, 2014; Kofinas *et al.*, 2014; Shabani *et al.*, 2016; Tiwari *et al.*, 2016).

Each accuracy metric has advantages and disadvantages. The MAE assigns the same importance to larger and smaller errors, as well as positive and negative errors. It solely provides an indication of the overall agreement between predicted and observed values (Tiwari *et al.*, 2016). The MAPE is independent of units and therefore can be used to compare results across different studies and utilities (Candelieri *et al.*, 2015). The RMSE is the square root of the mean square error (MSE) and is sensitive to larger errors (Tiwari *et al.*, 2016). The R^2 ranges from 0 to 1 and indicates the degree of association between modelled and observed values (Haque *et al.*, 2014). A wide range of accuracy metrics are available in the MLmetrics R package (Yan, 2016).

However, even if the model has good overall accuracy, it may fail to predict peak demands. ML algorithms assume that the distributions of the training set and test set are the same. Since extreme demand values are rare, the model is less likely to predict them. In the previous example of a water scarce area, accurate predictions are particularly important when a water utility may struggle to cover demand, that is on days and hours of peak consumption. Thus, it is important to ensure that the model performs well on those critical days. Improving data representation, as well as choosing the right model for the task and using methods that facilitate identifying rare events, can assist with improving model performance on days with peak demand (Xenochristou & Kapelan, 2020).

3.4.2 Model interpretability

ML model interpretability reflects the degree to which humans can understand the cause of algorithmic decisions (Miller, 2019). ML models can account for thousands or hundreds of thousands of features and learn complex, non-linear relationships between those features and one or multiple targets. Understanding these relationships and how they can influence a certain prediction can enhance the usability of these methods by informing planning and decision making, as well as instilling confidence in the model's decisions. This is particularly important in fields such as engineering and healthcare, where it is important to ensure that a ML model is making decisions based on true signal rather than data artifacts.

Interpretability methods can be model-specific when they apply only to a specific model type, or model-agnostic when they can be used with any model (Molnar, 2020). An example of a model-specific method is the interpretation of weights in linear models, where the target is modeled as a linear combination of a set of predictors. The higher the coefficient value of each predictor, the higher its importance. Examples of model-agnostic methods are Permutation Feature Importance (Breiman, 2001), Partial Dependence Plots (PDP) (Zhao & Hastie, 2021), Accumulated Local Effects (ALE) plots (Apley & Zhu, 2020), and Individual Conditional Expectation (ICE) curves (Goldstein *et al.*, 2015).

Permutation Feature Importance can be used with tabular data and is the reduction in predictive performance when a predictor is permutated. By shuffling the values of the predictor, we break the association with the target variable. Thus, the higher the drop in model performance, the higher the importance of that predictor. When using permutation feature importance, it is important to consider correlations between predictors, as these can lead to misleading results. If two (or more) features are highly correlated and therefore provide the same information to the model, removing one of them by permutating its values may not significantly affect the model's performance.

PDPs and ICEs visualize the model response for a certain change in the predictor. PDPs force a model feature to take the whole range of its values for each data instance and calculate the model response each time. For example, if the predictor is air temperature and the target is water consumption on a given day, PDPs will vary the values of air temperature within its range of possible values, while all other predictors are kept constant. The final plot consists of the mean water consumption among all days in the dataset for the corresponding temperature value. ICEs, on the other hand, demonstrate the model response for each data instance. In the same example, ICEs show the range of predicted water consumption for each day in the data, for the whole range of temperature values. Similar to Permutation Feature Importance, PDPs and ICEs assume independence between predictors. If the predictors are not independent, these methods may create instances with unrealistic combinations of feature values (e.g., an air temperature value of 35°C and soil temperature of 0°C).

ALEs are a faster, non-biased alternative to PDPs. Instead of forcing a predictor to take the whole range of its values, they analyze the variation of the model's response within a small window of the predictor's real value. Therefore, ALEs are robust to correlations among model features. For a detailed overview of ML interpretability methods, see Christoph Molnar's book on Interpretable ML (Molnar, 2020).

3.5 MODEL DEPLOYMENT

Deployment refers to incorporating the model as part of operations. For example, we could deploy an ML model for predicting water demand in real time with the aim to raise alerts for leakages or pipe bursts, when the prediction error is higher than a certain threshold. However, not all deployed models are required to run in real-time.

3.6 TOOLS AND SOFTWARE

3.6.1 Prerequisites

Working through the following examples requires installing R (R Core Team, 2019), a freely available programming language and software environment, and the RStudio Integrated Development

Environment (IDE) (RStudio Team, 2020). R offers a variety of packages that we will use in the following problems. Packages contain R code and reusable functions, as well as documentation that explains how to use them. R is available for Linux, Mac, and Windows.

3.6.2 Useful tools, packages, and APIs

In the following, we list some useful and popular R packages for ML:

- **Caret:** the 'caret' package (Classification and Regression Training) aims to simplify ML model training and hyperparameter tuning. It includes a variety of models as well as methods for data preprocessing, visualizations, and feature importance (Kuhn, 2008).
- **Keras:** Keras is a high-level, deep learning API (Application Modeling Interface), developed by Google, written on top of the Tensorflow ML platform. The 'keras' package provides an R interface for the Keras API. For more information, see https://tensorflow.rstudio.com/.
- **h2o:** the 'h2o' R package provides an R interface for the open-source AI platform h2o, built by the software company H2o.ai. The automl function of 'h2o' can automatically train and hyperparameter optimize several commonly used ML algorithms, as well as two stacked ensembles. A stacked ensemble is a combination of predictions from previously trained models. When training the stacked ensembles, h2o finds the best combination that minimizes prediction errors among (1) all previously trained models and (2) the best model (with the optimum set of hyperparameters) of each type.
- **randomForest:** 'randomForest' is an R package that uses Random Forests for classification and regression based on Breiman (2001).
- **ICEbox:** ICEbox (Goldstein *et al.*, 2015) is an R package that implements ICE curves for any supervised ML algorithm.
- **Ggplot2:** ggplot2 (Wickham, 2016) is one of the most popular data visualization libraries and it provides functionalities for a variety of graphs.
- **Plotly and Shiny:** plotly (Plotly Technologies Inc., 2015) and Shiny (Chang *et al.*, 2019) are popular R libraries for making interactive graphs.
- **MLmetrics:** the MLmetrics package contains a variety of metrics for ML that evaluate classification, regression, and ranking performance (Yan, 2016).
- **fpp2:** The 'fpp2' package (Hyndman, 2020) contains a set of datasets that are used within the book 'Forecasting: principles and practice' (Hyndman & Athanasopoulos, 2018). These datasets can be a great resource when you are experimenting with your first forecasting models!
- **dplyr:** 'dplyr' (Wickham *et al.*, 2020) is a popular R package used for various types of data manipulation.
- **iml:** the 'iml' (interpretable machine learning) R package (Molnar *et al.*, 2018) contains a selection of machine learning model interpretability methods, including ALE plots, PDP plots, and ICE curves.

We will use several of these packages in the following examples.

3.7 PRACTICAL EXAMPLES

3.7.1 Installation

(1) Instructions on how to download and install R are available from CRAN (https://cran.r-project.org/).
(2) RStudio Desktop is available to download for free under an open source license from (www.rstudio.com/products/rstudio/download/).

We run the following examples with R version 4.0.3, and R Studio version 1.2.5033.

3.7.2 Example 1: A simple model for demand forecasting

For this example, we will use electricity consumption at the household and daily level as the target variable, as well as temporal (day of the week) and weather (temperature) variables as predictors. We will use electricity instead of water consumption as this data is readily available and easy to load directly from the 'fpp2' R package.

First, open a new R Studio window (Figure 3.3). From your R studio window, create a new R script, by clicking on 'R Script' from the drop-down menu on the top left (Figure 3.4). You can use R scripts (or files) to write and save code. You can save the R file you created by clicking on 'File' and then 'Save as' at your menu bar on a Mac. We will name the file for this example 'Example_1.R'.

3.7.3 Installing and loading R packages

Next, you need to install the necessary R packages. You will do this only once (unless you uninstall them). You can install an R package by typing install.packages() and adding the package name in the brackets. For this example, we will install the packages 'fpp2', 'randomForest', 'MLmetrics', and 'dplyr' (Figure 3.5).

You can execute a line of code (command) by selecting it in the source editor window and either clicking the run icon on the top right menu of your script, or clicking control + enter. You can comment

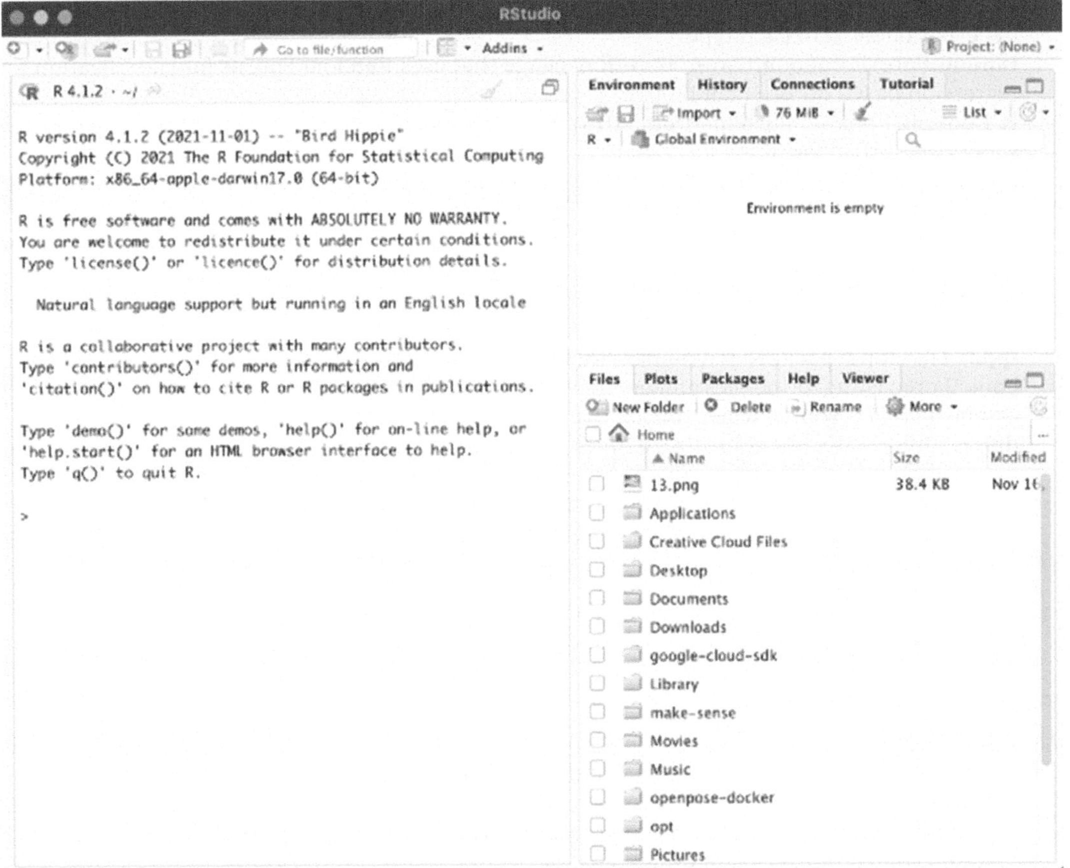

Figure 3.3 R Studio interface.

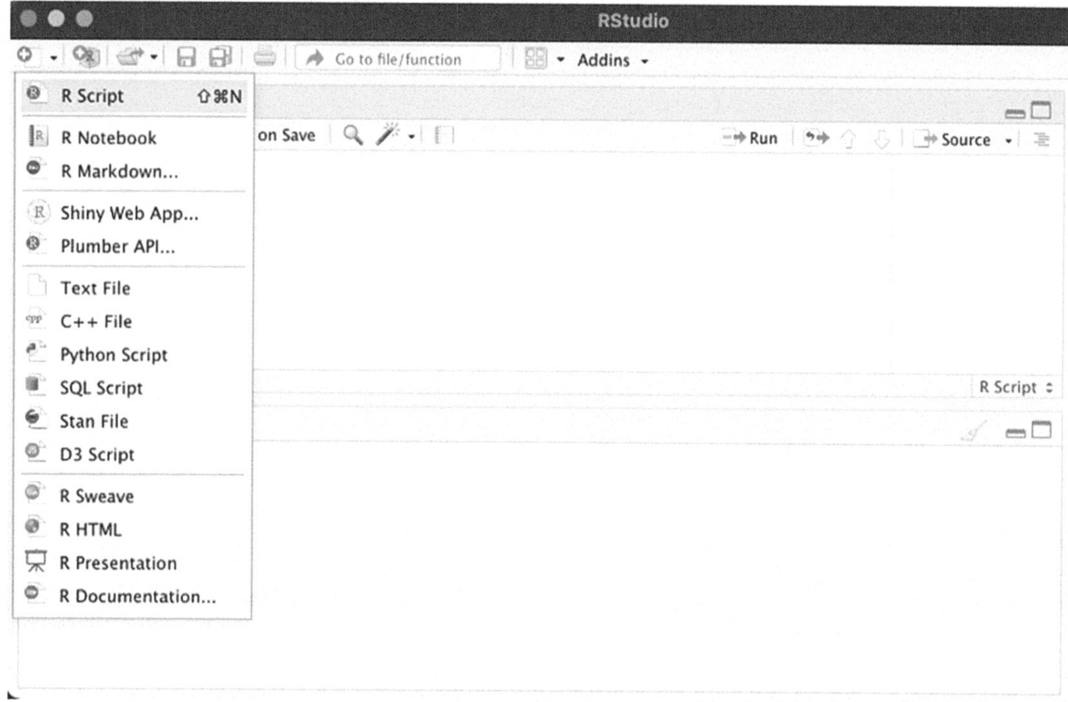

Figure 3.4 Create a new R script.

a line of code by using the # symbol at the start of thc line. Commented lines are not executed when you run your code. You can see the results of your command in the console window.

Unlike installing, you need to reload the necessary packages every time you start a new RStudio session. You can load an R package by typing library() and adding the package name in the brackets. For this example, we will load the packages we installed above, 'fpp2', 'randomForest', 'MLmetrics', and 'dplyr' (Figure 3.6).

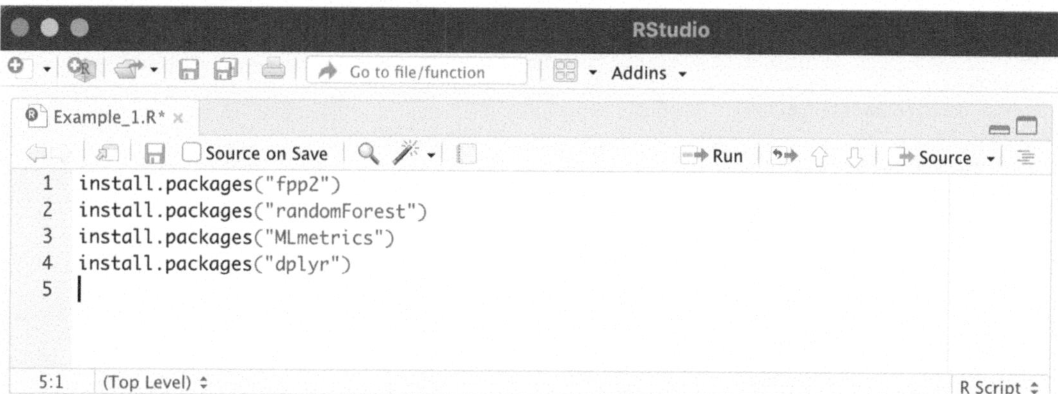

Figure 3.5 R Studio package installation.

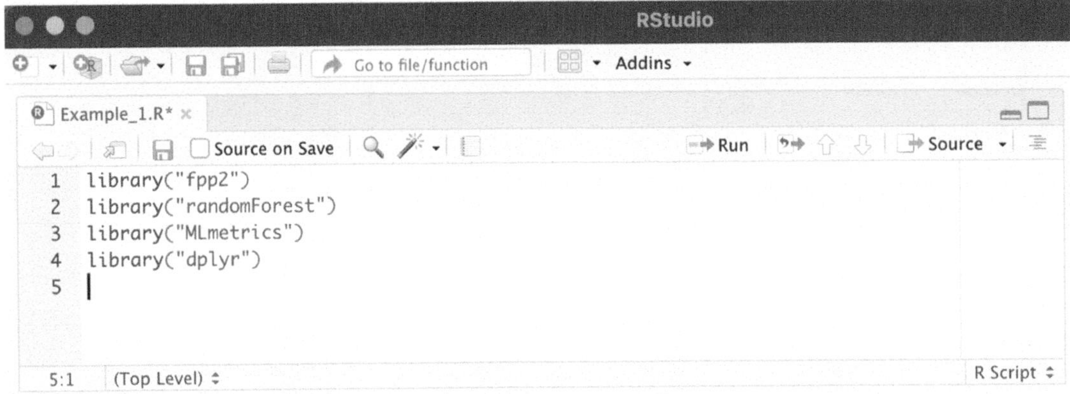

Figure 3.6 R Studio package loading.

3.7.4 Get and preprocess the data

Load the electricity demand dataset available from the 'fpp2' package (Figure 3.7). This dataset contains daily electricity demand, temperature, and type of day (working day or holiday/weekend), from 1/1/2014 to 31/12/2014. You can see the first six rows of the data frame by using the head() function (Figure 3.7). You can write, edit, and save your code as an.R script in the source pane (top window, Figure 3.7) and execute it in the console (bottom window, Figure 3.7).

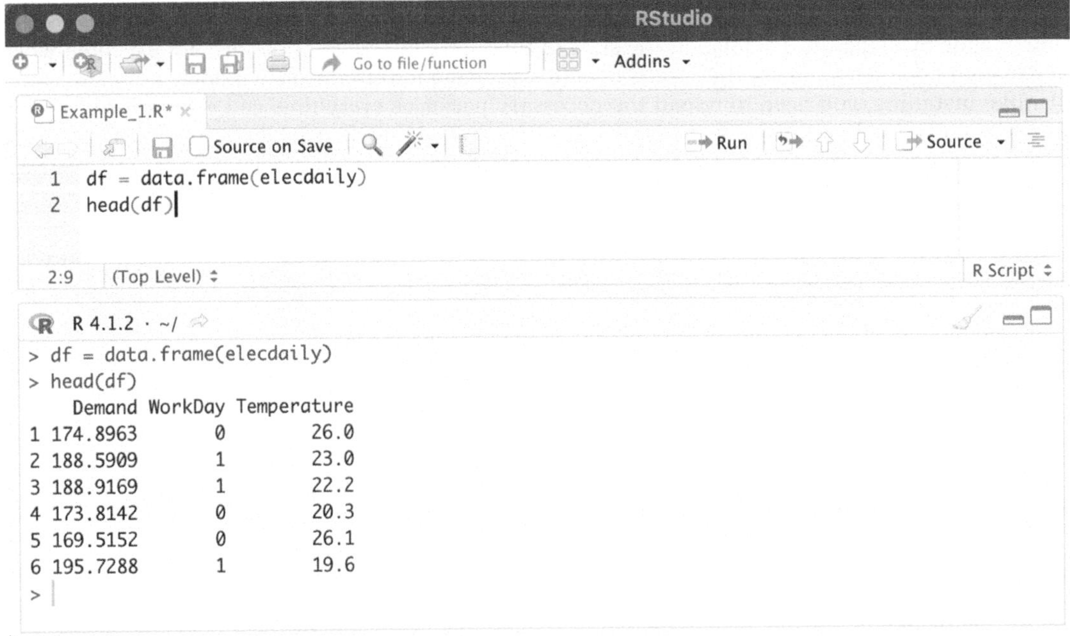

Figure 3.7 Load the data from the fpp2 package.

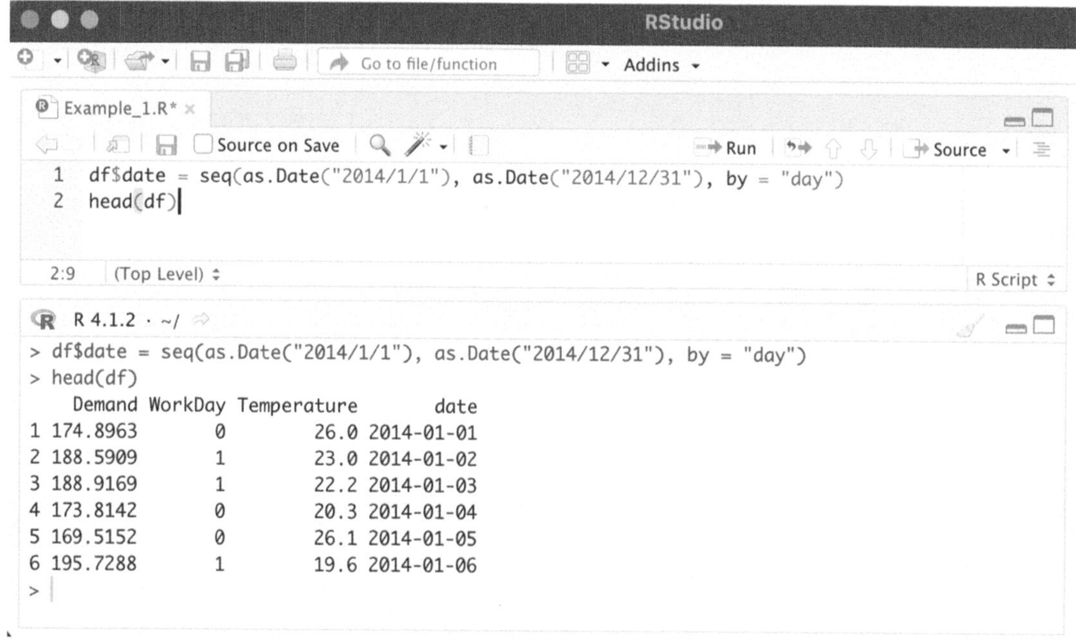

Figure 3.8 Create a date column.

Create a new data frame column called 'date', by defining a sequence with a start and end date (Figure 3.8). The elecdaily dataset in the fpp2 package contains daily electricity demand values for every day in 2014.

TIP: We are going to use a Random Forest model, so we can omit some data preprocessing steps.

Create seven additional columns with demand 1–7 days prior to each day. The following for-loop will run the statement in brackets for seven different values of the 'days_ahead' variable, from 1 to 7 (Figure 3.9). For example, demand on the 4th January 2014 (2014-01-04) was 173.8142 ('Demand' column), whereas demand on the previous day was 188.9169 ('Demand_1_days_prior'/'Demand' on 2014-01-03), and demand 2 days prior was 188.5909 ('Demand_2_days_prior'/'Demand' on 2014-01-02).

Next, create a new column from date with the day of the week (Figure 3.10).

Remove rows with missing values. You can inspect the number of rows before and after you remove missing values, using the nrow() function (Figure 3.11).

Define the predictors and the target. In this example, we use demand 1–7 days prior to the target day, as well as day of the week (Monday–Sunday) and temperature as predictors of demand (Figure 3.12).

3.7.5 Model training and testing

Divide your data chronologically into a training set (50%) and a test set (50%) (Figure 3.13). You can do this using the nrow() function to get the total number of samples (rows) in your dataset. You can get a subset of a data frame by defining the index of rows and columns you want your new data frame to have as:

df_new = df[start_row_index:finish_row_index, start_column_index:finish_column_index].

If either the column or row index is left blank, the new data frame will include the same rows or columns as the old data frame.

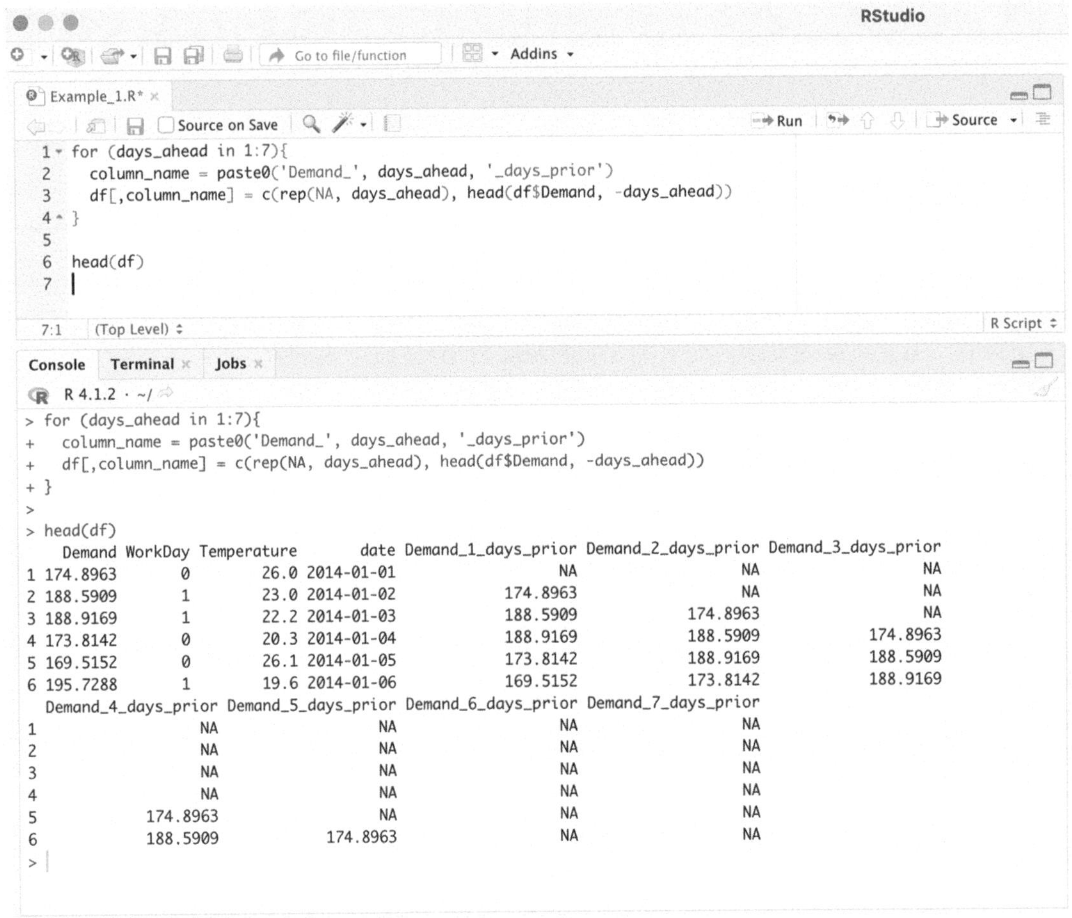

Figure 3.9 Create past demand columns to use as predictors of future demand.

In this case we create df_train as a subset of df, by selecting the first 50% of the rows of the original dataframe (1 to 0.50*k, where k is the number of rows of the original data frame). Similarly, we create df_test from the remaining 50% (rows 0.5*k+1 till k). Both the training and test set contain the same columns as the original data frame.

TIP: *Since we are not optimizing the model hyperparameters, we do not need a validation set.*

Train the model on the training set and make predictions on the train and test set (Figure 3.14). In the randomForest() function, we determine the target variable as the 'Demand' column, while all other columns are used as predictors. After we train and save the model, we use it to make predictions using the predict() function.

Evaluate your predictions using the R^2, RMSE, MAE, and MAPE. All metrics are available from the 'MLmetrics' package (Figure 3.15). You can compare the accuracy of the training set and the test set to assess how well the model is able to generalize on new, unseen data. If the accuracy of the training set is significantly higher than this of the test set, the model has overfitted on the training data. Tuning the model hyperparameters can assist with achieving the desired model fit.

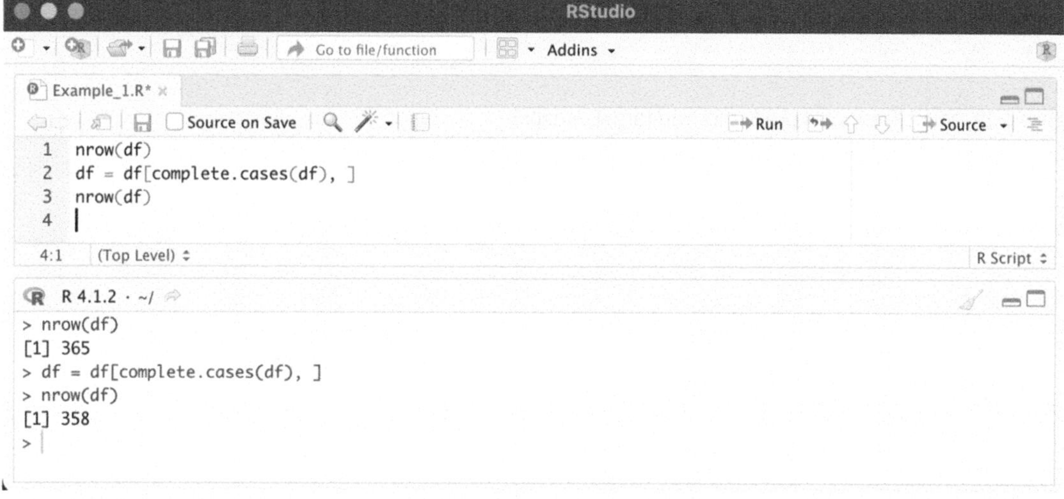

Figure 3.10 Create a weekday column to use as predictor of demand.

Figure 3.11 Remove missing values.

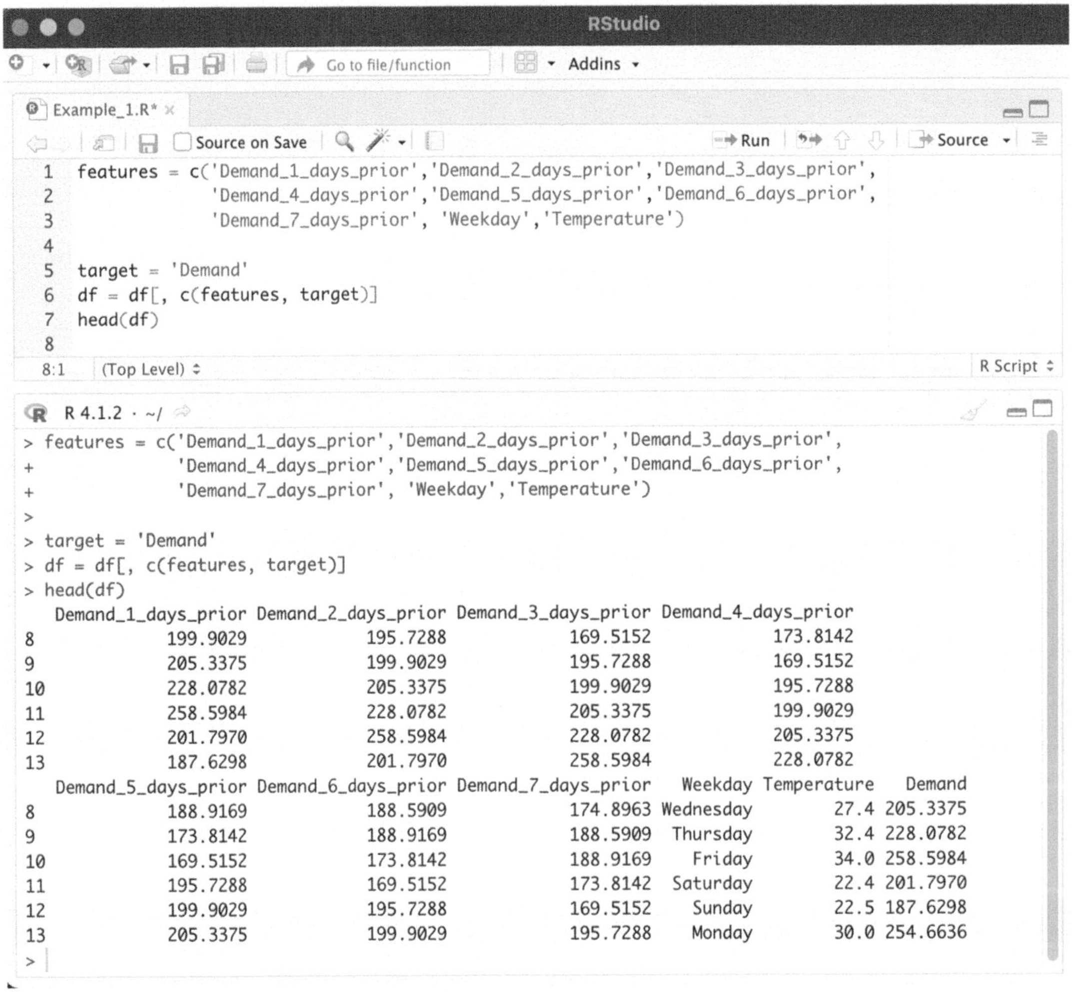

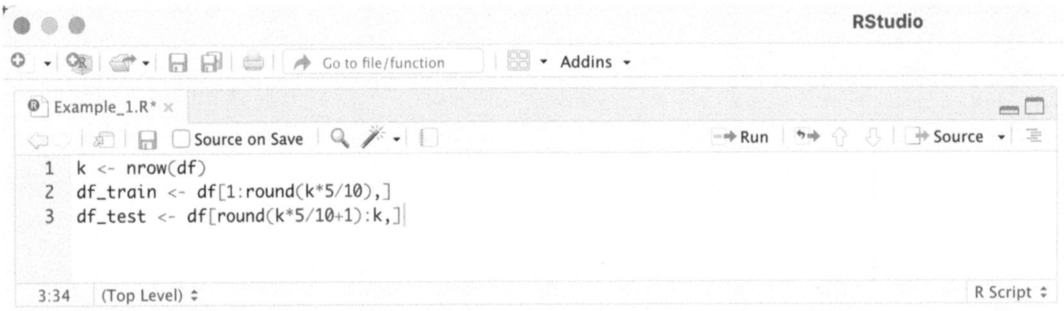

Figure 3.12 Define the model predictors and target variable.

Figure 3.13 Divide the data into a training set and a test set.

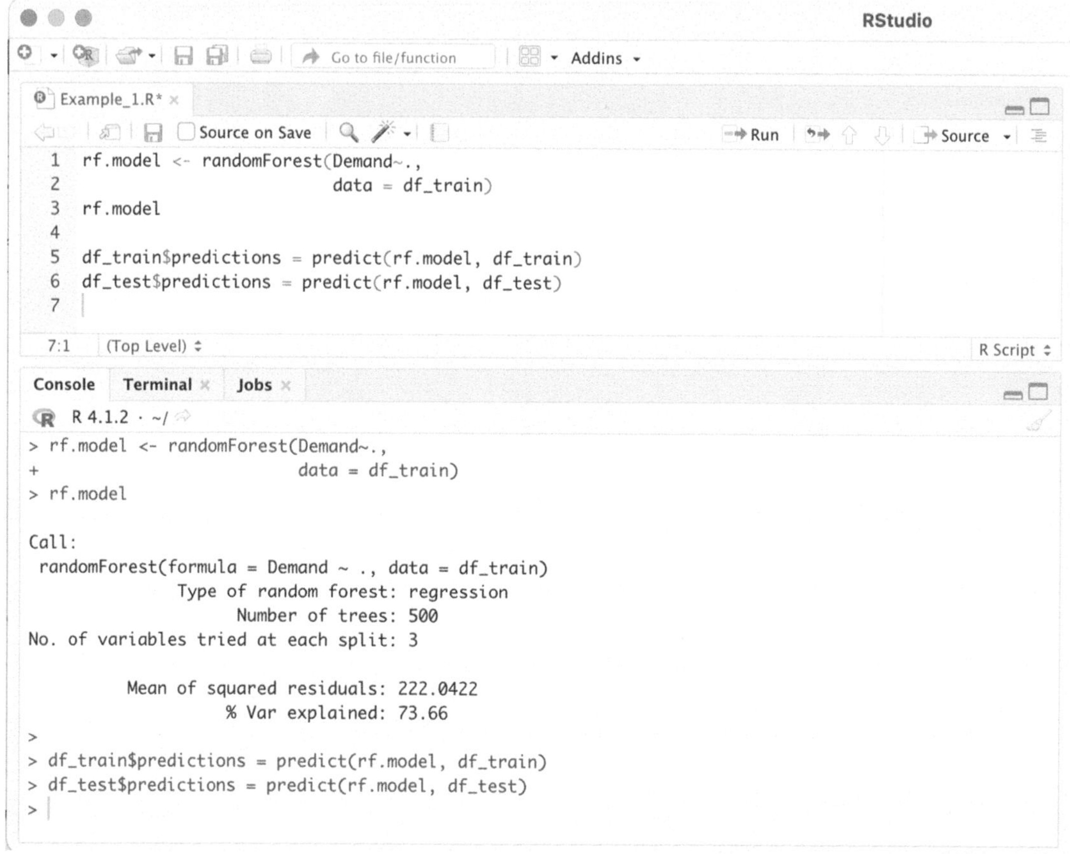

Figure 3.14 Train a Random Forest model on the training set and use it to make predictions on the training and test set.

3.7.6 Questions

1A) Visualize the results using the 'ggplot2' package. Plot the real demand on the x axis and the predicted demand on the y axis, and identify any patterns in the residual errors.

Solution 1A: Visualization of model predictions:

Install and load the ggplot2 package for data visualization. Define the input data ('df_test'), x axis ('Demand' column), and y axis ('predictions' column). Define the color ('brown3'), and size ('2') of the scatterplot points, the x axis ('Recorded demand') and y axis ('Predicted demand') labels, as well as the axis ranges (165–270). Add a straight line with slope 1 and 0 intercept (x=y) for reference (Figure 3.16).

If the model predictions were perfect, that is if the predicted demand matched exactly the recorded demand, all points in Figure 3.17 would fall on the gray line (x=y). The further away the points are from the gray line, the higher the model residual errors.

According to Figure 3.17, the model underestimates the highest recorded demand (Figure 3.17, points inside the green circle) and overestimates the lowest recorded demand (Figure 3.17, points inside the yellow circle). This systematic bias that is known to affect ensemble-tree machine learning regression models is particularly important in water demand forecasting due to the importance of

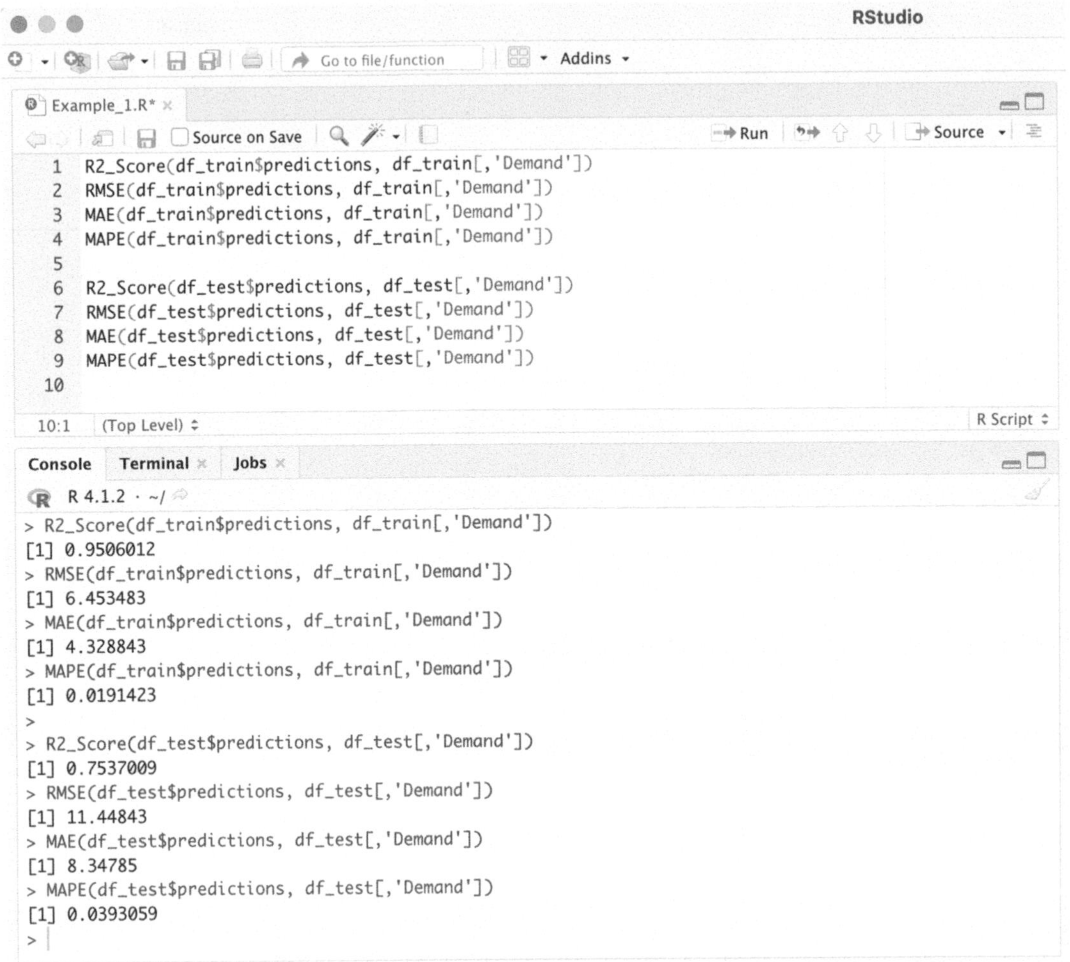

Figure 3.15 Calculate four evaluation metrics, R2, RMSE, MAE, and MAPE for the training and test set.

accurately predicting days with extreme consumption. For more information on this effect and a review of methods for correcting bias see Belitz & Stackelbers (2021).

Solution 1B) Use two model interpretability methods to identify the most important predictors and visualize the results.

TIP: Use the 'iml' package.

Solution 1B) Feature importance:

Install and load the 'iml' package for feature importance (Figure 3.18).

Define the model ('rf.model') and input data ('df_train') with the relevant columns, that is the ones used as model features and target, to create the predictor object ('mod'). Compute the feature importance for the prediction model using the predictor object, loss metric, comparison type between original model error and model error after permutation ('difference' or 'ratio'), and number of times the feature should be permutated – the higher the number of repetitions, the more stable the outcome.

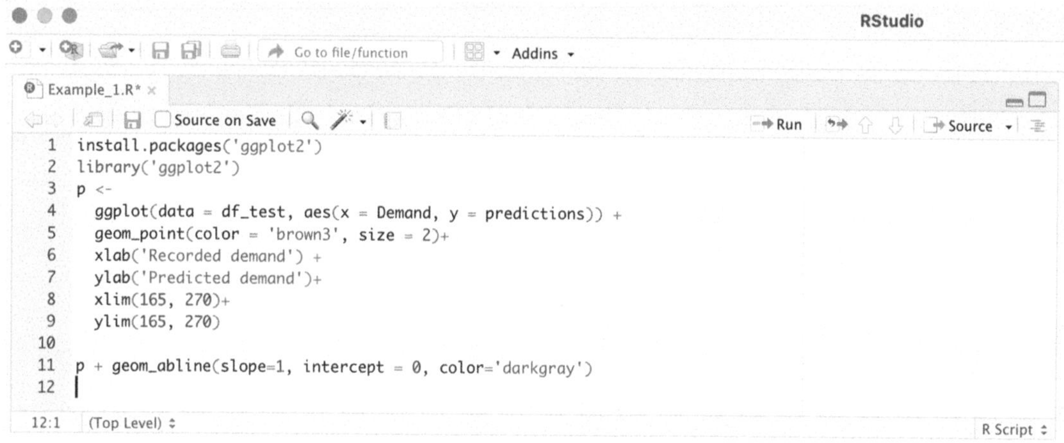


```r
1  install.packages('ggplot2')
2  library('ggplot2')
3  p <-
4    ggplot(data = df_test, aes(x = Demand, y = predictions)) +
5    geom_point(color = 'brown3', size = 2)+
6    xlab('Recorded demand') +
7    ylab('Predicted demand')+
8    xlim(165, 270)+
9    ylim(165, 270)
10
11 p + geom_abline(slope=1, intercept = 0, color='darkgray')
12
```

Figure 3.16 Code for solution 1A–Visualize data with the ggplot2 package.

Figure 3.19 depicts feature importance as a measure of MAE (loss = 'mae'). Specifically, it shows how many times the MAE increases (compare = 'ratio') if we permutate each one of the model features. Since this calculation is unstable, this process is repeated multiple times (n.repetitions = 20).

As mentioned earlier, permutating the values of a feature breaks the association between feature and target. The higher the predictive value of a feature, the higher the resulting increase in MAE, when the feature is not used as a predictor. In this case, demand 1 day prior, temperature, and demand 7

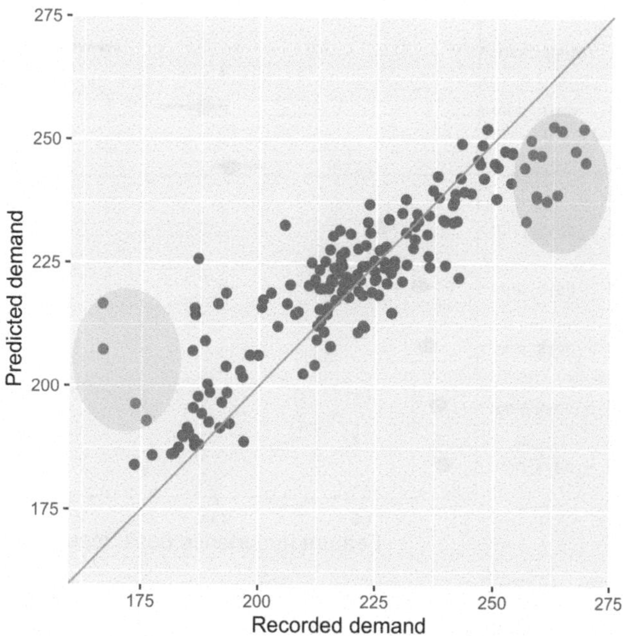

Figure 3.17 Solution 1A–Visualize predicted demand (y axis) vs recorded demand (x axis).

```
install.packages('iml')
library('iml')

mod <- Predictor$new(rf.model, data = df_train[,c(features,target)])
imp <- FeatureImp$new(mod, loss = "mae", compare = "ratio", n.repetitions = 20)
plot(imp) + theme_bw()
```

Figure 3.18 Code for solution 1B–Use the iml package to assess the permutation feature importance.

days prior are the most important predictors of demand. Demand on the previous day is an important predictor (MAE increases by ~2.7 times if demand 1 day prior is not included as a model predictor) due to autocorrelation between demand values, while demand 7 days prior carries the information of past demand on the same day of the week. The bar in Figure 3.19 shows the 5% and 95% quantile of importance values from all repetitions while the point shows the median importance. For more details see the documentation of the 'iml' R package (Molnar *et al.*, 2018) or the Interpretable ML book (Molnar, 2020).

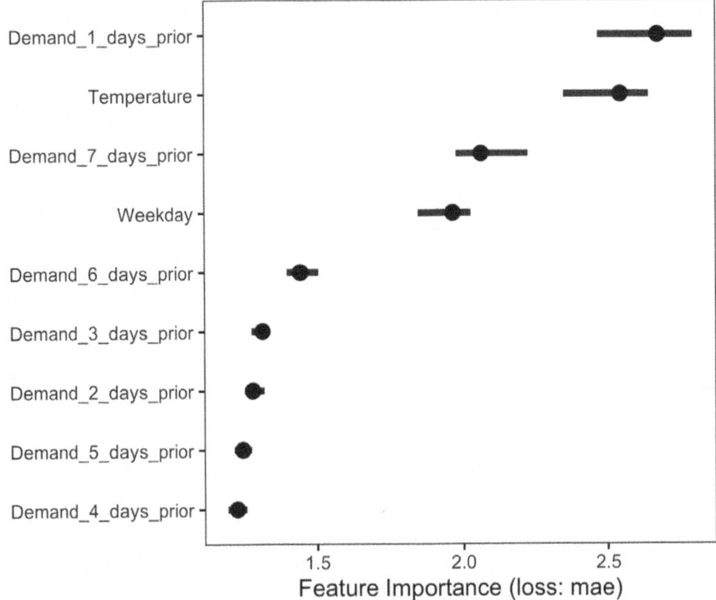

Figure 3.19 Solution 1B–Plot the feature importance as a measure of MAE. The x axis shows how many times the MAE increases if we permutate each model feature (y axis).

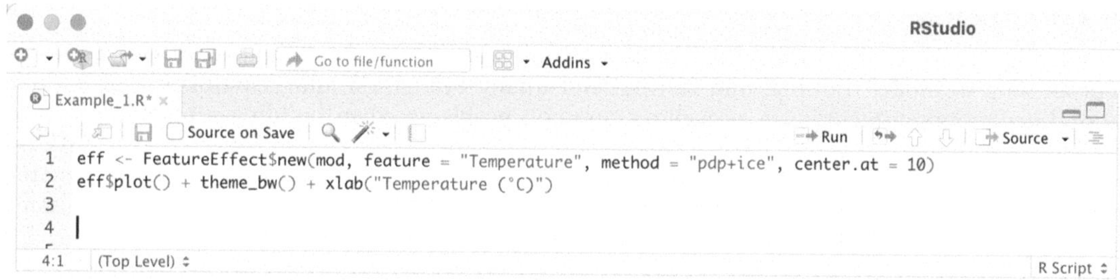

Figure 3.20 Code for solution 1B–Use the iml package to create the PDP and ICE plots.

3.7.7 PDP and ICE plots

Use the 'iml' R package to visualize a combined PDP and ICE plot (method = 'pdp + ice') for temperature (feature = 'Temperature') (Figure 3.20). Since demand predictions can vary for the same temperature, for different days or different customers, it can be difficult to compare ICE curves. For this reason, we centered the plot at 10 (center.at = 10), which means that the ICE curves show the difference in predicted demand for temperatures that are higher than 10°C for each day in the training data. The average of the ICE lines is a PDP plot.

According to Figure 3.21, demand remains relatively unaffected until temperature reaches values higher than 30°C. After this point, demand grows nearly exponentially (~50 GW increase in demand for a 12°C increase in temperature, from 30 to 42°C).

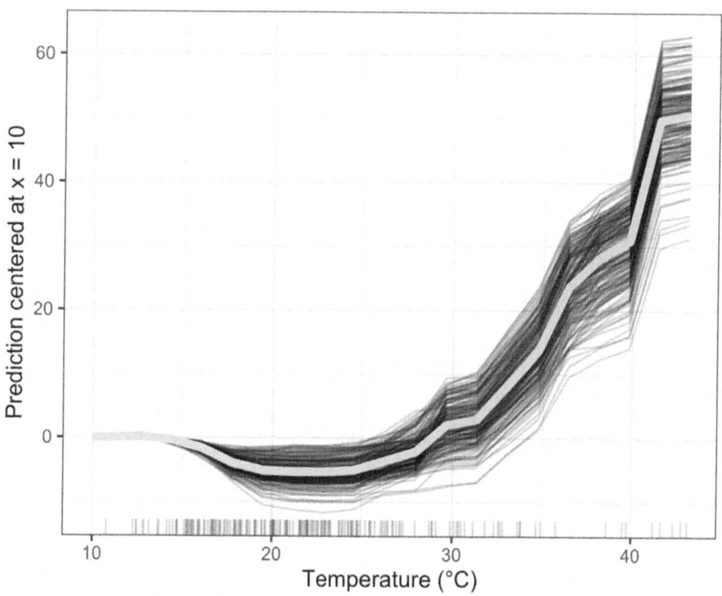

Figure 3.21 Solution 1B–Plot the ICE (black) and PDP (yellow) plots, centered around 10.

3.8 CONCLUSION

In this chapter, we covered the basics of a machine learning pipeline, from data collection and preprocessing to model training, and testing, and finally evaluation and visualization of findings. We outlined common techniques as well as common problems when building a machine learning pipeline. Even though these may vary depending on your dataset, aims, and problem constraints, this should be an iterative process that is constantly being checked, optimized, and updated. Ultimately, being confident in the accuracy of your predictions and at the same time understanding and sanity checking your results are important steps to building confidence in your model.

REFERENCES

Anele A. O., Hamam Y., Abu-Mahfouz A. M. and Todini E. (2017). Overview, comparative assessment and recommendations of forecasting models for shortterm water demand prediction. *Water (Switzerland)*, **9**(11), 887, https://doi.org/10.3390/w9110887

Antunes A., Andrade-Campos A., Sardinha-Lourenço A. and Oliveira M. S. (2018). Short-term water demand forecasting using machine learning techniques. *Journal of Hydroinformatics*, **20**(6), 1343–1366, https://doi.org/10.2166/hydro.2018.163

Apley D. W. and Zhu J. (2020). Visualizing the effects of predictor variables in black box supervised learning models. *Journal of the Royal Statistical Society: Series B (Statistical Methodology)*, **82**(4), 1059–1086, https://doi.org/10.1111/rssb.12377

Babel M. S., Gupta A. D. and Pradhan P. (2007). A multivariate econometric approach for domestic water demand modeling: An application to Kathmandu, 160 Nepal. *Water Resources Management*, **21**(3), 573–589, https://doi.org/10.1007/s11269-006-9030-6

Bai Y., Wang P., Li C., Xie J. and Wang Y. (2014). A multi-scale relevance vector regression approach for daily urban water demand forecasting. *Journal of Hydrology*, **517**, 236–245, https://doi.org/10.1016/j.jhydrol.2014.05.033

Bakker M., Van Duist H., Van Schagen K., Vreeburg J. and Rietveld L. (2014). Improving the performance of water demand forecasting models by using weather input. *Procedia Engineering*, **70**, 93–102, https://doi.org/10.1016/j.proeng.2014.02.012

Beal C. D. and Stewart A. S. (2014). Identifying residential water end uses underpinning peak day and peak hour demand. *Journal of Water Resources Planning and Management*, **140**(7), 04014008, https://doi.org/10.1061/(ASCE)WR.1943-5452.0000357

Belitz K. and Stackelberg P. E. (2021). Evaluation of six methods for correcting bias in estimates from ensemble tree machine learning regression models. *Environmental Modelling & Software*, **139**, 105006, https://doi.org/10.1016/j.envsoft.2021.105006

Breiman L. (2001). Random forests. *Machine Learning*, **45**(1), 5–32, https://doi.org/10.1023/A:1010933404324

Butler D. and Memon F. (2006). Water Demand Management. IWA Publishing, London.

Candelieri A., Soldi D. and Archetti F. (2015). Short-term forecasting of hourly water consumption by using automatic meter readers data. *Procedia Engineering*, **119**, 844–853, https://doi.org/10.1016/j.proeng.2015.08.948

Chang H., Parandvash G. H. and Shandas V. (2010). Spatial variations of single-family residential water consumption in portland, *Oregon. Journal of Urban Geography*, **31**(7), 953–972, https://doi.org/10.2747/0272-3638.31.7.953

Chang W., Cheng J., Allaire J., Xie Y. and McPherson J. (2019). *Shiny: web application framework for R.* Retrieved from: https://CRAN.R-project.org/package=shiny (last accessed 5 October 2022)

Cole G. and Stewart R. A. (2013). Smart meter enabled disaggregation of urban peak water demand: precursor to effective urban water planning. *Urban Water Journal*, **10**(3), 174–194, https://doi.org/10.1080/1573062X.2012.716446

Domene E. and Sauri D. (2006). Urbanisation and water consumption: influencing factors in the metropolitan region of Barcelona. *Urban Studies*, **43**(9), 1605–1623, https://doi.org/10.1080/00420980600749969

Dos Santos C. C. and Pereira F. A. J. (2014). Water demand forecasting model for the metropolitan area of São Paulo, Brazil. *Water Resources Management*, **28**(13), 4401–4414, https://doi.org/10.1007/s11269-014-0743-7

Fiorillo D., Kapelan Z., Xenochristou M., De Paola F. and Giugni M. (2021). Assessing the impact of climate change on future water demand using weather data. *Water Resources Management*, **35**, 1449–1462, https://doi.org/10.1007/s11269-021-02789-4

Goldstein A., Kapelner A., Bleich J. and Pitkin E. (2015). Peeking inside the black box: visualizing statistical learning with plots of individual conditional expectation. *Journal of Computational and Graphical Statistics*, **24**(1), 44–65, https://doi.org/10.1080/10618600.2014.907095

H$_2$O.ai 2020h2o: R Interface for H$_2$O. R package version 3.30.0.6. Available from: https://github.com/h2oai/h2o-3 (last accessed 7 March 2022)

Haque M., Hagare D., Rahman A. and Kibria G. (2014). Quantification of water savings due to drought restrictions in water demand forecasting models. *Journal of Water Resources Planning and Management*, **140**(11), 04014035, https://doi.org/10.1061/(ASCE)WR.1943-5452.0000423

Herrera M., Torgo L., Izquierdo J. and Pérez-García R. (2010). Predictive models for forecasting hourly urban water demand. *Journal of Hydrology*, **387**(1–2), 141–150, https://doi.org/10.1016/j.jhydrol.2010.04.005

Hyndman R. (2020). fpp2: Data for "Forecasting: Principles and Practice, 2nd edn. R package version 2.4. https://CRAN.Rproject.org/package=fpp2

Hyndman R. J. and Athanasopoulos G. (2018). Forecasting: Principles and Practice, 2nd edn. OTexts, Melbourne, Australia. Available at: OTexts.com/fpp2 (last accessed 8 September 2021)

Indyk P. and Motwani R. (1998). Approximate nearest neighbor: towards removing the curse of dimensionality. Proceedings of 30th Annual ACM Symposium on Theory of Computing, 24–26 May 1998, Dallas, pp. 604–613.

Kohavi R. (1995). A study of cross-validation and bootstrap for accuracy estimation and model selection. *IJCAI*, **14**(2), 1137–1145.

Kofinas D., Mellios N., Papageorgiou E. and Laspidou C. (2014). Urban water demand forecasting for the island of skiathos. *Procedia Engineering*, **89**, 1023–1030, https://doi.org/10.1016/j.proeng.2014.11.220

Kowalski M. and Marshallsay D. (2005). Using measured microcomponent data to model the impact of water conservation strategies on the diurnal consumption profile. *Water Supply*, **5**(3–4), 145–150, https://doi.org/10.2166/ws.2005.0094

Kuhn M. (2008). Building predictive models in R using the caret package. *Journal of Statistical Software*, **28**(5), 1–26, https://doi.org/10.18637/jss.v028.i05

Menapace A., Zanfei A. and Righetti M. (2021). Tuning ANN hyperparameters for forecasting drinking water demand. *Applied Sciences*, **11**(9), 4290, https://doi.org/10.3390/app11094290

Miller T. (2019). Explanation in artificial intelligence: insights from the social sciences. *Artificial Intelligence*, **267**, 1–38, https://doi.org/10.1016/j.artint.2018.07.007

Molnar C. (2020). *Interpretable Machine Learning*. Available at: https://christophm.github.io/interpretable-ml-book/

Molnar C., Bischl B. and Casalicchio G. (2018). lml: an R package for interpretable machine learning. _JOSS_, **3**(26), 786, https://doi.org/10.21105/joss.00786

Parker J. (2014). Assessing the sensitivity of historic micro-component household water-use to climatic drivers. Doctoral dissertation, Loughborough University, Loughborough, UK.

Pedregosa F., Varoquaux Ga"el Gramfort A., Michel V., Thirion B., Grisel O., Blondel M., Prettenhofer P., Weiss R., Dubourg V., Vanderplas J., Passos A., Cournapeau D., Brucher M., Perrot M. and Duchesnay E. (2011). Scikit-learn: machine learning in python. *Journal of Machine Learning Research*, **12**, 2825–2830.

Pesantez J. E., Berglund E. Z. and Kaza N. (2020). Smart meters data for modeling and forecasting water demand at the user-level. *Environmental Modelling and Software*, **125**, 104633, https://doi.org/10.1016/j.envsoft.2020.104633

Plotly Technologies Inc. (2015). Collaborative Data Science. Plotly Technologies Inc., Montréal, QC. Available at: https://plot.ly (last accessed 7 March 2022)

R Core Team. (2017). R: A Language and Environment for Statistical Computing. R Foundation for Statistical Computing, Vienna, Austria. Retrieved from: https://www.R-project.org/

Radakovich N., Nagy M. and Nazha A. (2020). Machine learning in haematological malignancies. *The Lancet Haematology*, **7**(7), e541–e550, https://doi.org/10.1016/S2352-3026(20)30121-6

Romano M. and Kapelan Z. (2014). Adaptive water demand forecasting for near real-time management of smart water distribution systems. *Environmental Modelling and Software*, **60**, 265–276, https://doi.org/10.1016/j.envsoft.2014.06.016

RStudio Team. (2020). RStudio: Integrated Development for R. RStudio, PBC, Boston, MA. http://www.rstudio.com/

Shabani S., Yousefi P., Adamowski J. and Gholamreza Naser G. (2016). Intelligent soft computing models in water demand forecasting. In *Water Stress in Plants*, IntechOpen, London, United Kingdom, 2016 [Online]. https://www.intechopen.com/chapters/51221 doi: 10.5772/63675 (accessed 5 October 2022)

Stekhoven D. J. and Buehlmann P. (2012). MissForest – nonparametric missing value imputation for mixed-type data. *Bioinformatics*, **28**(1), 112–118, https://doi.org/10.1093/bioinformatics/btr597

Tibshirani R. (1996). Regression shrinkage and selection via the lasso. *Journal of the Royal Statistical Society. Series B (Methodological)*, **58**(1), 267–288, https://doi.org/10.1111/j.2517-6161.1996.tb02080.x

Tiwari M., Adamowski J. and Adamowski K. (2016). Water demand forecasting using extreme learning machines. *Journal of Water and Land Development*, **28**(1), 37–52, https://doi.org/10.1515/jwld-2016-0004

Wickham H. (2016). Ggplot2: Elegant Graphics for Data Analysis. Springer-Verlag, New York. Available at: https://ggplot2.tidyverse.org (last accessed 7 March 2022)

Wickham H., François R., Henry L. and Müller K. (2020). *dplyr: A Grammar of Data Manipulation*. R package version 1.0.2. Available at: https://CRAN.R-project.org/package=dplyr (last accessed 7 March 2022)

Willis R. M., Stewart R. A., Panuwatwanich K., Capati B. and Giurco D. (2013). End-use water consumption in households: impact of sociodemographic factors and efficient devices. *Journal of Cleaner Production*, **60**, 107–115, https://doi.org/10.1016/j.jclepro.2011.08.006

Xenochristou M. (2019). Water demand forecasting using machine learning on weather and smart metering data. PhD thesis, College of Engineering, Mathematics, and Physical Sciences, Centre for Water Systems, University of Exeter.

Xenochristou M. and Kapelan Z. (2020). An ensemble stacked model with bias correction for improved water demand forecasting. *Urban Water Journal*, **17**(3), 212–223, https://doi.org/10.1080/1573062X.2020.1758164

Xenochristou M., Hutton C., Hofman J. and Kapelan Z. (2020a). Water demand forecasting accuracy and influencing factors at different spatial scales using a gradient boosting machine. *Water Resoures Research*, **56**(8), e2019WR026304, https://doi.org/10.1029/ 2019WR026304

Xenochristou M., Kapelan Z. and Hutton C. (2020b). Using smart demand metering data and customer characteristics to investigate the influence of weather on water consumption in the UK. *Journal of Water Resources Planning and Management*, **146**(2), 04019073, https://doi.org/10.1061/(ASCE)WR.1943-5452.0001148

Xenochristou M., Hutton C., Hofman J. and Kapelan Z. (2021). Short-term forecasting of household water demand in the UK using an interpretable machine learning approach. *Journal of Water Resources Planning and Management*, **147**(4), 04021004, https://doi.org/10.1061/(ASCE)WR.1943-5452.0001325

Yan Y. (2016). *MLmetrics: Machine Learning Evaluation Metrics*. R package version 1.1.1. Available at: https:// CRAN.R-project.org/package=MLmetrics (last accessed 7 March 2022)

Zhao Q. and Hastie T. (2021). Causal interpretations of black-box models. *Journal of Business & Economic Statistics*, **39**(1), 272–281, https://doi.org/10.1080/07350015.2019.1624293

doi: 10.2166/9781789062380_0075

Chapter 4

Water demand forecasting | time series data

*Jarai Sanneh[1], A. Di Mauro[2], S. Venticinque[2], G. F. Santonastaso[2], A. Di Nardo[2], Yi Wang[3] and Juneseok Lee[1]**

[1]*Civil & Environmental Engineering, Manhattan College, Riverdale, NY 10471, USA*
[2]*Università degli Studi della Campania Luigi Vanvitelli, Aversa 81031, Italy*
[3]*Electrical and Computer Engineering, Manhattan College, Riverdale, NY 10471, USA*
**Corresponding author: juneseok.lee@manhattan.edu*

LEARNING OBJECTIVES

At the end of this chapter, you will be able to:

(1) Apply ARIMA/SARIMA to forecast water demand in time-series data.
(2) Discuss the practical aspects and implications of using Machine Learning to water demand in time-series data.
(3) Build and run time series data using machinear learning techniques (MATLAB and Python).
(4) Interpret modeling results.

4.1 INTRODUCTION

Water demand forecasting is crucial in many aspects of Water Distribution Systems (WDS) because it helps minimize cost, optimize operations, and provide strategies for water conservation (Kofinas *et al.*, 2014). It plays a vital role in the planning, operations, and management of physical assets for water utilities such as pumping stations, treatment plants, tanks, and distribution networks, which rely on future consumption forecasts (Arandia *et al.*, 2015; Billings & Jones, 2008). For instance, water utilities need short-term water demand forecasting in order to provide a more stable urban freshwater supply that will be used in a timely manner 'by adjusting water supply to actual demand and consumption' (Kofinas *et al.*, 2014).

Traditional ***time series forecasting methods*** such as Auto-Regressive Moving Average (ARMA), Auto-Regressive Integrated Moving Average (ARIMA) and Seasonal Auto-Regressive Integrated Moving Average (SARIMA) have been used for decades to forecast water demand using time series historical data. Redondo *et al.* (2018) used ARIMA models to make operational analysis in a drinking water treatment plant by analyzing how the water quality is affected by rainfall. The results showed that the ARIMA models were more accurate for analyzing the water treatment operations using a weekly timescale compared to a daily timescale 'due to significant daily variations in the control parameters of water quality in the plant' (Redondo *et al.*, 2018). Lee and Chae (2016) developed seasonal ARIMA models to make hourly water demand forecasting for micro water grids (Lee & Chae,

2016). Arandia *et al.* (2015) forecasted short-term water demand using SARIMA models to make both offline and online forecasts. The offline forecasts were made using the most recent historical data to 're-estimate the models' while the online forecasts were made by combining the SARIMA models (state-space form) with data assimilation by applying a Kalman Filter (KF) to update the models efficiently (Arandia *et al.*, 2015).

In the past decade, artificial intelligence (AI) had a rapidly growing presence in many applications, including the water sector. Machine learning (ML) techniques are an artificial intelligence approach that has drawn serious attention in water-demand forecasting. Machine learning techniques have the advantage of being able to forecast nonlinear relationships between response variables and their predictors in time series models with the presence of noisy data. The increasing use of smart water metering in the water sector has made available a great amount of data which cannot be processed with traditional methods (Cominola *et al.*, 2015). Therefore, the need has emerged to identify new data analysis techniques able to extract valuable information from available data and support water utilities in their decision systems. Analytics in the Drinking Water Industry support improvements in demand side management and water distribution network efficiencies, lead significant water savings, promote customers' sustainable behaviours, identify peak hours of use, and facilitate water forecast demand modelling (Monks *et al.*, 2019).

In this context, machine learning techniques (MLT) represent the key to many challenges. In the literature, especially in the last five years, various MLT for water demand analysis and forecasting have been proposed showing how they can also be applied in the water sector (Pesantez *et al.*, 2020; Rahim *et al.*, 2020; Villarin & Rodriguez-Galiano, 2019; Xenochristou *et al.*, 2018).

4.2 TIME SERIES DATA ANALYSIS

A time series, consisting of a sequence of numerical observations recorded successively in time, has an intrinsic feature of dependence between adjacent observations, which is analyzed using time series analysis (Box *et al.*, 2016). ARIMA and SARIMA models utilize historical time series data and consist of a three-step iterative process: identification, estimation, and diagnostics checking (Box *et al.*, 2016).

4.2.1 ARIMA model

An ARIMA model is denoted as ARIMA(p,d,q) and is expressed using the mathematical formulations given in Equations (4.1)–(4.4) (Lee & Chae, 2016):

$$Y_t = \mu + \sum_{k=1}^{p} \varnothing_k Y_{t-k} + \epsilon_t \tag{4.1}$$

$$Y_t = C + \epsilon_t + \sum_{k=1}^{q} \varnothing_k \epsilon_{t-k} \tag{4.2}$$

$$Y_t = \mu + \sum_{k=1}^{p} \varnothing_k Y_{t-k} + \epsilon_t + \sum_{k=1}^{q} \theta_k \epsilon_{t-k} \tag{4.3}$$

$$\varnothing_p(B)(1-B)^d Y_t = \theta_q(B)\epsilon_t \tag{4.4}$$

where $\varnothing$ = autoregressive or damping parameter; θ = moving average parameter; μ = mean value of the process; ϵ_t = forecast error at time t, in which ϵ_t is assumed to follow a normal $(0, \sigma)$ distribution, σ = standard deviation of the process (Lee & Chae, 2016). Equation (4.1) defines an autoregressive process of order p, AR(p), 'which predicts values from previous values'; Equation (4.2) defines a moving average process of order q, MA(q), 'which accounts for previous random trends'; Equation (4.3) defines an autoregressive moving average process of order (p,q), ARMA(p,q); and Equation (4.4) defines an autoregressive integrated moving average process of order (p,q) differenced by order d, ARIMA(p,d,q) (Lee & Chae, 2016).

4.2.2 SARIMA model

A SARIMA or seasonal ARIMA model is obtained when an ARIMA model has a seasonal component (periodic pattern). It is denoted as ARIMA(p,d,q)x(P,D,Q)$_s$ and is expressed using Equation (4.5) (Arandia *et al.*, 2015):

$$\Phi_P(B^s)\varnothing(B)(1-B^s)^D(1-B)^d Y_t = \delta + \Theta_Q(B^s)\theta(B)\epsilon_t \tag{4.5}$$

$$\Phi_P(B^s) = 1 - \Phi_1 B^s - \Phi_2 B^{2s} - \ldots - \Phi_P B^{Ps} \tag{4.6}$$

$$\Theta_Q(B^s) = 1 + \Theta_1 B^s + \Theta_2 B^{2s} + \ldots + \Theta_Q B^{Qs} \tag{4.7}$$

$$\varnothing(B) = 1 - \varnothing_1 B - \varnothing_2 B^2 - \ldots - \varnothing_p B^p \tag{4.8}$$

$$\theta(B) = 1 + \theta_1 B + \theta_2 B^2 + \ldots + \theta_q B^q \tag{4.9}$$

$$\delta = \mu(1 - \varnothing_1 - \ldots - \varnothing_p)(1 - \Phi_1 - \ldots - \Phi_P) \tag{4.10}$$

$$B^k Y_t = Y_{t-k} \tag{4.11}$$

where Equations (4.6)–(4.11) give the seasonal autoregressive polynomial, seasonal moving average polynomial, ordinary (non-seasonal) autoregressive (AR) polynomial, and the ordinary (non-seasonal) moving average (MA) polynomial respectively; B is the backshift operator as defined in Equation (4.11); P is the seasonal AR polynomial order, Q is the seasonal MA polynomial order, p is the non-seasonal AR polynomial order, q is the non-seasonal MA polynomial order, D is the seasonal differencing order, d is the non-seasonal differencing order, s is the seasonal period, Y_t is the water demand time series, μ = mean value of the process; ϵ_t = forecast error at time t, in which ϵ_t is assumed to follow a normal (0, σ) distribution, and σ = standard deviation of the process.

4.2.3 Creating ARIMA/SARIMA models using econometric toolbox

This example shows how to use MATLAB's Econometric Modeler App to create ARIMA and SARIMA models for time series analysis using the following 36-months hypothetical water demand data, with each time step corresponding to one month:

[266.0, 145.9, 183.1, 119.3, 180.3, 168.5, 231.8, 224.5, 192.8, 122.9, 336.5, 185.9, 194.3, 149.5, 210.1, 273.3, 191.4, 287.0, 226.0, 303.6, 289.9, 421.6, 264.5, 342.3, 339.7, 440.4, 315.9, 439.3, 401.3, 437.4, 575.5, 407.6, 682.0, 475.3, 581.3, 646.9]

You can download the Econometrics toolbox in MATLAB by clicking on Apps → Get More Apps → and then search for 'Econometrics Toolbox' in the Add-On Explorer Search bar. You can run the example by using the following procedures:

Step 1. Save the water demand data as an excel file with each data value in a row so that you have one column of data (you can write the 'water demand' header in column A and row 1 and the data values in column A from rows 2 to 37. Import it to MATLAB's workspace by clicking on Home → Import Data.

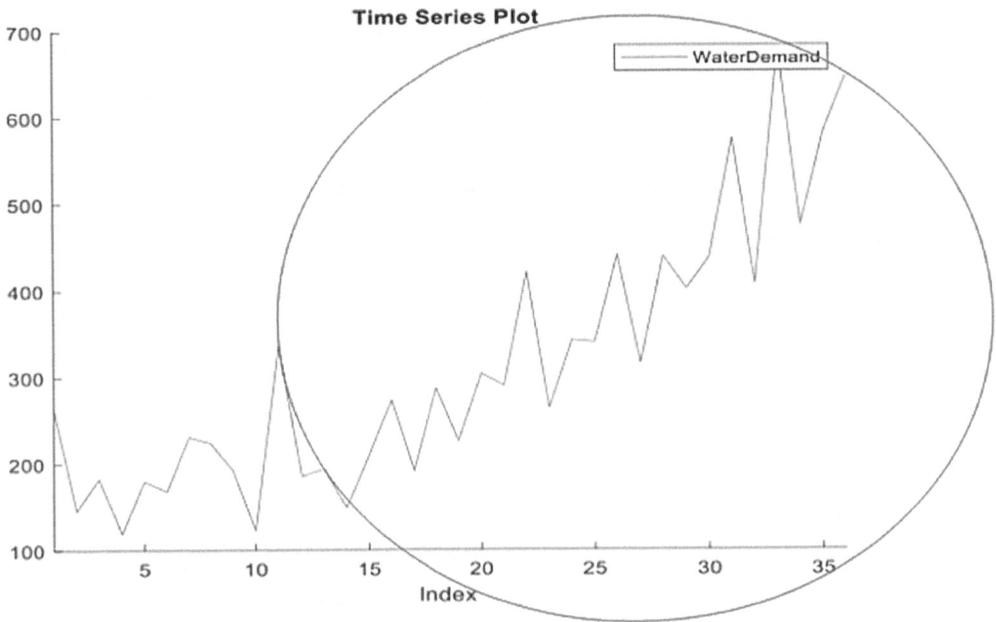

Figure 4.1 Time series plot of water demand.

Step 2. Open the Econometric Modeler app and click on Import → Import from Workspace to import and load the water demand time series data.

Step 3. The time series is plotted automatically and is shown in Figure 4.1. From the time series plot, the presence of a linear trend and seasonality (cyclic pattern) is evident, which means that the time series is non-stationary. Box–Jenkins models can only be applied to stationary time series, therefore, the nonstationary time series needs to be differenced to make it stationary.

Step 4. Click on the time series tab in the data browser (see Figure 4.2) and click on the time series variable that was just loaded. You can right-click to rename the variable 'Water Demand.'

Step 5. Click on 'ACF' and 'PACF' in the plots tab (see Figure 4.3) to plot the autocorrelation function (ACF) and partial autocorrelation function (PACF) of the time series as shown in Figures 4.4 and 4.5 respectively. ACF, which 'gives the correlation of time-series data with its previous time-series data,' and PACF, which 'correlates the time-series with its own lagged values separated by certain time units,' are analytical tools that are used to assess the 'reliability of time-series analysis' (ArunKumar *et al.*, 2021).

The presence of a trend can also be noticed by looking at the ACF plot, which is indicated by continuing large autocorrelations even after several lags (NCSS). The first five lags in the ACF plot shown in Figure 4.4 are significant, which indicates the presence of a trend.

Step 6. Click on 'difference' in the econometric modeler tab to perform a first order non-seasonal difference operation ($d=1$) to remove the trend. A new differenced time series shown in Figure 4.6 was created with 'Diff' automatically added next to the variable name, for example WaterDemandDiff. It is clear that there is no trend present anymore, however, if trend was still present, a second order difference operation ($d=2$) would have been applied by clicking on 'WaterDemandDiff' and clicking on 'difference' to get a new time series with the variable name 'WaterDemandDiffDiff' – the two 'Diff' words after the name of the variable means that the time series was differenced twice ($d=2$).

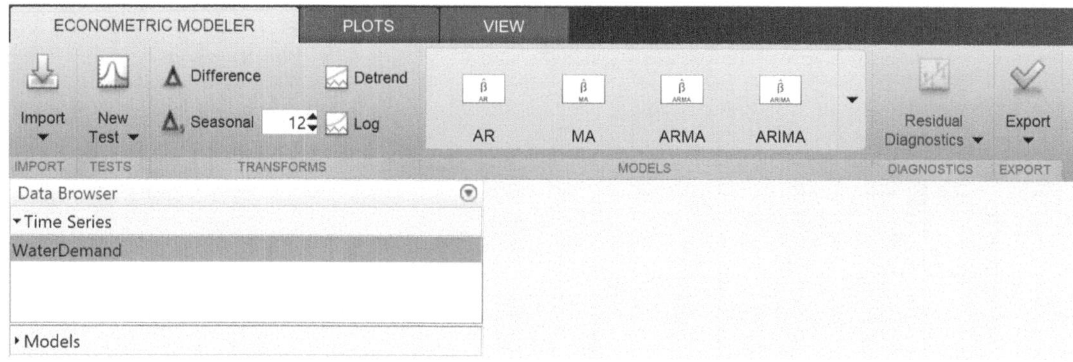

Figure 4.2 Data browser.

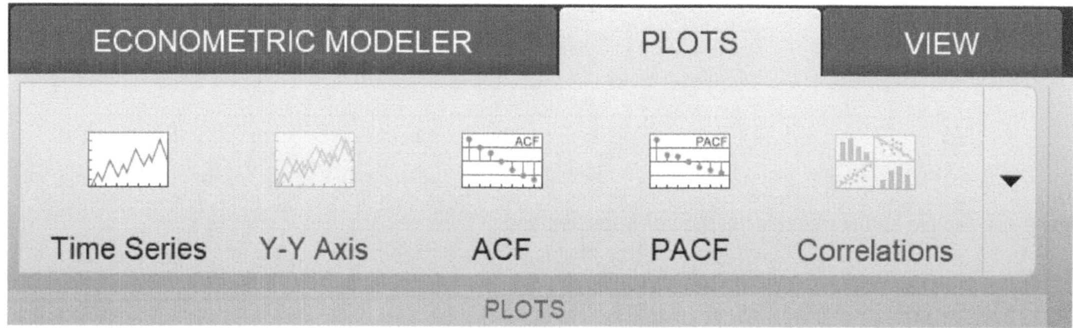

Figure 4.3 Plots tab.

Step 7. Click on 'WaterDemandDiff' in the time series tab, and then click on 'ACF' and 'PACF' to plot the autocorrelation function and partial autocorrelation function respectively of the first order differenced time series, which are shown in Figures 4.7 and 4.8. From the ACF plot, the autocorrelations attenuate quickly, which means that there is no more trend, and a suitable value of d has been attained ($d=1$) (Kofinas *et al.*, 2014). We will refer back to the ACF and PACF plots of 'WaterDemandDiff' in Step 9.

Step 8. The value of p and q are found from the PACF and ACF respectively of the appropriately differenced time series (Kofinas *et al.*, 2014). We have an AR model if the partial autocorrelations of the appropriately differenced time series cut off after a small number of lags, where the value of p is the last lag with a large value, and we have an MA model if the autocorrelations of the appropriately differenced time series cut off after a small number of lags, where the value of q is the last lag with a large value (NCSS). However, if the partial autocorrelation or autocorrelation plots of the appropriately differenced time series do not cut off, that means that we either have a mixed ARIMA model with p and q values greater than zero, or that we have an AR model with $p=0$ when only the partial autocorrelation plot does not cut off, or that we have a MA model with $q=0$ when only the autocorrelation plot does not cut off. If both partial autocorrelation and autocorrelation plots of the appropriately differenced time series do not cut off, we have a mixed ARIMA model with positive p and q values that can be estimated by using trial and error until the autocorrelations are minimal (NCSS).

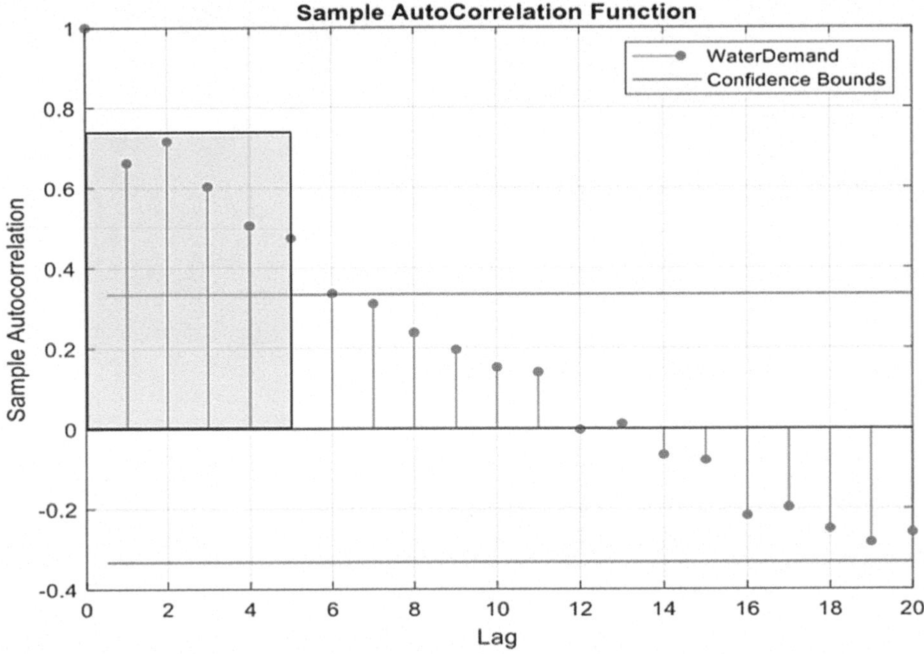

Figure 4.4 Sample autocorrelation function of WaterDemand.

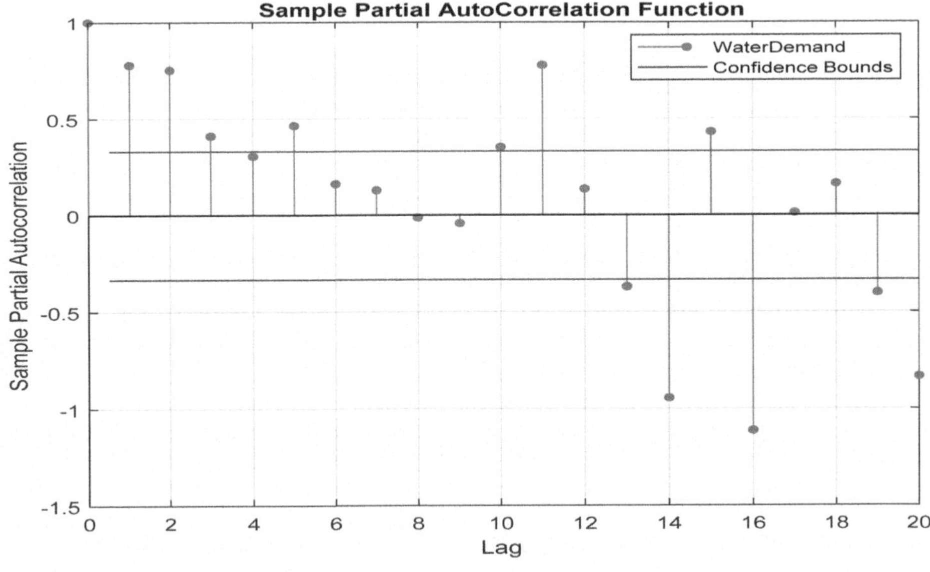

Figure 4.5 Sample partial autocorrelation function of WaterDemand.

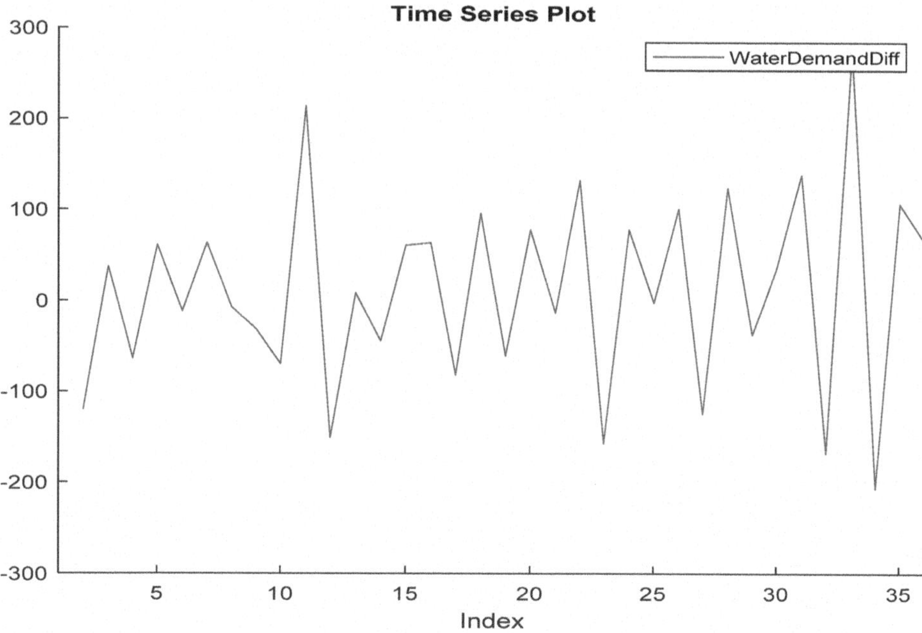

Figure 4.6 Time series plot of WaterDemandDiff.

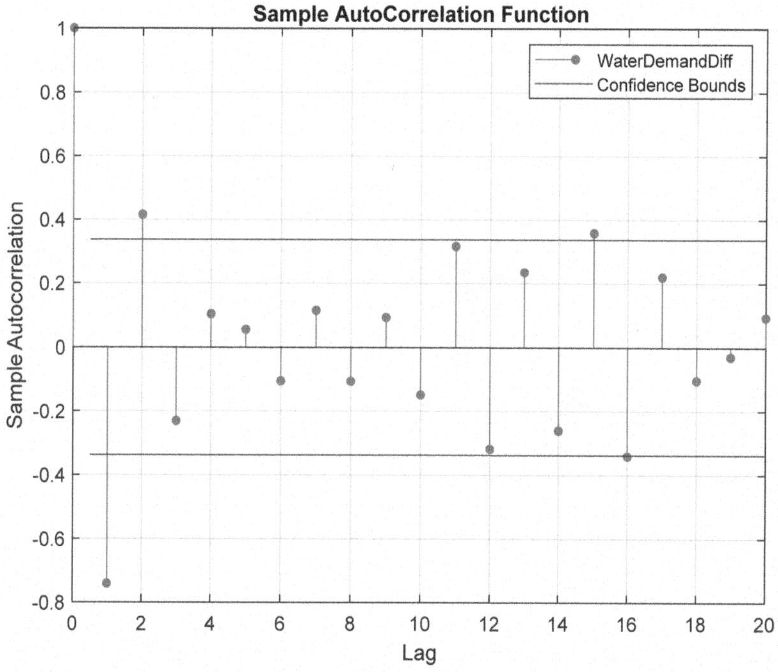

Figure 4.7 Sample autocorrelation function of WaterDemandDiff.

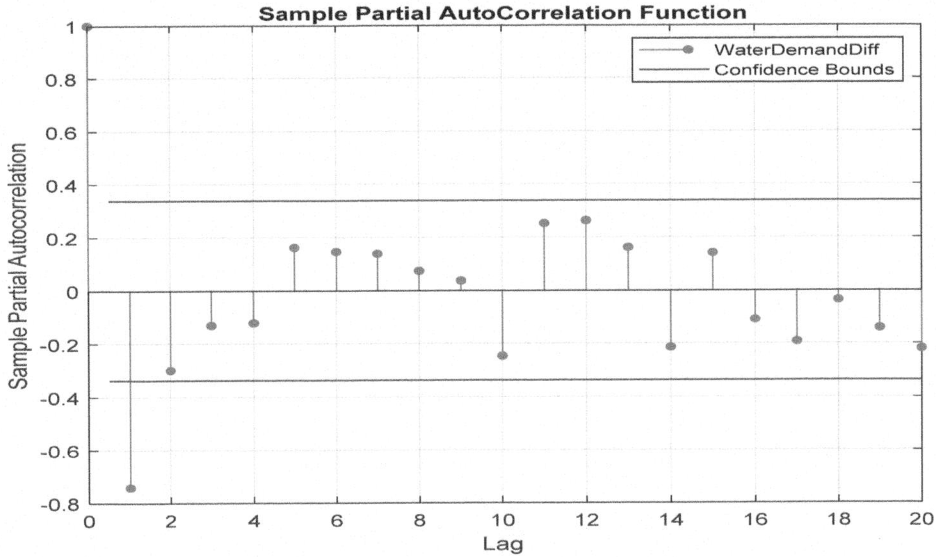

Figure 4.8 Sample partial autocorrelation function of WaterDemandDiff.

Step 9. By looking at the ACF plot of 'WaterDemandDiff' in Figure 4.7, the autocorrelation cuts off shortly after lag 2, therefore q can be chosen as 2. Similarly, by looking at the PACF plot of 'WaterDemandDiff' in Figure 4.8, the partial autocorrelation cuts off shortly after lag 1, therefore p can be chosen as 1. Therefore, we could fit the water demand time series data to an ARIMA (11,2) model where $p = 1$, $d = 1$, and $q = 2$ and then check if the model is a good fit.

Step 10. Click on 'WaterDemandDiff' in the time series tab and then click on the econometric modeler tab. Click on ARIMA and enter the degree of integration or d as 1, autoregressive order or p as 1, moving average order or q as 2, and then click on 'Estimate' to create the ARIMA model as shown in Figure 4.9. The created model is put under the models tab and has the variable name 'ARIMA_WaterDemandDiff.' A model summary as shown in Figure 4.10 is automatically created and it features the model fit plot to compare the differenced time series and the ARIMA model, the estimated ARIMA model parameters and their associated standard errors and p-values, the residual plot, and the goodness of fit using Akaike Information Criterion (AIC) and Bayesian Information Criterion (BIC) to assess the model reliability. The p-values for the constant, AR and MA parameters are used to determine whether the terms in the model are statistically significant by comparing them to the level of significance, α, which is usually taken as 0.05 – a parameter is considered statistically significant if its p-value is less than or equal to $\alpha = 0.05$. AIC and BIC are analytical tools that are used to assess the quality or reliability of time-series models by determining 'how well a model explains the relationships between the variables' – the lower AIC and BIC values are, the more a model is 'likely to be considered as a true model' (ArunKumar *et al.*, 2021).

Step 11. As mentioned earlier, the water demand time series had both trend and seasonality, and the trend was removed after it was differenced with $d = 1$ to get 'WaterDemandDiff.' Now, the seasonality will be removed, and the time series will be fitted to a SARIMA model. Click on 'WaterDemandDiff' in the time series tab and enter '12' next to 'Seasonal' since the water demand data is monthly, and then click on 'Seasonal' to perform a seasonal difference ($D = 1$) to remove the seasonality (see Figure 4.11).

A new seasonal differenced time series with the name 'WaterDemandDiffSeasonalDiff' shown in Figure 4.12 was created with 'SeasonalDiff' automatically added to the name 'WaterDemandDiff.'

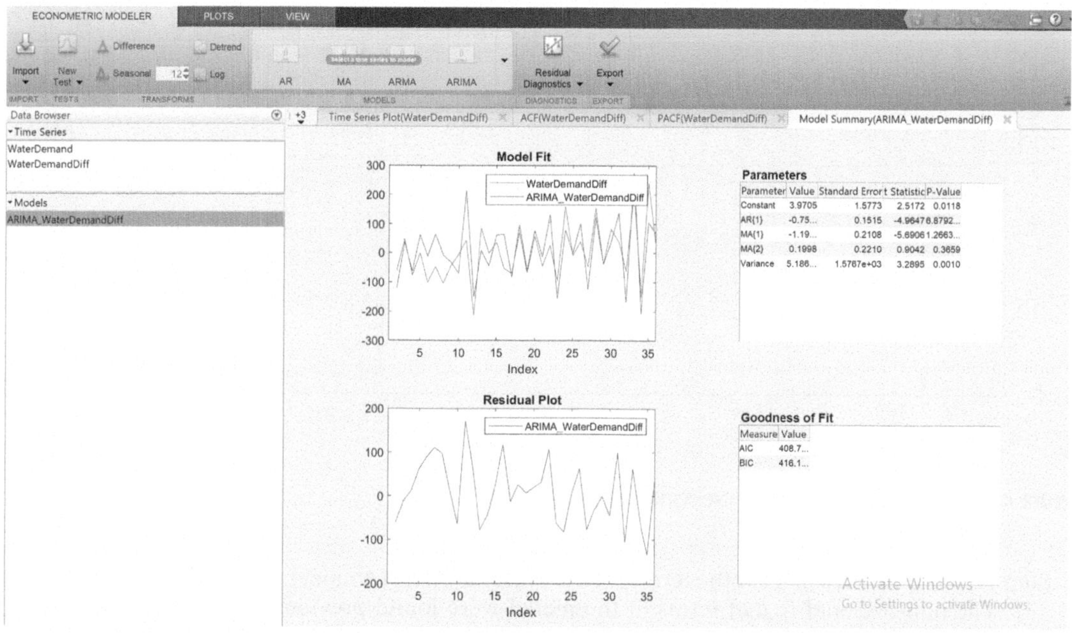

$$(1 - \phi_1 L)(1 - L)y_t = c + (1 + \theta_1 L + \theta_2 L^2)\varepsilon_t$$

Figure 4.9 ARIMA model parameters.

Figure 4.10 Summary results for ARIMA_WaterDemandDiff.

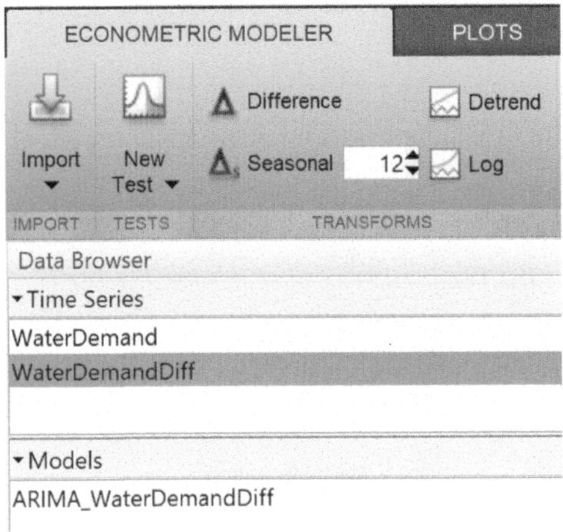

Figure 4.11 Performing seasonal difference.

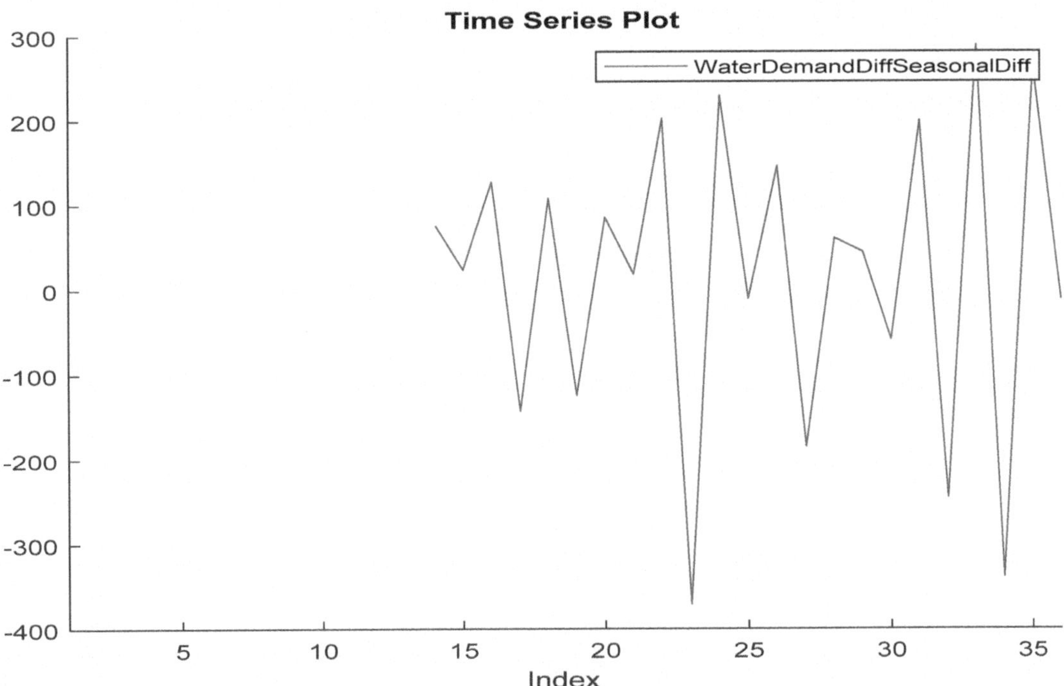

Figure 4.12 Time series plot of WaterDemandDiffSeasonalDiff.

Step 12. We now have most of the terms for the seasonal ARIMA model or ARIMA$(p,d,q) \times$ (P,D,Q) s. The non-seasonal (p,d,q) terms of the model were found previously ($p=1$, $d=1$, and $q=2$), $s=12$, $D=1$, and we can try $P=0$ and $Q=1$. Therefore, we could fit the water demand time series data to an ARIMA $(11,2) \times (01,1)12$ model and then check if the model is a good fit.

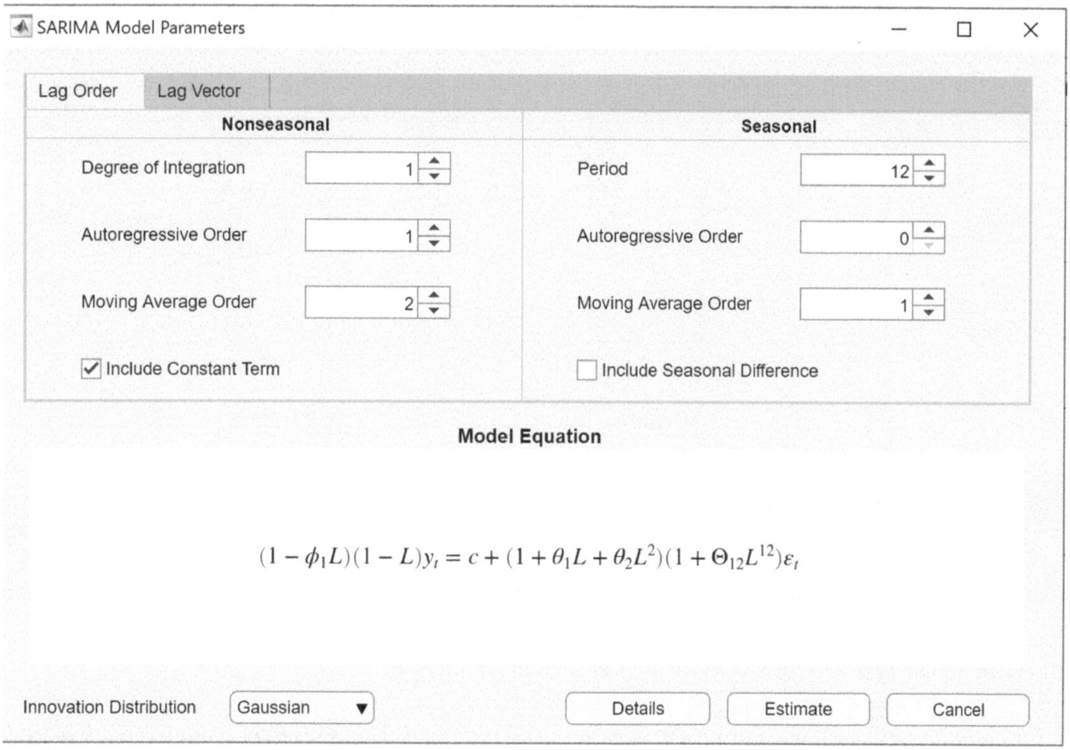

$$(1 - \phi_1 L)(1 - L)y_t = c + (1 + \theta_1 L + \theta_2 L^2)(1 + \Theta_{12}L^{12})\varepsilon_t$$

Figure 4.13 SARIMA model parameters.

Step 13. Click on 'WaterDemandDiffSeasonalDiff' in the time series tab and then click on the econometric modeler tab. Click on the arrow next to ARIMA to show all of the available models and then click on SARIMA and enter the non-seasonal degree of integration or d as 1, non-seasonal degree autoregressive order or p as 1, non-seasonal degree moving average order or q as 2, seasonal period or s as 12, seasonal degree autoregressive order or P as 0, seasonal degree moving average order or Q as 1, and then click on 'Estimate' to create the SARIMA model as shown in Figure 4.13. Normally, you should click on the checkbox next to 'Include Seasonal Difference' to include the seasonal difference term, however, checking that box for this example causes an error since the water demand data size is small – we will include the seasonal difference term manually when we do the forecast in the next section.

Step 14. The created model is put under the models tab and has the variable name 'SARIMA_WaterDemandDiffSeasonalDiff.' The automatically created model summary is shown in Figure 4.14. The AIC and BIC of the ARIMA $(11,2) \times (01,1)12$ model are 286.9 and 293.1 respectively, which are about half of the values for the ARIMA $(11,2)$ model, which has an AIC of 408.7 and BIC of 416.2. Therefore, the SARIMA model has a better fit than the ARIMA model for this monthly water demand data, which makes it more reliable.

Step 15. Click on the econometrics modeler tab and then click on 'ARIMA_WaterDemandDiff' in the model tab followed by 'Export' → 'General Function' to generate a MATLAB code for creating the selected ARIMA model. A new MATLAB file with the model code will be automatically opened. Go back to the Econometric Modeler app and do same for the SARIMA model: click on 'SARIMA_WaterDemandDiffSeasonalDiff' in the model tab followed by 'Export' → 'General

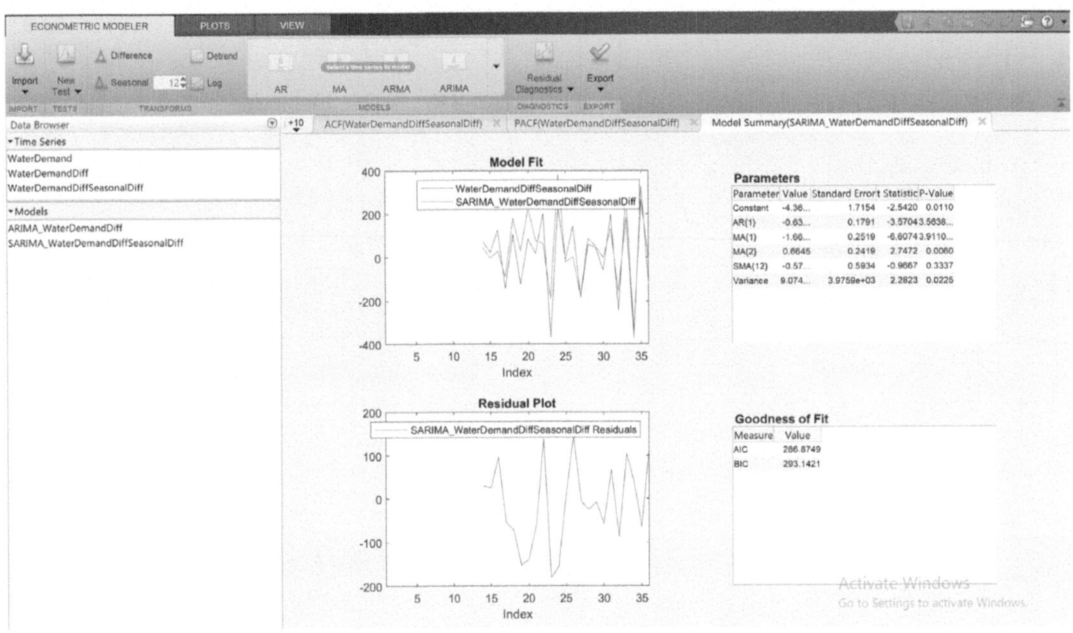

Figure 4.14 Summary results for SARIMA_WaterDemandDiffSeasonalDiff.

Function' to generate a MATLAB code for creating the selected SARIMA model. Save the two MATLAB files since we will use them in the forecasting section.

Step 16. Click on 'Export' → 'Generate Report' to generate a report summarizing the results of what we did using the econometrics modeler app. The report can be either in pdf, docx, or html format, and you can click on the check box next to the name of the time series or models that that you would like to include in the report (see Figure 4.15).

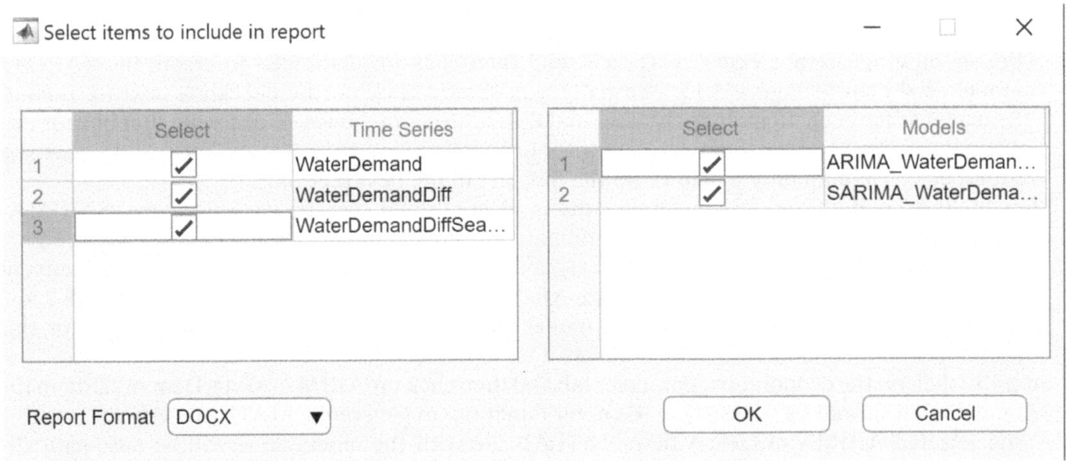

Figure 4.15 Generating a report.

4.2.4 Forecasting

MATLAB's forecast function uses an observed time series as a presample data (to initialize the forecasts) and a fitted regression model such as an ARIMA or SARIMA model to generate minimum mean square error (MMSE) forecasts denoted in Equation (4.12):

$$\hat{y}_{t+1} = E(y_{t+1}|H_t, X_{t+1}) \tag{4.12}$$

where H_t is the history of the process up to time t *and* X_{t+1} is the exogenous covariate series up to time $t+1$ (Mathworks, 2021a).

Equation (4.13) shows an s-step ahead forecast mean square error (MSE) corresponding to the MMSE forecasts (Mathworks, 2021b):

$$MSE = E(y_{t+s} - \hat{y}_{t+s} \mid H_{t+s-1}, X_{t+s})^2 \tag{4.13}$$

The performance of ARIMA and SARIMA models can be evaluated using either the MSE or the root mean squared errors (RMSE) given in Equation (4.14):

$$RMSE = \sqrt{\sum_{i=1}^{n} \frac{(Yt - Yo)^2}{n}} \tag{4.14}$$

where Y_t is the forecasted observation, Y_o is the actual observation, and n is the number of observations. The ARIMA and SARIMA models obtained in the example were used respectively to make a 12-months future forecast using the following procedures given in the two MATLAB codes:

ARIMA FORECAST MATLAB CODE:

```
% Forecast ARIMA Model
% This example shows how to forecast an ARIMA (11,2) model for a
% hypothetical water demand data using MATLAB's forecast function.

% Step 1: Load the water demand data and prepare it for analysis.

[~, ~, data] = xlsread('C:\Users\User\Documents\ waterdemand.xlsx'); % change this to your file location
data = data(:,1); % corresponds to the 1st column in the excel file (column A)
data = data(2:37); % corresponds to the data range from row 2 to 37 in the excel file
data = [data{:}];
data = data';
y = data;
T = length(y);

% Step 2: Estimate an ARIMA (11,2) model for the water demand time series
% data.

Mdl = arima('Constant',NaN,'ARLags',1,'D',1,'MALags',1:2,'Distribution','Gaussian'); % the ARIMA model
function on the right hand side of the equal to sign was copied directly from the model estimate equation
given in the saved MATLAB function that was generated from the Econometric Modeler.

EstMdl = estimate(Mdl,y,'Display','off');

% Step 3: Forecast future water demand for the next 12 months using
% the fitted ARIMA model and the observed water demand time series as
% presample data to generate MMSE forecasts and their corresponding MSE and RMSE
```

```
[yF,yMSE] = forecast(EstMdl,12,'Y0',y);
upper = yF + 1.96*sqrt(yMSE);
lower = yF − 1.96*sqrt(yMSE);

mse = mean((lower-yF).^2) % calculate the MSE
rmse = sqrt(mse) % calculate the RMSE

figure
plot(y,'Color',[.75,.75,.75])
hold on
h1 = plot(T + 1:T + 12,yF,'r','LineWidth',2);
h2 = plot(T + 1:T + 12,upper,'k—','LineWidth',1.5);
plot(T + 1:T + 12,lower,'k—','LineWidth',1.5)
xlim([0,T + 12])
title({'Forecast and 95% Forecast Interval using ARIMA (11,2)', 'RMSE = ' + rmse})

legend([h1,h2],'Forecast','95% Interval','Location','NorthWest')
xlabel('Month')
ylabel('Water Demand')
hold off
```

The results of the ARIMA forecast are shown in Figure 4.16.

SARIMA FORECAST MATLAB CODE:

```
% Forecast SARIMA Model
% This example shows how to forecast a seasonal ARIMA (11,2) × (01,1)12 model
% for a hypothetical water demand data using MATLAB's forecast function.
```

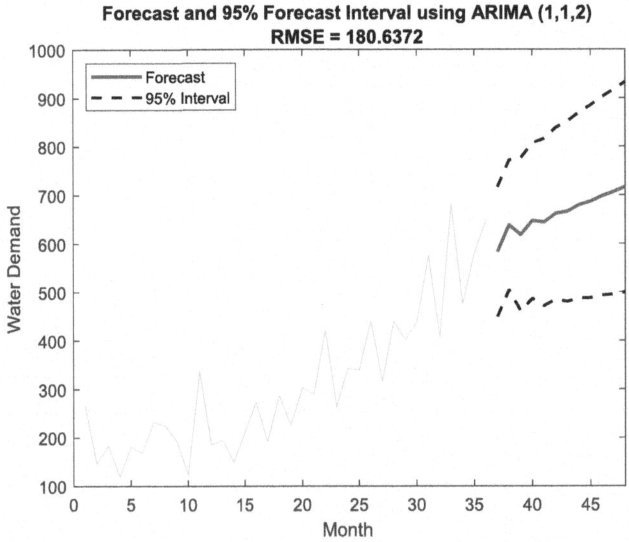

Figure 4.16 Forecast and 95% forecast interval using ARIMA (1,1,2).

```
% Step 1: Load the water demand data and prepare it for analysis.

[~, ~, data] = xlsread('C:\Users\User\Documents\NYC 311 Water Complaints\waterdemand.xlsx'); %
change this to your file location

data = data(:,1); % corresponds to the 1st column in the excel file (column A)
data = data(2:37); % corresponds to the data range from row 2 to 37 in the excel file
data = [data{:}];
data = data';
y = data;
T = length(y);
% Step 2: Estimate an ARIMA (11,2) × (01,1)12 model for the water demand time series data.

Mdl = arima('Constant',NaN,'ARLags',1,'D',1,'MALags',1:2,'SARLags',[],'Seasonality',12,'SMALags',12,
'Distribution','Gaussian'); % the seasonal ARIMA model function on the right hand side of the equal to
sign was copied directly from the model estimate equation given in the saved MATLAB function that was
generated from the Econometric Modeler. However, the seasonality term was changed from '0' to '12'
to include the seasonal difference, which was not included in the estimation as discussed Step 15 in the
previous section.
EstMdl = estimate(Mdl,y,'Display','off');
% Step 3: Forecast future water demand for the next 12 months using
% the fitted ARIMA model and the observed water demand time series as
% presample data to generate MMSE forecasts and their corresponding MSE and RMSE.

[yF,yMSE] = forecast(EstMdl,12,'Y0',y);
upper = yF + 1.96*sqrt(yMSE);
lower = yF - 1.96*sqrt(yMSE);

mse = mean((lower-yF).^2) % calculate the MSE
rmse = sqrt(mse) % calculate the RMSE

figure
plot(y,'Color',[.75,.75,.75])
hold on
h1 = plot(T + 1:T + 12,yF,'r','LineWidth',2);
h2 = plot(T + 1:T + 12,upper,'k--','LineWidth',1.5);
plot(T + 1:T + 12,lower,'k--','LineWidth',1.5)
xlim([0,T + 12])
title({'Forecast and 95% Forecast Interval using ARIMA (11,2) × (01,1)12', 'RMSE = '+rmse})

legend([h1,h2],'Forecast','95% Interval','Location','NorthWest')
xlabel('Month')
ylabel('Water Demand')
hold off
```

The results of the SARIMA forecast are shown in Figure 4.17.

4.2.5 Limitations

Although ARIMA and SARIMA can be used to model a wide range of time series problems, one of
the major limitations of these models is their inability to capture nonlinear patterns due to their linear
structure (Kofinas *et al.*, 2014). Machine learning-based time series models such as artificial neural
networks (ANNs) can capture both linear and non-linear patterns, therefore hybrid ARIMA and ANN
models have been proposed to tackle the nonlinearity deficiencies (Kofinas *et al.*, 2014). Faruk (2010)

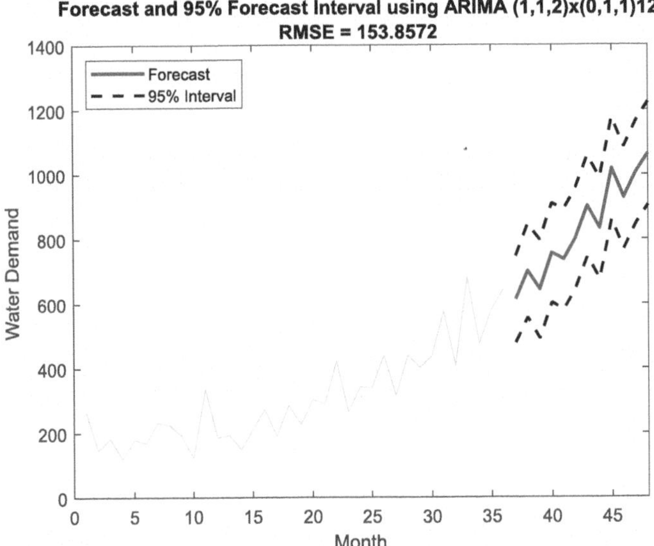

Figure 4.17 Forecast and 95% Forecast Interval using ARIMA (1,1,2)x(0,1,1)12.

used a hybrid neural network and ARIMA model for water quality time series prediction by using water quality data such as water temperature, and boron and dissolved oxygen concentrations collected at the Buyuk Menderes river in Turley from 1996 to 2004. The hybrid model provided accurate results by tackling both the linear and nonlinear patterns of the complex water quality time series (Faruk, 2010).

4.3 MACHINE LEARNING TIME SERIES

4.3.1 Machine learning

4.3.1.1 Artificial neural network

Artificial neural networks (ANNs) mimic the biological neural structure of the brain and form interconnected groups of artificial neurons which are organized in layers. It is a supervised machine learning technique that can be used to forecast water demand patterns over time. ANNs consist of three layers: input layer, hidden layer, and output layer. The inputs or predictors are inserted into the input layer as the bottom layer. The hidden layer is an intermediate layer with hidden neurons. The output layer forms the top layer as forecasts. Among the various architecture of ANN, the feedforward, back propagation (BP) neural network is the most popular, effective model to recognize patterns. A multilayer feedforward network is shown in Figure 4.18. There are four inputs, one hidden layer with three hidden neurons. Each layer of nodes receives inputs from previous layers.

Suppose the input of an ANN is $x = [x_1, x_2, ..., x_n]'$ and its output is $y(x) = [y_1, y_2, ..., y_n]'$. There exists a mapping M from the input space $X:\{x \in X | x$ is the input to the system$\}$ to output space $Y:\{y \epsilon Y | y$ is the output of the system for given input $x\}$. So, the mapping M is as follows:

$$M : X \rightarrow Y \tag{4.15}$$

The training process can be considered a process of gradually adjusting the network internal parameters, for example, the weight w in the weight space ω, that is $w \in \omega$, so that the error between the expected outputs $\hat{y}(x, w)$ and the real outputs $y(x)$ of the network are minimal:

$$error = \min \|\hat{y}(x, w) - y(x)\|^2 \tag{4.16}$$

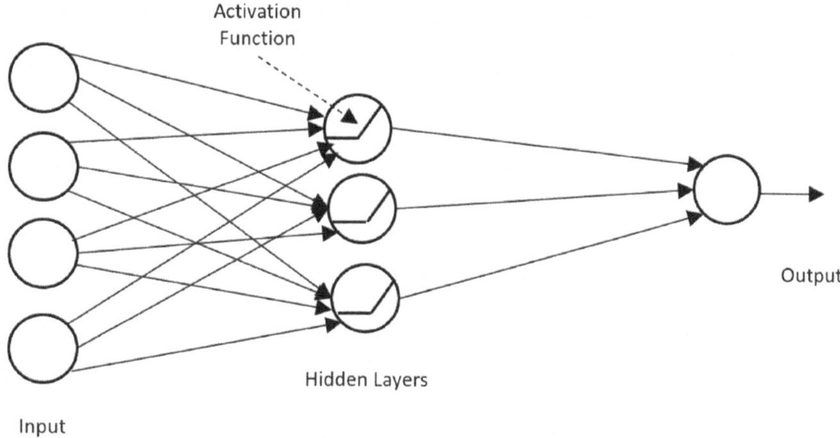

Figure 4.18 Artificial neural networks (ANNs).

The activation of the artificial neuron is conducted through the following equations:

$$\phi(\mathbf{z}) = \phi\left(\sum_i w_i x_i + b\right) \tag{4.17}$$

where i stands for the independent variables that we are considering. The activation function is a non-linear function. Three activation functions that we will consider are the sigmoid function (sigmoid), the hyperbolic tangent function (tanh) and the rectified linear function (ReLU) shown below:

$$sigmoid(z) = \frac{1}{1 + e^{-z}} \tag{4.18}$$

$$\tanh(z) = \frac{e^{2z} - 1}{e^{2z} + 1} \tag{4.19}$$

$$\text{ReLU}(\mathbf{z}) = \max(0, z) \tag{4.20}$$

The training process of feedforward backpropagation ANN is summarized as follows: (1) *Initialize*: construct the feedforward neural network by choosing the input units and output units; (2) *Feedforward*: the input value is propagated from the input layer via the hidden layer to the output layer using the weight and offset value of the network. Compute the output and the error until a stopping criterion is met; (3) *Backpropagation*: the weight is continuously updated and modified so that the error is minimized.

4.3.1.2 Support vector machine

SVM is a supervised machine learning algorithm (Candelieri, 2017; Msiza *et al.*, 2007; Sengupta *et al.*, 2018). The goal of SVM is to separate a given set of binary labeled training data with a hyperplane that is maximally distant from them, that is with maximized margin. However, a hyperplane cannot separate the training data if they are non-linearly separable. Hence, kernel function is introduced to map the training data from its original input space to a high dimensional space where a linear separation can be achieved. In this case, the hyper-plane found by the SVM in the feature space corresponding to a non-linear decision boundary in the original input space. Several common kernel functions are linear kernel, Gaussian radial basis kernel and Sigmoid kernel, and so on.

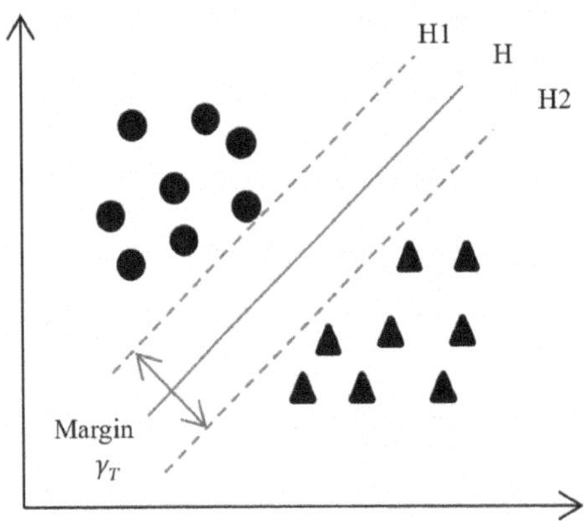

Figure 4.19 Support vector machines (SVMs).

As shown in Figure 4.19, the decision boundary of SVMs is a hyperplane H: (w, b), where w is a normal vector, or a weight vector, perpendicular to the hyperplane with initial value $w_0 = 0$. It is adjusted iteratively each time when training examples are misclassified by current w. b is intercept or bias. The hyperplane equation is defined as:

$$w^T x_i + b = 0 \tag{4.21}$$

To assign class labels to each class for test data, another two hyperplane H1 and H2 are used to determine their classification labels:

$$\begin{cases} H1 : w^T x_i + b \geq 1, & \textit{if } y_i = +1 \\ H2 : w^T x_i + b \leq -1, & \textit{if } y_i = -1 \end{cases} \tag{4.22}$$

Therefore, the final goal is to find the hyperplane with the largest margin. The points on H1 and H2 are called support vectors. Margin of the hyperplanes are the distance from support vectors to the hyperplane γ_T, namely the distance between H1 and H2. To solve the minimization problem, Lagrange multiplier method and Karush-Kuhn-Tucker (KTT) conditions are used to transform this problem to its dual problem. An equivalent dual problem of minimizing $\|w\|_2$ is a maximization problem solving by QP (Quadratic Programming) below:

$$maximize\, W(\alpha) = \sum_{i=1}^{m} \alpha_i - \frac{1}{2} \sum_{i=1}^{m} \sum_{j=1}^{m} \alpha_i \alpha_j y_i y_j x_i \cdot x_j$$

$$subject\, to\quad \begin{aligned} & \sum_{i=1}^{m} y_i \alpha_i = 0 \\ & W = \sum_{i=1}^{m} \alpha_i y_j x_i \end{aligned} \tag{4.23}$$

$$0 \leq \alpha_i \leq C, i = 1, \dots, m.$$

where $\alpha_1, \dots, \alpha_m$ is the Lagrangian multiplier associated with each training example (x_i, y_i). The Lagrangian multipliers are bounded by C, called a box constraint. α_i is the Lagrangian multipliers for support vectors.

The training process of SVM is summarized as follows: (1) Initialize: construct the SVM by entering input and output pairs of the training data sets. Compute the support vectors. (2) Sequential minimal optimization (SMO) is used to solve the QP problem. The goal of this problem to find the hyperplane with the largest margin.

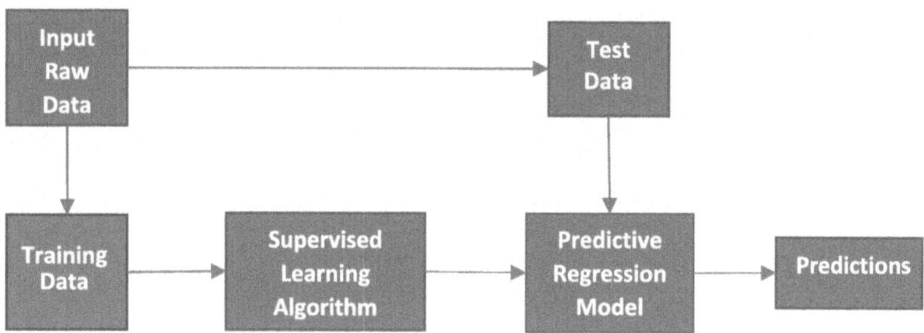

Figure 4.20 Machine learning for water demand forecasting.

4.3.1.3 Forecasting

The water demand forecasting problems can be formalized as supervised machine learning tasks. Supervised learning builds a predictive model that relies on the availability of a finite set of observations. These observations are the mapping or relation between a set of input variables and one or more output variables of the forecast problem.

The flow of a supervised machine learning forecasting task is presented in Figure 4.20. A raw dataset is divided into two subsets: a training set and test set. Data points in the training set are excluded from the test set. The training set is a collection of the input and output pairs. The training set is fed to a supervised learning algorithm to build a predictive regression model. Then, the test set validates the model using its output, that is predictions. In this case, the test set can also be referred to as the validation set. In some literatures, validation set is different from test set. Validation set is a third part of raw data which is used to tune the model's parameters to minimize the overfitting.

Water demand forecast can be solved using machine learning regression models. The input of the model is non-linear water demand time series. The output is real values depicting the water demand on a specific date. The regression problem will find a function $f(x)$ that can map the training inputs to the training outputs.

4.3.2 Practice problems

In this section, we present a simple forecasting problem using SVM regression. The data set we used is from hourly inflow/outflow data of production and storage facilities of the south-central water distribution network in Hillsborough County, FL, Apr 2012–Dec 2012 (Chen, 2018). The first 500 data points were selected for our example below for illustration purposes.

Step 1: Import the data. Separate the data as training and test set. Plot the training set as shown in Figure 4.21.

```
%Import the data from the data file 'Water demand data set 2_Unit_MLD.mat'. This file includes 500 data
points, where 450 data points (90% of the data) is chosen as training data set. The 50 data points are chosen
as the test data set. Plot the training datasets.

rawdata=importdata('Water demand data set 2_Unit_MLD.mat');
rawdata=rawdata';

data1=rawdata(1:450,:);
data1=data1'
figure
plot(data1)
xlabel('Hour')
ylabel('Million gallons')
title('System-wide water demands aggregated in 1-hour intervals in million gallons per day')
```

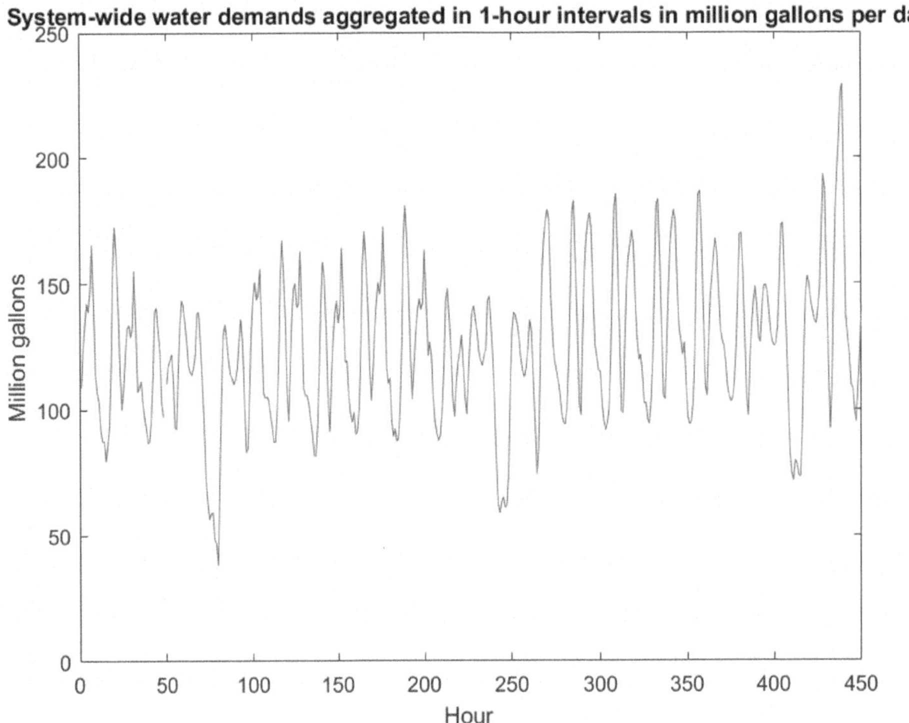

Figure 4.21 SVM Training data set.

Step 2: Construct training and testing data sets. Ninety per cent of the data (450 data points) is chosen as training data set. The remaining 10% of the data (50 data points) is chosen as the test data set.

```
data = rawdata(1:500,:);

numTimeStepsTrain = 450;

dataTrain = data(1:numTimeStepsTrain + 1);
dataTest = data(numTimeStepsTrain + 1:end);

numTimeStepsTest = numel(dataTest(1:end−1));

%XTrain is training data set
%YTrain is the response values of the training data set

XTrain = dataTrain(1:end−1);
YTrain = dataTrain(2:end);

YTest = dataTest(2:end);
```

Step 3: Configure and train the SVM.

%Use 'fitrsvm' function to train the SVM. List the kernel function as 'gaussian' kernel, and set the 'standardize' as true. The function will standardize the training data set.

svm_Mdl = fitrsvm(XTrain,YTrain, 'KernelFunction','gaussian','Standardize',true);

Step 4: Validate the trained SVM model. The forecasting results are showed in Figure 4.22 and compared with the observed results shown in Figure 4.23. The RMSE (root mean square error) values for SVM forecast model are shown in Figure 4.24.

%Use 'predict' function to validate the SVM predictive model svm_Mdl, with input test data set YTest. YPred stores the forecast results.

YPred = predict(svm_Mdl,YTest);

%Plot the forecast results
figure
plot(dataTrain(1:end−1))
hold on
idx = numTimeStepsTrain:(numTimeStepsTrain + numTimeStepsTest);
plot(idx,[data(numTimeStepsTrain) YPred'],'.-')
hold off
xlabel('Hourly water demands')
ylabel('Million gallons')
title('Forecast 50 red data points in the future')
legend(['Observed' 'Forecast'])

%Plot the forecast results versus observed results

figure

plot(YTest)
hold on
plot(YPred,'.-')
hold off
legend(['Observed' 'Forecast'])
ylabel('Million gallons')
title('Forecast vs Observed')

% Quantitative evaluation of forecast results using RMSE

rmse = sqrt(mean((YPred-YTest).^2));
figure(),

stem(YPred − YTest);
xlabel('Hourly water demands')
ylabel('Error')
title('RMSE =' + rmse)

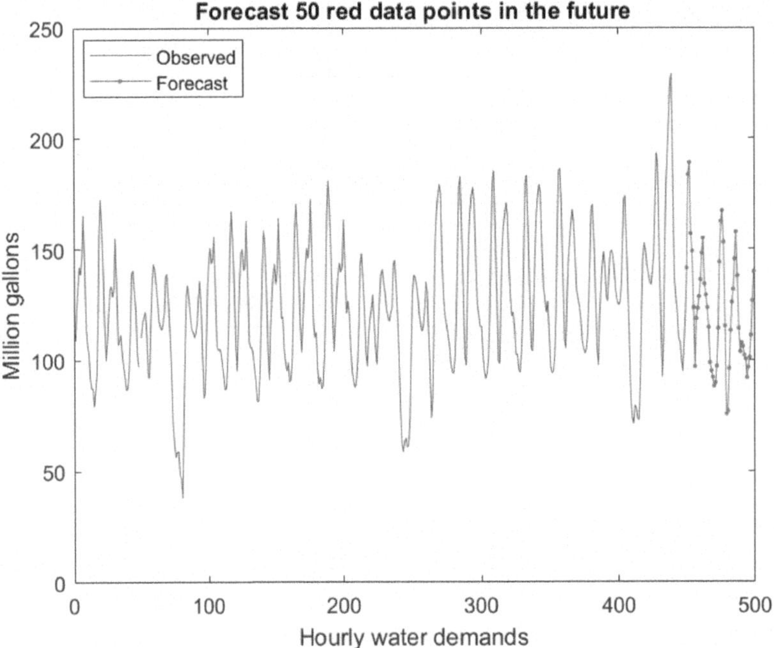

Figure 4.22 SVM forecast results.

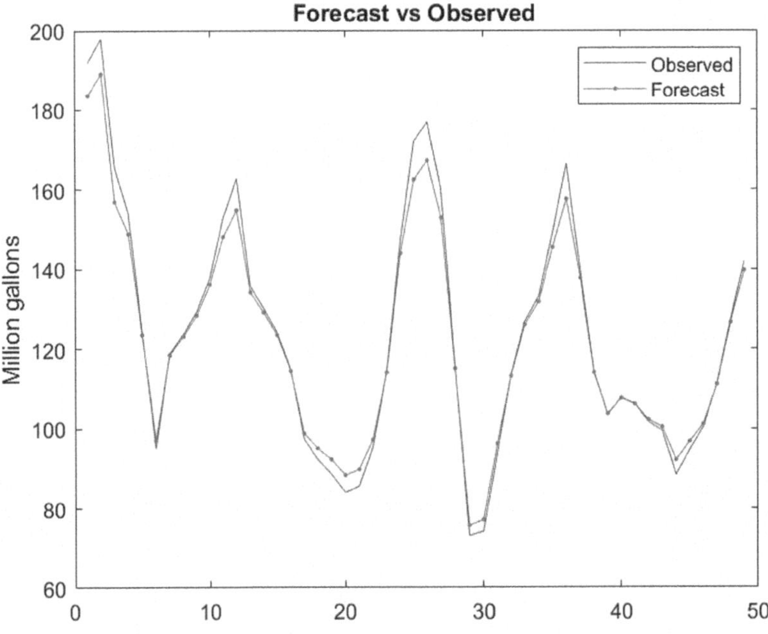

Figure 4.23 SVM forecast (testing) results compared with observed results.

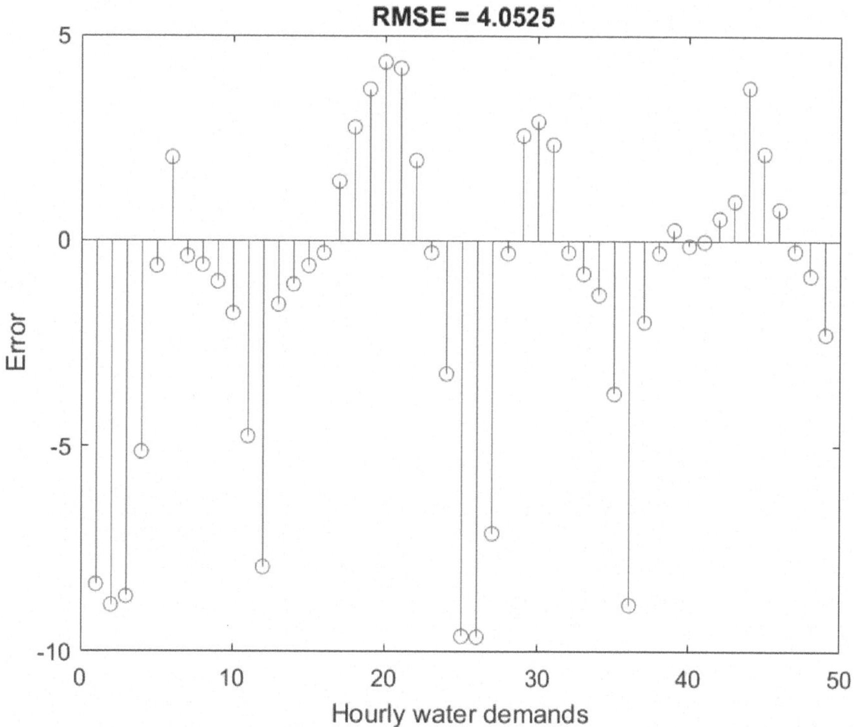

Figure 4.24 RMSE for SVM forecast model.

4.4 DEEP LEARNING TIME SERIES

Deep learning is a promising type of machine learning technique that has attracted much attention over the past few years. Deep learning has the advantages of processing big data, feature learning and strong generalization capability compared to shallow machine learning models. The deep learning time series model exhibits attractive performance in terms of accuracy, stability, and effectiveness (Bedi & Toshniwal, 2019; Du *et al.*, 2021; Guo *et al.*, 2018). We introduce two deep learning time series forecasting models in this section: Convolutional neural networks (CNN) and recurrent neural networks (RNN).

4.4.1 Deep learning models
4.4.1.1 Convolutional neural network
Convolutional neural network (CNN) is a neural network that has been successfully applied in image classification and feature mining. The main advantage of CNN is that it enables the most important features from the input to be extracted (Goldberg, 2016). CNN consists of three types of layers as building blocks: convolution layer, subsampling or pooling layer, as well as a fully connected layer as shown in Figure 4.25.

The convolution layer is a two-layer feed-forward neural network that includes a convolution operation that is designed to extract features from the input. CNN is designed to accept two-dimensional (2D) image data for feature extraction. Time series is one dimensional (1D) data in time domain, so a conversion from 1D to 2D data needs to be carried out before feeding into CNN for

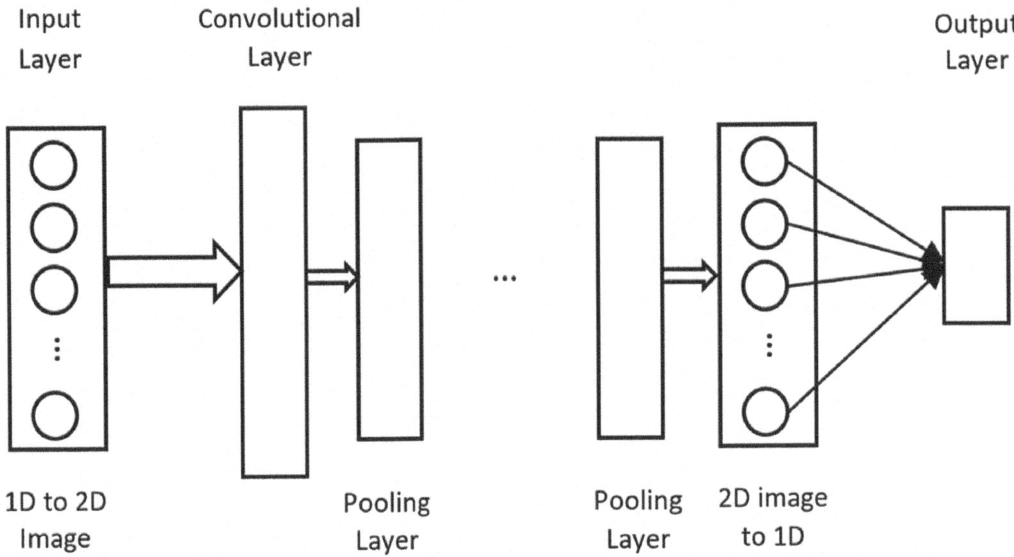

Figure 4.25 Convolutional neural networks (CNNs).

forecasting. Specifically, the input features x_i are convolved with shared weight w and bias term b and get the output y_j in the next layer as follows:

$$y_j = f(\textstyle\sum x_i \otimes w_{i,j} + b_j) \tag{4.24}$$

where $\otimes$ is a convolutional operation and f is a sigmoid function.

The pooling layers are connected to convolutional layers to build up the high-level invariant structures in data. The pooling layer aims to reduce the dimensions of the data and create a down-sampled version of the input. The pooling operations include the max pooling and average pooling.

4.4.1.2 Recurrent neural network

Recurrent neural networks (RNNs) are designed to use the previous information in the sequence to produce the current input and gained popularity in time series forecasting with the recent advances of AI. Unlike ANN, it has forwarding connections in between the neuros and feedback loops. The main advantage of RNN is its acquisition of the internal sequential nature that remembers information through many timesteps, making it a powerful tool in forecasting long term trends from time series data. RNN is comprised of single rolled RNN units as shown in Figure 4.26.

Three kinds of RNN units are most popular for sequence modelling. They are the Elman RNN (ERNN) cell (Elman, 1990), the gated recurrent unit (GRU) cell (Cho *et al.*, 2014) and the long short-term memory (LSTM) cell (Hochreiter & Schmidhuber, 1997). The LSTM RNN network has been applied in time series prediction as a special kind of deep learning model.

The structure of RNN includes hidden state h, input X and an optional output Y. Given a time series input sequence $X = \{x_1, x_2, \ldots, x_t\}$, at time step t, RNN learns a mapping from x_t to h_t depending on the hidden state at h_{t-1}:

$$h_t = f(h_{t-1}, x_t), \tag{4.25}$$

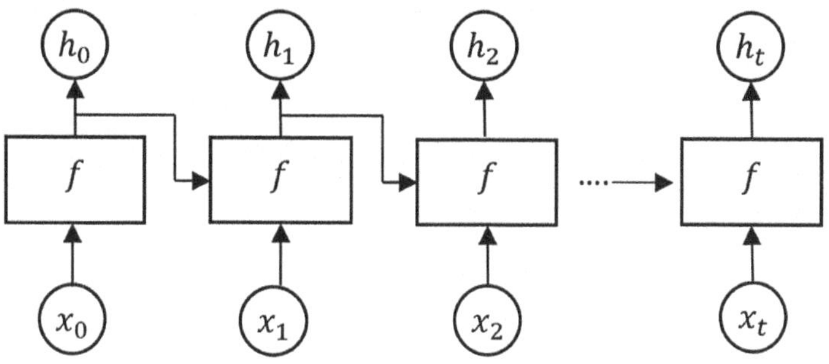

Figure 4.26 Recurrent neural networks (RNNs).

where f is a non-linear activation function. This function can be ERNN, GRN or LSTM, or as simple as a logic sigmoid function.

The training process of RNN suffers from problems of vanishing or exploding gradients which occur when backpropagating errors across many time steps. LSTM was introduced to overcome the above problem by replacing the hidden layer in the standard RNN by a memory cell. Each memory cell contains several gates and four interactive layers: forget gate layer, input gate layer, Tanh layer, and output gate layer.

4.4.2 Practice problems

In this section, we present a simple forecasting problem using LSTM regression. The data set we used is from hourly sewer flows monitored at Station S2 in Columbus, OH, Jun 1998–Dec 2013 (Chen, 2018). The first 500 data points was selected for our example below for illustration purpose. The task is to forecast the sewer flow in the 1-hour intervals.

Step 1: Import the data. Separate the data as training and test set. Plot the training set as shown in Figure 4.27.

```
%Import the data from the data file 'sewer_hourly.mat'. This file includes 500 data points, where 450 data
points (90% of the data) is chosen as training data set. The 50 data points are chosen as the test data set.
Plot the training datasets.

rawdata = importdata('sewer_hourly.mat');
data1 = rawdata(1:450,:);
data1 = data1'
figure
plot(data1)
xlabel('Hour')
ylabel('Million gallons')
title('Hourly Sewer flow aggregated in million gallons per day')
```

Step 2: Construct training and testing data sets. 90% of the data (450 data points) is chosen as training data set. The remaining 10% of the data (50 data points) is chosen as the test data set.

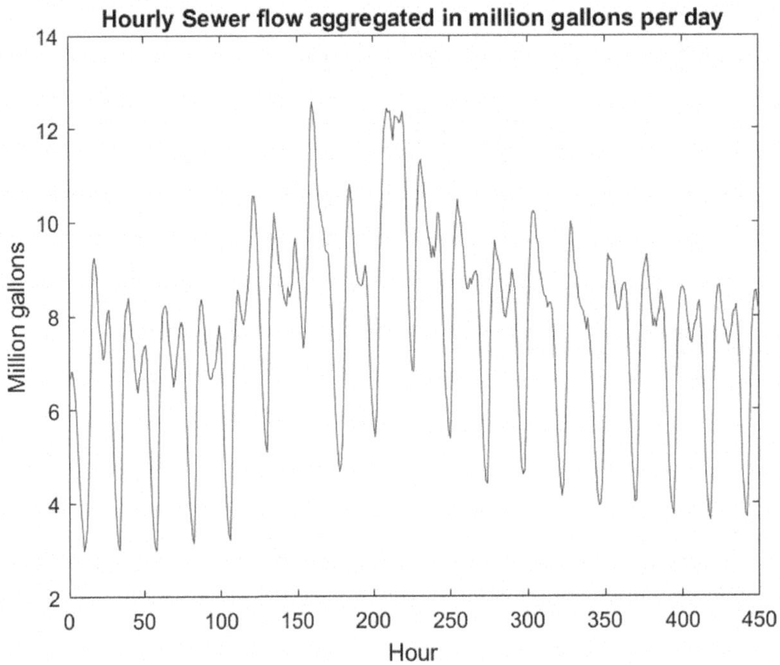

Figure 4.27 LSTM training data set.

```
data = rawdata(1:500,:);
data = data';

numTimeStepsTrain = 450;

% The data with index 1 to numTimeStepsTrain + 1 will be training set
% The data with index numTimeStepsTrain + 1 to end will be test set
dataTrain = data(1:numTimeStepsTrain + 1);
dataTest = data(numTimeStepsTrain + 1:end);

% Standardize the data by putting different data on the same scale. We calculate the mean and standard
deviation for each variable. Then, for each observed data, we subtract the mean and divide by the standard
deviation.

mu = mean(dataTrain);
sig = std(dataTrain);

dataTrainStandardized = (dataTrain – mu)/sig;

%XTrain is training data set
%YTrain is the response values of the training data set

XTrain = dataTrainStandardized(1:end–1);
YTrain = dataTrainStandardized(2:end);
```

Step 3: Configure the LSTM neural network.

```
% Set the LSTM regression network training option as follows: 250 hidden units.
numFeatures = 1;
numResponses = 1;
numHiddenUnits = 250;

layers = [...
  sequenceInputLayer(numFeatures)
  lstmLayer(numHiddenUnits)
  fullyConnectedLayer(numResponses)
      regressionLayer];

% Set the maximum epochs to 250. Gradient threshold to 1. Learn rate determines the step size at each
iteration while moving toward a minimum of a loss function. Initial learn rate 0.005. After 125 epochs, the
learn rate will be multiplied by a factor of 0.2.
options = trainingOptions('adam',...
  'MaxEpochs',250,...
  'GradientThreshold',1,...
  'InitialLearnRate',0.005,...
  'LearnRateSchedule','piecewise',...
  'LearnRateDropPeriod',125,...
  'LearnRateDropFactor',0.2,...
  'Verbose',0,...
  'Plots','training-progress');
```

Step 4: Train the LSTM neural network.

```
% Generate a trained recurrent neural network model in variable 'net'
net = trainNetwork(XTrain,YTrain,layers,options);
```

Step 5: Validate the trained LSTM model. The forecasting results are showed in Figure 4.28 and
compared with the observed results shown in Figure 4.29. The RMSE values for LSTM forecast
model are shown in Figure 4.30.

```
dataTestStandardized = (dataTest – mu)/sig;
XTest = dataTestStandardized(1:end−1);

% predictAndUpdateState function: Predict responses using a trained recurrent neural network 'net' and
update the network state

net = predictAndUpdateState(net,XTrain);

% YPred variable stores the forecast results of 50 data points

[net,YPred] = predictAndUpdateState(net,YTrain(end));

numTimeStepsTest = numel(XTest);
for i = 2:numTimeStepsTest
  [net,YPred(:,i)] = predictAndUpdateState(net,YPred(:,i−1),'ExecutionEnvironment','cpu');
end
```

```
YPred = sig*YPred + mu;

% YTest variable stores the observed results of 50 data points

YTest = dataTest(2:end);

% Plot the forecast results

rmse = sqrt(mean((YPred-YTest).^2));

figure
plot(dataTrain(1:end−1))
hold on
idx = numTimeStepsTrain:(numTimeStepsTrain + numTimeStepsTest);
plot(idx,[data(numTimeStepsTrain) YPred],'.-')
hold off
xlabel('Hourly sewer flows')
ylabel('Million gallons')
title('Forecast 50 red data points in the future')
legend(['Observed' 'Forecast'])

% Compare the forecast results with the observed results.

figure
plot(YTest)
hold on
plot(YPred,'.-')
hold off
legend(['Observed' 'Forecast'])
ylabel('Million gallons')
title('Forecast vs Observed')

%% Quantitative evaluation of forecast results using RMSE.

figure(),
stem(YPred − YTest)
xlabel('Hourly sewer flows')
ylabel('Error')
title('RMSE = ' + rmse)
```

4.5 OTHER POPULAR ML TECHNIQUES

4.5.1 Ensemble learning

In this section, we demonstrate how ensemble methods may be used to combine multiple MLT to improve the solution of regression and classification problems, with practical applications to a real case study, using high-resolution water-flow measures. All the applications reported in this paragraph are made available in the Github repository (https://github.com/Water-End-Use-Dataset-Tools/EL-WaterDemandTS). An ensemble includes a number of learners called base learners, usually generated from training data by a base learning algorithm which can be a decision tree, neural network or other kinds of learning algorithms. They try to build a set of learners from training data and combine them (Dong *et al.*, 2020). The use of ensemble methods is related to the possibility of achieving higher predictive performance than using an individual algorithm by itself (Zhou, 2012). In this section, the example code is given in Python for the variety of coding capacities (and it is also free!)

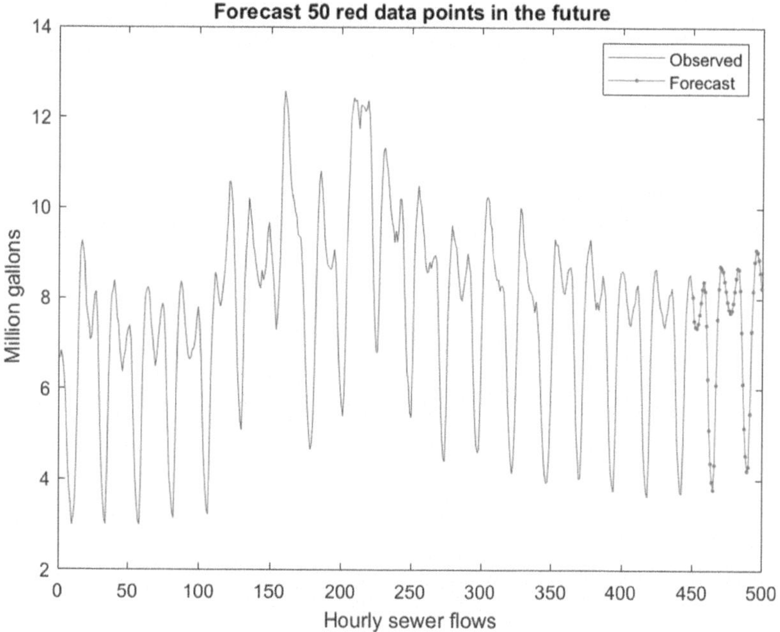

Figure 4.28 LSTM forecast results.

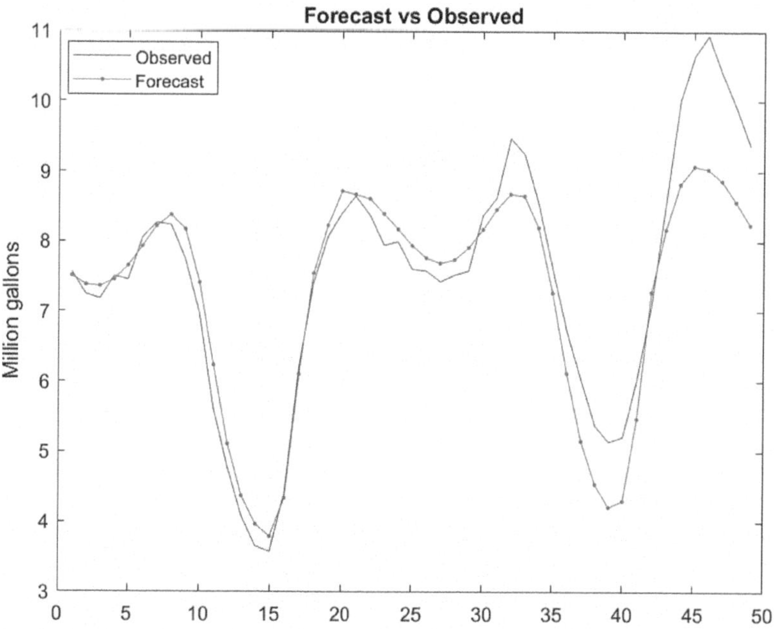

Figure 4.29 LSTM forecast (testing) results compared with observed results.

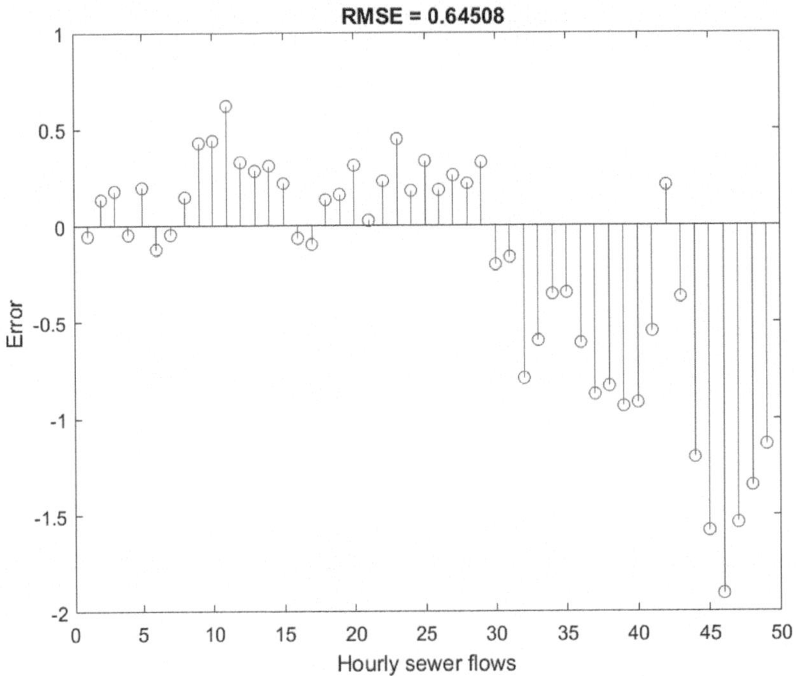

Figure 4.30 RMSE for LSTM forecast model.

4.5.1.1 *Water end use dataset*

For the applications reported in this paragraph, a dataset of water end use consumption is used. The dataset has been generated processing the water consumption measured at different fixtures of a domestic pilot and collected as water flow time-series. Each time-series contains the water-flow data in ml/sec with a sample period of 1 sec (Di Mauro *et al.*, 2019).

The water_usages dataset is a list of records provided as a CSV (comma separated values). Each record characterizes the occurrence of a water usage and is described by the following parameters:

- *start_date_time*: long [sec] it is the starting date-time of the usage as Unix epoch
- *duration*: int [ms], how long lasts the usage
- *liters*: int [mL], how many liters of water have been consumed
- *month*:int, month of occurrence
- *hour*:int, hour of the day
- *day*: int, day of the week {0,...,6}
- *max_flow*: int [mL/sec], maximum flow rate measured during the usage
- *av_flow_rate*: float [mL/sec], the average flow rate calculates for the usage
- *sec_from_midnight*: int, the number of seconds after the midnight
- *fixture*: string, the lable that identifies the fixture (e.g., shower, washbasin, etc.)
- *num_fixture*: int, an integer that identifies the fixture (e.g., 0: shower, 1: washbasin, …)

The original time-series have been split to identify every single usage, and then the usages have been clustered to identify similar water consumption profiles (e.g. hand washing, teeth brushing). The individual time-series excerpts will be also used later in this chapter. The complete dataset is available in a different GitHub repository (https://github.com/Water-End-Use-Dataset-Tools/WEUSEDTO)

4.5.1.2 Bootstrapping

Bootstrapping is a statistical method that resamples a single dataset to create many simulated samples. Applying the bootstrap method works like collecting many datasets. Increasing the dataset and computing the mean of the means estimates will eventually lead to a zero bias. In other words, it aims at computing an unbiased estimator of the population mean. The bootstrapping process allows us to evaluate statistics on a population which is obtained by sampling a dataset with replacement in order to make the selection procedure completely random. Bootstrapping is commonly useful to evaluate statistics such as the mean, standard deviation, construct confidence intervals and perform hypothesis testing for different types of statistics samples. It is used in applied machine learning to value the ability of machine learning models when making predictions on data not included in the training dataset. The importance of bootstrap sampling is related to their use as a basic step for several modern MLT, as for example the bagging technique used in various ensemble machine learning algorithms like random forests, gradient boost, and so on. Moreover, bootstrapping can be used to estimate the parameters of a population when the data sample available is not large enough to assume that the sampling distribution is normally distributed. Bootstrapping uses the distribution of the sample statistics among the simulated samples as the sampling distribution. The application reported below shows an example of mean evaluation on resampled datasets.

Bootstrap method formulation: Let there be a sample X of size N. We can make a new sample from the original sample by drawing N elements from the latter randomly and uniformly, with replacement. In other words, we select a random element from the original sample of size N and do this N times. All elements are equally likely to be selected, thus each element is drawn with the equal probability $1/N$. More details on the bootstrap method can be found in (Efron & Tibshirani, 1993).

Bootstrapping example application and code: The water_usages data-set has been used here to demonstrate how bootstrapping works. The amount of water consumed on each usage is the feature that is used in the model. Let us visualize in Figure 4.31 the data and look at the distribution of this feature for two fixtures, which are the washbasin and the kitchen faucet.

```python
import numpy as np
import pandas as pd
import seaborn as sns

sns.set()
from matplotlib import pyplot as plt

#Graphics in retina format are more sharp and legible %config InlineBackend.figure_format = 'retina'
water_data = pd.read_csv('.data/dataset.csv', delimiter = ' ')

water_data.loc[water_data['fixture'] == 'washbasin', 'liters'].hist(label = 'Washbasin')

water_data.loc[water_data['fixture'] == 'kitchen faucet', 'liters'].hist(label = 'kitchen faucet')

plt.xlabel('mL')
plt.ylabel('Density')
plt.legend();
```

It is straightforward to observe that the washbasin is used more often than the kitchen faucet. Moreover, a higher percentage of usages consume less water in the case of the washbasin with respect to the kitchen faucet. Now, it may be a good idea to estimate the average amount of water consumed for each fixture for designing predictive strategies of water management. Since our dataset is small,

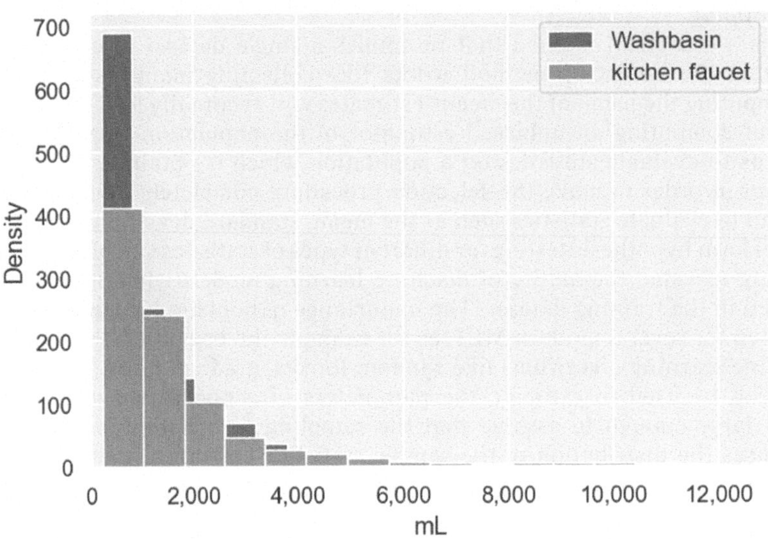

Figure 4.31 Data distribution of washbasin and kitchen faucet fixtures.

and the number of samples is different for the two fixtures (washbasin: 1354, kitchen faucet: 895), we would not get a good estimate by simply calculating the mean of the original sample. With a small dataset the estimation of the mean value could be different from the mean value of the population. Such a difference is called bias. We will be better off applying the bootstrap method. Let us generate 5000 new bootstrap samples from our original population and produce an interval estimate of the mean.

```python
def get_bootstrap_samples(data, n_samples):
    '''Generate bootstrap samples using the bootstrap method.''' indices =
    np.random.randint(0, len(data), (n_samples, len(data))) samples =
    data[indices]
    return samples

def stat_intervals(stat, alpha):
    '''Produce an interval estimate.'''
    boundaries = np.percentile(stat, [100 * alpha / 2.0, 100 * (1 – alpha / 2.0)])
    return boundaries

#Save the data about the washbasin and kitchen faucet to split the dataset
wb_liters = water_data.loc[water_data['fixture'] == 'washbasin', 'liters' ].values
kit_liters = water_data.loc[ water_data['fixture'] == 'kitchenfaucet', 'liters'].values

#Set the seed for reproducibility of the results
np.random.seed(0)

#Generate the samples using bootstrapping and calculate the mean for each of them
wb_liters_mean_scores = [

    np.mean(sample) for sample in get_bootstrap_samples(wb_liters, 5000)]
kit_liters_mean_scores = [
```

```
np.mean(sample) for sample in get_bootstrap_samples(kit_liters, 5000)]

#Print the resulting interval estimates
print('mliters consumed by washbasin: mean interval',
stat_intervals(wb_liters_mean_scores, 0.05))
print('mliters consumed by kitchenfaucet: mean interval',
        stat_intervals(kit_liters_mean_scores, 0.05))
```

As a result, the mean interval for the milliliters consumed by washbasin and kitchen faucet are respectively: [1344, 1547] and [1645, 1953].

In Figure 4.32, the same procedure is applied to compare different fixtures as kitchen faucet and shower. It is straightforward to observe that the shower is used less often than the kitchen faucet. Moreover, shower usages usually consume more water respect to the kitchen faucet, with a reduced variance.

As a result, the mean interval for the milliliters consumed by washbasin and shower are respectively: [20056, 24218] and [1344, 1547])

4.5.1.3 Bagging

The bagging is a machine learning ensemble algorithm realized to improve the stability and accuracy of algorithms used for statistical classification and regression. Bagging, also called bootstrap aggregating, trains multiple models of the same learning algorithm on bootstrapped samples of the original dataset, and then aggregates their individual predictions to produce a final prediction as shown in Figure 4.33. Bagging is typically used with decision trees and this kind of MLT prevents overfitting, reducing the variance of a classifier by decreasing the difference in error when the model is trained on different datasets. Besides the use of the bagging technique to reduce model overfitting, it is used in the case of high-dimensional data due to its good performance. Furthermore, possible missing values in the dataset do not alter the execution of the algorithm. More details on the bagging method can be found in Bühlmann and Yu (2002).

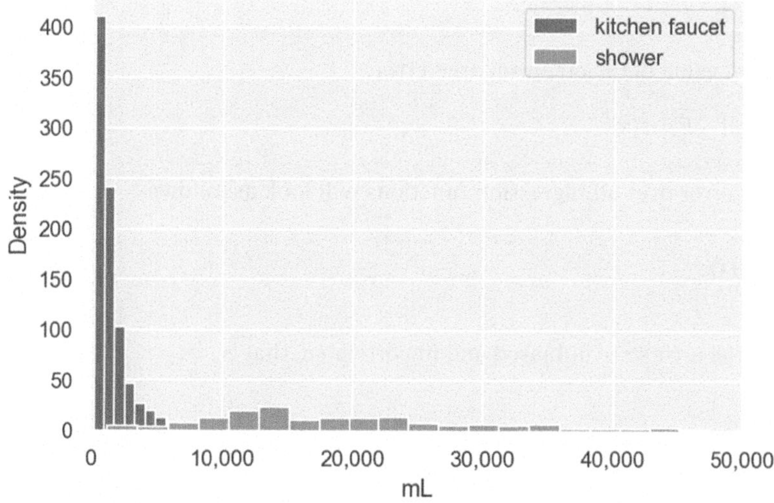

Figure 4.32 Data distribution of kitchen faucet and shower fixtures.

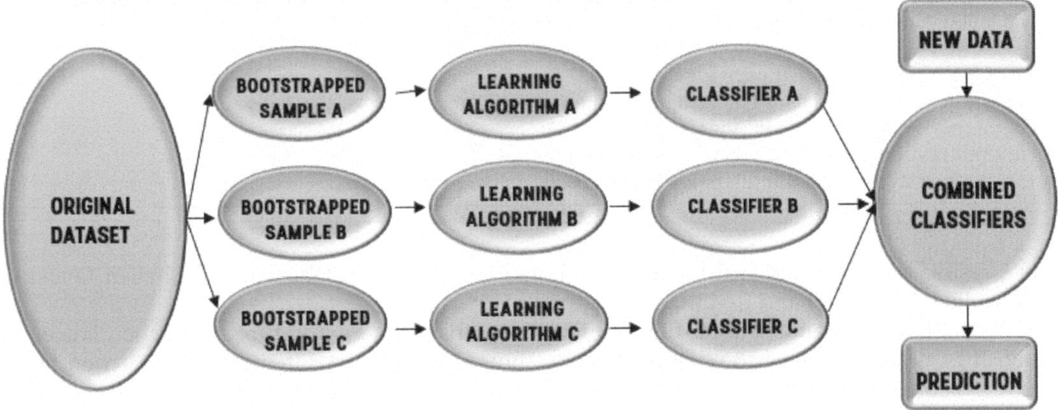

Figure 4.33 Diagram of bagging technique.

Bagging method formulation: Bagging method formulation was presented by Breiman as reported in Breiman (1996). Consider a training set X then $X_1, ..., X_M$ are generated using bootstrapping. Now, a classifier $a_i(x)$ is trained for each bootstrap sample. The final classifier will average the outputs from all these individual classifiers:

$$a(x) = \frac{1}{M} \sum_{i=1}^{M} a_i(x) \tag{4.26}$$

Figure 4.33 illustrates the bagging algorithm.

Let us consider a regression problem with base algorithms $b_1(x), ..., b_n(x)$. Assume that there exists an ideal target function of true answers $y(x)$ defined for all inputs and that the distribution $p(x)$ is defined. Then the error can be expressed for each regression function as follows:

$$\varepsilon_i(x) = b_i(x) - y(x), \quad i = 1, ..., n \tag{4.27}$$

and the expected value of the mean squared error:

$$E_x[(b_i(x) - y(x))^2)] = E_x\left[\varepsilon_i^2(x)\right] \tag{4.28}$$

Then, the mean error over all regression functions will look as follows:

$$E_1 = \frac{1}{n} E_x\left[\sum_{i}^{n} \varepsilon_i^2(x)\right] \tag{4.29}$$

Assuming that the errors are unbiased and uncorrelated, that is:

$$E_x\left[\varepsilon_i(x)\right] = 0, \tag{4.30}$$

$$E_x\left[\varepsilon_i(x)\varepsilon_j(x)\right] = 0, \quad i \neq j \tag{4.31}$$

Now, let us construct a new regression function that will average the values from the individual functions:

$$a(x) = \frac{1}{n} \sum_{i=1}^{n} b_i(x) \tag{4.32}$$

Let us find its mean squared error:

$$E_n = E_x \left[\frac{1}{n} \sum_{i=1}^{n} b_i(x) - y(x) \right]^2 \tag{4.33}$$

$$= E_x \left[\frac{1}{n} \sum_{i=1}^{n} \varepsilon_i \right]^2 \tag{4.34}$$

$$= \frac{1}{n^2} E_x \left[\sum_{i=1}^{n} \varepsilon_i^2(x) + \sum_{i \neq j} \varepsilon_i(x)\varepsilon_j(x) \right] \tag{4.35}$$

$$= \frac{1}{n} E_1 \tag{4.36}$$

Thus, by averaging the individual answers, the mean squared error can be reduced by a factor of n. Let us recall the components that make up the total out-of-sample error:

$$Err(\vec{x}) = E\left[(y - \hat{f}(\vec{x}))^2\right] \tag{4.37}$$

$$= \sigma^2 + f^2 + Var(\hat{f}) + E[\hat{f}]^2 - 2fE[\hat{f}] \tag{4.38}$$

$$= (f - E[\hat{f}])^2 + Var(\hat{f}) + \sigma^2 \tag{4.39}$$

$$= Bias(\hat{f})^2 + Var(\hat{f}) + \sigma^2 \tag{4.40}$$

Bagging example application and code: In this example, similarly to what is presented in section 7.3 of Hastie *et al.* (2009), we show and compare the variance of the expected mean squared error of a single estimator against a bagging ensemble in a regression problem applied to time-series of real data. A cluster of washbasin usages has been used as they were noisy measures of the same water consumption profile (e.g., hand washing), and the spline that approximates all measures of the cluster as the true profile. Figure 4.34 shows the results of the application: the upper left figure illustrates the predictions (in dark dashed line) of a single decision tree that has been trained over a down-sampled time-series of one usage profile. It also illustrates the predictions (in light dashed line) of other single decision trees trained over the down-sampled consumption profiles of the cluster. The variance term in this application corresponds to the width of the bundle of predictions (in light dashed line) of the individual estimators. The predictions for x are more sensitive. The lower left figure plots the pointwise decomposition of the expected mean squared error of a single decision tree. It shows the variance in the rectangular marker line and also illustrates the noise part of the error which, as expected, appears to be comparable to the variance as we considered real profiles as a noisy version of the cluster centroid. The figures on the right reported to the same plots using a bagging ensemble of decision trees. In terms of variance, the bundle of predictions is narrower, which indicates that the variance is lower. Moreover, as shown by the lower right figure, the variance term (rectangular marker line) is lower than for single decision trees.

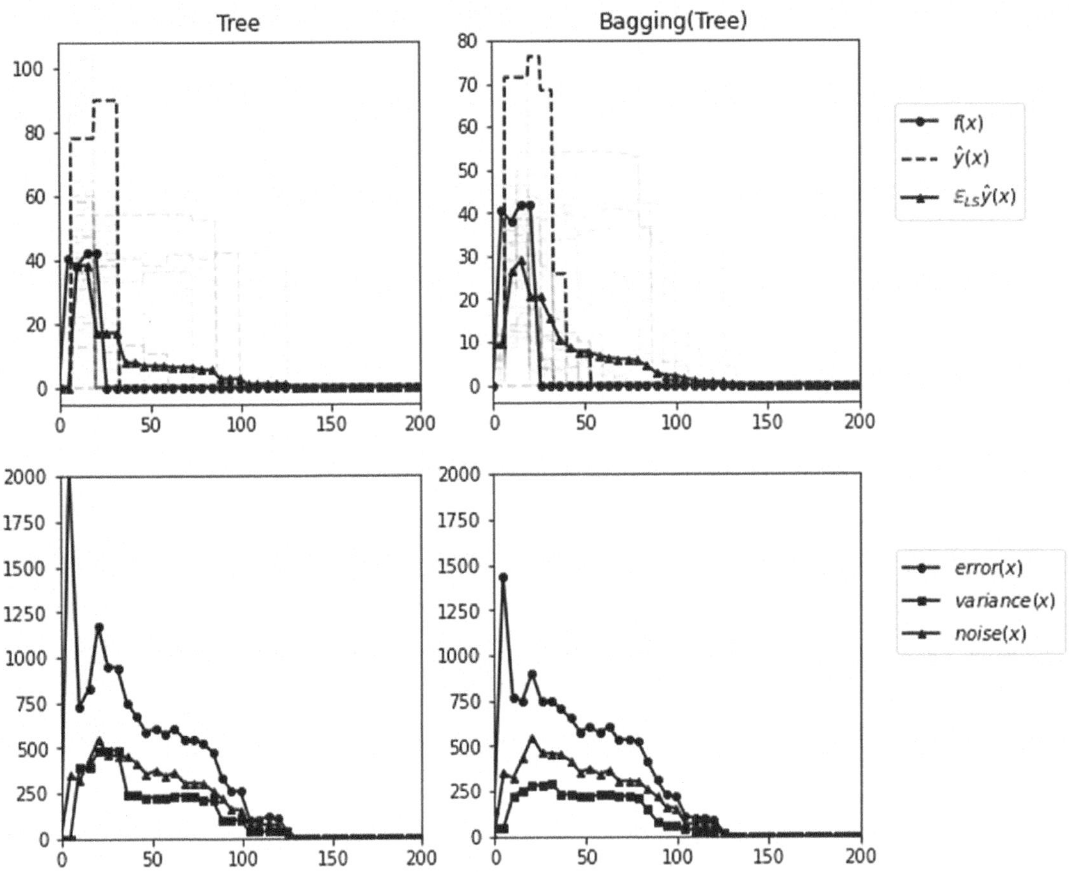

Figure 4.34 Bagging application to washbasin time series.

Comparing the circle marker line in the lower graphics, it is worth noticing that for bagging the overall mean squared error is lower. It depends on the fact that for bagging, averaging several decision trees fit on bootstrap copies of the dataset allows for a reduction of the variance.

The total error of the bagging ensemble is a wee bit lower than the total error of a single decision tree, this difference hinges on the reduced variance.

In Figure 4.34, $f(x)$ is the true function $y(x)$ are the estimators, $\mathcal{E}l_s\hat{y}(x)$ is the average of the estimators, error(x) is the mean square error between the true value and one estimator, noise(x) is the variance of the measured timeseries (it is evaluated on the test set that represent noisy measures).

4.5.1.4 Random forest

Random forest (RF) is one of the most popular machine learning algorithms. It was introduced by Breiman as an ensemble tree learner (Breiman 2001). The algorithm consists of many decision trees, each with the same nodes, built using a different bootstrap sample of the data from the original training dataset. RF merges the prediction result from every decision tree in order to find an answer, which represents the average of all the decision trees. It selects the best solution by means of voting, the most voted is chosen as the final prediction, as shown in Figure 4.35. One of the advantages of random forest is its flexibility, in fact, it is used to solve both regression and classification problems. It is used mostly because it is not influenced by noise, and due to the presence of several trees in the

```python
from scipy.interpolate import interp1d
import numpy as np
import matplotlib.pyplot as plt
from sklearn.ensemble import BaggingRegressor
from sklearn.tree import DecisionTreeRegressor
import glob

# Settings
n_repeat = 50            # Number of iterations for computing expectations
n_train = 50             # Size of the training set
n_test = 1000            # Size of the test set
np.random.seed(0)

estimators = [('Tree', DecisionTreeRegressor()),
                    ('Bagging(Tree)', BaggingRegressor(DecisionTreeRegressor()))]

n_estimators = len(estimators)
ts_files = glob.glob('data/csv_Washbasin/cluster_0/*.csv')
f_true = 'data/csv_Washbasin/1_spline.csv'

def f(x, iteration):
  if iteration ==-1:
   ts = np.genfromtxt(ts_files[4], delimiter = ' ')
  else:
   ts = np.genfromtxt(ts_files[iteration], delimiter = ' ')
   start_time = ts[0,0]
   ts[:,0] -= start_time
   ts[0,1] = 0
   if ts[-1,0] < 650:
                    ts = np.vstack((ts,[ts[-1,0] + 1, 0]))
                    ts = np.vstack((ts,[650, 0]))
          for i in range(1,len(ts)-1):
                    if ts[i,1] ==0 and ts[i+1,1]!=0:
                     ts[i,1] = (ts[i-1,1] + ts[i+1,1])*0.5
   linfunc = interp1d(ts[:,0], ts[:,1])
  return linfunc(x)

def generate(n_samples, n_repeat = 1):
 max_duration = 650
 X = np.linspace(0, 650, n_samples)
 if n_repeat == 1:
  y = f(X, np.random.randint(1,len(ts_files)))
 else:
  y = np.zeros((n_samples, n_repeat))
 for i in range(n_repeat):
  y[:, i] = f(X, np.random.randint(1,len(ts_files)))
 X = X.reshape((n_samples, 1))
 return X, y

X_train []
y_train = []
```

```python
for i in range(n_repeat):
    X, y = generate(n_samples = n_train)
    X_train.append(X)
    y_train.append(y)
X_test, y_test = generate(n_samples = n_test,n_repeat = n_repeat)
plt.figure(figsize=(10, 8))
# Loop over estimators to compare
for n, (name, estimator) in enumerate(estimators):
    # Compute predictions
    y_predict = np.zeros((n_test, n_repeat))
    for i in range(n_repeat):
        estimator.fit(X_train[i], y_train[i])
        y_predict[:, i] = estimator.predict(X_test)

    y_error = np.zeros(n_test)
    for i in range(n_repeat):
        for j in range(n_repeat):
            y_error += = (y_test[:, j] - y_predict[:, i]) ** 2

    y_error /= (n_repeat * n_repeat)
    y_noise = np.var(y_test, axis = 1)
    y_var = np.var(y_predict, axis = 1)

# Plot figures
plt.subplot(2, n_estimators, n + 1)
plt.plot(X_test, f(X_test,-1), 'b', label = '$f(x)$')

for i in range(n_repeat):
 if i == 0:
  plt.plot(X_test, y_predict[:, i], 'r', label = r'$\^y(x)$')
 else:
  plt.plot(X_test, y_predict[:, i], 'r', alpha = 0.05)

plt.plot(X_test, np.mean(y_predict, axis = 1), 'c', label = r'$\mathbb{E}_{LS} \^y(x)$')
plt.xlim([0, 350])
plt.title(name)

if n == n_estimators - 1:
 plt.legend(loc = (1.1, 0.5))

plt.subplot(2, n_estimators, n_estimators + n + 1) plt.plot(X_test,
y_error, 'r', label = '$error(x)$')        plt.plot(X_test,  y_var, 'g',
label = '$variance(x)$'),  plt.plot(X_test,          y_noise,   'c',
label = '$noise(x)$')
pltxlim([0, 350])
plt.ylim([0, 2000])
if n == n_estimators - 1:
 plt.legend(loc = (1.1, 0.5))

plt.subplots_adjust(right = .75)
plt.show()
```

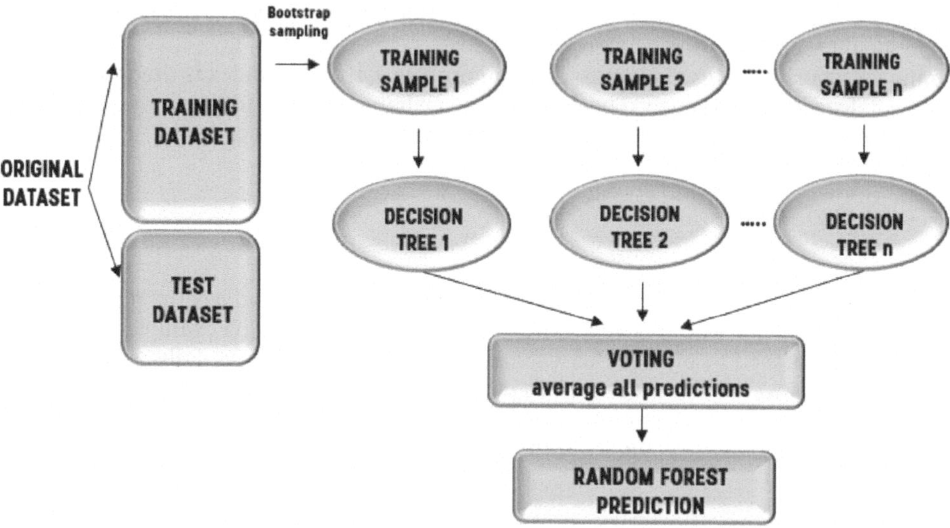

Figure 4.35 Diagram of random forest technique.

forest, it will not overfit the model. RF also has some limitations in terms of computation, which becomes slower as the number of trees in the model is larger. More details on random forest method can be found in Hastie *et al.* (2009).

***Random forest method formulation*:** The random forest method formulation was presented by Breiman as reported in Breiman (2001) and Hastie *et al.* (2009).

The algorithm for building a random forest of N trees goes as follows:

For each $b = 1, \ldots, N$;

- Draw a bootstrap sample Xb;
- Build a decision tree T_b on the bootstrap sample Xb repeating the following steps:
  - Pick the best feature according to the given criteria. Split the sample by this feature to create a new tree level. Repeat this procedure until the sample is exhausted;
  - Building the tree until any of its leaves contains no more than n_{min} instances;
  - For each split, first randomly pick m features from the original ones and then search for the next best split only among the subset.

Output the ensemble of trees $\{T_b\}_1^N$

The final prediction at a new point x is defined:

For Regression by: $f(x) = \dfrac{1}{N} \displaystyle\sum_{b=1}^{N} T_b(x)$

For Classification by: Let $C_b(X)$ be the class prediction of the bth random forest tree. Then $c(x) = majorityvote\{C_b\}_1^N$

When the RF algorithm is used for regression problems, the mean squared error (MSE) is used to evaluate the distance of each node from the predicted value in order to select which branch represents the best decision for the forest:

$$MSE = \frac{1}{N} \sum_{i=1}^{N} (f_i - y_i)^2 \tag{4.41}$$

where N is the number of data points, f_i is the value returned by the decision tree and y_i is the value of the data point you are testing at a certain node.

When the RF algorithm is used for classification problems, the Gini index is used to determine how nodes are on a decision tree branch. The class and probability are used to determine the Gini of each branch on a node, establishing which of the branches is more likely to occur:

$$Gini = 1 - \sum_{i=1}^{C} (p_i)^2 \qquad\qquad 4.42$$

where pi is the relative frequency of the class observed in the dataset and c is the number of classes.

Random forest regression example application and code: In the following example machine learning techniques have been used to predict the water consumption profiles of a daily usage from a down-sampled time-series of water-flow measures. In particular, the examples start from the high-resolution time-series of the daily water consumption of a kitchen faucet. The training set is composed of 100 samples, which are randomly selected from the original time-series.

```python
import numpy as np
from scipy.interpolate import interp1d
from matplotlib import pyplot as plt

import seaborn as sns
from sklearn.datasets import make_circles
from sklearn.ensemble import (BaggingClassifier,
    BaggingRegressor, RandomForestClassifier, RandomForestRegressor)
from sklearn.model_selection import train_test_split
from sklearn.tree import DecisionTreeClassifier, DecisionTreeRegressor
import glob
import random

n_train = 100
n_test = 779

#Generate data
def generate(n_samples):
        ts = np.genfromtxt('data/oneday_kitchen.csv', delimiter = ' ')
        start_time = ts[0,0]
        ts[:,0] -= start_time
        X = random.sample(range(0, len(ts)), n_samples)
        X.sort()y = ts[X,1]
        X = np.reshape(X, (n_samples, 1))
        return X, y

X_train, y_train = generate(n_samples = n_train)
X_test, y_test = generate(n_samples = n_test)
```

In the first example, a decision tree is used to predict all 779 samples of the original one, as shown in Figure 4.36.

```
# One decision tree regressor
dtree = DecisionTreeRegressor().fit(X_train, y_train)
d_predict = dtree.predict(X_test)
plt.figure(figsize = (10, 6))
plt.plot(X_test, y_test, color = '0.5',linestyle = 'dashed')
plt.scatter(X_train, y_train, c = 'b', s = 20)
plt.plot(X_test, d_predict, 'g', lw = 2)
plt.title('Decision tree, MSE = %.2f' % np.divide(np.sum((y_test – d_predict) ** 2),n_test))
```

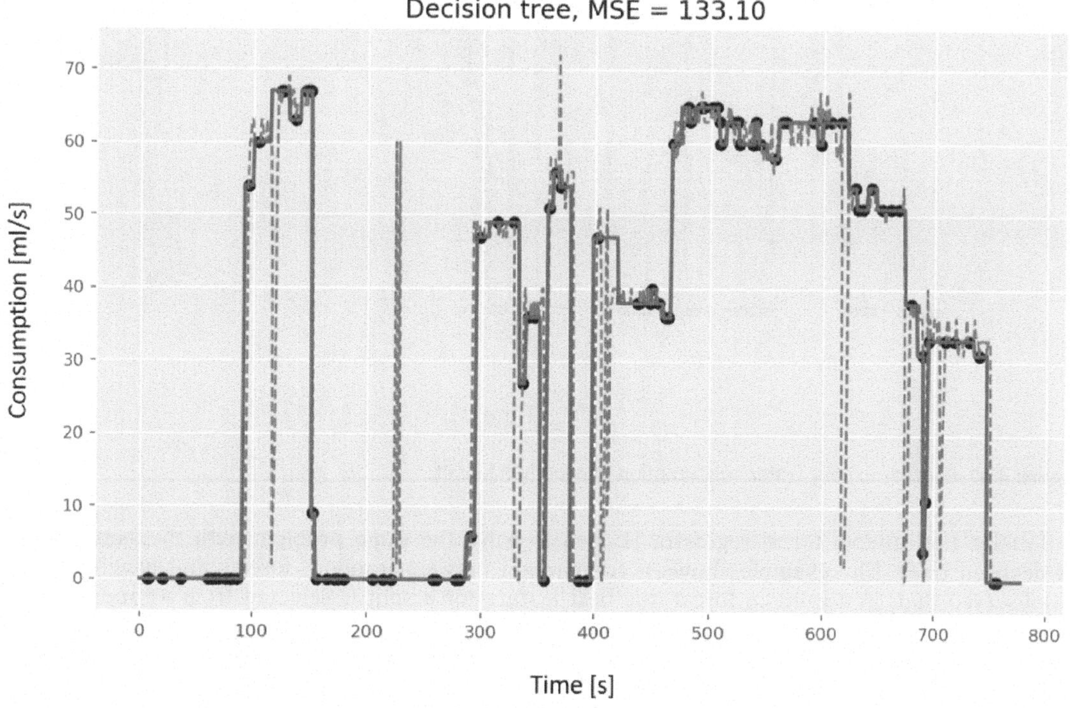

Figure 4.36 Decision tree to daily water consumption of a kitchen faucet.

In the second example, the bagging regressor uses ten trees to generate the solution, presenting a lower MSE, as shown in Figure 4.37.

```
# Bagging with a decision tree regressor
bdt = BaggingRegressor(DecisionTreeRegressor()).fit(X_train, y_train)
bdt_predict = bdt.predict(X_test)

plt.figure(figsize = (10, 6))
plt.plot(X_test, y_test, color = '0.5',linestyle = 'dashed')
plt.scatter(X_train, y_train, c = 'b', s = 20)
plt.plot(X_test, bdt_predict, 'y', lw = 2)
plt.title('Bagging for decision trees, MSE = %.2f'
% np.divide(np.sum((y_test – bdt_predict) ** 2),n_test));
```

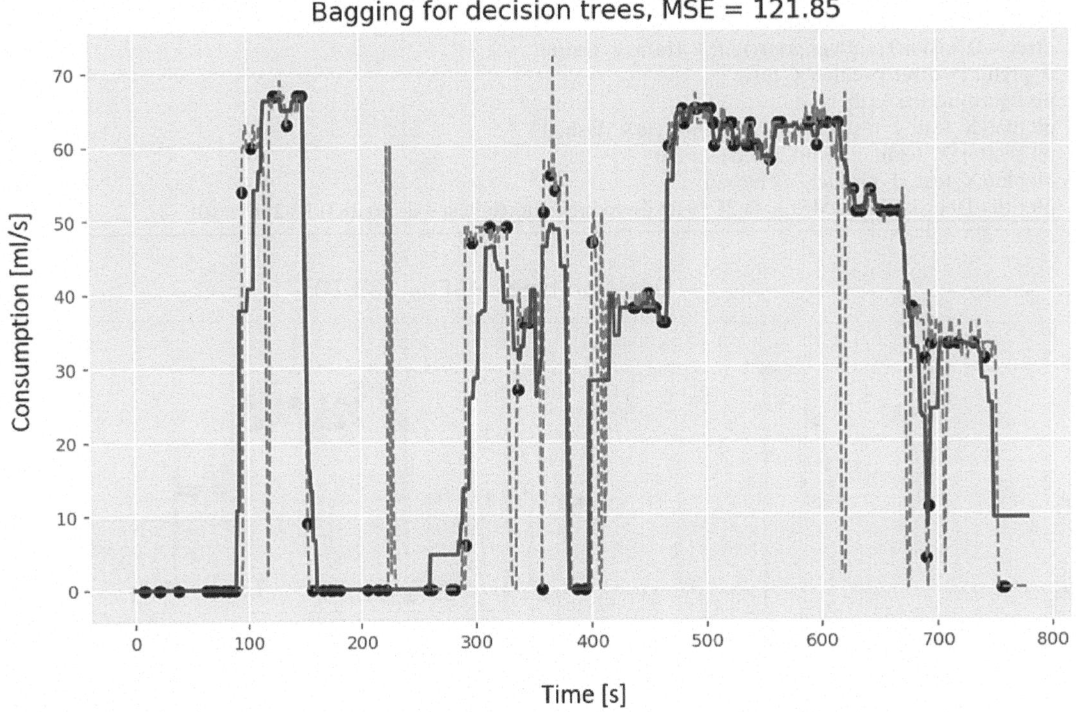

Figure 4.37 Bagging to daily water consumption of a kitchen faucet.

Finally, the random forest regressor is used to solve the same problem with the same number of decision trees. The example shows a comparison between random forests and bagging. It can be observed that, in a random forest, the best feature for a split is selected from a random subset of the available features, while in bagging all features are considered for the next best split. This represents the main difference between the two methods. The effect is, at least in this example, a slight improvement of the MSE as shown in Figure 4.38.

```
# Random Forest
rf = RandomForestRegressor(n_estimators = 10).fit(X_train, y_train)
rf_predict = rf.predict(X_test)


plt.figure(figsize = (10, 6))
plt.plot(X_test, y_test, color = '0.5',linestyle = 'dashed')
plt.scatter(X_train, y_train, c = 'b', s = 20)
plt.plot(X_test, rf_predict, 'r', lw = 2)
plt.title('Random forest, MSE = %.2f' % np.divide(np.sum((y_test – rf_predict) ** 2),n_test));
```

Random forest classification example application and code: The following example looks at the advantages of random forests and bagging in classification problems. The goal is to classify the corresponding fixture of each water usage using two features. In this example, random forest classification has been applied to three fixtures: washbasin, shower and kitchen faucet. In particular, the following code reads from the dataset of water usages, the time of the day (seconds), the volume in liters and the related fixture for washbasin, shower and kitchen faucet.

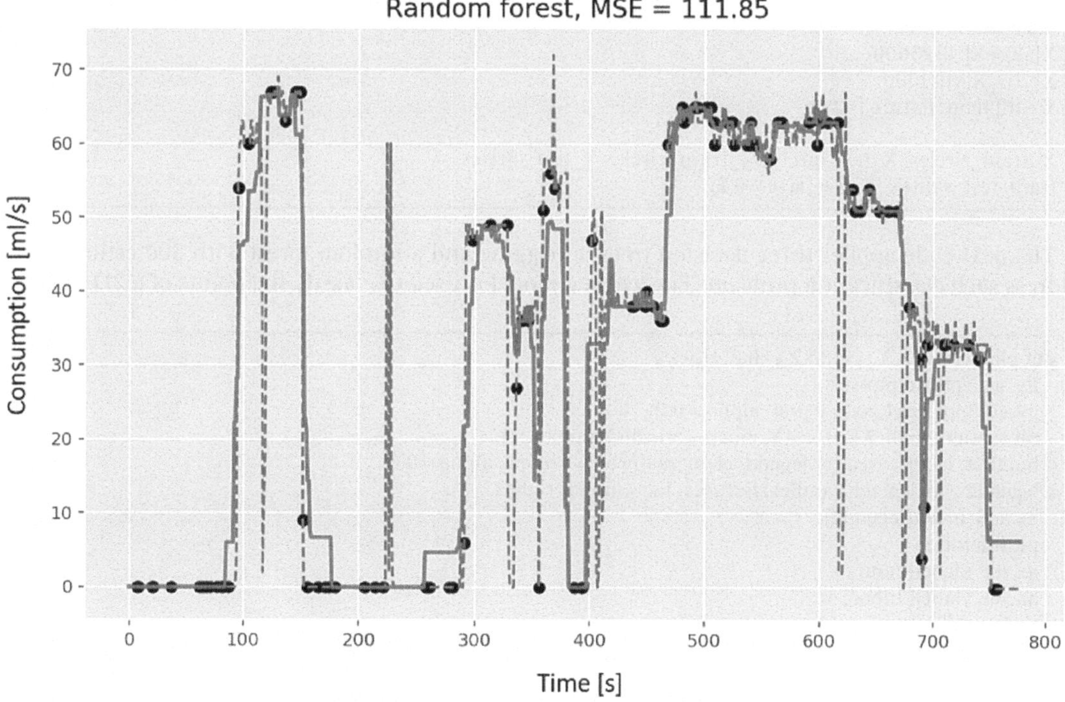

Figure 4.38 Random forest to daily water consumption of a kitchen faucet.

The fixtures are represented as an integer from 0 to 2. Twenty per cent of usages are used as the training set.

```python
from matplotlib import pyplot as plt
import seaborn as sns
from sklearn.datasets import make_circles
from sklearn.ensemble import (BaggingClassifier, BaggingRegressor,
        RandomForestClassifier, RandomForestRegressor)
from sklearn.model_selection import train_test_split
from sklearn.tree import DecisionTreeClassifier, DecisionTreeRegressor
import pandas as pd
import sklearn

# Load data
df = pd.read_csv('./data/dataset.csv', delimiter=' ')
fixtures = ['washbasin', 'shower', 'kitchenfaucet']
# Choose the numeric features
df = df[['sec_from_midnight','liters','fixture', 'num_fixture']]
df = df[(df['fixture']==fixtures[0]) | (df['fixture']==fixtures[1]) | (df['fixture']==fixtures[2])] plt
df.head()
df = sklearn.utils.shuffle(df)
X = np.asarray(df[['sec_from_midnight','liters']],dtype=float)
max_dur = max(X[:,0])
```

```
max_lit = max(X[:,1])
X[:,0] = X[:,0] /3600
X[:,1] = X[:,1]/1000
Y = df['num_fixture']−2

X_train_circles, X_test_circles, y_train_circles, y_test_circles = \
train_test_split(X, Y, test_size = 0.2)
```

The next code applies three decision trees, a bagging and a random forest with 300 estimators to address such classification problem. The learned model is used to classify the points of a 2D grid.

```
def plot_class (X,Y, xx1,xx2,y_hat, title):
    fig, ax = plt.subplots()
    plt.contourf(xx1, xx2, y_hat, alpha = 0.2)
    plt.scatter(X[:,0], X[:,1], c = Y, cmap = 'viridis', alpha = .7)
    handles, labels = scatter.legend_elements(prop = 'colors', alpha = 0.6)
    legend2 = ax.legend(handles, fixtures, loc = 'upper right')
    ax.add_artist(legend2)
    plt.title(title)
    ax.set_xlabel('hours')
    ax.set_ylabel('liters')
    ax.legend()
    plt.show()

x_range = np.linspace(X[:,0].min(), X[:,0].max(), 100)
y_range = np.linspace(X[:,1].min(), X[:,1].max(), 100)
xx1, xx2 = np.meshgrid(x_range, y_range)

dtree = DecisionTreeClassifier()
dtree.fit(X_train_circles, y_train_circles)

y_hat = dtree.predict(np.c_[xx1.ravel(), xx2.ravel()])
y_hat = y_hat.reshape(xx1.shape)
plot_class(X,Y,xx1,xx2,yhat, 'Decision tree')

dtree = BaggingClassifier(DecisionTreeClassifier(),
        n_estimators = 300, random_state = 42)
b_dtree.fit(X_train_circles, y_train_circles)
y_hat = b_dtree.predict(np.c_[xx1.ravel(), xx2.ravel()])
y_hat = y_hat.reshape(xx1.shape)
plot_class(X,Y,xx1,xx2,yhat, 'Bagging (Decisione tree)')

rf = RandomForestClassifier(n_estimators = 300, random_state = 42)
rf.fit(X_train_circles, y_train_circles)
y_hat = rf.predict(np.c_[xx1.ravel(), xx2.ravel()])
y_hat = y_hat.reshape(xx1.shape)

plot_class(X,Y,xx1,xx2,yhat, 'Random Forest')
```

Figure 4.39 shows the results of the classification problems performed with the three methods described before. The models are used to classify a mesh-grid of 100×100 points in a 2D space whose dimensions range from the minimum to the maximum values of the two features of the training set.

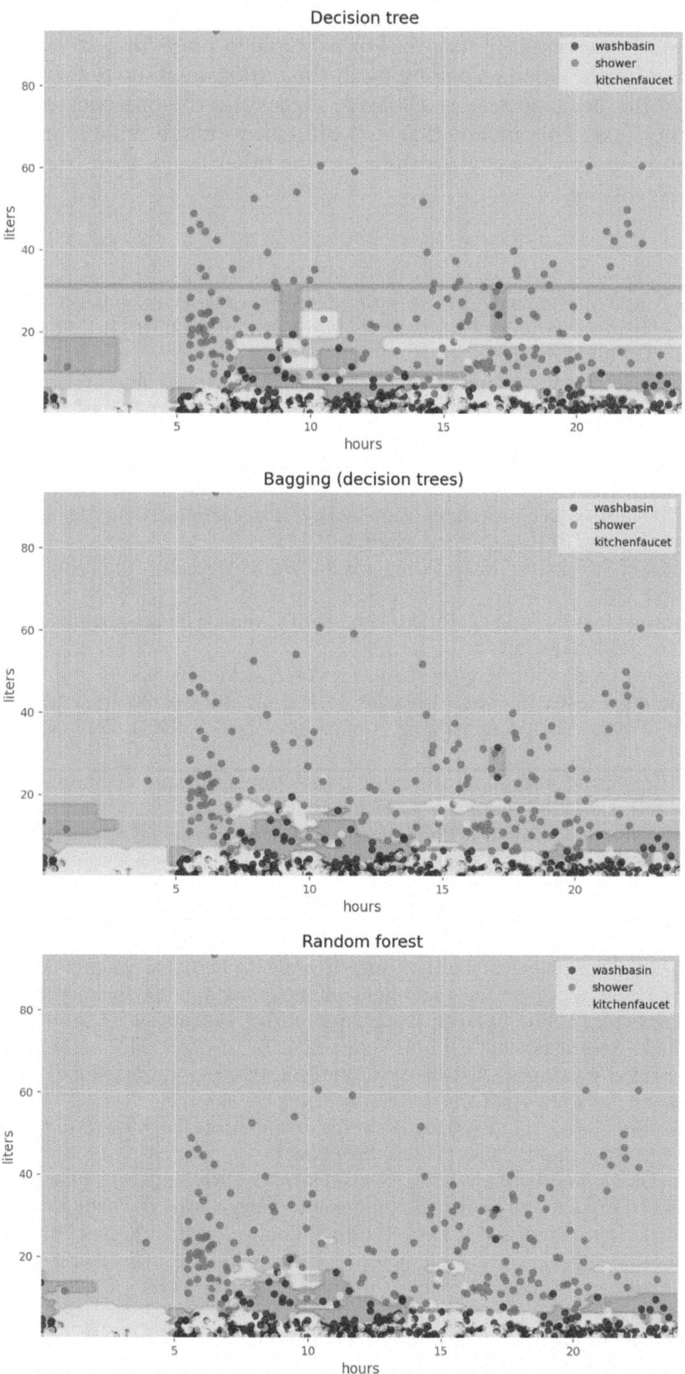

Figure 4.39 Random forest classification of washbasin, shower and kitchen faucet fixtures.

Each chart shows how the points of the mesh-grid have been clustered, coloring the associated region with a gradient of the corresponding fixture. For example, a point in a strip represented by darker gray shade then it will be classified as belonging to the dark gray fixture. Figure 4.39 shows that the decision boundary of the decision tree is serrated, suggesting the presence of overfitting and a not clear definition of the class. This means that it is difficult to make reliable predictions for new test data. The bagging and random forest algorithms, on the other hand, show more regular bounds and no evident signs of overfitting.

REFERENCES

Arandia E., Ba A., Eck B. and McKenna S. 2015 Tailoring seasonal time series models to forecast short-term water demand. *Journal of Water Resources Planning and Management*, **142**(3), 04015067, https://doi.org/10.1061/(ASCE)WR.1943-5452.0000591

ArunKumar K. E., Kalaga D. V., Mohan Sai Kumar C., Chilkoor G., Kawaji M. and Brenza T. M. (2021). Forecasting the dynamics of cumulative COVID-19 cases (confirmed, recovered and deaths) for top-16 countries using statistical machine learning models: auto-regressive integrated moving average (ARIMA) and seasonal auto-regressive integrated moving average (SARIMA). *Applied Soft Computing Journal*, **103**, 107161, https://doi.org/10.1016/j.asoc.2021.107161

Bedi J. and Toshniwal D. (2019). Deep learning framework to forecast electricity demand. *Applied Energy*, **238**, 1312–1326, https://doi.org/10.1016/j.apenergy.2019.01.113

Billings R. B. and Jones C. V. (2008). Forecasting Urban Water Demand, 2nd edn. American Water Works Association, Denver, CO.

Box G., Jenkins G., Reinsel G. and Ljung G. (2016). Time Series Analysis: Forecasting and Control, 5th edn. John Wiley and Sons, Inc., Hoboken, NJ.

Breiman L. (1996). Bagging predictors. *Machine Learning*, **24**(2), 123–140.

Breiman L. (2001). Random forests. *Machine Learning*, **45**(1), 5–32, https://doi.org/10.1023/A:1010933404324

Bühlmann P. and Yu B. (2002). Analyzing bagging. *Annals of Statistics*, **30**(4), 927–961, https://doi.org/10.1214/aos/1031689014

Candelieri A. (2017). Clustering and support vector regression for water demand forecasting and anomaly detection. *Water*, **9**(3), 224, https://doi.org/10.3390/w9030224

Chen J. (2018). *Water Demand Datasets* (Publication no. 10.17632/4yhprsgjrf.1). Retrieved from: https://data.mendeley.com/datasets/4yhprsgjrf/1 (last accessed July 2021)

Cho K., Van Merriënboer B., Gulcehre C., Bahdanau D., Bougares F., Schwenk H. and Bengio Y. (2014). Learning phrase representations using RNN encoder-decoder for statistical machine translation. *arXiv preprint arXiv*, 1406.1078.

Cominola A., Giuliani M., Piga D., Castelletti A. and Rizzoli A. E. (2015). Benefits and challenges of using smart meters for advancing residential water demand modeling and management: a review. *Environmental Modelling and Software*, **72**, 198–214, https://doi.org/10.1016/j.envsoft.2015.07.012

Di Mauro A., Di Nardo A., Santonastaso G. F. and Venticinque S. (2019). An IoT system for monitoring and data collection of residential water end-use consumption. Proceedings – International Conference on Computer Communications and Networks (ICCCN), IEEE, pp. 1–6.

Dong X., Yu Z., Cao W., Shi Y. and Ma Q. (2020). A survey on ensemble learning. *Frontiers of Computer Science*, **14**(2), 241–258, https://doi.org/10.1007/s11704-019-8208-z

Du B., Zhou Q., Guo J., Guo S. and Wang L. (2021). Deep learning with long short-term memory neural networks combining wavelet transform and principal component analysis for daily urban water demand forecasting. *Expert Systems with Applications*, **171**, 114571, https://doi.org/10.1016/j.eswa.2021.114571

Efron B. and Tibshirani R. J. (1993). An Introduction to the Bootstrap. Springer, Berkeley, CA.

Elman J. L. (1990). Finding structure in time. *Cognitive Science*, **14**(2), 179–211, https://doi.org/10.1207/s15516709cog1402_1

Faruk D. (2010). A hybrid neural network and ARIMA model for water quality time series prediction. *Engineering Applications of Artificial Intelligence*, **23**, 586–594, https://doi.org/10.1016/j.engappai.2009.09.015

Goldberg Y. J. J. o. A. I. R. (2016). A primer on neural network models for natural language processing. **57**, 345–420.

Guo G., Liu S., Wu Y., Li J., Zhou R. and Zhu X. (2018). Short-term water demand forecast based on deep learning method. *Journal of Water Resources Planning and Management*, **144**(12), 04018076, https://doi.org/10.1061/(ASCE)WR.1943-5452.0000992

Hastie T., Tibshirani R. and Friedman J. (2009). Elements of statistical learning. *The Mathematical Intelligencer*, **27**(2), 267–268.

Hochreiter S. and Schmidhuber J. (1997). Long short-term memory. *Neural Computation*, **9**(8), 1735–1780.

Kofinas D., Mellios N., Papageorgiou E. and Laspidou C. (2014). Urban water demand forecasting for the island of skiathos. *Procedia Engineering*, **89**, 1023–1030, https://doi.org/10.1016/j.proeng.2014.11.220

Lee J. and Chae S. (2016). Hourly water demand forecasting for micro water grids. *Journal of Water Supply Research and Technology – AQUA*, **65**(1), 12–17.

Mathworks. (2021a). Forecast Multiplicative ARIMA Model. Available at: https://www.mathworks.com/help/econ/forecast-airline-passenger-counts.html?searchHighlight=forecast%20multiplicative&s_tid=srchtitle (last accessed 9 March 2022)

Mathworks. (2021b). MMSE Forecasting of Conditional Mean Models. Available at: https://www.mathworks.com/help/econ/mmse-forecasting-for-arima-models.html#btbqqom (last accessed 9 March 2022)

Monks I., Stewart R. A., Sahin O. and Keller R. (2019). Revealing unreported benefits of digital water metering: literature review and expert opinions. *Water (Switzerland)*, **11**(4), 838–870, https://doi.org/10.3390/w11040838

Msiza I. S., Nelwamondo F. V. and Marwala T. (2007). Artificial neural networks and support vector machines for water demand time series forecasting. Paper presented at the 2007 IEEE International Conference on Systems, Man and Cybernetics, IEEE, pp. 638–643.

NCSS Statistical Software. The Box Jenkins Method, pp. 470-4 to 470-9. Available at: https://ncss-wpengine.netdna-ssl.com/wp-content/themes/ncss/pdf/Procedures/NCSS/The_Box-Jenkins_Method.pdf (last accessed 9 March 2022)

Pesantez J. E., Berglund E. Z. and Kaza N. (2020). Smart meters data for modeling and forecasting water demand at the user-level. *Environmental Modelling and Software*, **125**, 104633, https://doi.org/10.1016/j.envsoft.2020.104633

Rahim M. S., Nguyen K. A., Stewart R. A., Giurco D. and Blumenstein M. (2020). Machine learning and data analytic techniques in digital water metering: a review. *Water (Switzerland)*, **12**(1), 294–320.

Redondo E., Zafra-Mejía C. and Gutiérrez-Malaxechebarría A.-M. (2018). Operational analysis in a drinking water treatment plant using ARIMA models. *International Journal of Applied Engineering Research*, **13**, 16093–16099.

Sengupta A., Hawley R. J. and Stein E. D. (2018). Predicting hydromodification in streams using nonlinear memory-based algorithms in southern California streams. *Journal of Water Resources Planning and Management*, **144**(1), 04017079, https://doi.org/10.1061/(ASCE)WR.1943-5452.0000853

Villarin M. C. and Rodriguez-Galiano V. F. (2019). Machine learning for modeling water demand. *Journal of Water Resources Planning and Management*, **145**(5), 04019017, https://doi.org/10.1061/(ASCE)WR.1943-5452.0001067

Xenochristou M., Kapelan Z., Hutton C. and Hofman J. (2018). Smart water demand forecasting: learning from the data. *EPiC Series in Engineering*, **3**, 2351–2358, https://doi.org/10.29007/wkp4

Zhou Z. H. (2012). Ensemble methods: foundations and algorithms. In: Ensemble Methods: Foundations and Algorithms. Chapman and Hall/CRC Press, Taylor & Francis Group, New York, pp. 1–218.

doi: 10.2166/9781789062380_0123

Chapter 5

Use of cost-benefit analysis (CBA) in water infrastructure

*Anita M. Chaudhry PhD**

Department of Economics, California State University, Chico
**Corresponding author: achaudhry@csuchico.edu*

LEARNING OBJECTIVES

At the end of this chapter, you will be able to:

(1) Understand the policy context of CBA in United States.
(2) Understand the importance of using marginal benefits and costs in water infrastructure planning.
(3) Understand the components of CBA such as benefits, costs, and discount rate
(4) Learn how to organize the components of CBA in Excel
(5) Learn how to build a basic CBA

5.1 INTRODUCTION

Cost-Benefit Analysis (CBA) is one of the most prominent and widely used policy evaluation and decision-making tools in public policy. CBA has played a key role in water infrastructure project analysis, and at the same time, application of CBA tools and methods in water industry have also contributed to the development and refinement of tools and approaches now used in CBA. This chapter gives an overview of the methods CBA, with a brief outline of the history and the regulatory requirements of using CBA in water industry.

5.2 CONTRIBUTION OF CBA TO WATER POLICYMAKING

5.2.1 Imperatives of water scarcity: demand management or supply enhancement?

Figure 5.1 shows hypothetical demand and supply curves of water in a region. The downward sloping demand curve (D) represents the behavior of water users or buyers, which could include municipal, industrial, and agricultural users. A demand curve shows the quantity of water demanded at various prices. The downward-sloping demand curve shows that quantity demanded increases at lower prices. In other words, the additional or marginal benefit of water to users declines at higher volumes of water. For example, at smaller water volume, Q_1, more necessary water uses with higher marginal benefit such as drinking and washing, and at higher volume, Q_2 marginal benefit to users is lower which captures uses such as lawn irrigation or washing cars. This property of diminishing the marginal benefit of water can be leveraged by policy makers to incentivize water conservation by increasing

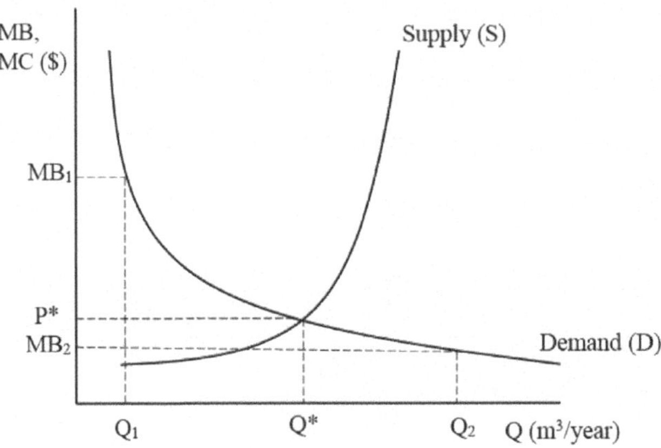

Figure 5.1 A hypothetical demand and supply of water in a region.

water rates to encourage users to reduce or eliminate uses of water that have a lower marginal benefit. The demand curve of water, *D*, can also be called the marginal benefit or willingness-to-pay curve. The supply curve (*S*) represents the behavior of water providers, for example a water utility. The upward sloping supply curve captures the idea that the additional or marginal cost of water supplies rises at higher water volume.

In practice, *water is often not priced at marginal cost of supply and a true market of water hardly ever exists*. However, this basic supply-demand framework can be useful for us to understand some basic economics of water, and how CBA can be helpful to measure the marginal benefit and marginal cost of water supply in a region.

Broadly speaking, there are two general methods to address the problem of water scarcity. We can either undertake projects that increase water supply or pursue approaches to control or reduce water demand. Supply enhancement projects may include projects such as building new or enlarging existing dams and reservoirs, drilling or deepening groundwater wells, building inter-basin water transfer facilities, repair deteriorating water infrastructure, building desalinization plants, or capturing and reusing rainwater. Costs of each of these options vary by location due to environmental, geographic, economic or regulatory reasons. For example, Ziolkowska (2015) reports that in 2010 the price of desalinated water ranged between $0.2–1.2/m³ ($0.8–4.5/kgal) for desalinated brackish groundwater and $0.3–3.2/m³ ($1.1 12.1/kgal) for desalinated seawater depending on location, local capital and operational costs and environmental regulations. Similarly, for some regions accessing deeper groundwater may be cheaper than accessing inter-basin water transfer to enhance supply. For the region under study, we can rank different supply augmentation options from lower to higher marginal cost and the result may look like the upward-sloping supply (*S*) curve in Figure 5.1.

Demand management options may include policies to reduce water use, such as raising water rates, that is a movement up the demand curve at a higher price and lower volume of water is consumed. Policy makers often pursue non-price options such as improving plumbing codes or educating water users about conservation options, or drought awareness messaging to nudge persuading users to lower water use, which would not increase the price of water but would shift the demand back curve and lower water demand.

A related concept is the price elasticity of demand, which is defined as the percentage change in quantity demanded that will occur for a percentage change in price. Since demand curve is downward-sloping, price elasticity is negative. Many studies have been conducted estimating price elasticity of urban water demand (Dalhuisen *et al.*, 2003). Water demand functions are generally found to be inelastic meaning that elasticity estimates are between −1 and 0, that is a 1% increase in price will affect a less than 1% decline in water demand. Price elasticity of water demand can vary within a year, it is found that it is more elastic (closer to −1) during summer months when lawn irrigation use is higher.

Water policy makers often want to choose from amongst the myriad options in both demand management and supply enhancement to achieve larger benefits and lower costs, that is the highest net benefit. CBA is very useful tool for evaluating different supply augmentation or demand management options. CBA can help make projections for different future scenarios and calculate the net benefit in terms of present value. This information can allow policymakers to not only assess whether a project provides enough benefits to warrant investing limited resources in it, CBA can also provide a measuring stick to help choose among alternative uses of limited resources.

5.2.2 CBA as a decision-making tool

In principle, CBA covers the full range of benefits and costs of a project, whether they have market prices or not. Many projects generate intangible benefits, which may be difficult to monetize. In such cases, CBA uses techniques to value unpriced benefits, both current and future, in present-dollar terms. On many of the water infrastructure projects, placing a dollar value on intangible, indirect, and unintended benefits could be crucial. For instance, construction of a reservoir provides additional water supplies that could be easily monetized and quantified based on measurable units by using water demand and market prices, but the additional recreational benefits provided by reservoirs such as boating, fishing, swimming, and wildlife observation etc. may be harder to monetize because they are not traded in a market. In such cases, economists rely on non-market valuation methods such as the travel cost method, or contingent valuation methods to construct a demand curve of additional water supplies as shown in Figure 5.1. In some cases, recreational benefits may be a very significant part of total economic value on the reservoir, as was seen in in a case study of the Cumberland River system of the southern United States where Bonnet *et al.* (2015) found that recreational benefits were the greatest economic benefit on the river even though no reservoir was built for that purpose. An in-depth presentation of discussion of non-market valuation methods and details is beyond the scope of this chapter (see Chapter 17 in this book). Please refer to Champ *et al.* (2017) for those who are interested.

5.2.3 Policy background

CBA was pioneered in the pursuit of a better framework for decision making about national water projects in the US (Griffin, 2012). Although CBA is clearly applicable for a wide range of public investment decisions, its growth as a useful tool in policy decision making is closely tied to construction of large water projects in the United States in the 20th century. Several water development agencies, and a few States, have made CBA a required step in project evaluation processes. US government rules stipulate that water projects that making use of federal dollars must be subjected to CBA, and the project approval is contingent on the findings of the CBA (Griffin, 2012).

The beginning of the central role of CBA started with the Flood Control Act of 1936 which stated that water projects were economically acceptable, *'if the benefits to whomsoever they accrue are in excess of the estimated costs'*, which clearly refers to requiring positive net benefits as a benchmark for project approval. Since the Flood Control Act, the federal requirement for water project CBAs have evolved considerably, with major rules established in 1952, 1958, 1962, 1973, 1979, 1983, and 2013 (Griffin, 2012). The most recent rules for federal water project analysis were set in 2013 by the US Council on Environmental Quality (CEQ), which is part of the President's executive branch of government. In these rules the agencies subjected to CBA have been extended from the traditional four that included the US Army Corp of Engineers, US Bureau of Reclamation, Tennessee Valley Authority, and the Natural Resources Conservation Service, to Environmental Protection Agency, National Oceanic and Atmospheric Administration, Federal Emergency Management Agency, and the Office of Management and Budget. Thus, CBA retains an influential place in the conversation around the decision making of water infrastructure.

5.3 CBA METHODS

In this section, let us now turn to the methods in building a CBA analysis of a project.

5.3.1 Building a spreadsheet of the CBA model

Figure 5.2 shows an example of a CBA spreadsheet. The goal of the CBA is to arrive at the sum of discounted net benefits, the value at the bottom right corner of the spreadsheet, also referred to as the net present value (NPV). This example is the case of a household's CBA of installing a 1893 liter rainwater barrel and using the captured rainwater for domestic uses, and reduce the use of piped water. This example is taken from Dallman *et al.* (2021) who conducted a CBA of rainwater capture and reuse in Los Angeles. In CBA analysis, net present value (NPV) of the discounted net benefits is to be calculated as follows:

$$NPV = \sum_{t=0}^{T} \frac{B_t - C_t}{(1+d)^t} \tag{5.1}$$

Equation (5.1) compresses or summarizes all the benefits and costs in one number, which is a very useful and powerful contribution of CBA to decision-making. To arrive at the NPV of the project, the analysts must measure the benefits (B) and costs (C) of the project. The analyst also must decide on the time horizon of analysis (T) which could last from a few months to several years. Typically, water infrastructure projects have long time horizons from a few to several decades. Therefore, the choice of the discount rate (d) becomes really important. The next three subsections address the main issues involved in identifying and monetizing the benefits and costs.

5.3.2 Identifying and measuring the benefits

Water infrastructure projects can be accompanied by a diverse set of benefits. From a water scarcity perspective, the main benefit is the increased water supply. Other major benefits may include enhancement in recreation, flood control, hydropower, water quality improvements. A first step in using the spreadsheet in Figure 5.1 is to identify whose benefits will be considered, and for this relevant population, which benefits will be included. In the rainwater harvesting example given in Figure 5.2, the relevant population is the single household and the only benefits being considered are the reduction in the household's monthly water bill, which is reduced by the amount of rainwater captured. Assuming this household is an average water consumer in Los Angeles, its water charges are $1.68 per cubic meter, based on typical water rates charged by the utilities in the watershed in 2018. Dallman *et al.* (2021) estimated that one 1893 liter barrel captures 15.8 cubic meters of water per year. Therefore, each year the household had a benefit of ($1.68×15.8) about $27 per year. In general, the population of interest is broader than one household and the benefit could include non-market benefits as mentioned in section 5.2.2. The analyst should identify all the benefits to the relevant population, and once identified, prioritize the ones that are more significant and will be monetized and included in the CBA.

5.3.3 Identifying and measuring the costs

Main cost categories for water infrastructure projects include all the planning and construction costs, such as design services, materials, equipment, land, and labor costs involved in construction, and interest rates on funds if borrowed funds are being used for construction. Also, it is extremely important to measure the losses to recreation or environmental resources from the diversion of water from one place to another. As Griffin (2016) puts it aptly *'The key aspect of water infrastructure*

Year	Benefits (B_t)	Costs (C_t)	Net Benefits (NB_t) $=B_t-C_t$	Discounted Net Benefits $= (NB_t)/(1+d)^t$
0	0	500	-500	-500.00
1	26.68	0	26.68	25.91
2	26.68	0	26.68	25.15
3	26.68	0	26.68	24.42
4	26.68	0	26.68	23.71
5	26.68	0	26.68	23.02
6	26.68	0	26.68	22.35
7	26.68	0	26.68	21.70
8	26.68	0	26.68	21.06
9	26.68	0	26.68	20.45
10	26.68	0	26.68	19.85
11	26.68	0	26.68	19.28
12	26.68	0	26.68	18.72
13	26.68	0	26.68	18.17
14	26.68	0	26.68	17.64
15	26.68	0	26.68	17.13
16	26.68	0	26.68	16.63
17	26.68	0	26.68	16.14
18	26.68	0	26.68	15.67
19	26.68	0	26.68	15.22
20	26.68	0	26.68	14.77
21	26.68	0	26.68	14.34
22	26.68	0	26.68	13.93
23	26.68	0	26.68	13.52
24	26.68	0	26.68	13.13
25	26.68	0	26.68	12.74
26	26.68	0	26.68	12.37
27	26.68	0	26.68	12.01
28	26.68	0	26.68	11.66
29	26.68	0	26.68	11.32
Sum of Discounted Net Benefits (NPV) =				**12.01**

Figure 5.2 An example of a CBA spreadsheet: A household CBA of rainwater harvesting.

development is to understand that water resources are being redirected and consumed in a different time, place and manner, not increased'. Like benefits, costs also need to be prioritized, and then the main cost categories should be monetized. When costs include items available in the market, such as materials, equipment, labor etc., market prices can be used to measure the costs. In some cases, the costs pertain to goods not traded in the market. Again, lost recreational services or environmental costs from diverted waters from receiving water bodies are important examples of non-market costs of a water infrastructure project. In the example above, the costs of the barrel and its installation were estimated at $500 using market prices. This was a one-time cost paid upfront in the first year of installation, and we assumed no costs incurred afterwards (although some maintenance/operations cost may incur!).

5.3.4 Time horizon and discount and interest rates

Projects that involve multiple years or decades cannot be evaluated without the use of a discount rate that makes dollar figures comparable across time. Discounting means placing a lower value on benefits and costs the further away in time they occur. There are two reasons why future values are discounted: (i) opportunity cost of capital and (ii) preferences. Capital is scarce, just like water, and when it is used in a water infrastructure project, its return from investing in another project is given up. This is known as the opportunity cost of capital and is typically measured by the prevailing interest rate in the economy. Another motivation for discounting future costs and benefits is typical human preference for benefits to come sooner rather than later. This human impatience makes today's rewards (costs) more (less) preferred to the same reward tomorrow. Let us consider a simple example to demonstrate this: Suppose you are given an offer of receiving $100 today or $100 a year from today. Suppose both amounts are tax free, there is no inflation, that is the purchasing power of $100 remains the same a year later, and there is no risk of not receiving the amount. Given this choice, most people will choose to receive $100 today. Why? This is because most people's individual rate of time preference is such that today is more important than tomorrow. Put simply, most people are impatient. What if the choice is between $100 today and $150 a year from today? There is a 50% reward for waiting a year. Given this choice, some people may choose to wait for a year to receive $150, while some may still choose $100 today. The key insight here is that people vary in their individual rate of time preference. Relatively patient people will choose to wait, but for relatively impatient people, this reward is not enough to compensate them for the 'pain of waiting', and they will prefer to be paid today and forgo the 50% reward for waiting.

The basic insight from this discussion on discounting is that when water projects, or any project for that matter, involve multiple years, it is essential to consider the rate of time preference. Water projects that affect a lot of people for many decades, tend to use *social rate of time preference*. A rate of time preference that applies to an individual may be different from that applied for society. It has been argued that governments should not base their social discount rates on individual impatience, because water infrastructure projects tend to generate public benefits (i.e., shared by all in the society).

In the example above, Dallman *et al.* (2021) used a time horizon of 30 years and discount rate of 3%. Using Equation (5.1) to calculate NPV was $12. A positive net present value means this project can be pursued.

5.4 CBA IN PRACTICE

5.4.1 CBA of reservoir construction

The Applewhite Project consisted of a dam and reservoir on the Medina River about 19 km (12 miles) south of San Antonio, Texas, United States. This project was approved by a San Antonio city council resolution in July 1979. Griffin and Chowdhury (1993) performed a CBA of this project; the section below is based on Griffin and Chowdhury (1993) and Griffin (2015).

Year	total annual benefit from water saved ($/m3)	Benefits	Costs	discounted net benefits
0	0	0	342,576,150	-342576150
1	9,124,706.08	9,961,605.40	0	9671461.551
2	9,580,941.38	10,417,840.70	0	9819814.028
3	10,059,988.45	10,896,887.77	0	9972195.956
4	10,562,987.87	11,399,887.19	0	10128652.12
5	11,091,137.27	11,928,036.59	0	10289229.14
6	11,645,694.13	12,482,593.45	0	10453975.5
7	12,227,978.84	13,064,878.16	0	10622941.53
8	12,839,377.78	13,676,277.10	0	10796179.43
9	13,481,346.67	14,318,245.99	0	10973743.3
10	14,155,414.00	14,992,313.32	0	11155689.11
11	14,863,184.70	15,700,084.02	0	11342074.74
12	15,606,343.93	16,443,243.26	0	11532959.98
13	16,386,661.13	17,223,560.45	0	11728406.57
14	17,205,994.19	18,042,893.51	0	11928478.17
15	18,066,293.90	18,903,193.22	0	12133240.41
16	18,969,608.59	19,806,507.91	0	12342760.91
17	19,918,089.02	20,754,988.34	0	12557109.28
18	20,913,993.47	21,750,892.79	0	12776357.14
19	21,959,693.15	22,796,592.47	0	13000578.14
20	23,057,677.80	23,894,577.12	0	13229848.01
21	24,210,561.69	25,047,461.01	0	13464244.53
22	25,421,089.78	26,257,989.10	0	13703847.6
23	26,692,144.27	27,529,043.59	0	13948739.23
24	28,026,751.48	28,863,650.80	0	14199003.58
25	29,428,089.05	30,264,988.38	0	14454727
26	30,899,493.51	31,736,392.83	0	14715998.02
27	32,444,468.18	33,281,367.50	0	14982907.41
28	34,066,691.59	34,903,590.91	0	15255548.19
29	35,770,026.17	36,606,925.49	0	15534015.67
	Sum of discounted net benefits (NPV)=			14,138,576.26

Figure 5.3 An example of a CBA spreadsheet: Watershed-scale adoption of rainwater harvesting.

The Applewhite reservoir was expected to increase San Antonio's water supply by 59.2 m^2 per year. Griffin and Chowdhury (1993) decided to restrict the benefits to municipal water users, and assumed the non-market benefits of recreation to the users from the reservoir were zero in the baseline model. Even though non-market recreational benefits could be significant, analysts may explicitly acknowledge and not undertake elaborate non-market valuation techniques to assess if just including municipal water benefits is sufficient for positive NPV. What would be the demand of the additional water supplies by the municipal water users? Griffin and Chowdhury (1993) used the demand curve for municipal water (as shown in Figure 5.1) and estimates from studies of water demand in other municipal areas to estimate the *marginal value* of the water supplies from the reservoir. Previous analyses had shown that the monthly price elasticity of water demand vary from -0.31 to -0.4. The analysts also accounted for population growth in San Antonio. An increase (decrease) in water users means an increase (decrease) in total benefits of the project.

The cost of the project was estimated to be around $180 million. At the time, the cost was to be financed by the city issuing municipal bonds and the city planners decided to increase water prices for the water users every five years to repay the bonds. Griffin and Chowdhury (1993) assumed this payment schedule until all the bonds are repaid. They assumed inflation of 2%. Griffin and Chowdhury assumed that bond buyers would receive periodic payments until the purchase price of the bond was returned to them as well as some interest for lending their money. At the time of this proposal the interest payments were expected to be around 7.5% per year. They also assumed no environmental change in the watershed from the diversion of Medina River into a reservoir. They chose a discount rate of 4% and a time period of 1991–2040, until all bonds were fully repaid.

Griffin and Chowdhury (1993) found that discounted net benefits of the project were negative. The net benefits were initially negative and become more negative during the first few years because the water rate increases were harming the consumers. The increase in water rate reduced water demand, which meant that water users in San Antonio were not making any use of the additional supplies from the reservoir. Eventually, in about 100 months or about eight years, the city collects enough revenues to fully pay back all bonds, which lowers water rates for consumers, and they begin using the additional water supplies from the reservoir. Population growth also helps in increasing the benefits from the reservoir, but thus the conclusion of this CBA was that this reservoir should not be constructed. The NPV was negative, $-$86 million, and for numerous reasons this project was never undertaken. CBA showed that this project was economically undesirable. Sensitivity analyses were also performed by altering the rate of population growth or changing the pattern of bond repayments.

5.4.2 CBA of rainwater harvesting systems (RWHS)

Rainwater harvesting provides a potential source of supplemental water supply to meet increasing urban demand. Dallman *et al.* (2021) undertook a CBA of using captured rainwater as a substitute for a share of municipal water supplied to residential and commercial buildings in the densely urbanized Ballona Creek watershed in Los Angeles, California. This research developed a framework that organizes the diverse variables that may affect the benefits and costs of RWHS to answer several questions: how high are the net benefits of RWHS to replace potable water supplies? What quantity of 'new' water supply can be realized? What is the scale of RWHS that maximizes net benefits affected by the key parameters, such as weather patterns, tank size, use style (indoor/outdoor), or the cost of other water supplies?

The CBA model of an RWHS was developed for the two possible uses of captured rainwater: outdoor use only, and both indoor and outdoor use. Also, the CBA was evaluated at increasing scales of adoption in the watershed, that is 20, 40 and 60% of residential and commercial buildings in the study area to evaluate the benefits of a coordinated policy to encourage RWHS adoption. Also, cisterns of varying sizes were considered for each scale, 208, 1893 and 7571 liters, commonly used and available in the market.

The main benefit identified was the economic value of captured water, energy and carbon saved from the use of captured rainwater rather than piped municipal water supply. In order to monetize this benefit, the volume of water saved and price per unit of water were needed. Dallman *et al.* (2021) estimated the quantity of water saved by using the Environmental Protection Agency (EPA) Storm Water Management Model (SWMM) model based on historic patterns of rainfall, land use, and irrigation demand (based on evapotranspiration). The monetary value of a unit of saved water was calculated from the wholesale rate of the urban water utility that supplied water to the study area. In this study, Dallman *et al.* (2021) valued the additional supplies at the same marginal rate as the current water use.

The costs included were those required to implement and maintain the RWHS infrastructure. To estimate the cost of RWHS equipment and installation, purchase price and insallation costs of the cisterns were collected from various vendors of cisterns and associated equipment, such as hardware stores and specialty vendors. The CBA model used average market prices for southern California. A time frame of 30 years, and a discount rate of 3%, was consistent with the Regulatory Impact Analysis conducted for EPA's Clean Power Plan. Figure 5.3 shows the CBA calculations for the scenario of 60% of the watershed participating in installing 7.57 m^3 (2000 gallon) barrels and using this water for outdoor irrigation. The price of water increases at the rate of 5% every year. The total cost of barrels including installation was a one-time fixed cost of $342.5 million, and the benefits from saved municipal water occurred each year. The NPV of $14.1 m showed that this project was desirable.

It is important to recognize that Dallman *et al.* (2021) assumed certain benefits and costs as zero. For example, they argued that reductions in the peak runoff (flooding risk) resulting from rainwater capture in this watershed are minimal and thus were not included as a benefit in this analysis, although this could be a significant component in other regions. Similarly, Dallman *et al.* (2021) recognized that installing RWHS by a homeowner that is visible to neighbors may be a source of pride for the homeowner and yield psychological 'warm glow' benefits. Such non-market benefits may be important for certain homeowners but are difficult to measure and so were not monetized, although they were acknowledged. There may also be non-market benefits of RWHS, if participants believe saving water is important even if the monetary benefits are minimal. This analysis does not include such non-market benefits or benefits from water quality improvements (e.g., due to non-point source pollution).

In their CBA model, Dallman *et al.* (2021) found that the discounted net benefits (NPV) of RWHS were positive for outdoor use of captured rainwater. RWHS NPV rise as cistern size and participation rates increase. For example, for the smallest cistern discounted net benefits range from $4 to 12 million but for the largest cistern discounted net benefits range from about $32 to 100 million for the 30-year project life. Installing RWHS for only outdoor use is likely to be an economically efficient policy for the region if the price of water will rise. One of the key insights of this analysis was the value of saved water for outdoor use was the largest component of the benefits of the RWHS and hence the most important consideration in the decision to implement RWHS. The conomic value of saved water contributes 63% of the total benefits of RWHS, whereas energy and carbon savings respectively constitute 30 and 7% of the benefits of RWHS. They assumed that the water price would increase at the historic rates of 5%. The results are dependent on the annual rate of increase in water rates as well. If the wholesale price of water increases at a higher rate (than 5%), RWHS for outdoor and indoor water use could potentially achieve positive net discounted benefits.

This case study showed how a CBA model can help delineate the primary driver of benefits and costs and help guide the policy maker to the relatively important benefit and costs.

5.5 CONCLUSION

CBA is an economic tool for helping decision-makers assess the economic efficiency of a policy or a project. As this chapter showed, CBA does this by quantifying all the benefits and costs of the project for the relevant population. Although it seems straightforward to fill in the spreadsheet cells in the benefits and costs columns, and determine the NPV, it is important to remember that a CBA is more than NPV for several reasons: First, it can be quite hard to reduce all of the impacts (costs or benefits) of a project to a single metric. For practical reasons, an NPV will not include all important project consequences. However, a well-done CBA includes determination and disclosure of *all* project impacts, not just those that can be readily quantified in dollar terms. Therefore, the researcher often must make decisions on which impacts to include in the calculation of NPV and which to leave aside. Also, the choice of the discount rate to convert future benefits and costs to present values is an important choice. These decisions can make a substantial impact on the calculated NPV. It is imperative that the researchers/practitioners should clearly disclose all assumption and make modeling decisions transparent, and so that the audience understands the true scope of the analysis.

REFERENCES

Bonnet M., Witt A., Stewart K., Hadjerioua B. and Mobley M. (2015). The Economic Benefits of Multipurpose Reservoirs in the United States – Federal Hydropower Fleet. National Technical Information Service, Springfield, VA, USA.

Champ P. A., Boyle K. J. and Brown T. C. (2017). A primer on nonmarket valuation. In: The Economics of Non-Market Goods and Resources, I. J. Bateman (ed.), 2nd edn. Springer, Switzerland.

Dallman S., Chaudhry A. M., Muleta M. K. and Lee J. (2021). Is rainwater harvesting worthwhile? A benefit–cost analysis. *Journal of Water Resources Planning and Management*, **147**(4), 04021011, https://doi.org/10.1061/(ASCE)WR.1943-5452.0001361

Dalhuisen J. M., Florax R. J., De Groot H. L. and Nijkamp P. (2003). Price and income elasticities of residential water demand: a meta-analysis. *Land Economics*, **79**(2), 292–308, https://doi.org/10.2307/3146872

Griffin R. C. (2012). The origins and ideals of water resource economics in the United States. *Annual Review of Resource Economics*, **4**(1), 353–377, https://doi.org/10.1146/annurev-resource-110811-114517

Griffin R. C. (2016). Water Resource Economics: The Analysis of Scarcity, Policies, and Projects. MIT Press. https://mitpress.mit.edu/books/water-resource-economics-second-edition

Griffin R. C. and Chowdhury M. E. (1993). Evaluating locally financed reservoir: the case of Applewhite. *Journal of Water Resources Planning and Management*, **119**(6), 628–644, https://doi.org/10.1061/(ASCE)0733-9496(1993)119:6(628)

Ziolkowska J. R. (2015). Is desalination affordable? Regional cost and price analysis. *Water Resources Management*, **29**(5), 1385–1397, https://doi.org/10.1007/s11269-014-0901-y

Part II
Operations

doi: 10.2166/9781789062380_ 0135

Chapter 6

Water quality modeling and analysis

*Maria A. Palmegiani[1] and Juneseok Lee[2]**

[1] *Lyles School of Civil Engineering, Purdue University, West Lafayette, IN 47907, USA*
[2] *Department of Civil and Environmental Engineering, Manhattan College, Riverdale, NY 10471, USA*
**Corresponding author: juneseok.lee@manhattan.edu*

LEARNING OBJECTIVES

At the end of this chapter, you will be able to:

(1) Explain water quality modeling in water distribution systems.
(2) Apply EPANET and EPANET-MSX for water quality modeling.
(3) Calibrate modeling results to observed field data.
(4) Interpret modeling results.

6.1 INTRODUCTION

Water quality within water distribution and plumbing systems is a highly complex and rapidly changing issue that is intuitively difficult to predict. This is because it is affected by many factors such as the materials, layout, level of disinfectant, water demand, corrosion levels, and other hydraulic factors. Seasonal changes in temperature can also affect the water quality as higher temperature is known to increase chemical reaction rates (Courtis *et al.*, 2009). Many opportunistic pathogens (OPs) and complex chemical species can exist within a plumbing or distribution system, which can expose communities to waterborne diseases such as Legionnaire's disease and cause outbreaks (Falkinham *et al.*, 2015; Kusnetsov *et al.*, 2003). Issues often occur as the water ages due to low demand (Rhoads *et al.*, 2016). The amount of time that it takes water to exit a system after entering the system is referred to as the water age. Drinking water is often treated with a chlorine disinfectant to prevent growth of harmful chemical and microbial contaminants, as well as corrosion control inhibitors to prevent metal leaching from the pipes. However, as the water age increases, the system experiences decay of both the disinfectant and the corrosion control inhibitors, allowing for contaminants and pathogens to grow inside the system and biofilm (Ley *et al.*, 2020; Salehi *et al.*, 2018, 2020).

Several federal and state laws such as The Safe Drinking Water Act in the United States exist to define the maximum contaminant levels for various parameters within water distribution systems (The Safe Drinking Water Act, 2000; USEPA, 2016a, 2021). However, the water quality is only reported at selected sampling locations within the system (USEPA, 2013, 2014). Studies have shown that chemical and microbial water quality can vary over the course of a day at different locations in a distribution system so sampling at select locations does not capture the complete and changing water quality characteristics of the system (Clark *et al.*, 1999). The maximum contaminant level for free chlorine in water distribution systems is 4 mg/L as Cl_2 disinfectant concentration. However, levels that reach buildings are often much lower, sometimes nonexistent. Also, as chlorine navigates through the pipes, it often reacts with other materials in the bulk and wall phase that further reduce the chlorine concentration. Many physical, chemical, and biological activities occur during the transport of water in a distribution system that contribute to the reactions of chlorine and deteriorate the water quality (Munavalli & Kumar, 2004). Without models, extensive sampling and analysis during different times of the day for every water distribution system is necessary to ensure that safe water is delivered to homes.

In contrast to distribution systems, plumbing systems are different because they often contain fluctuating temperatures, smaller pipe diameters, lower disinfectant residual, and intermittent water demand which increases the residence time and can promote greater chemical and microbial growth (Lautenschlager *et al.*, 2010). Because water quality in these systems is so complex, methods to mitigate the risk of occurrences of opportunistic pathogens and growth of chemical species are often done inside buildings by continuous flushing and installing in home treatment (Hozalski *et al.*, 2020; Lothrop *et al.*, 2015). However, it is often difficult to know what the water quality conditions are at certain points within a home plumbing system or water distribution system without routine sampling and analysis. This collection of data is highly time consuming and expensive. To account for seasonal variations and periods of low water use, lengthy study periods are necessary. Even then, because most water systems are unique in their geometry and water demand, sampling would need to occur at each system. To further complicate matters, efforts are currently being made to improve water conservation by reducing flow (U.S. Green Building Council, 2015; USEPA, 2016b). These water conservation practices could have adverse effects on water quality by increasing the water age.

Models are important because they serve as tools to aid in water infrastructure design, and instantly identify health risks associated with various water use patterns for different water systems and scenarios. The goal of water quality models is to combine hydraulic and water quality parameters to predict the concentration of various species that exist within water infrastructure over time (Palmegiani *et al.*, 2022).

This section describes the methods to model the hydraulic and water quality components of an example water distribution network using the EPANET and EPANET-MSX software. The section briefly explains how to build a network on EPANET, how to extract the configuration properties of the pipes, nodes, tanks, and reservoirs when given a network, and how to use it to perform hydraulic and water quality calibrations and analysis.

6.2 EPANET AND EPANET-MSX SOFTWARE

The EPANET software can be used to model the hydraulic components of drinking water distribution systems when configuration and flow demand is known. It can also model water quality but is limited to the transport and fate of one water quality species. The water is transferred in pipes through advective transport and mixing at pipe junctions and storage nodes. It is often assumed that complete mixing occurs at the junctions and storage tanks.

Most water quality problems in water distribution systems involve many species, as well as species that interact with one another. The EPANET-MSX software can model the chemical and microbial contaminants of multiple species within the distribution system when used alongside the EPANET software. Inside a distribution system, a mobile bulk phase and a fixed pipe wall phase exist. Bulk phase species are chemical or biological contaminants that are transported through the pipe with an average velocity. Wall phase species are attached to the pipe wall and do not move but react with the other species. Both wall and bulk phase species are considered for modeling purposes.

The equations for flow distribution in each pipe and head can be calculated as follows for each instant of time.

(1) Continuity equation at each node:

$$\sum_{k\in J} Q_{k,J} - \sum_{l\in J} Q_{J,l} = \pm qJ \tag{6.1}$$

where $q_J = (+)$ for outflow from node J and $(-)$ for inflow into node J, $(k,J)=$ a pipe entering node J, and $(J, l)=$ a pipe leaving node J.

(2) Energy equation for each link:

$$H_i - H_j = f\left(Q_{i,j}\right) \tag{6.2}$$

where $H_i=$ head at node i, $H_j=$ head at node j, and $Q_{i,j}=$ flow in pipe (i,j).

(3) Rate of change of volume equal to inflow rate minus outflow rate:

$$\frac{dH_T}{dt} = \frac{1}{A_T}\left(Q_{k,T} - Q_{T,l}\right) \tag{6.3}$$

where $A_T=$ surface area of tank T, $H_T=$ water level in tank T, $t=$ time, $Q_{k,T}=$ flow into tank T, and $Q_{T,\#}=$ flow out of tank T.

The calculation of constituent concentration propagation through a pipe network is described below. The notation for this includes:

$$T = \min_i\left(\frac{V_i}{Q_i}\right) = \text{Water quality time step (sec)}$$

$C_i=$ concentration (Kg/m³) in the ith pipe
$m_i^{'k} =$ reacted mass within segment k for pipe i (Kg) $= m_i^k e^{\alpha\tau}$; $m_i^k = C_i Q_i\tau\,(kg)$
$Q_i=$ flow into pipe i
$R[.] =$ Reaction rate expression $= \alpha\, C_i$; $\alpha =$ parameter
$u_i=$ velocity of flow in the ith pipe (m/s)
$V_i=A_iL_i$; $V_i=$ volume of pipe i (m³); $A_i=$ area of pipe i (m²); $L_i=$ length of pipe i (m)
$x=$ distance along pipe i (m);

The advection of constituent concentration is given by:

$$\frac{\partial C_i\left(x,t\right)}{\partial t} + u_i\frac{\partial C_i\left(x,t\right)}{\partial x} - R\left[C_i\left(x,t\right)\right] = 0 \tag{6.4}$$

where $R(C_i)=\alpha C_i$

The solution for this equation is:

$$C_i\left(x,t+\tau\right) = C_i\left(x-u_i\tau,t\right)e^{\alpha\tau}; \forall_\tau \leq \frac{x}{u_i} \tag{6.5}$$

Concentration at distance x at time $t=$ [concentration at distance $(x-u_i\tau)$ at time $(t-\tau)$]$e^{\alpha\tau}$.

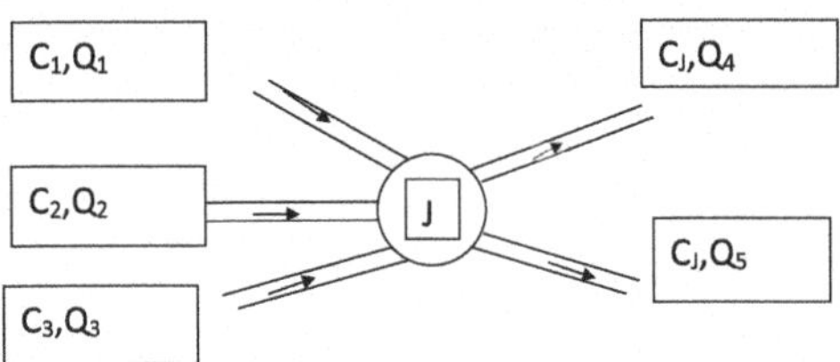

Figure 6.1 Water quality at junction.

Assuming complete and instantaneous mixing at node k (the head node of outgoing pipe i) due to the incoming pipes j of lengths L_j with flow Q_j and concentration C_j at time t, the constituent concentration for pipe I is as follows (see Figure 6.1):

$$C_i(0,t) = C_k(t) = \frac{\displaystyle\sum_{j \in \{k\}}^{\beta_k} Q_j C_j(L_j,t)}{\displaystyle\sum_{j \in \{k\}}^{\beta_k} Q_j} \tag{6.6}$$

If k were to be a tank T, the concentration at the tank is as follows:

$$C_T(t+\tau) = \frac{1}{V_T + Q_{j\tau}}\left[c_j(L_j,t)Q_{j\tau} + V_T(t)C_T(t)\right] \tag{6.7}$$

and for any outgoing pipe i from the tank T, the concentration is obtained as:

$$C_i(0,t+\tau) = C_T(t) \tag{6.8}$$

To accommodate the mixing that takes place at intervening junctions, each pipe must be considered as a whole but separately in solving the advection equation. A single water quality time step is utilized as follows:

$$\tau = \min_i\left(\frac{V_i}{Q_i}\right) \tag{6.9}$$

where $V_i = A_i L_i$, A_i = area of pipe i (m²), L_i = length of pipe i (m), V_i = volume of pipe i (m³). Such a water quality time step is utilized to split every other pipe into an integer number of segments as given in the following: $\eta_i = Int[V_i/Q_i\tau]$ = largest integer number of segments of pipe i for water quality computations smaller than $[V_i/Q_i\tau]$. Once all the pipes are partitioned into volume segments and the initial concentration distribution is computed, the propagation of mass through the network over each water-quality time step proceeds in four steps: a kinetic reaction step, in which the mass in each segment undergoes a kinetic concentration change; a nodal mixing step, in which incoming masses are mixed together and divided by the total incoming flows to obtain an average outgoing concentration; an advective step, in which mass is moved to the next segment within the same pipe; and an allocation step, in which the mixed, average concentration is assigned the first (head) segment of each outgoing link. These steps are shown in Figure 6.2 and are summarized in the following:

Step 1. (*Reaction within a segment*) m'^k_i = reacted mass within segment k for pipe i (Kg) = $m^k_i e^{\alpha\tau}$ and $m^k_i = C_i Q_i \tau$ (Kg).

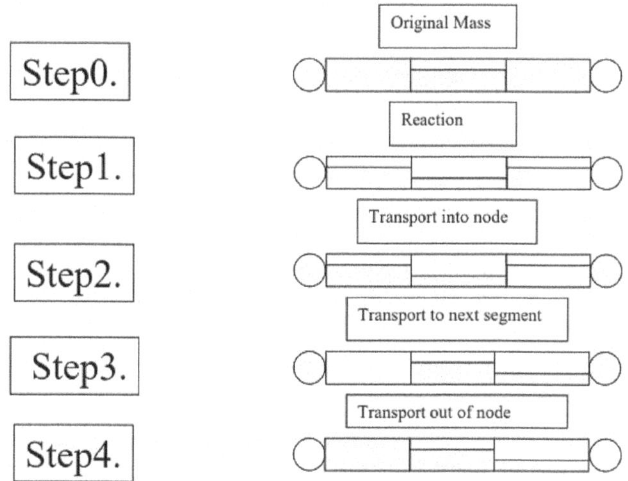

Figure 6.2 Water quality transport steps.

Step 2. (*Transport mass from last segment into head node*)

$$M_J = \sum_{j \in \{J\}} m_i^{'\eta_i} \quad \text{and} \quad V_J = \sum_{i \in \{J\}} v_i^{\eta_i}$$

Note the role of η_i, which says that the transport is only from the end segment of each pipe into the head node of the downstream pipe J. Also, M_J and V_J are the mixed masses and volumes at node J; and $V_i^{\eta_i} =$ volume from the end segment of pipe i. We compute the mixed, average concentration C_J as:

$$C_J = \frac{M_J}{V_J}$$

Step 3. (*Move contents to the next segment within a pipe*)

$$m_i^{'1} = C_J Q_i \tau$$

Initial Concentration Distribution: To compute initial masses within segments, initial concentrations are necessary. These are obtained by linear interpolation between head node and tail node concentrations. The initial masses are therefore computed as follows:

$$m_i^k = C_i^0 v_i + \frac{(k-1)\left(C_i^{\eta_i} - C_i^0\right) v_i}{(\eta_i - 1)}; k = 1, \ldots, \eta_i \tag{6.10}$$

where: C_i^0 and $C_i^{\eta_i}$ are the head and tail node concentrations. If the entire pipe is made up of just one segment, that is $\eta_i = 1$, we compute the initial mass as given below:

$$m_i^1 = \frac{\left(C_i^0 + C_i^1\right)}{2} \tag{6.11}$$

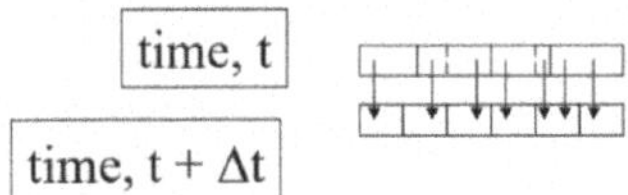

Figure 6.3 Interpolation between hydraulic time steps.

The flow distribution changes for every iteration of flow simulation. Associated with each flow distribution change there is a change in the water quality time step length, which alters the segmental division for each pipe. Consequently, an interpolation of concentrations among the previous segments should be carried out to fit the current segmentation, as shown in Figure 6.3. The number of segments increases for the next time step.

6.3 CREATING AN EPANET NETWORK FILE

When a project is created on EPANET and a network system is made, it will be saved as a network file. To create this file, open the EPANET software and select File>>New. A blank project will appear on the screen with a ribbon at the top with various options. The icons on the right of the ribbon allow the user to add certain features to the network such as pipes, junctions, reservoirs, valves, pumps, and tanks. Before starting a project, it is important to set default values. To do so, select Project>>Defaults. The pop-up dialog that appears is especially useful for hydraulic components (Figure 6.4). However, the ID label tab is useful when setting the ID increment value which is at a default of 1. This means that each new junction, pipe, tank, and so on. that is added to the tank will be labeled numerically, increasing

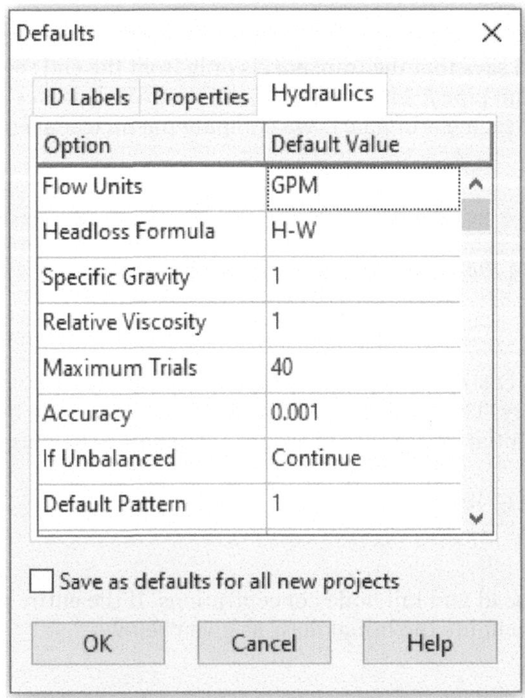

Figure 6.4 Pop up dialog for setting project defaults.

Figure 6.5 Project toolbox.

by 1. Many of the other properties can be adjusted on an individual basis so all options in this tab can be cleared other than that 1 which will allow for easy generation of the network. In the hydraulics tab, the default flow units, head loss formula, and other options are set to default values and can be changed according to the project. For this example, the default values will be used so flow will be in units of gallons per minute (GPM; 1 GPM = 3.79 liters per minute (LPM)) and the head loss will be calculated using the Hazen–Williams formula. By selecting the units of GPM for flow, the rest of the units will be adjusted to US Customary units as well. For example, by selecting GPM as units for flow, one is also automatically selecting units of US units, in this case feet for pipe lengths, inches for pipe diameters, psi for pressure, and so on. Likewise, if the user chooses LPM, then all units will automatically be in SI units. If changes were made to the defaults, the user can either select ok, or check the box that says 'Save as defaults for all new projects' to avoid having to set the defaults for each new project.

To draw the network, the toolbar at the top of the page is used. If this toolbox is not there, the user can select View>>Toolbars>>Map to display the toolbox on the map (Figure 6.5). This toolbox allows users to add pipes, junctions, tanks, valves, pumps and reservoirs to the project, as well as labels. More information on what each icon does can be accessed by hovering the mouse over the button. The black arrow enables selection mode. In selection mode, the properties of each object can be set by simply double clicking on the objects for a pop-up window to appear (Figure 6.6). Other methods of opening this window include right clicking the object and selecting properties, or selecting the object from the data page and then clicking edit. Using this property editor, each object can be given properties consistent with that of a given water distribution system. Knowledge of these properties is required.

Junction 129			Pipe 137		
Property	Value		Property	Value	
*Junction ID	129	^	*Pipe ID	137	^
X-Coordinate	30.320		*Start Node	129	
Y-Coordinate	26.390		*End Node	131	
Description			Description		
Tag			Tag		
*Elevation	51		*Length	6480	
Base Demand	0		*Diameter	16	
Demand Pattern			*Roughness	130	
Demand Categorie	1		Loss Coeff.	0	
Emitter Coeff.			Initial Status	Open	
Initial Quality			Bulk Coeff.		
Source Quality			Wall Coeff.		
Actual Demand	#N/A	v	Flow	#N/A	v

Figure 6.6 Property editor.

Once the network is completed the file can be exported to an input file. This is done by selecting project>>run. Once the analysis is run, select File>>Export>>Network. The user can export the map or scenario depending on individual needs but for this analysis the network works accordingly as we just need the hydraulic layout of the system. This will save the previous network (.net) file into an input file (.inp) format. The input file is what is used to analyze the hydraulic and water quality components of a model.

Consider the sample water distribution network that was created using the EPANET software (Figure 6.7). This distribution system describes how the percentage of lake water in a dual-source system changes over time and consists of a series of pipes, nodes or junctions, tanks, and two reservoirs which are the lake and river. This network resembles that of a functioning water distribution system and will be referred to as Net3 for this example. The junctions that will be focused on for this analysis are labeled in the figure. Water distribution systems are typically sampled at several locations in the network, and these are the locations assumed to be sampled. Net3 was exported into an input file as explained in the previous paragraph which will be used to add key components to the network. The

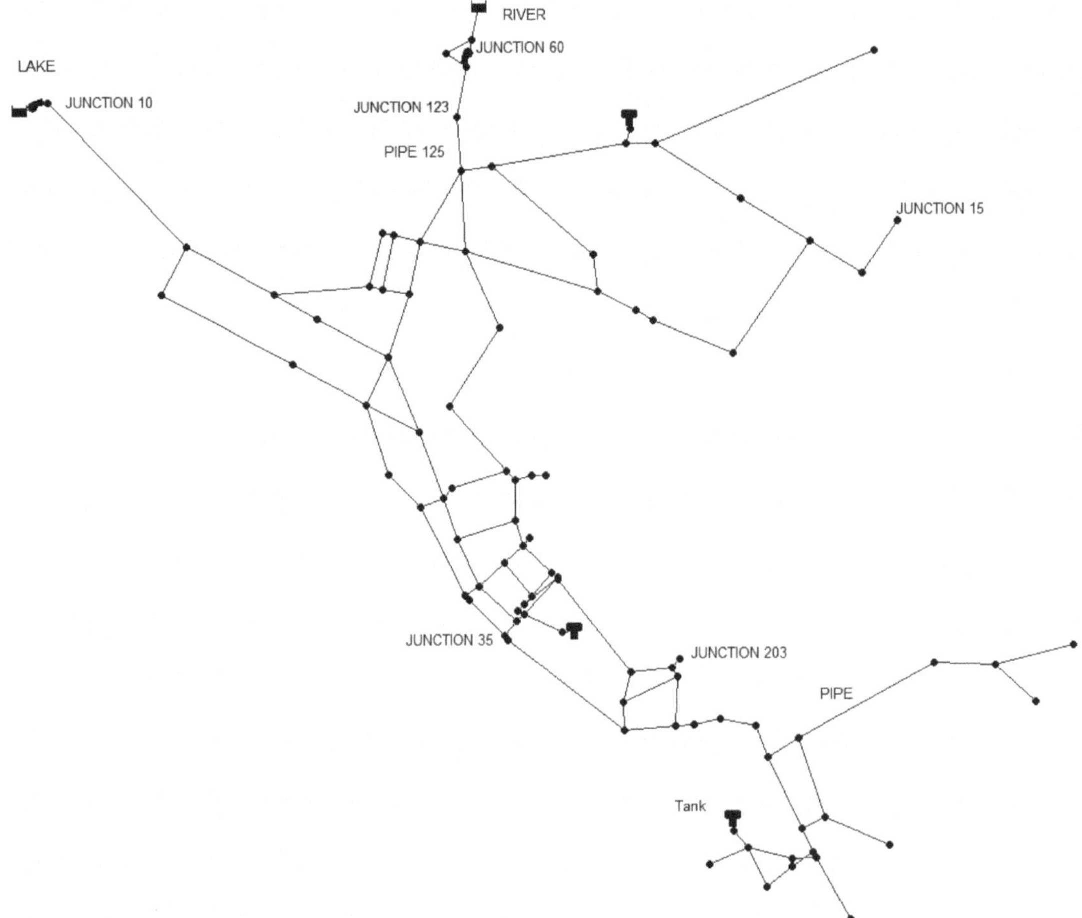

Figure 6.7 Example water distribution network, Net3.

```
Net3 - Notepad

File   Edit   Format   View   Help
[TITLE]
EPANET Example Network 3
Example showing how the percent of Lake water in a dual-source
system changes over time.

[JUNCTIONS]
;ID                      Elev            Demand          Pattern
 10                      147             0                               ;
 15                      32              1               3               ;
 20                      129             0                               ;
 35                      12.5            1               4               ;
 40                      131.9           0                               ;
 50                      116.5           0                               ;
 60                      0               0                               ;
 601                     0               0                               ;
 61                      0               0                               ;
 101                     42              189.95                          ;
 103                     43              133.2                           ;
```

Figure 6.8 Net3 input file on Notepad.

input file can be opened in the Notepad app which provides the descriptive summary of the network that is seen on the EPANET software (Figure 6.8).

To model water as it flows through the distribution system, the duration of the simulation as well as the hydraulic timestep must be set. For this example, the simulation will run for 1 week, and the hydraulic timestep will be 1 hour. That means that calculations will be done for every hour during a 1-week period (168 times). There are two ways to do this, on the EPANET software before the input file is exported, and on the EPANET input file using Notepad. There are advantages to both methods but modifying on Notepad is preferred because it can be modified more efficiently as all the data is in the same place and there is no need to continue to export the data from EPANET if adjustments need to be made. This is especially true for calibration purposes when adjustments are made often. However, using the EPANET software has advantages because there is less room for error as one can visually see where values are being entered more clearly. To adjust the duration and timestep on EPANET, select project>>Analysis Options>>Times. Then set hydraulic time step to 1 and total duration to 168 which is 1 week (Figure 6.9). On Notepad, the timesteps can simply be changed by scrolling down to the [Times] section and changing the values (Figure 6.9). The Quality Time Step is unimportant since the EPANET-MSX software will handle the microbial and chemical species and will be explained in a later section. Regardless of the method, make sure to save the project. An asterisk will appear next to the file name on Notepad if adjustments have been made to the file and have not been saved.

An important parameter that is necessary to model the flow of water throughout the system is the demand at certain junctions which is equal to the flowrate at that junction over time. The demand can be set to a constant value, 0, or assigned a multiplier value of 1 with an associated pattern (Figure 6.10). A pattern is made when the demand at a junction changes over time and when the hourly demand values are known, usually because of sampling. Although patterns can be assigned using EPANET, this pattern will involve 168 hours of data, which is easier to create on Excel or other sheet software, and then copied and pasted to Notepad. To create a pattern on EPANET, one should select the node that they want to assign a pattern to. For this network, the patterns are assigned to the outlet

Times Options	
Property	Hrs:Min
Total Duration	168:00
Hydraulic Time Step	1:00
Quality Time Step	0:05
Pattern Time Step	1:00
Pattern Start Time	0:00
Reporting Time Step	1:00
Report Start Time	0:00
Clock Start Time	12 am
Statistic	NONE

Net3 – Notepad
File Edit Format View Help

```
[TIMES]
Duration                168:00
Hydraulic Timestep      1:00
Quality Timestep        0:05
Pattern Timestep        1:00
Pattern Start           0:00
Report Timestep         1:00
Report Start            0:00
Start ClockTime         12 am
Statistic               None
```

Figure 6.9 Time adjustments for model setup using EPANET and Notepad.

points as that is where water is discharged. The base demand should be set to 1 because it will serve as a multiplier to the demand pattern and the Demand pattern option should be assigned a number correlating to the pattern. The multiplier can be set to different values such as 2 or 0.5 if the modeler wants to observe the output when increases or decreases to the demand occur. Then, in the browser dropdown, select patterns>>specified pattern number (double click). This will give a pop-up window of the pattern for each hour at that node which can be modified (Figure 6.10). Again, this option is only useful for small duration patterns. For this example, the pattern will be entered into Notepad,

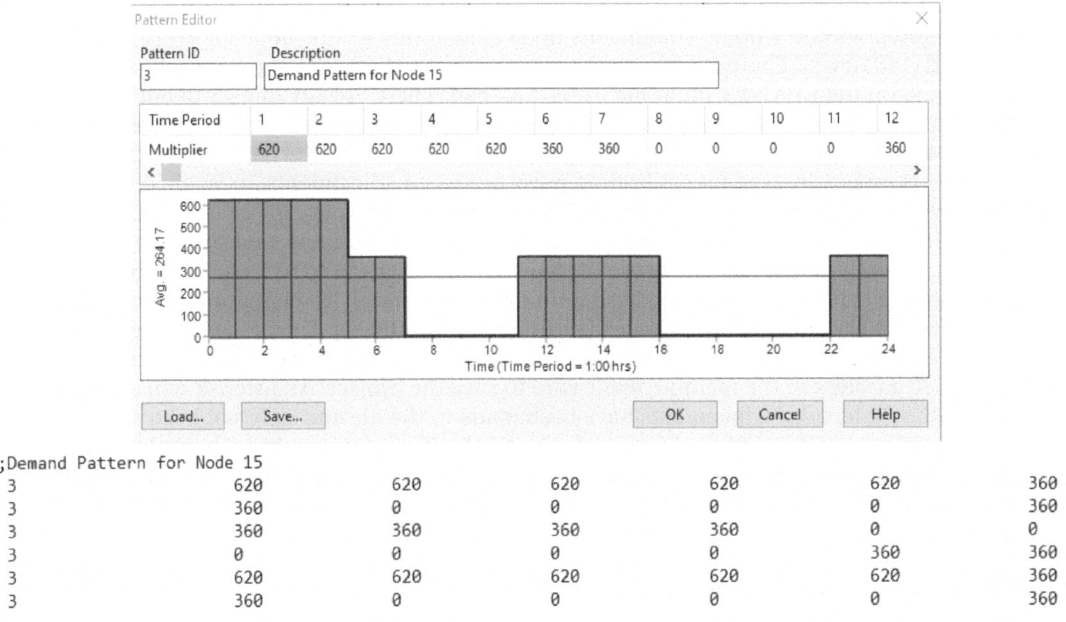


```
;Demand Pattern for Node 15
3          620          620          620          620          620          360
3          360          0            0            0            0            360
3          360          360          360          360          0            0
3          0            0            0            0            360          360
3          620          620          620          620          620          360
3          360          0            0            0            0            360
```

Figure 6.10 Modification of node demand using EPANET software (top) and using Text Editor (bottom).

and formatted using Excel. On notepad, the pattern comes out in the form of seven columns per row (Figure 6.10). The first column always has the pattern number, then columns 2–6 provide the pattern for hours 1–6. The next row also has the pattern number in the first column, followed by hours 7–12. For large patterns consisting of many hours, or for patterns consisting of small hydraulic time steps such as 1 minute, Excel can be used to ensure column/row consistency, then copied and pasted to Notepad. Note that the Notepad figure only includes some of the hours in the pattern. The true notepad pattern consists of 168 entries (number of hours in a week) or 28 rows (six entries per row). Once the demand has been determined for each of the nodes, the EPANET file is ready for use.

6.4 MODELING WATER AGE AND SINGLE-SPECIES WATER QUALITY ON EPANET

EPANET is capable of modeling both water age and water quality for one single species when the plumbing flow demand and initial concentration of the species are known. Water age is correlated to the hydraulic retention time of the water in the system, or the amount of time that the water remains in the pipe system. When the average hydraulic retention time increases, likely due to periods of decreased water use or lower plumbing demand, as well as reduced flowrate at the outlet points, the water age increases. EPANET 2.2 has a function built into the software that calculates the water age for every second when the water is in the system, and for every second during stagnant conditions. For a given system, water age continues to increase until the next water use event in which water leaves the system occurs. This simple method of calculating water age is useful to monitor the water quality in a system (Güngör-Demirci *et al.*, 2020) because high water age is associated with loss of disinfectant residual as well as growth of chemical and microbial contaminants (Hozalski *et al.*, 2020; Ji *et al.*, 2015).

For this example, the water age and free chlorine concentration will be modeled for arbitrary junction 115. Free chlorine is a rapidly decaying species that is used to treat water and inactivate chemical and microbial contaminants that may be present (Nguyen *et al.*, 2012). Chlorine disinfectant in drinking water is crucial to avoid waterborne disease outbreaks (Falkinham *et al.*, 2015). It will be assumed that the initial chlorine concentration at the reservoirs is 2.0 mg/L and that it follows a first order decay rate as it flows through the pipes, meaning that the free chlorine decay is equal to $(dC/dt) = -KC$ where C is the free chlorine concentration, and K is the free chlorine decay rate coefficient which will be assumed to be 0.05 hr^{-1}.

First, like the hydraulic analysis option, in the browser menu, once again select Options>>Times and make the Quality Time Step equal to 5 minutes (0:05). This is because we are often interested in the various water quality fluctuations that could occur within a given hour. Then, select Options>>Quality then select age in the parameter row. Once these options have been selected, the user can select Project>>Run Analysis. Once the project has completed its run, a pop up will appear saying that the run was successful. To view the water age hourly plot, select Report>>Graph. In the graph selection window, select time series for graph type, age for parameter, and nodes for object type. We often prefer to get concentration values for nodes rather than links (pipes) because they are more representative of the water that leaves the system and can be compared to measured data. Then, click on the node to be plotted and press add (Figure 6.11). Note that multiple nodes/links can be plotted at once. Once all options are selected, press ok to see the plot (Figure 6.12). The graph properties can be edited by right clicking anywhere on the plot.

The free chlorine species is modeled on EPANET with a similar approach as the water age. To model the species, select Options>>Quality in the browser menu. Then, select chemical in the parameter row. Free chlorine is typically measured in units of mg/L so select that option in the mass units row. For this example, we will focus on bulk reactions as opposed to both bulk and wall reactions as we assume that we do not have any information regarding the wall species in the system. Select reactions and specify the bulk reaction order as 1, wall reaction order as zero, global bulk coefficient as −0.05, global wall coefficient as 0, and leave the rest of the options as 0. After setting the reaction and quality options, the initial conditions must be set. In this system, water enters from the two reservoirs which

  Embracing Analytics in the Drinking Water Industry

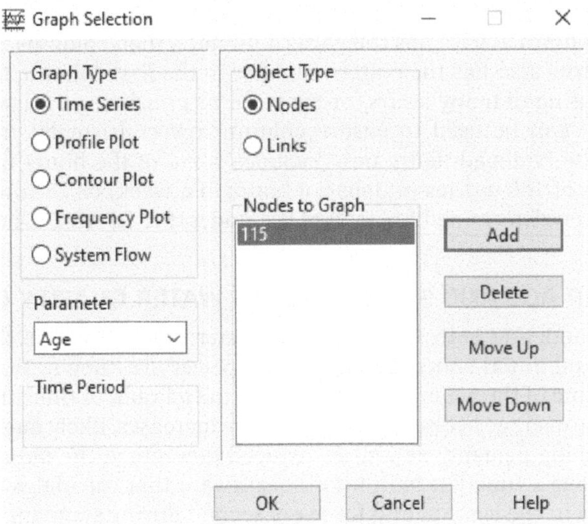

Figure 6.11 Graph options for node 115.

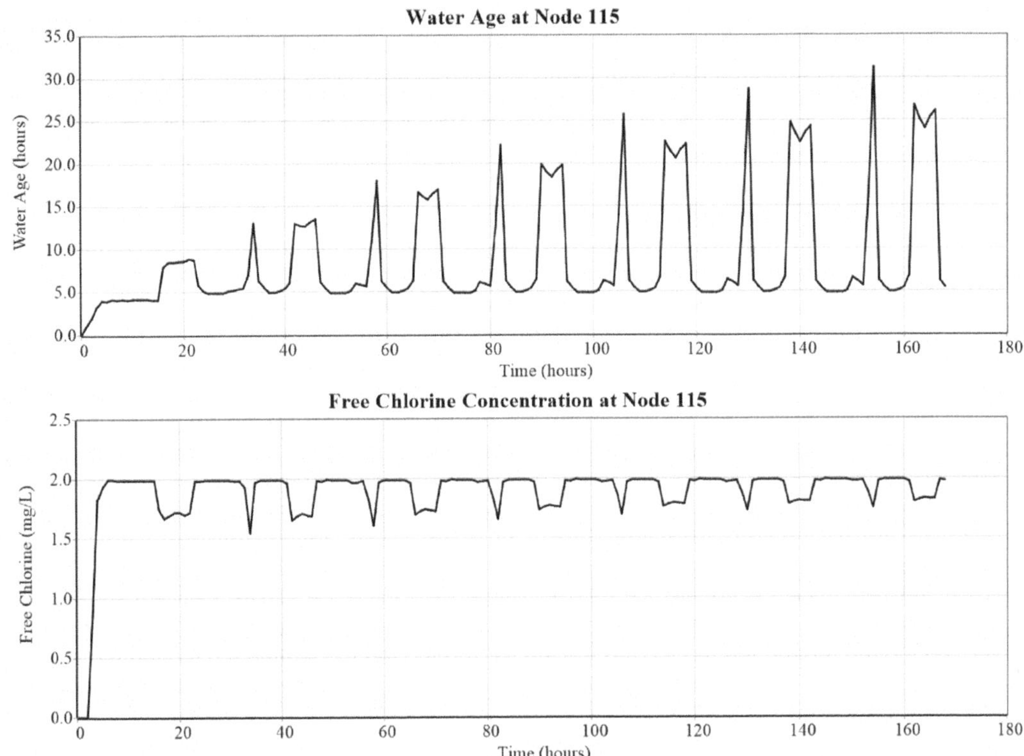

Figure 6.12 Time-series plots of water age and free chlorine for a sample water distribution system using EPANET.

are the lake and the river. Double click on one of the two reservoirs and insert a value of 2.0 in the quality row. No units are necessary since they have already been specified for the quality options. Run the analysis for the project and open the graph for node 115 in the same way that we did for water age. This time, select chemical in the parameter option. The resulting plots of water age and free chlorine concentration for the 1-week (168 hour) time interval are shown in Figure 6.12.

6.5 MODELING MULTIPLE SPECIES USING EPANET-MSX

Unlike the EPANET file, the EPANET-MSX file is created entirely on Notepad or another text editor and run using the computer's terminal or command prompt in unison with the input file. The file should be saved as an MSX file (.msx). Figure 6.13 describes the contents of the MSX file. A semicolon is used to add comments to the file and will not be read by the EPANET-MSX software. Not every section in the file needs to be included, just the ones that are necessary for the individual program.

For this example, the title of the file will be 'Net3 MSX Program' and the options will set the area units as US Customary to match the EPANET input file options, rate units as concentration per hour and quality time step to 300 seconds (5 minutes). The rest of the options are recommended for most programs but can be changed on an individual need basis. The solver option will be set to a 5th order Runge–Kutta integrator, and the relative, as well as absolute concentration tolerance will be set to 0.1 (Figure 6.14).

The species that will be modeled for this example include free chlorine and total trihalomethanes (TTHM). As previously mentioned, free chlorine is useful at preventing waterborne disease outbreaks. However, chlorination results in the formation of disinfection biproducts such as TTHM which are harmful to human health (Brown *et al.*, 2011).

Water quality models on EPANET assume advective-reactive transport with no dispersion effects. Therefore, the governing equations often consider reactions in the bulk flow and at the pipe wall. It is not always possible to obtain separate data for bulk and wall species concentrations to calibrate the models, so the bulk and wall reactions are often grouped into one governing equation and instantaneous mixing of water at the nodes, junctions, and storage facilities is assumed (Seyoum *et al.*, 2013). For this example, we will assume that all reactions occur in the bulk phase and that first order TTHM growth depends on free chlorine concentration. Equations (6.12) and (6.13) were used as the governing equations for the EPANET-MSX file:

$$\frac{dC_{TTHM}}{dt} = k_{TTHM1} - k_{TTHM2} \cdot C_{TTHM} \cdot C_{FCL} \tag{6.12}$$

$$\frac{dC_{FCL}}{dt} = -kC_{FCL} \tag{6.13}$$

where the k values are the kinetic coefficients and C values are the concentrations of free chlorine and TTHM.

The governing equations will be used in the MSX file under both the pipes and tanks section (Figure 6.14). since we have both in our network. Also, the coefficients section will include selected coefficient values. Often, these values are unknown, so the selected coefficients are an educated guess. Once the program is run, the output concentrations will be compared to known, collected data at that location and the coefficient values will be adjusted accordingly (whether the output concentrations are too high or too small) and the program will be run again until the values match as well as possible. This process will be explained in more detail in the calibrations section. To start, let us set a value of 0.05 for bulk free chlorine constant, 10 for TTHM bulk constant 1, and 4 for TTHM bulk constant 2. As mentioned, free chlorine is measured in units of mg/L so we are setting its units to MG (representing mg/L because the program already takes the flow into account). TTHM is typically measured in units

```
Net3 - Notepad                                                    —    □    >

File  Edit  Format  View  Help
[TITLE]
;Add a title to the network

[OPTIONS]
;Computational options

[SPECIES]
;Define the species to be evaluated, whether they are in the bulk or wall phase, and
their units

[COEFFICIENTS]
;Defines the parameters and the constant values for the rate/equilibrium expressions.

[TERMS]
;If neccessary, include this section for intermediate terms/equations that are used in
rate/equilibrium expressions (that need a separate equation)

[PIPES]
;Include the rate or equilibrium expressions that describe the behavior of the species
in the water distribution pipes.

[TANKS]
;Include the rate or equilibrium expressions that describe the behavior of the species
in the water distribution tanks. This section must be included when program contains
both bulk and wall species even if there are no tanks.

[SOURCES]
;If neccesssary, the input source for selected species can be included in the form of an
incoming time pattern.

[QUALITY]
;Provide a constant initial condition/initial concentration for species throughout the
pipe network for the entire simulation duration.

[PARAMETERS]
;If neccessary, parameters can be assigned on a pipe by pipe basis

[PATTERNS]
;This section is neccessary when there is a pattern that is assigned to a species in the
sources section. The pattern is included here. |

[REPORT]
;indicates what species and at what locations the report will be included
```

Figure 6.13 EPANET-MSX file input sections with descriptions.

```
Net3 - Notepad
File  Edit  Format  View  Help
[TITLE]
  Net3 MSX Program

[OPTIONS]
  AREA_UNITS FT2                            ;Surface concentration is mass/ft2
  RATE_UNITS HR                            ;Reaction rates are concentration/hour
  SOLVER     RK5                           ;5-th order Runge-Kutta integrator
  TIMESTEP   300                           ;300 sec (5 min) solution time step
  RTOL       0.1                           ;Relative concentration tolerance
  ATOL       0.1                           ;Absolute concentration tolerance

[SPECIES]
  BULK FCL      MG                         ;Free chlorine in the bulk phase
  BULK TTHM     UG                         ;TTHM in the bulk phase

[COEFFICIENTS]
  PARAMETER Kfcl      0.05                 ;Free Chlorine rate bulk coefficient
  PARAMETER Ktthm1    10                   ;TTHM bulk rate coefficient 1
  PARAMETER Ktthm2    4                    ;TTHM bulk rate coefficient 2

[TERMS]
  ;N/A for this example

[PIPES]
  RATE     FCL       -Kfcl*FCL             ;Rate of bulk free chlorine in the pipes
  RATE     TTHM      Ktthm1-Ktthm2*TTHM*FCL   ;Rate of bulk TTHM in the pipes

[TANKS]
  RATE     FCL       -Kfcl*FCL             ;Rate of bulk free chlorine in the tanks
  RATE     TTHM      Ktthm1-Ktthm2*TTHM*FCL   ;Rate of bulk TTHM in the tanks

[SOURCES]
  ;N/A for this example

[QUALITY]
  ;Initial conditions (= 0 if not specified here)
  NODE Lake  FCL    2.0
  NODE River FCL    2.0

[PARAMETERS]
;N/A for this example

[PATTERNS]
;N/A for this example

[REPORT]
  NODES       115                          ;Report results for node 115
  SPECIES     FCL      YES                 ;Report results for bulk FCL
  SPECIES     TTHM     YES                 ;Report results for bulk TTHM
```

Figure 6.14 EPANET-MSX file for Net3.

of μg/L so the units are set to UG (Figure 6.14). We will assume that the coefficients will remain constant among all pipes. If we wanted to have different coefficient values at different pipes, this could be done in the parameters section by specifying the pipe number, the coefficient, and the value at that specific pipe.

Finally, let us set the initial conditions for each parameter. For this example, we will use the quality section which assumes a constant incoming concentration of each parameter from the reservoirs. If water quality was not constant, an incoming water quality pattern could be set using the sources and pattern section. For this network, there are two sources of water to the system, the lake, and the river, so there are two incoming water quality conditions. At the quality section, add an initial concentration of 2.0 for free chlorine, and 0 for TTHM since it is assumed that TTHM begins to grow in the plumbing system and is not present in the reservoir. If no initial concentrations are specified, the model assumes an initial value of 0. It is important to note that each action in both the input and MSX files begins with a space. There will be an error on the program if the space is not added before each row so a space must be included before each line that is not a section label or does not have a semicolon.

6.6 RUNNING EPANET-MSX SOFTWARE AND CALIBRATING RESULTS TO SAMPLED DATA

To run the EPANET-MSX program, the user should first ensure that both the EPANET input file and EPANET-MSX file are stored in the 'C' drive under the user's username. Also, the EPANET-MSX program, that can be downloaded from the USEPA website, must also be stored in the 'C' drive for the program to run. Once all the files are in the correct location, the user can open a terminal, or the command prompt on the computer and type 'epanetmsx [name of input file].inp [name of MSX file].msx [defined name of output report file].rpt' and press enter. For this example, the command window is as follows:

C:\Users\username>epanetmsx Net3.inp Net3.msx Net3.rpt

The user must make sure that the input,.msx, and.rpt files are separated by a space, and that the file names do not have any spaces themselves as it will confuse the terminal into thinking that they are two separate files. Once entered, the program should run and the window will inform the user when the program run is complete (Figure 6.15), and the new report file will appear in the 'C' drive. Several errors could occur in this process and there are a few ways to troubleshoot them. One of the two main kinds of errors is a file error where the program cannot run because there is an error in either the msx file or the EPANET input file (Figure 6.16). When this occurs, first make sure that there is no error on the command line. Often, there could be an issue in which the names of the files were not entered exactly as they are saved, or that there is no space between the file names, or that the type of file (.inp,.msx,.rpt) are not specified. If naming is not the issue, a.rpt file will be formed in the 'C' drive. This file will specify why the inputs could not be read, and on what line the error occurred.

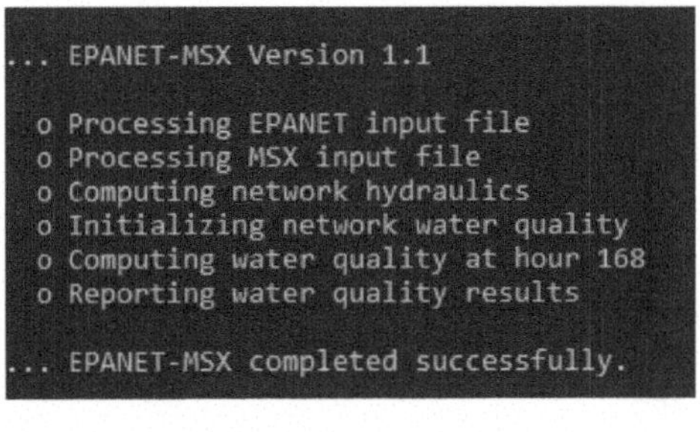

Figure 6.15 Successful EPANET-MSX run for 168 hours or 1 week.

```
...  EPANET-MSX Version 1.1

 o Processing EPANET input file

...  Cannot read EPANET file; error code = 302

...  EPANET-MSX Version 1.1

 o Processing EPANET input file
 o Processing MSX input file

...  Cannot read EPANET-MSX file; error code = 508
```

Figure 6.16 Input errors.

This file will be overwritten once the error is resolved, and the program runs as expected or it can be deleted by the user so that a new file is made. The second kind of error can occur while the program is running (Figure 6.17). This occurs because the EPANET solver cannot solve the differential equations based on the given inputs. This is often either because the RTOL/ATOL values are too high, the quality timestep is too low, or the kinetic coefficients are too high for certain differential equations, specifically ones that have exponential functions. All the error codes that occur can be found in the EPANET-MSX manual or EPANET manual depending on where the errors occur. These manuals should be located to troubleshoot other less common errors that may occur.

Once the.rpt file is made, the files can be opened using Microsoft Excel, MATLAB, or other software to calibrate results. For this example, Microsoft Excel will be used. Before opening the.rpt file, the user must have the sample data from the notes that have been modeled so that model values can be compared to the sample data. For this example, Link 125 will be used so that both wall and bulk concentrations appear. Table 6.1 demonstrates the measured data at each of the points collected every 8 hours from the start of the sample period at pipe 125.

To open the.rpt file on Excel, open Excel, then press File>>Open>>Browse. Go to the C drive where the.rpt file is stored and press the file dropdown and change from 'All Excel Files' to 'All Files'. The.rpt file should appear in the C drive. Open the file, check the delimited option, then press next. In the next page, check the box that says tab, and the box that says space. This should create a new Excel column for each column separated by a space on the.rpt file. Once that is done, press next>>Finish. The file will appear in Excel format. Delete any unnecessary information from the file to make it easier to read (Figure 6.18).

On that Excel file, the user can create another column to turn the hours into date format for easier reading and plotting. Let us assume that node 115 was sampled twice a day from January 1 to January 7. We will assume that the samples were taken at different hours of the day to capture different water use patterns, and periods of high and low demand. Table 6.1 shows the measured TTHM and FCL values for this example. In the working Excel file that has columns of hours, date time, and simulated

```
...  EPANET-MSX Version 1.1

 o Processing EPANET input file
 o Processing MSX input file
 o Computing network hydraulics
 o Initializing network water quality
 o Computing water quality at hour 17

...  EPANET-MSX runtime error; error code = 513
```

Figure 6.17 Runtime errors.

Table 6.1 Sample data at node 115.

Date/Time	Measured FCL (mg/L)	Measured TTHM (μg/L)
1/1 8:00 AM	1.60	6.90
1/1 10:00 PM	1.00	22.56
1/2 4:00 AM	1.50	7.80
1/2 10:00 AM	1.00	27.80
1/3 1:00 AM	1.50	7.750
1/3 10:00 AM	1.00	27.20
1/4 3:00 AM	1.50	7.75
1/4 6:00 PM	1.08	22.00
1/5 3:00 AM	1.50	7.76
1/5 10:00PM	1.00	23.10
1/6 10:00 AM	0.96	27.25
1/7 3:00 AM	1.50	7.75
1/7 10:00 AM	0.98	27.25
1/7 4:00 PM	1.45	8.50

TTHM and FCL values, add another two columns and insert the sample data for FCL and TTHM at the correct row that corresponds with the hour in which the sample was collected. Now we can plot simulated and measured values of TTHM as well as FCL on two separate graphs (one for FCL one for TTHM). The measured and simulated values can now be compared visually (Figure 6.19).

By inspecting the figures, it is clear that the chlorine decay rate coefficient is too small as the simulated values are larger than the actual values. Also, the TTHM concentrations are much lower than the expected values. We must therefore go back to the EPANET-MSX file and adjust the coefficient values until the simulated results match the actual concentrations as closely as possible. This is achieved by ensuring that the maximum and minimum values in the dataset are captured.

<<<	Node	115	>>>
Time	FCL	TTHM	
hr:min	MG/L	UG/L	
-------	----------	----------	
0:00	0	0	
1:00	0	10	
2:00	0	20	
3:00	0.78	17.66	
4:00	1.58	1.72	
5:00	1.67	1.5	
6:00	1.66	1.5	
7:00	1.7	1.46	
8:00	1.64	1.52	

Figure 6.18 RPT file transferred to Microsoft Excel.

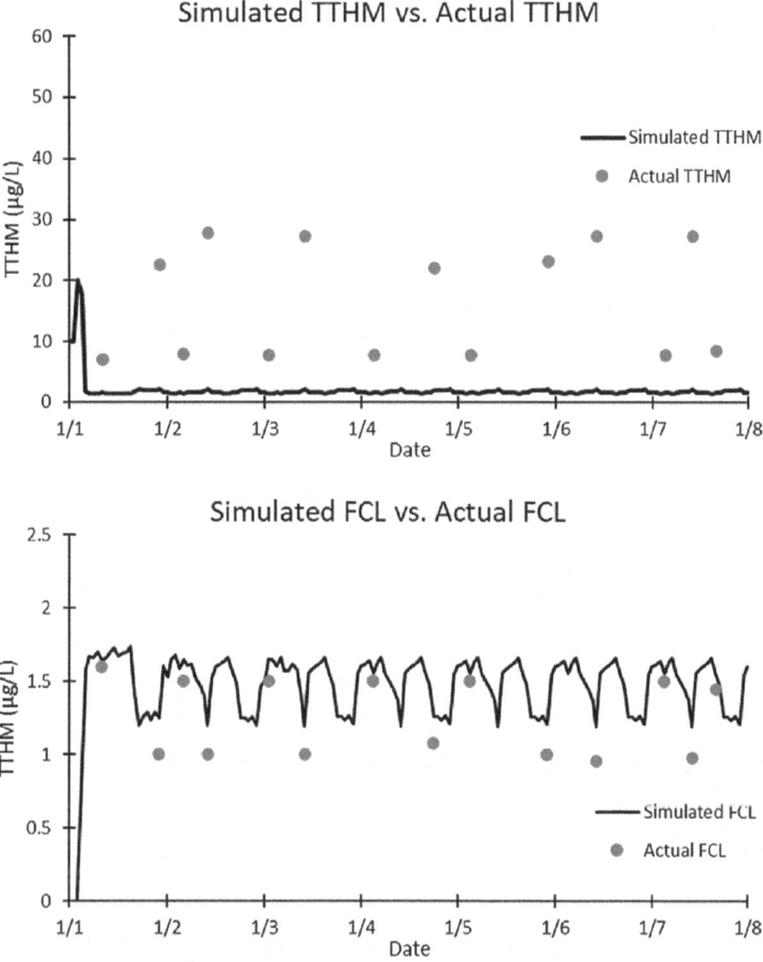

Figure 6.19 First attempt simulated vs. actual concentrations of TTHM and FCL.

For the example, we will change the chlorine decay rate (K_{FCL}) to 0.1 and the second TTHM rate coefficient ($KTTHM_2$) to 0.8. Figure 6.20 shows the resulting calibrated plots.

By visually inspecting the plots, it is seen that the simulated values match the expected TTHM and FCL values. We can see a relatively even distribution of maximum and minimum peaks that are captured. We can also notice that the TTHM concentrations tend to peak when the FCL concentrations are at a minimum. This is expected from our governing equation that states that TTHM grows under low chlorine conditions.

6.7 MODEL STATISTICAL VERIFICATION

There are many statistic methods to verify that the model works correctly in the form of equations. One of the most common methods is the RMSE method. This method represents the mean error not affected by cancellation and is given in the same units as the model outputs. The RMSE values can range from zero to infinity, with the ideal value being zero or as close to zero as possible (Bennett

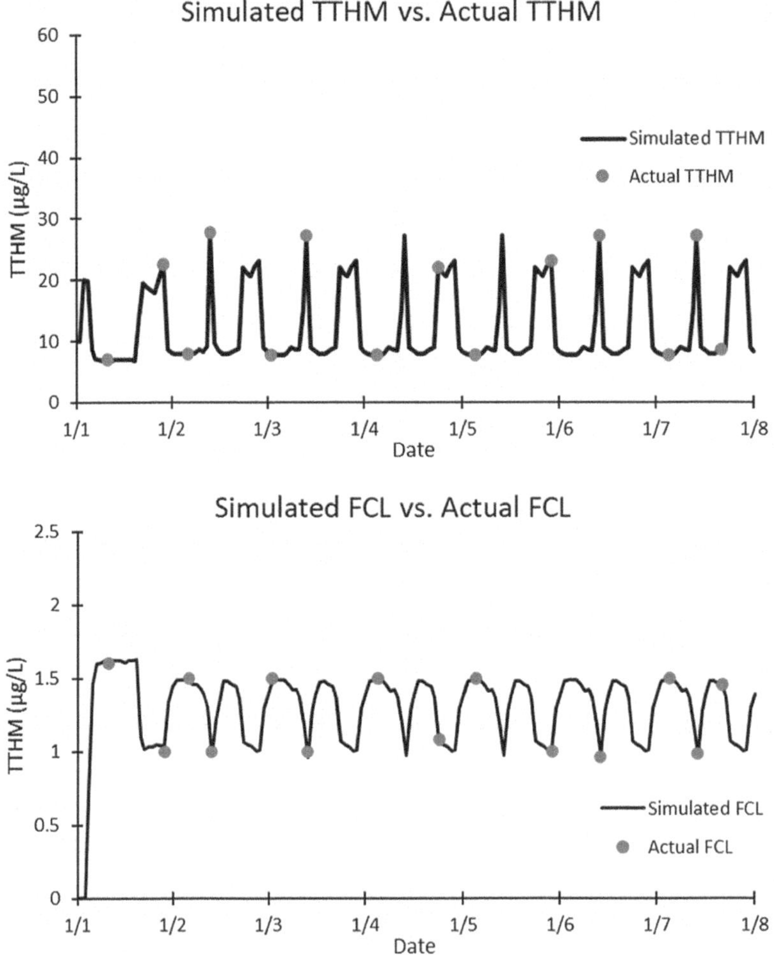

Figure 6.20 Calibrated simulated vs. actual concentrations of TTHM and FCL.

et al., 2013). The normalized root mean squared error (NRMSE) normalizes the RMSE values so that species with different units (such as TTHM FCL) can be compared accurately (Bennett *et al.*, 2013). The Nash–Sutcliffe (NSE) criterion compares the effectiveness of the models to one that uses only the mean of the observed data. NSE values range from negative infinity to 1, with the ideal value being 1. For this equation, negative values indicate poor model performance (Bennett *et al.*, 2013). The percent bias (PBIAS) is the average tendency of the simulated data to be larger or smaller than the observed data. PBIAS values are typically expressed as a percentage, ranging from negative infinity to infinity, with the optimal value being 0 (Gupta *et al.*, 1999). Equations (6.14)–(6.17) represent the selection criteria that will be used for this example to evaluate the model performance:

$$\text{RMSE} = \sqrt{\frac{1}{n}\sum_{i=1}^{n}\left(y_i - \hat{y}_i\right)^2} \tag{6.14}$$

Table 6.2 TTHM statistical analysis of model vs. observed data.

Time hr:min	TTHM Observed UG/L	TTHM Model							
		UG/L	Obs.-model	Obs.-mean	RMSE	NRMSE	NSE	PBIAS
–	–	–			0.024	0.002	1.000	0.058
0:00:00		0						
1:00:00		10.00						
2:00:00		20.00						
3:00:00	19.90	19.89	0.010	5.757				
4:00:00		8.48						
5:00:00		7.16						
6:00:00		6.93						
7:00:00		6.83						
8:00:00	6.90	6.91	−0.010	−7.243				
9:00:00		6.85						
10:00:00		6.90						

where y_i = observed values, $\hat{y}_i$ = simulated values and N = number of observed values.

$$NRMSE = RMSE / \bar{y} \tag{6.15}$$

where $\bar{y}$ = average model value.

$$NSE = 1 - \frac{(1/n)\sum_{i=1}^{n}(y_i - \hat{y}_i)^2}{(1/n)\sum_{i=1}^{n}(y_i - \bar{y})^2}\mathbb{F} \tag{6.16}$$

$$PBIAS = \frac{\sum_{i=1}^{n}\left(Y_i^{obs} - Y_i^{sim}\right)}{\sum_{i=1}^{n}\left(Y_i^{obs}\right)} \tag{6.17}$$

where Y_i^{obs} = observed values and Y_i^{sim} = simulated values.

To evaluate the RMSE, NRMSE, PBIAS, and NSE values, the measured values in Table 6.1 will be used and will be compared to the model values that correlate to the exact time as the measured values. Tables 6.2 and 6.3 demonstrate the table with calculation of RMSE, NRMSE, NSE, and PBIAS. Although the tables only show 24 hours of data, the calculations are for the entire 168-hour duration. The resulting values are representative of a good model. Both the RMSE and NRMSE values are near zero for both TTHM and FCL, NSE is 1 or close to 1, and PBIAS is expressed as a per cent so a decimal per cent is also very close to zero. It is important to note that this was a hypothetical example with hypothetical numbers so depending on the model, assumptions, and measurements, it may be unrealistic to have a model with near perfect performance as is shown here. This is especially true because chemical reactions in water systems are highly complex. The goal is to get the values as close to as possible to these metrics.

6.8 CONCLUSION

This chapter describes the tools and methods for modeling hydraulic and water quality for water distribution systems. The EPANET software is used to model the system's hydraulic characteristics,

Table 6.3 FCL statistical analysis of model vs. observed data.

Time hr:min	FCL Actual MG/L	FCL Model							
		MG/L	Obs.-model	Obs.-mean	RMSE	NRMSE	NSE	PBIAS
–	–	–			0.029	0.023	0.985	0.242
0:00:00		0						
1:00:00		0						
2:00:00		0						
3:00:00	0.70	0.72	−0.020	−0.556				
4:00:00		1.46						
5:00:00		1.6						
6:00:00		1.61						
7:00:00		1.62						
8:00:00	1.60	1.61	−0.010	0.344				
9:00:00		1.62						
10:00:00		1.62						

as well as a single parameter. The system inputs such as demand can be added on the software itself, or by using a text editor to edit the system's input file. For models of high duration, or small timesteps that require a large pattern, it is recommended that the system layout be edited in the text editor rather than on the software. However, for patterns with few inputs, it is recommended to use the software as it is visually easy to see and has less room for error. The EPANET-MSX software is used alongside the EPANET software to model the behavior of multiple species in the system, particularly when they react with each other. This software is only applied by using a text editor. Many options are given for the calculation of the species. Using the EPANET software with the EPANET-MSX software allows for realistic water quality modeling scenarios.

The models are calibrated by fitting the collected sample data to the model's outputs. To validate the calibrated model, several different criteria are given. The user can select one or compare all criteria together to verify the best fit. A hypothetical validated calibration is given in the chapter as an example but in reality, model performance may be less exact due to error in sampling, or because species are highly complex in their behavior within systems.

Although plumbing systems differ from water distribution systems, the methods outlined in this model can also be used to model smaller home plumbing systems. These results will depend on the ability to simulate actual demand within the building, and consideration of temperature changes, especially due to the hot water plumbing that needs to be modeled separately. Overall, water distribution systems and plumbing systems can both be modeled with the steps that are outlined in this chapter.

REFERENCES

Bennett N. D., Croke B. F., Guariso G., Guillaume J. H., Hamilton S. H., Jakeman A. J. and Andreassian V. (2013). Characterising performance of environmental models. *Environmental Modelling & Software*, **40**, 1–20, https://doi.org/10.1016/j.envsoft.2012.09.011

Brown D., Bridgeman J. and West J. R. (2011). Predicting chlorine decay and THM formation in water supply systems. *Reviews in Environmental Science and Bio/Technology*, **10**(1), 79–99, https://doi.org/10.1007/s11157-011-9229-8

Clark R. M., Rizzo G. S., Belknap J. A. and Cochrane C. (1999). Water quality and the replacement and repair of drinking water infrastructure: the Washington, DC case study. *Journal of Water Supply: Research and Technology–AQUA*, **48**(3), 106–114, https://doi.org/10.2166/aqua.1999.0011

Courtis B. J., West J. R. and Bridgeman J. (2009). Temporal and spatial variations in bulk chlorine decay within a water supply system. *Journal of Environmental Engineering*, **135**(3), 147–152, https://doi.org/10.1061/(ASCE)0733-9372(2009)135:3(147)

Falkinham J. O., Hilborn E. D., Arduino M. J., Pruden A. and Edwards M. A. (2015). Epidemiology and ecology of opportunistic premise plumbing pathogens: *Legionella pneumophila, Mycobacterium avium*, and *Pseudomonas aeruginosa*. *Environmental Health Perspectives*, **123**(8), 749–758, https://doi.org/10.1289/ehp.1408692

Güngör-Demirci G., Lee J. and Keck J. (2020). Optimizing pump operations in water distribution systems: energy cost, greenhouse gas emissions and water quality. *Water and Environment Journal*, **34**, 841–848, https://doi.org/10.1111/wej.12583

Gupta H. V., Sorooshian S. and Yapo P. O. (1999). Status of automatic calibration for hydrologic models: comparison with multilevel expert calibration. *Journal of Hydrologic Engineering*, **4**(2), 135–143, https://doi.org/10.1061/(ASCE)1084-0699(1999)4:2(135)

Hozalski R. M., LaPara T. M., Zhao X., Kim T., Waak M. B., Burch T. and McCarty M. (2020). Flushing of stagnant premise water systems after the COVID-19 shutdown can reduce infection risk by Legionella and Mycobacterium spp. *Environmental Science & Technology*, **54**(24), 15914–15924, https://doi.org/10.1021/acs.est.0c06357

Ji P., Parks J., Edwards M. A. and Pruden A. (2015). Impact of water chemistry, pipe material and stagnation on the building plumbing microbiome. *PloS One*, **10**(10), e0141087.

Kusnetsov J., Torvinen E., Perola O., Nousiainen T. and Katila M. L. (2003). Colonization of hospital water systems by legionellae, mycobacteria and other heterotrophic bacteria potentially hazardous to risk group patients. *Apmis*, **111**(5), 546–556, https://doi.org/10.1034/j.1600-0463.2003.1110503.x

Lautenschlager K., Boon N., Wang Y., Egli T. and Hammes F. (2010). Overnight stagnation of drinking water in household taps induces microbial growth and changes in community composition. *Water Research*, **44**(17), 4868–4877, https://doi.org/10.1016/j.watres.2010.07.032

Ley C. J., Proctor C. R., Singh G., Ra K., Noh Y., Odimayomi T., Salehi M., Julien R., Mitchell J., Nejadhashemi A. P., Whelton A. J. and Aw T. G. (2020). Drinking water microbiology in a water-efficient building: stagnation, seasonality, and physicochemical effects on opportunistic pathogen and total bacteria proliferation. *Environmental Science: Water Research & Technology*, **6**(10), 2902–2913, https://doi.org/10.1039/D0EW00334D

Lothrop N., Wilkinson S. T., Verhougstraete M., Sugeng A., Loh M. M., Klimecki W. and Beamer P. I. (2015). Home water treatment habits and effectiveness in a rural arizona community. *Water*, **7**(3), 1217–1231, https://doi.org/10.3390/w7031217

Munavalli G. R. and Kumar M. M. (2004). Modified lagrangian method for modeling water quality in distribution systems. *Water Research*, **38**(13), 2973–2988, https://doi.org/10.1016/j.watres.2004.04.007

Nguyen C., Elfland C. and Edwards M. (2012). Impact of advanced water conservation features and new copper pipe on rapid chloramine decay and microbial regrowth. *Water Research*, **46**(3), 611–621, https://doi.org/10.1016/j.watres.2011.11.006

Palmegiani M. A., Whelton A. J., Mitchell J., Nejadhashemi A. P. and Lee J. (2022). New developments in premise plumbing: integrative hydraulic and water quality modeling. *AWWA Water Science*, **4**(2), e1280.

Rhoads W. J., Pruden A. and Edwards M. A. (2016). Survey of green building water systems reveals elevated water age and water quality concerns. *Environmental Science: Water Research & Technology*, **2**(1), 164–173, https://doi.org/10.1039/C5EW00221D

Salehi M., Abouali M., Wang M., Zhou Z., Nejadhashemi A. P., Mitchell J., Caskey S. and Whelton A. J. (2018). Case study: Fixture water use and drinking water quality in a new residential green building. *Chemosphere*, **195**, 80–89.

Salehi M., Odimayomi T., Ra K., Ley C., Julien R., Nejadhashemi A. P., Hernandez-Suarez J. S., Mitchell J., Shah A. D. and Whelton A. (2020). An investigation of spatial and temporal drinking water quality variation in green residential plumbing. *Building and Environment*, **169**, 106566, https://doi.org/10.1016/j.buildenv.2019.106566

Seyoum A. G., Tanyimboh T. T. and Siew C. (2013). Assessment of water quality modelling capabilities of EPANET multiple species and pressure-dependent extension models. *Water Science and Technology: Water Supply*, **13**(4), 1161–1166, https://doi.org/10.2166/ws.2013.118

The Safe Drinking Water Act. (2000). U.S. Government Printing Office, Washington, DC.

USEPA. (2013). Revised Total Coliform Rule (RTCR) 78 FR 10269, February 13, Vol. 78, No. 30. USEPA, Washington, DC.

USEPA. (2014). The Revised Total Coliform Rule (RTCR) State Implementation Guidance – Interim Final. USEPA, Washington, DC.

USEPA. (2016a). *Revised Total Coliform Rule Seasonal Startup Checklist*. Retrieved from: https://www.epa.gov/region8-waterops/revised-total-coliform-rule-seasonal-startup-checklist (last accessed 10 May 2022)

USEPA. (2016b). *WaterSense Accomplishments 10 Years of Saving Water Together*. Retrieved from: https://www.epa.gov/sites/production/files/2017-07/documents/ws-aboutus-2016accomplishments-report-spreads.pdf (last accessed 10 May 2022)

USEPA. (2021). *EPANET – Application for Modeling Drinking Water Distribution Systems*. Available at: https://www.epa.gov/water-research/epanet (last accessed 2 March 2022)

U.S. Green Building Council. (2015). *LEED v4.1*. Washington, DC. Retrieved from: LEED v4.1 | U.S. Green Building Council (usgbc.org)

doi: 10.2166/9781789062380_0159

Chapter 7

Calibration and uncertainty analysis of hydraulic models

Adell Moradi Sabzkouhi[1]*, Juneseok Lee[2] and Jonathan Keck[3]

[1]Department of Hydraulic Engineering, ASNRUKH, Mollasani, Iran
[2]Department of Civil and Environmental Engineering, Manhattan College, Riverdale, NY, USA
[3]Founder/Principal, Water First, LLC, Naperville, IL, USA
*Corresponding author: adellmoradi@asnrukh.ac.ir

LEARNING OBJECTIVES

At the end of this chapter, you will be able to:

(1) Understand the basic definition of calibration and uncertainty analysis.
(2) Define the most important calibration parameters prone to uncertainty in water supply networks.
(3) Explain the steps and the procedure of a hydraulic model calibration.
(4) Employ MATLAB Optimization Tool to solve the problem arising from a pipe network calibration example.
(5) Understand the difference between hydraulic analysis under the most pessimistic parameters versus uncertainty analysis.
(6) Implement a simple method to quantify the uncertainty of a pipe network results as a function of input known uncertainties.

7.1 INTRODUCTION

Today, hydraulic models play an undeniable facilitating role in various stages of design/development, rehabilitation, operation and management of urban water distribution networks. Models represent an estimate of the behavior of Water Distribution Networks (WDNs), not their entire reality, and this is because hydraulic models are prone to different sources of uncertainty. Uncertainties due to incomplete understanding of the dynamics of phenomena, uncertainties in the structure of models and uncertainties in data and parameters are the most important types of uncertainty associated with modeling WDNs, among which, in this chapter, we are going to discuss the latter.

In WDN modeling, parameters are unknowns (constants or non-constant) that appear in the governing equations describing the system dynamics, mainly as coefficients or exponents that can be spatiotemporal variable. Roughness coefficients of pipes, nodal demand patterns, bulk and wall reaction rate coefficient of chemicals and so on, are examples of parameters in WDNs modeling. Parameters may be estimated by laboratory tests (e.g., new pipe roughness coefficients) or by analysis of field measurements (e.g., demand patterns or pipe roughness coefficients for systems under operation) or by a combination of them.

Calibration of water distribution models is a process that adjusts network parameters to minimize the differences between simulation results in the model and real measurements in the network (Zanfei *et al.*, 2020). Any parameter calibration is prone to inaccuracy since we just have to make an estimate of the parameters. Hence, parameter calibration is generally accompanied by an uncertainty analysis. Uncertainty analysis is performed to quantify to what extent the inaccuracies of parameter estimation would make the model results imprecise (e.g., nodal heads, velocity in pipes, concentration of chemicals etc.). Such analysis is called parameter 'uncertainty quantification' or 'uncertainty analysis' (UA). An important function of UA for operators could be awareness of the expected range of fluctuations in model results. Obviously, using an UA, we will be able to recognize less reliable model outputs and plan to minimize such uncertainties (e.g., by appropriate modification in the network) to have a more robust system in real-life operation conditions. In this chapter we are going to review the concepts of WDNs calibration and UA, and represent how to apply these concepts on practical examples.

Another term that is sometimes confused with UA is sensitivity analysis (SA). SA is the process of recognizing the effects of parameter variation on model results. UA tries to find the variability features of the model results (or responses) against parameters' variability. To put it more simply, SA is performed to distinguish the most important model parameters (i.e., that has the highest impacts on the model outcomes/variations), while UA is conducted to determine the most dependent model responses (results). SA is mainly relevant when creating and calibrating models, whereas UA is performed when models are employed to predict the actual behavior of the system under specific operation conditions.

Following the recognition of the effective parameters of the model, the calibration, as a process to determine the approximate values of the parameters by tuning them, is performed to attain the least squares of differences between the system responses (results) predicted by the model and measured in the field. The general method to parameter calibration in WDNs is minimization of the above mentioned least square function using optimization techniques. The procedure will be explained in detail through the following sections.

7.2 UNCERTAIN PARAMETERS IN PIPE NETWORK ANALYSIS

In this chapter, a distinction should be made between *design variables* (or decision variables) and *system parameters*. A design variable is basically a factor whose actual value can be changed by the system analyst. Each combination of design variables creates a design alternative. For example, diameters of the pipes are decision variables in the network sizing problem. The actual value of the system parameter, however, is not under control of the analyst, unlike stated above as for the design variable. In other words, when the actual value of a parameter was determined (or estimated) through the calibration process, the user is no longer allowed to change this value in analysis of the system. However, the user may change the parameter's value if by redefining the problem, the scenario expressing the status of the parameter is changed. As an example, imagine that to promote the water conveyance capacity of an aged water transmission pipeline by an update in pump station specification (scenario A), we need to estimate the pipe roughness parameter. After determining the roughness of the aged pipe, for scenario A, we are not allowed to change the value obtained for the pipe roughness. However, if as scenario B, it is intended to be study the possibility of pipe replacement, we can modify the roughness according to the different options of pipe materials commercially available.

Depending on the network under consideration and how the problem is defined, the system parameters will be different. The most important parameters in modeling water distribution networks are as discussed below.

7.2.1 Pipe roughness coefficients

In steady-state hydraulics of pipe networks, the Hazen–William roughness coefficient appears in the head loss equation with the exponential power of -1.852. It is not possible to directly measure the parameter of pipe roughness coefficient. In this regard, past experience and engineering judgment can provide an acceptable rough guess for the possible range of pipe roughness. Depending on the type of pipe material, operation condition, the quality of the water inside the pipe and so on, roughness parameter could have been significantly affected by pipe aging (Lamont, 1981; Sharp & Walski, 1988), which should be considered in dynamic design of WDNs (Creaco *et al.*, 2014; Minaei *et al.*, 2020).

Since pipe material and age are the most effective factors influencing roughness, it has been proposed to categorize the pipes with the same material and age into a separate group having identical roughness in the calibration process (Ormsbee & Lingireddy, 1997).

7.2.2 Nodal demands

For Extended Period Simulation (EPS) of WDNs, nodal base demand and demand pattern coefficients are considered as known values, while the way base demand is allocated to nodes as well as the distribution of demand pattern coefficients over the simulation period are exposed to uncertainties.

Assuming data availability, the simplest way for allocation of the base demands to the nodes could be dividing the total outflow from the pipe into two parts and equally loading them on the pipe upstream and downstream ends (Ormsbee & Lingireddy, 1997). Hence, the demand pattern coefficients for the nodes, as unknown parameters, should be determined in the calibration.

7.2.3 Pipe diameters

Pipe diameters are rarely considered as parameters in the analysis of WDNs. However, in modeling aged systems where reducing the inner diameter of pipes (due to sediment deposition, scale, or tuberculation) is likely to exist, the pipe diameters may also be considered as unknown parameters. Also, in some cases, due to changes in the network layout over time and the lack of as-built drawings, we may be unsure about the existence, the material or the size of some pipes in specific sites. In such case, if it is not possible to employ any detection facilities, the diameter of uncertain pipes may also be considered as the calibration parameters.

7.2.4 Leakage parameters

For aged water distribution networks suffering from high levels of leakage, network calibration without considering leakage generally does not lead to reliable results. In this case, leak parameters including the number of leaks, leak location and leak area size should also be considered as the calibration parameters. Since leak outflow from pipe systems is a function of pressure head at leak locations, in most practical application related to leakage simulation, a pressure driven approach is required for modeling hydraulics of the network.

7.2.5 Boundary conditions, tanks, valves and pump characteristics

These parameters are especially the case for the networks being under operation. The performance of tanks, pumps and valves may change due to 'wear-and-tear' over time. For example, the Head (H) and Flow rate (Q) characteristic curve of a pump or Q-head loss (H_{loss}) relationship of a valve due to mechanical depreciation may be subject to variability. In such cases, the original characteristic curves cannot be used in modeling the system. These sources of uncertainty, however, can be largely eliminated by using field measurements. As an example, to reproduce the modified H-Q curve for a pump, it is possible to measure flow passing through the pump (using the flow meter usually available on the discharge line of the pump station) and the differential pressure between pump suction and discharge, at several different openings of the regulating valve installed on the pump discharge line. This technique seems to be hard to implement in the case of network's regulation or isolation valves due to the large number of valves and difficulties related to flow measurement within the network. Hence, the H_{loss} coefficients of network valves are sometimes considered as unknown parameters in the calibration process.

7.3 REVIEW ON CALIBRATION STEPS

Ormsbee and Lingireddy (1997) divided the calibration process into the seven basic steps outlined below.

7.3.1 Identifying the intended use of the model

The objective of using the model determines the type of calibration parameters and the method of collecting field data. For example, in a model developed for network water quality management, in addition to hydraulic parameters (e.g., pipe roughness and nodal demands), reaction rate coefficient of the intended chemical should also be considered as calibration parameters. Moreover, if the model is created to use in water/energy management, due to the dependence of system results on hourly and daily varying demands, it is necessary to use the EPS approach of modeling and a weekly-basis field data collection scheduling. Obviously, for a network with pressure deficit, a pressure driven modeling approach is needed for calibration purposes.

7.3.2 Determining initial estimation of model parameters

In this step we need to have an initial rough estimate of the calibration parameters. For this purpose, useful tabular and diagrammatic information represented in references and standards provide a good initial estimate of the values of pipe roughness coefficients according to pipe material, diameter and age (Lamont, 1981; Wood, 1991). In addition, from standard field measurements (e.g., fire hydrant flow test) reliable data for specific pipe roughness calibration could be achieved (McEnroe *et al.*, 1989; Walski *et al.*, 2003).

Additionally, nodal demands can be roughly estimated by identifying the region influencing each node, identifying the types of demand units in the service area, and multiplying the number of each type by an associated demand factor. Alternatively, the estimate can be obtained by first identifying the area associated with each type of land use in the service area and then multiplying the area of each type by an associated demand factor (Ormsbee & Lingireddy, 1997).

7.3.3 Collecting calibration data

The accuracy of the initial estimated values for the model parameters must be evaluated in some way. To calculate the accuracy of the parameters estimation, it is required to collect field measurements (observations) from a number of available sites of nodal pressures and/or pipe flows, and compare them with the corresponding results predicted by the model.

Although the routine flow/pressure data collected at the pump station outlet and the water level in the storage tanks inside WDNs provides very useful information to be used in the calibration process, it is not sufficient in practical applications. Hence, calibration of WDNs usually needs a special measurement site design and establishment of measurement devices (like sensors) to collect more flow/pressure/water quality data required to achieve reliable results. From this point of view, the optimal design of measurement sites is an important and challenging issue, considering the expected accuracy of the collected data and the cost associated with sensors' purchase and installation. To study about optimal measurement site design, interested readers are referred to Kapelan *et al.* (2003), Vítkovský *et al.* (2003) and Ranginkaman *et al.* (2019).

7.3.4 Evaluating model results

To evaluate the results of parameter estimation, it is required to compare the measured data with the corresponding calculated result. Simulated flow in hydrant tests or water level in storage tanks are examples of model results to evaluate the accuracy of the parameter estimation. To make quantifiable the assessment of accuracy, various criteria can be considered. The absolute error of estimation and the squared error of estimation are two popular criteria to be considered as evaluation of the accuracy of parameter estimation. Ormsbee and Lingireddy (1997) have suggested that for planning and development purposes, a maximum deviation of 10% for nodal pressures, pipe flows and storage tank water levels and for design, operation and water quality management purposes, a maximum deviation of 5% for the model results and field measurements can be suitable criteria for accepting calibration results.

7.3.5 Performing macro-level calibration

Macro calibration is performed when the difference between the measurements in the field and the calculations in the model are greater than 30%. It is most likely that the large difference cannot be merely due to the parameter estimation error. In such cases it is necessary to further investigate the true status of the elements, the boundary conditions and so on, to find the cause of the large disagreement between the observed and predicted results. In this regard, possible dissimilarities between the model and prototype data may be due to closed or partially-closed valves, different pump performance curve, unseen links and network configuration, different pipe diameters and length, incorrect information of tanks, incorrect network zones, and so on.

7.3.6 Performing sensitivity analysis

Sensitivity analysis (SA) is a useful tool to determine the parameters to which the model results are more sensitive. Through SA time and effort are more efficiently spent in determining the parameters being more important in modeling. For example, if there is a limited labor force to do a hydrant flow test (a test whose results are very useful for estimating roughness coefficient of a specific water main) it is sensible to have more focus on the pipes with more sensitivity to model results.

7.3.7 Micro-calibration

The calibration process in the previous steps led to the determination of the possible range of variation and rough estimation of the model parameters. In micro-calibration, as the final step of the procedure, accurate values of the parameters are achieved by fine tuning so that the difference between the measured and calculated results reaches a minimal value. The fine-tuning process has experienced a gradual evolution in past years. Different approaches of micro-calibration include manual methods, trial-and-error methods, analytical methods, and optimization-based methods (also called automatic calibration). In the following section the optimization-based method, which is capable of implementation for practical purposes, is described.

7.4 AUTOMATIC CALIBRATION

This approach is an implicit method in which the problem of estimating the unknown parameters of the model is defined as an inverse problem to match some referenced condition. In such an inverse problem, the independent variables are initialized so that a number of selected dependent variables match, as much as possible, to the predetermined (or measured) values. Assuming that the coefficients of pipe roughness and nodal demands are network parameters, since the number of unknowns (network parameters) is greater than the number of equations (number of field measurements) a unique combination cannot be explicitly found for network parameter values. One practical way to solve such problems is to employ simulation-optimization techniques. In this approach, the calibration of WDNs can be defined as a nonlinear optimization problem so that unknown parameters are considered the decision variables and an equation related to '*the difference between the calculated results (in the model) and their corresponding measured ones (in the field)*' is taken as the objective function. Moreover, the problem consists of the hydraulic constraints of mass balance in the nodes and energy conservation in the pipes as well as the predefined ranges of variation of unknown parameters. Such an optimization problem can be solved using efficient optimization techniques like meta-heuristic algorithms.

7.4.1 Conceptual framework

Figure 7.1 shows the conceptual framework of the simulation-optimization model for micro-calibration of water distribution networks. The framework has two main cores including the simulation model

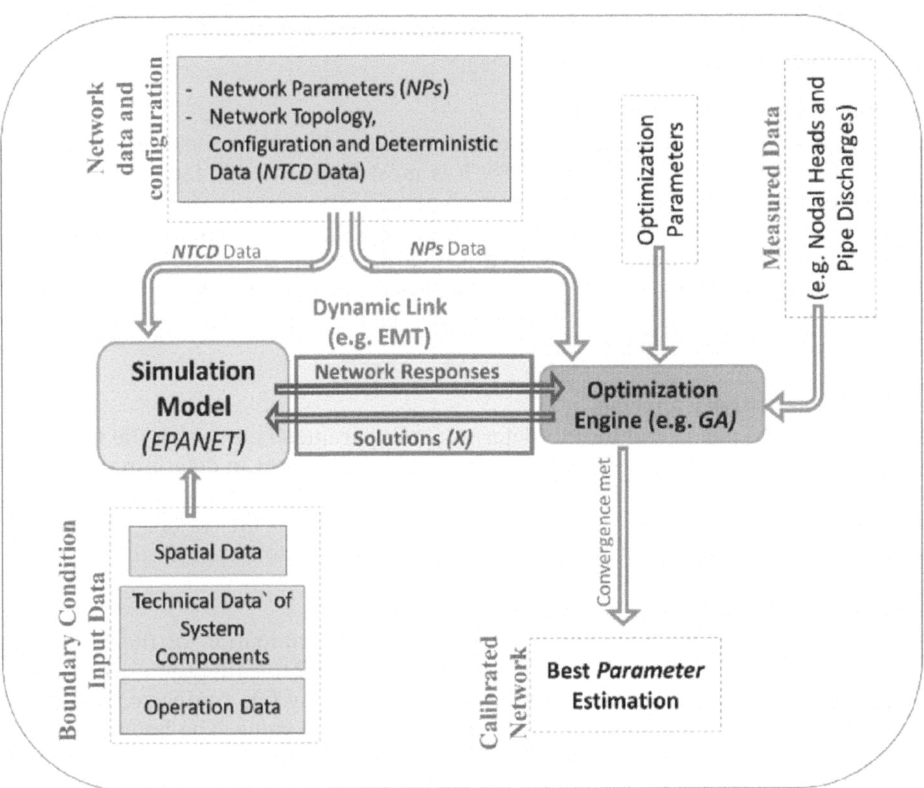

Figure 7.1 Conceptual framework for automatic parameter calibration of WDNs.

(e.g., EPANET) and the optimization engine (e.g., Genetic Algorithm). First, through the box labelled '*Network data and configuration*' and the box labelled '*Boundary Condition Input Data*' the known *Network Topology, Configuration and Deterministic Data (NTCD Data)* and the information related to the boundary conditions and other specifications, are introduced to the EPANET to build the network hydraulic model *inp* file. The '*Network data and configuration*' box also contains the information associated with *network parameters* (e.g., initial estimates and possible ranges) which enters the GA. Second, the GA randomly generates a population of solutions based on the initial information received (Network Parameters (*NPs*) Data in Figure 7.1), and sends them to the EPANET for hydraulic analysis. By execution of the EPANET, the computational results of the model are produced for each solution of the population, and then the results return back to the GA box as labelled 'Network Responses' in Figure 7.1. In addition to the previous inputs, GA also receives the 'Measured Results' from the field which enables the GA to evaluate and sort the solutions in regard to 'how much their calculated results match the corresponding ones measured in the field' (i.e. minimum error based on the objective function).

In the third step, according to the sorted population and the predefined *Optimization Parameters* (see Figure 7.1) the GA tries to eliminate the less fitted individuals (solutions) and preserve the more fitted ones, and produces a new generation of the solutions by imposing its operators (e.g., selection, cross over, migration, mutation, etc.) on the previous generation. The above procedure is repeated in a similar way on the next generations until the GA finally converges and the best fitted values for the calibration parameters are achieved.

7.4.2 Dynamic link for the simulation-optimization model

During the simulation-optimization process, there are a huge number of 'sending solutions to' and 'receiving responses from' the simulation model (EPANET), making the process impossible to manually handle. Therefore, any simulation-optimization model requires a *dynamic link* to make the interaction between simulation and optimization model automatic (see Figure 7.1). The major role of the dynamic link is to get a solution data (encoded in GA) and translate (decode) it into the language being meaningful to EPANET, and then return the calculated results to the GA. A widely popular dynamic link for WDN problems modeling is the EPANET-MATLAB Toolkit (EMT) developed by KIOS Research Center for Intelligent Systems and Networks of the University of Cyprus (Eliades *et al.*, 2016) which is a public domain programming interface to bridge MATLAB (the optimization environment) and EPANET (the simulation model).

To use *EMT*, it is required to first install *MinGW-w64* compiler in the MATLAB environment. To this end, in MATLAB from *APPS Menu Bar*, click on *Get More Apps*, then in *Add-On Explorer* search for *MinGW-w64* and install it. After *MinGW-w64* installation, from https://github.com/ OpenWaterAnalytics/EPANET-Matlab-Toolkit you should download source code and save it within the MATLAB current folder

For introducing an EPANET model to EMT, you should export the model to an *inp* file and move it to the same folder as EMT source code location in MATLAB. By executing *start_toolkit.m* and *epanet('filename.inp')* an EPANET object is created in which all the properties of the input *inp* file for the network model are provided (see Figure 7.2).

By creation of the EPANET object, EMT can read all available data in the inp file and update the properties of the system elements. There are many commands provided for the user to retrieve the initial data in the inp file, many others to overwrite the network model properties, and run hydraulic and water quality simulation as well. A complete list of these functions has been presented in the file README.md available for different network design, operation and management problems. Detailed examples of how to use different functions of EMT can be found in Eliades *et al.* (2016).

7.4.3 Mathematical statement of the problem

In addition to the need to properly understand the conceptual relationship among different parts of the simulation-optimization model (Figure 7.1), to put the problem into a computer code, it is better to state the relationships in the form of mathematical expressions. These expressions appear in two forms

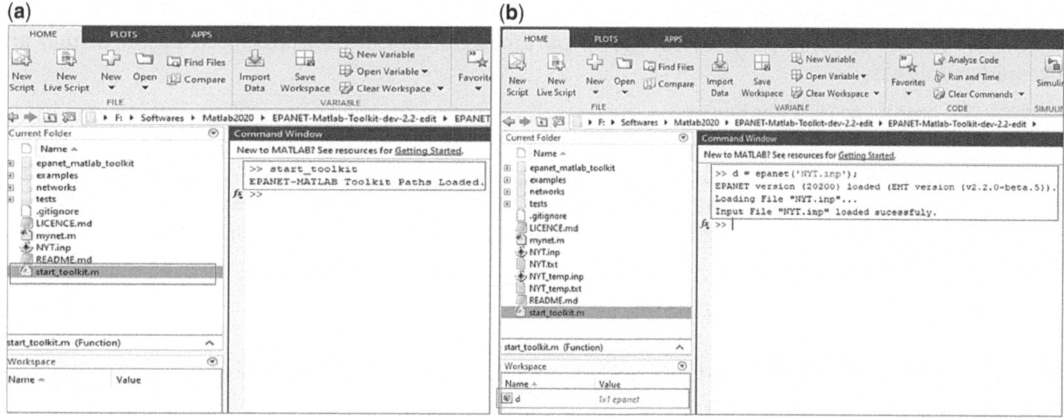

Figure 7.2 EMT in MATLAB environment: (a) start_toolkit.m command, and (b) EPANET object (d) creation for NYT. inp network model.

including the relationship stating the *Objective Function* and the ones stating the *Constraints*, both directly or indirectly are dependent on the values of independent variables (so called decision variables).

7.4.3.1 Actual decision variables

In the case of optimization resulting from an automatic calibration, the unknown parameters of the model (calibration parameters) will be the actual decision variables (or independent variables) of the optimization problem. Therefore, for a network with M groups of pipes, each group has distinct unknown Hazen–William's roughness coefficient (C_H), and N groups of junctions each has distinct unknown hourly-demand pattern coefficients, for a T-hour hydraulic simulation model, the string that expresses a generic solution (called a decision vector) is expressed as follows:

$$X_{1 \times n_{var}} = \left[x_1, x_2, \ldots, x_M, x_{M+1}, x_{M+2}, \ldots, x_{n_{var}} \right] \tag{7.1}$$

where M and N are respectively the number of pipe and junction groups, and n_{var} is the total number of variables (calibration parameters) in which $n_{var} = M + N \times T$.

7.4.3.2 Objective function

The objective function in the problem of optimization arising from micro-calibration of WDNs is the minimization of the sum of squared error (SSE) between calculated and observed pipe flows and nodal heads which could be mathematically expressed as follows:

$$Minimize\ SSE = \sum_{t=1}^{T} \sum_{i=1}^{N_{fms}} w_i \left(Q_{i,t}^c - Q_{i,t}^m \right)^2 + \sum_{t=1}^{T} \sum_{j=1}^{N_{hms}} w_j \left(H_{j,t}^c - H_{j,t}^m \right)^2 \tag{7.2}$$

where N_{fms} and N_{hms} are respectively the number of flow and head measurement sites, $Q_{i,t}^c$ and $Q_{i,t}^m$ are the calculated and measured flow in the ith flow measurement site at the tth time step, and $H_{j,t}^c$ and $H_{j,t}^m$ are the calculated and measured head on the jth head measurement site at the tth time step. Since the degree of importance of measurements may not be the same for all sites, the weighting coefficients w_i and w_j allow us to define different weights for each measurement.

7.4.3.3 Constraints

Constraints of the calibration problem are divided into two general categories. The first set of constraints is related to the possible ranges of parameters (decision variables) variation, which are expressed as follows:

$$x_i^{\min} \leq x_i \leq x_i^{\max} \quad \text{for } i = 1 \text{ to } n_{var} \tag{7.3}$$

The above set of constraints can be automatically handled while the GA works with normal decision variables (i.e., the value changes over the interval [0, 1]). In such a case, using the following equation, a normal decision variable x_i^{nor} is decoded into meaningful decision variable x_i with no extra imposing constraint:

$$x_i = x_i^{\min} + x_i^{nor} \left(x_i^{\max} - x_i^{\min} \right) \quad \text{for } i = 1 \text{ to } n_{var} \tag{7.4}$$

x_i^{nor} is initially generated in GA with a random function (e.g. the continuous uniform distribution), and updated as the GA advances in successive generations.

Another set of constraints are related to the hydraulic equations of the network (i.e., the mass conservation in junctions and energy balance in closed loops) which are automatically satisfied by the simulation model (i.e., EPANET), and we do not have to explicitly define any constraint in the GA.

7.5 EXAMPLE 7.1: CALIBRATION OF ANYTOWN MODIFIED NETWORK

To implement different automatic calibration procedures, the Any Town Modified (ATM) benchmark network is considered in this section. The Any Town network has been used as a benchmark in previous research with different configurations for different purposes (Ahmadian *et al.*, 2019; Cimorelli *et al.*, 2020; Costa *et al.*, 2016; Kapelan *et al.*, 2007; Rao & Salomons, 2007). The network, shown in Figure 7.3, consists of 19 demand nodes, 41 pipes, and three elevated storage tanks. The system is fed by a fixed head reservoir at an elevation 3.048 m and a pumping station with three identical centrifugal pumps in parallel. The data of system components and characteristic features of the pumps are presented in Table 7.1.

In this example, the roughness of pipes and the demand pattern are the calibration parameters. It is supposed that the 'true' but 'unknown' demand pattern multipliers of nodes and Hazen–Williams roughness coefficients of pipes are the ones presented in Table 7.1. As a matter of fact, the automatic calibration is going to find these true values using the systematic search engine (i.e., the Genetic Algorithm). Table 7.2 shows the results of field measurements of water level in tanks, pressure head in Node 100 and Node 50, and flow in Pipe 32 and inflow to the network as well. Obviously, if in the EPANET hydraulic model of the network we set the pipe roughness coefficients and water demand

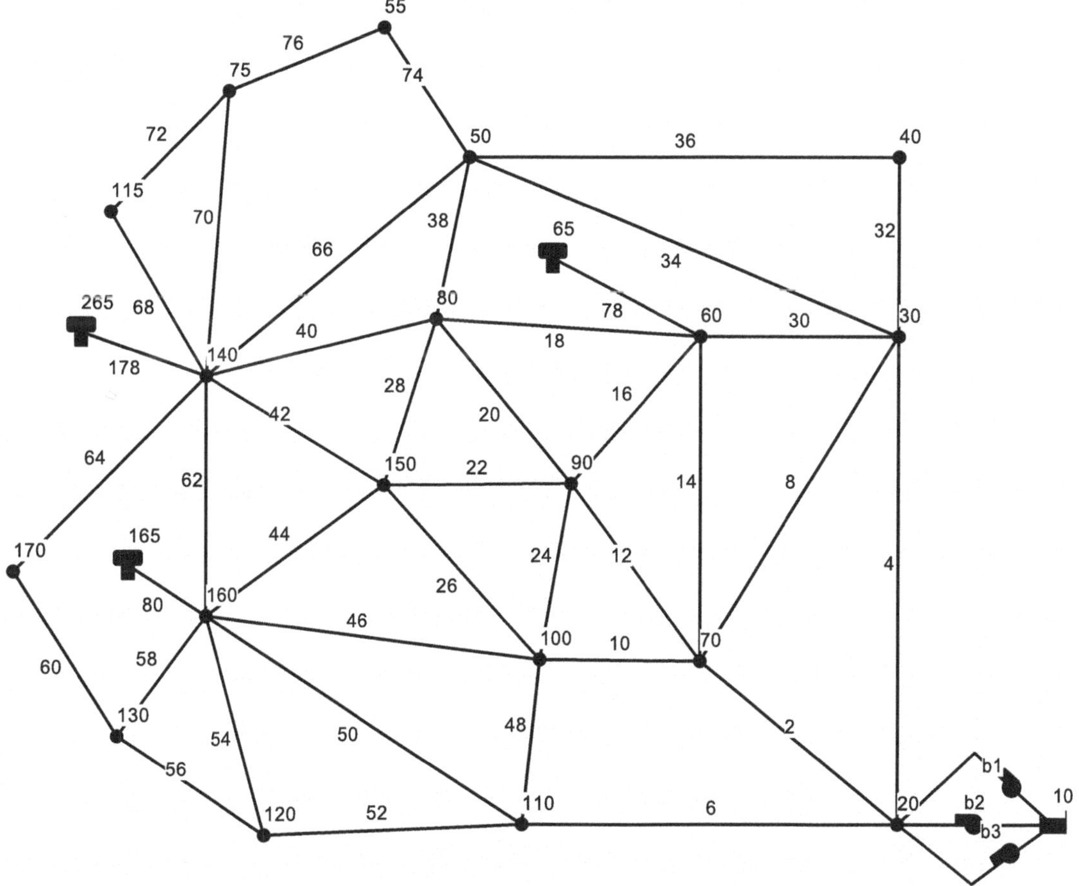

Figure 7.3 Layout of Anytown Modified Network.

Table 7.1 Pipe, node, tank and pump data of ATM.

Pipe Data

ID	Length (m)	Diameter (mm)	Roughness	ID	Length (m)	Diameter (mm)	Roughness
4	3657.6	609.6	109.2	64	3657.6	304.8	123.6
30	1828.8	508	111.3	60	1828.8	304.8	123.9
16	1828.8	152.4	108.8	58	1828.8	406.4	124.3
14	1828.8	406.4	110.1	44	1828.8	355.6	125.1
12	1828.8	254	109.5	50	1828.8	304.8	126.5
2	3657.6	609.6	110.9	52	1828.8	355.6	126.1
6	3657.6	457.2	116.1	56	1828.8	304.8	125.8
48	1828.8	304.8	114.6	62	1828.8	508	128.8
24	1828.8	508	113.8	46	1828.8	457.2	131.4
10	1828.8	609.6	115.2	34	2743.2	355.6	131.1
32	1828.8	203.2	114.9	78	30.48	508	131.8
36	1828.8	304.8	114.2	80	30.48	406.4	131.6
38	1828.8	203.2	119.2	8	2743.2	355.6	130.1
18	1828.8	457.2	118.7	74	1828.8	304.8	135.8
20	1828.8	406.4	118.9	76	1828.8	304.8	135.3
66	3657.6	203.2	119.5	72	1828.8	304.8	133.7
40	1828.8	304.8	119.9	68	1828.8	304.8	133.3
28	1828.8	457.2	119.3	70	1828.8	304.8	134.7
22	1828.8	152.4	118.5	178	30.48	406.4	136.6
26	1828.8	609.6	120.9	54	2743.2	304.8	135.1
42	1828.8	508	121.1				

Node Data

ID	Elevation (m)	Base Demand (L/S)
20	6.10	31.55
30	15.24	12.62
110	15.24	31.55
70	15.24	31.55
60	15.24	31.55
90	15.24	63.09
100	15.24	31.55
40	15.24	12.62
50	15.24	12.62
80	15.24	31.55
150	36.58	12.62
140	24.38	12.62
170	36.58	12.62
130	36.58	12.62
160	36.58	50.47
120	36.58	12.62
55	24.38	6.31
75	24.38	6.31
115	24.38	6.31

Pattern Multipliers

Time Step	Demand Pattern	Energy Price Tariff
1	0.7	0.1814
2	0.7	0.1814
3	0.7	0.1814
4	0.6	0.1814
5	0.6	0.1814
6	0.6	0.1814
7	1.2	0.1814
8	1.2	0.3528
9	1.2	0.3528
10	1.3	0.3528
11	1.3	0.3528
12	1.3	0.3528
13	1.2	0.3528
14	1.2	0.3528
15	1.2	0.3528
16	1	0.3528
17	1	0.3528
18	1	0.8097
19	0.9	0.8097
20	0.9	0.8097
21	0.9	0.8097
22	0.7	0.1814
23	0.7	0.1814
24	0.7	0.1814

Tank Data

ID	Elevation/ Diameter (m)	Initial Level (m)	Min./Max. Water Levels (m)
65	0/21.55	66.93	66.53/71.53
165	0/21.55	66.93	66.53/71.73
265	0/21.55	66.93	66.53/72.53

Pump data

Q (L/S)	TDH (m)	Efficiency%
0	91.4	0
126.9	89	50
252.4	82.3	65
378.5	70.1	55
504.7	55.2	40

Table 7.2 Field measurement data for Example 7.1

| Time Steps | Pump Station Outlet Data | | Site Measurement Data | | | | | |
| | Total Head at Node 20 (m) | Total Inflow to the Network (lps) | Water Level in Tanks (m) | | | Pressure Heads in Nodes (m) | | Flow Rate (lps) |
			Tank 65	Tank 165	Tank 265	Node 50	Node 100	Pipe 32
1	75.25	674.53	66.93	66.93	66.93	52.82	52.26	10.99
2	71.28	390.99	69.15	67.8	67.58	53.17	52.99	9.07
3	71.2	384.93	68.69	68	68.77	53.58	53.38	6.45
4	77	645.76	69.32	69.61	67.42	54.61	54.53	11.89
5	73.23	383.86	71.07	68.46	70.69	55.62	55.15	7.11
6	73.42	366.43	71.08	71.38	69.02	55.38	55.76	8.61
7	73.16	377.52	71.41	70.07	71.1	55.66	55.45	8.01
8	72.75	375.33	70.77	71.41	69.12	54.61	55.19	9.74
9	72.47	400.78	70.41	68.07	71.51	55.24	54.51	6.22
10	71.91	367.28	70.13	71.09	67.73	53.57	54.41	10.83
11	71.59	416.97	69.39	67.01	70.74	54.24	53.47	6.48
12	80.89	843.35	69.09	70.04	66.71	54.08	54.54	16.08
13	82.73	870.23	70.94	67.48	70.32	56.35	55.19	14.84
14	70.25	50.93	71.53	70.45	69.99	54.65	54.81	6.58
15	76.43	668.88	68.79	68.32	70.36	54.84	54.19	9.68
16	77.32	636.74	70.88	70.67	67.52	54.87	55.24	13.64
17	78.46	636.09	71.53	68.42	71.23	56.92	55.99	12.34
18	73.67	385.13	71.53	71.35	70.35	55.67	55.9	8.78
19	71	91.71	71.43	70.6	70.88	55.48	55.43	4.92
20	69.52	0	69.66	70.72	69.68	54.17	54.35	3.79
21	71.18	386.15	68.73	67.66	69.9	54	53.46	4.82
22	76.79	636.53	69.42	70.07	66.86	54.37	54.67	12.48
23	73.2	396.19	71	68.02	70.7	55.49	54.98	7.35
24	70.08	0	70.87	71.51	68.33	54.17	54.78	7.11

pattern coefficients equal to the values presented in Table 7.1, and the head of the network inlet point (Node 20) equal to the values in Table 7.2, the exact time series illustrated in Table 7.2 for nodal pressures (Node 50 and Node 100), tank water levels and the Pipe 32 flow rate must be attained. In other words, a reliable calibration optimization model should approximately find the true values of the calibration parameters stated in Table 7.1.

7.5.1 Optimization model: genetic algorithm

For automatic calibration of the ATM network, we will exploit the Genetic Algorithm (GA) to find the true roughness coefficients of the pipes and the demand pattern multipliers. The GA is an evolutionary optimization algorithm method inspired by natural genetics. As briefly explained earlier, this method starts with creating a population of solutions (called chromosomes in GA terminology) in order to systematically search the space of decision variables (called genes). Over successive generations, GA gradually alters the solutions to reach the solution that globally has optimal objective function. GA randomly generates the initial population and calculates (evaluates) the value of the objective function or the fitness function for each individual. Then, based on the evaluation results, the population is sorted, and relatively better individuals are preserved while the rest are eliminated. In the next step, by applying the GA operators having random internal processes, children (called off-springs) are produced from parents previously preserved, and added to the main population to form the next generation. Similar to the previous generation, the next generation of solutions are evaluated and altered by applying GA operators, and this iterative process continues until the algorithm converges.

The most important GA operators include *Selection, Crossover* and *Mutation. Selection* is responsible for the process of selecting parents for mating, in which individuals with better fitness are given a higher chance of participating in reproduction of the next generation. *Crossover* controls the intersection of parent chromosomes and the way of exchanging the genes to create new off-springs. Crossover produces children who receive some characteristics from the father chromosome and some characteristics from the mother chromosome. Moreover, *Mutation* is an operator that creates new individuals (solutions) by randomly changing a very limited number of genes which may result in new chromosomes with completely different characteristics compared to the population. Mutation is a mechanism to free the search process from being trapped in local optima. Without the mutation, the diversity of the population decreases rapidly causing the individuals to become too similar to each other. This makes the search process to be premature leading to stopping criteria without convergence to the global optima.

7.5.2 Optimization model setting

Based on previous information, we also suppose that in terms of the factors affecting Hazen–Williams roughness coefficient (e.g., material, age, etc.) all ATM pipes could be categorized into six separate groups of pipes: Group 1 (including Pipe 4, 30, 16, 14, 12 and 2), Group 2 (including Pipe 6, 48, 24, 10, 32 and 36), Group 3 (including Pipe 38, 18, 20, 66, 40, 28, 22, 26 and 42), Group 4 (including Pipe 64, 60, 58, 44, 50, 52 and 56), Group 5 (including Pipe 62, 46, 34, 87, 80 and 8) and Group 6 (including Pipe 74, 76, 72, 68, 70, 178 and 54). The above grouping has fictitiously been done in this example. In real applications, however, we probably have some evidence (e.g., the same pipe installation time, material, etc.) to group the pipes. This information leads us to the key point that we can consider the same grouping in the pipe roughness coefficients so that only six decision variables may be required to represent the roughness of all 41 pipes of the network. Although this point may affect the accuracy of calibration, it will reduce the dimensions of the optimization problem (number of decision variables) likely causing the whole procedure to be much more efficient.

According to the above details, the number of decision variables in this example is 30 which includes six variables for the roughness coefficient of groups of pipes and 24 demand multipliers (i.e., 24-hour demand patterns). We take w_i and w_j in the objective function (Equation 7.2) to be respectively $\left(1\big/Q_{i,t}^m\right)^2$ and $\left(1\big/H_{i,t}^m\right)^2$ based on the suggestion made by Do *et al.* (2016). The weights are squared since they should have the right same order with the squared errors, and are inversed as larger flow rates and pressure heads should have relatively small weights. Moreover, based on Equation (7.3), it is required to introduce upper and lower bounds for each decision variable. Prior experience and engineering judgment could effectively help accurately estimate these bounds. For example, by comparing the minimum night flow and maximum daily flow entering the network and comparing them with the 24-hour average inflow to the network, an acceptable estimate of the upper and lower bounds of demand pattern multipliers could be made. Also, for the roughness coefficient of pipes, the results of hydrant tests, pipe material and age, technical specifications by manufacturers, and so on give a good insight into the roughness coefficient upper and lower bounds. For the ATM example, we take 0.2 and 2 respectively as the lower and upper bounds of demand pattern multipliers, and 95 and 140 for the lower and upper bounds of roughness coefficient of all pipes.

The simplest way to access different optimization engines available in MATLAB environment, is the Optimization Tool (OT). To access OT, type 'optimtool' in Command Window. In OT (see Figure 7.4) we have two main parts as follows:

(1) The section *'Problem Setup and Results'* in which we define and address the information related to the objective function(s), decision variables and the constraints, and also we run the optimization model and see the final solutions found by the optimization engine. In this section, set the *Solver* as Genetic Algorithm. Below the Solver at the section *Problem*, we must define the *Fitness function* as a function handle of the form *@ObjFunc*, where ObjFunc is the name of the objective function we are going to minimize. To create ObjFunc, using EMT we must develop a MATLAB code as a *Function File* in which the input is a normal decision vector having 30 variables and the output is a variable defined for the value corresponding to the fitness of the input decision vector. In the section *Constraints*, we must only enter the bounds of decision variables. For *Lower* and *Upper*, type *zeros(1,30)* and *ones(1,30)*.

(2) The section *Options* in which we make different settings of GA. Since we defined the decision variables as real numbers, in *Population type*, choose *Double vector* for the type of the decision variable digits. In this section, also specify the *'Population size'* as 200. In *'Stopping criteria'* for the field *'Generation'* specify 1000. Furthermore, in *'Plot Function'* tick *'Best Fitness'* and *'Best Individual'* to check how GA converges to the best solution over generations. Leave the rest of the fields in 'Options' blank, meaning that we have accepted the default operators and parameters of the GA for the rest of the setting. You may come back later to change the defaults of different GA operators' settings (e.g., Mutation, Selection, Crossover, etc.) to check to what extent different values and types of operator are effective in GA progress to find the best solution.

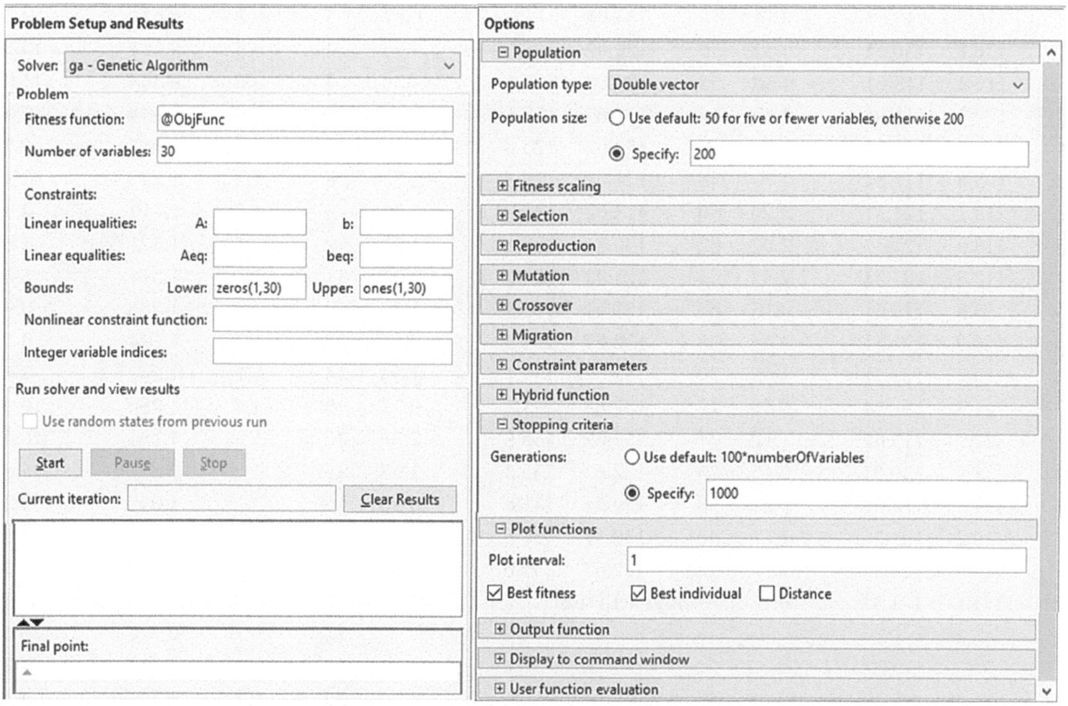

Figure 7.4 GA settings in Optimization Tool for calibration of the ATM example.

7.5.3 Model execution and results

According to the settings mentioned in the previous section, the optimization model was implemented. Table 7.3 shows the results of calibrated roughness of pipes and demand patterns multipliers and the comparison of the results with their true values. As can be seen, the average relative error of estimation (REE) for pipes roughness parameters and demand patterns multipliers are respectively 1.69 and 5.77%, while the maximum REEs are 4.53 and 18.43%. The agreement between calculated and measured data is shown in Figure 7.5, which indicates that the GA has a successful performance in searching the decision space to minimize the objective function. As Figure 7.6 also shows, the REEs between the measured data and calculated results by the calibrated model for a 24-hour extended period simulation, are below 10% indicating that the accuracy of results is acceptable (see section 7.3.4). As can be seen in these bar charts, the REE for the flow in the pipes is much greater than the REE for nodal pressure heads. Generally, in the calibration process, REE for flow measurements is higher than REE for pressure measurement. Increasing the number of flow measurement sites and performing a sensitivity analysis can reduce the REE. The trade-off between improvement of the accuracy of calibration and the costs associated with more measurement sites has always been a challenging issue for the planning and development of calibration models.

Table 7.3 Results of parameter calibration and the relative error of estimation (REE) for Example 7.1

Pipe ID	Pipe Roughness			Pipe ID	Pipe Roughness			Time Steps	Demand Multipliers		
	True value	Calibrated Value	REE (%)		True Value	Calibrated Value	REE (%)		True Value	Calibrated Value	REE (%)
2	110.9	111.84	0.85	44	121.02	125.1	3.37	1	0.7	0.673	3.90
4	109.2	111.84	2.41	46	130.51	131.4	0.68	2	0.7	0.731	4.39
6	116.1	113.61	2.14	48	113.61	114.6	0.87	3	0.7	0.631	9.80
8	130.1	130.51	0.32	50	121.02	126.5	4.53	4	0.6	0.693	15.57
10	115.2	113.61	1.38	52	121.02	126.1	4.20	5	0.6	0.534	11.05
12	109.5	111.84	2.13	54	132.67	135.1	1.83	6	0.6	0.648	7.94
14	110.1	111.84	1.58	56	121.02	125.8	3.95	7	1.2	1.167	2.75
16	108.8	111.84	2.79	58	121.02	124.3	2.71	8	1.2	1.223	1.94
18	118.7	121.31	2.20	60	121.02	123.9	2.38	9	1.2	1.140	4.98
20	118.9	121.31	2.02	62	130.51	128.8	1.31	10	1.3	1.321	1.60
22	118.5	121.31	2.37	64	121.02	123.6	2.13	11	1.3	1.314	1.09
24	113.8	113.61	0.16	66	121.31	119.5	1.49	12	1.3	1.288	0.93
26	120.9	121.31	0.34	68	132.67	133.3	0.47	13	1.2	1.194	0.50
28	119.3	121.31	1.68	70	132.67	134.7	1.53	14	1.2	1.242	3.48
30	111.3	111.84	0.48	72	132.67	133.7	0.77	15	1.2	1.080	10.00
32	114.9	113.61	1.12	74	132.67	135.8	2.36	16	1	1.184	18.43
34	131.1	130.51	0.45	76	132.67	135.3	1.98	17	1	0.822	17.84
36	114.2	113.61	0.51	78	130.51	131.8	0.98	18	1	1.092	9.23
38	119.2	121.31	1.77	80	130.51	131.6	0.83	19	0.9	0.864	3.98
40	119.9	121.31	1.17	178	132.67	136.6	2.96	20	0.9	0.890	1.16
42	121.1	121.31	0.17	Average REE (%)			1.69	21	0.9	0.913	1.44
								22	0.7	0.704	0.61
								23	0.7	0.712	1.70
								24	0.7	0.729	4.10
								Average REE (%)			5.77

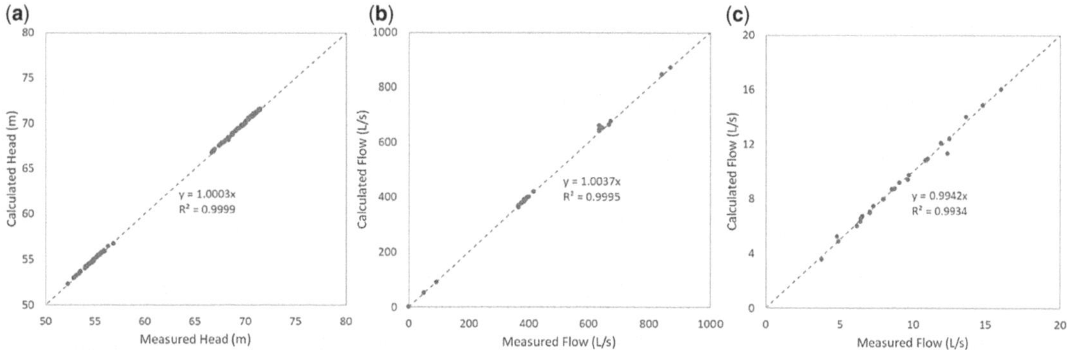

Figure 7.5 Agreement between the calculated and measured data for: (a) pressure head in the tanks and nodes 50 and 100; (b) inflow to the network; and (c) flow in pipe 32.

Figure 7.7 depicts the extent to which the calibrated roughness of pipes (Figure 7.7a) and demand pattern multipliers (Figure 7.7b) match the corresponding values considered as the true figures for the parameters.

Due to the random nature of the GA, different runs may not lead to exactly the same solution. Moreover, since the calibration is typically an under-determined problem especially in real world projects, we may find solutions representing other states of the system. To avoid such consequences, it is better to test the models with new measured data to approve the results obtained in the calibration process.

7.6 PARAMETER UNCERTAINTY ANALYSIS IN PIPE NETWORK MODELING

As seen in the results of the previous example, estimating the parameters of a hydraulic model of a WDN is accompanied with error. In addition, some modeling parameters of water distribution networks have inherent variability with respect to time and space. For example, water consumption in a DMA at 10 a.m. on a weekday may be different from the consumption of the same DMA at the exact same time on the next weekday, or the roughness of a particular pipe deteriorates over time due to pipe materials, environmental factors and operating conditions. Therefore, the values of the parameters of a WDN can be subject to uncertainty.

In a WDN, through the equations governing system dynamics, the network responses (e.g., nodal pressures) relate to the network parameters (e.g., roughness of pipes and nodal demands). As a result, *in modeling a WDN, the uncertainties associated with parameters propagate through the governing equations and cause the network responses to be uncertain too*. The process of calculating 'variability features of the responses as a result of variability of the parameters' is called uncertainty analysis (UA) or uncertainty quantification. Simply, in UA we are going to quantify the effect of uncertainty of independent variables (system parameters or inputs) on the system dependent variables (responses or outputs). For any system that includes *Inputs → Process → Outputs*, the deterministic (also called Crisp) outputs are achieved only if both the *Inputs* and the *Process* are deterministic, otherwise the *Outputs* will be uncertain (Tung & Yen, 2005). Hence, it is quite realistic to assume that hydraulic responses in WDNs are subject to uncertainty.

What the variability of the network parameters perceptibly brings to the system operators is the variability in tangible network responses such as flow velocity in pipes and especially nodal pressure heads. Therefore, it would be very desirable for system operators if network analysts specify, due to uncertainty in network parameters, what to expect from the statistical features of the responses (e.g., average, variance and range of the responses). With UA application in operation and especially in design projects, we can wisely detect the system's weak points, critical elements, and plan to make modifications where/when needed.

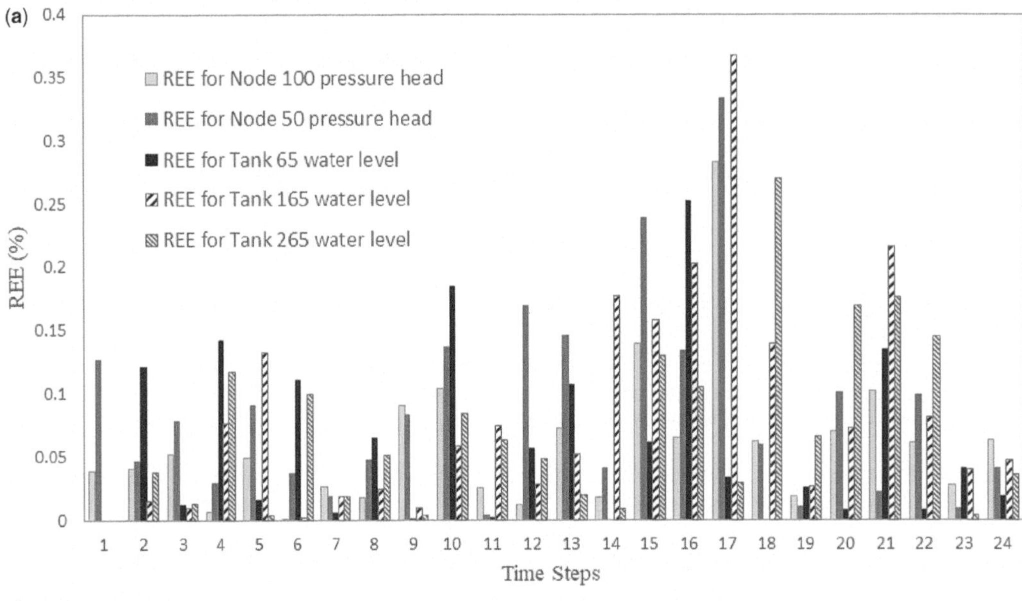

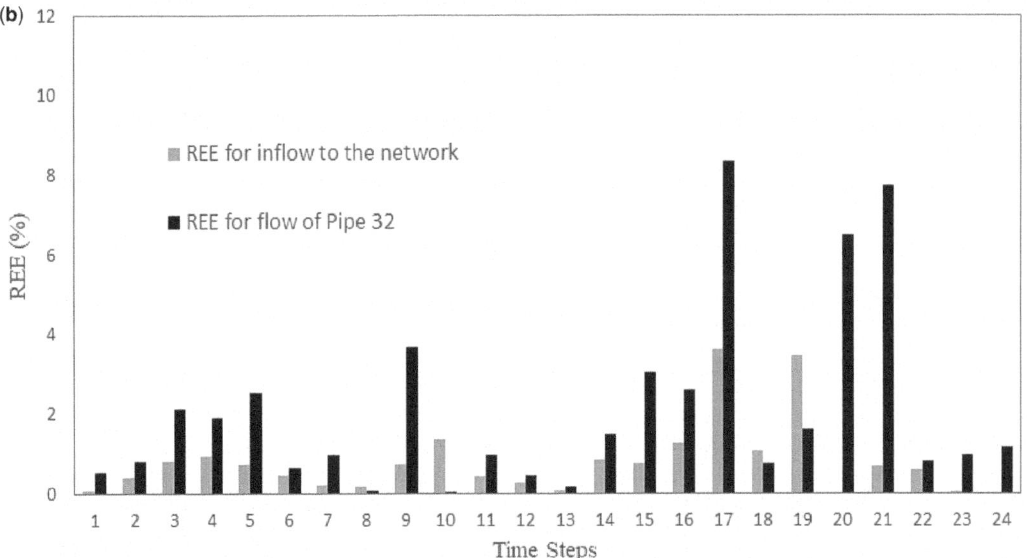

Figure 7.6 Relative Error of Estimation (REE) with respect to the measured data for: (a) calculated pressure heads; and (b) calculated pipe flow rate for Example 7.1.

7.6.1 Does UA search for the most pessimistic combination of parameters?

Many prudent engineers imagine that instead of the parameter UA, if the most pessimistic values of system parameters are simultaneously introduced to a deterministic model, the most pessimistic values would accordingly be captured for system responses. The following example reveals that how the above statement will not be true under some conditions, especially regarding looped pipe networks.

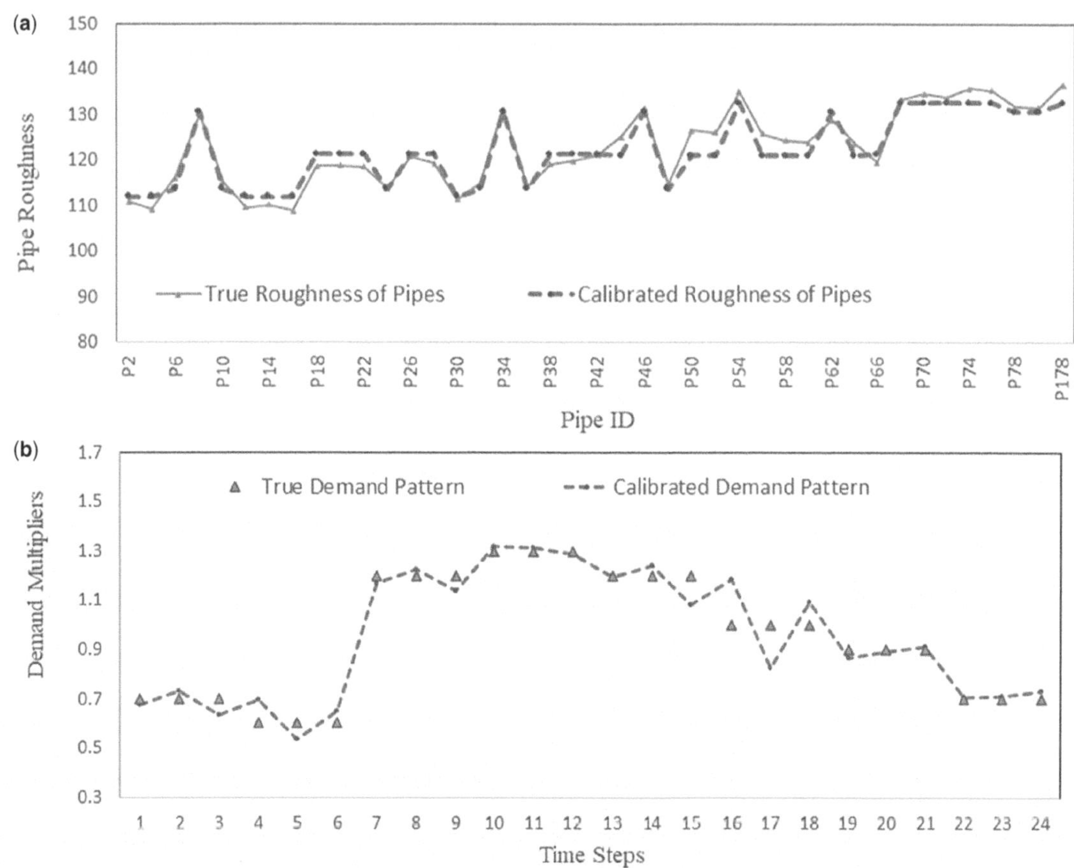

Figure 7.7 Comparison of the parameter calibration results with the true values in Example 7.1 for: (a) roughness of pipes; and (b) demand pattern multipliers.

The example is adapted from Sabzkouhi *et al.* (2017). As illustrated in Figure 7.8, the example includes a simple looped network with five pipes and four demand nodes supplied by a reservoir (via gravity flow) with a fixed water level at elevation 78 m. The system's physical and hydraulic properties are shown in Figure 7.8. Imagine that except for Hazen–Williams (HW) pipe roughness, all other parameters are deterministic with no uncertainty. Let us also assume that the HW roughness of pipes would average to 125 for all pipes while, based on technical knowledge and engineering experience, they are exposed to ±20% uncertainty. This means that the maximum and minimum possible HW roughness are expected to be 150 and 100, respectively. Based on the principles of pipe hydraulics, the less the HW pipe roughness becomes, the more the head loss along the pipe occurs. Hence, if we set all the HW pipe roughness equal to 100 (the most pessimistic value), we should apparently expect the lower pressure head values for all nodes. Let us see whether this assumption is true for all nodes or not. In Figure 7.8a, the results for nodal pressure heads are shown for this scenario. As can be seen, the pressure head for Node 3 is 18 m. As another scenario, in Figure 7.8b, we have kept all H-W pipe roughness fixed at 100 (as Figure 7.8a) except for Pipe 3 in which it has changed into 150 (the most optimistic value). The results in Figure 7.8b demonstrate that the minimum nodal pressure for Node 3 has been produced with the most optimistic HW roughness for Pipe 3 (i.e., 150) where it drops

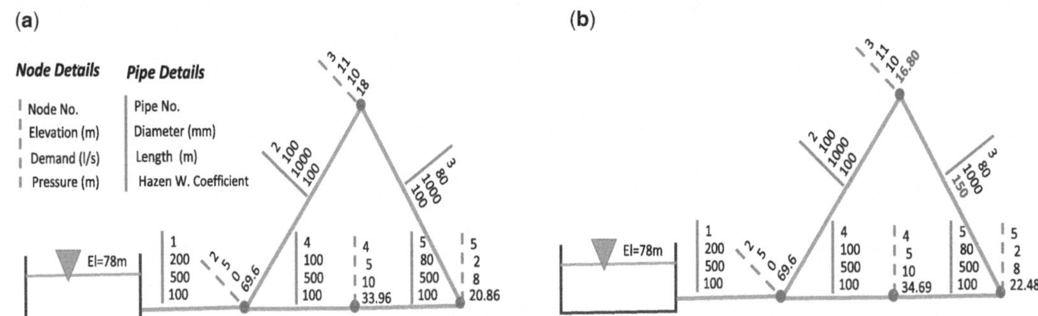

Figure 7.8 Uncertain nodal pressures corresponding to: (a) the most pessimistic pipes roughness; and (b) the critical combination of pipes roughness for the lowest pressure (Sabzkouhi *et al.*, 2017).

to 16.8 m. As a matter of fact, compared to the previous scenario, the more hydraulic transmission capacity for Pipe 3 results in more flow rate and more energy loss along Pipe 2 which causes a 1.2 m decrease in the pressure head of Node 3 compared to Figure 7.8a.

Clearly speaking, the statement that 'the most pessimistic combination of system parameters always leads to the most pessimistic values of network hydraulic responses' cannot be true for all cases. Therefore, evaluating the most pessimistic combination of parameters cannot replace UA for the system as a whole.

7.6.2 Approaches for parameter UA

The theory of probability as well as the fuzzy sets theory are the two major approaches of UA in WDNs, for which different methods and application have been proposed in the literature (Bao & Mays, 1990; Duan *et al.*, 2010; Haghighi and Asl, 2014; Hwang *et al.* 2017; Sabzkouhi and Haghighi, 2016, 2018a; Sabzkouhi *et al.* 2017; Seifollahi-Aghmiuni *et al.*, 2013; Tsakiris and Splilotis, 2017). Monte Carlo Simulation (MCS) is the most well-known probabilistic method for UA. This method requires numerous sampling stochastic input parameter(s) having known probability density function (PDF) and successive model execution to derive the PDF of the dependent random output(s). The MCS method has been a benchmark for assessing the accuracy of probability-based methods accepted by researchers for UA of engineering systems (Kang & Lansey, 2009).

As an alternative approach for UA, using fuzzy sets theory a WDN is considered as an uncertain system in which the fuzzy membership functions (FMFs) of the system output(s) are determined, given the known FMFs of the input parameter(s).

Implementing both the above approaches requires sufficient information about the parameters' statistical properties (e.g., PDFs) or fuzziness features (e.g., FMFs) which may not be simply accessible in most real-world problems. Both approaches also require a huge number of network simulation model runs which in the case of large networks makes the UA a very time-consuming and burdensome procedure. Hence, by taking simplifying assumptions, alternative approximation methods have been developed to reduce the cost of analysis in terms of calculations (Braun *et al.*, 2016; Gupta & Bhave, 2007; Kang & Lansey, 2011). In the next section, we are going to describe a simple method named Interval Analysis (Rao & Berke, 1997; Sabzkouhi & Haghighi, 2019) to practically perform UA in WDNs.

7.6.3 Interval analysis (IA) for parameter UA

In the IA approach for carrying out UA, the least information about system parameters is needed – the known parameter ranges. If q_i is the ith independent parameter (e.g., demand in the ith node) of the system, then a known interval $[\underline{q}_i, \bar{q}_i]$ could be considered for its range of variation. Additionally, let us take H_j the jth system dependent output (e.g., pressure head in the jth node) whose relationship with independent parameters is established as $H_j = \varphi(q_1, q_2, \ldots, q_i, \ldots q_{np})$. Here np is the number of independent parameters and φ is the set of equations governing pipe network hydraulics that returns H_j as a function of network parameters. As a matter of fact, since we employ EPANET hydraulic model to calculate the system responses (i.e., H_j), here EPANET takes the role of φ.

Corresponding to the interval $[\underline{q}_i, \bar{q}_i]$, an interval $\left[\underline{H}_j, \bar{H}_j\right]$ is conceivable for the variation of H_j. In other words, we can simply consider the UA as $\left[\underline{q}_i, \bar{q}_i\right] \overset{UA}{\Rightarrow} \left[\underline{H}_j, \bar{H}_j\right]$. As a desired outcome of IA, we need to capture the two extreme values for H_j including the upper bound ($\bar{H}_j$) and the lower bound ($\underline{H}_j$). However, systematically each extreme happens in a distinct critical combination of system parameters as follows:

$$\begin{cases} \underline{H}_j = \varphi\left(q_1, q_2, \ldots, q_i, \ldots q_{np}\right)_{\underline{H}_j} \\ \bar{H}_j = \varphi\left(q_1, q_2, \ldots, q_i, \ldots q_{np}\right)_{\bar{H}_j} \end{cases}, \quad q_i \in \left[\underline{q}_i \le q_i \le \bar{q}_i\right] \quad \text{for } i = 1 \text{ to } np \tag{7.5}$$

where $\left(q_1, q_2, \ldots, q_i, \ldots q_{np}\right)_{\underline{H}_j}$ and $\left(q_1, q_2, \ldots, q_i, \ldots q_{np}\right)_{\bar{H}_j}$ are respectively the combinations of system parameters corresponding to $\underline{H}_j$ and $\bar{H}_j$.

According to the nonlinear nature of the equations governing pipe network hydraulics, finding $\underline{H}_j$ and $\bar{H}_j$ generally requires an implicit procedure in which determining the upper and lower bounds are formulated as a mathematical optimization problem (a minimization type for $\underline{H}_j$ and a maximization type for $\bar{H}_j$). In each of the above optimization problems, $\underline{H}_j$ or $\bar{H}_j$ is the objective function, and q_1, $q_2, \ldots, q_i, \ldots$ and q_{np} are decision variables. Accordingly, in terms of UA of nodal pressure heads for the whole system, the number of optimization problems to be solved is two times the number of demand nodes. Hence, due to the large number of optimization runs required to perform UA of nodal pressures, this classical approach is very time consuming, especially for real networks. Although much effort has been made to develop specific optimization methods to reduce computational time (Haghighi & Asl, 2014; Sabzkouhi & Haghighi, 2016) these methods still do not have high efficiency in the case of real networks. Therefore, in this chapter, to carry out the UA of nodal pressure heads with IA approach, an approximate deterministic method developed by Gupta and Bhave (2007) is presented.

7.6.3.1 Impact Table Method

To better understand the Impact Table Method (ITM) (Gupta & Bhave, 2007), we first need to well understand how uncertainty in dependent variables may differ between monotonic and non-monotonic variations. In Figure 7.9, if we take Y as a function of X, the difference between monotonic and non-monotonic variation of Y is easily understood, where over the interval $[X_{\min}, X_{\max}]$, Y changes monotonically in the left chart and non-monotonically in the right chart. As we can see, in a monotonic variation, the extreme values of the dependent variable (i.e., $Y_{\min}$ and $Y_{\max}$) definitely occur in the extreme points of the independent variable (i.e. $X_{\min}$ and $X_{\max}$). In a non-monotonic variation, however, $Y_{\min}$ and $Y_{\max}$ do not occur in $X_{\min}$ and $X_{\max}$.

Accordingly, assuming a monotonic behavior of the responses to the system parameters, Gupta and Bhave (2007) proposed the ITM method for UA of a WDN responses including nodal pressures and pipe flows. The steps of this algorithm are as follows:

(1) Assuming the system parameters have been already determined, run the hydraulic model with the Crisp (i.e., the most likely value) parameters (q_i^c) and extract the values of model responses (H_j^c). To avoid confusion between the above information and other ones in the next steps, we label the calibrated data and the corresponding responses as Crisp.

(2) Consider that each system parameter q_i^c is exposure to uncertainty in the interval $\left[q_i^c - \Delta q_i, q_i^c + \Delta q_i\right]$. By keeping all parameters constant in their crisp values, for $i = 1$ allow only

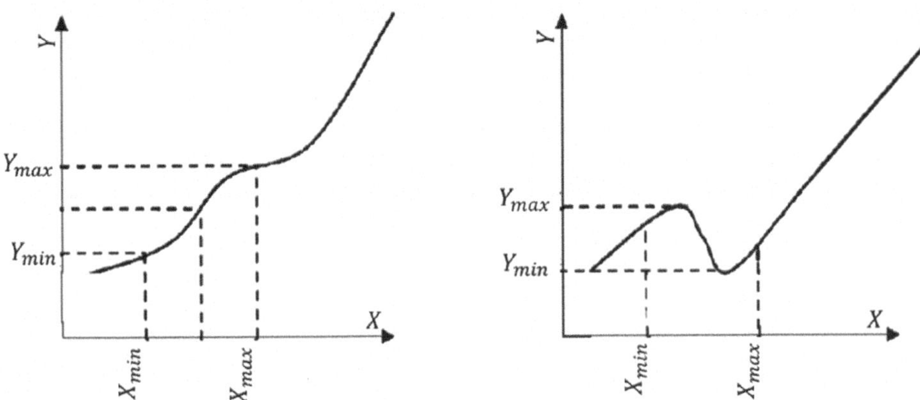

Figure 7.9 Monotonic and non-monotonic function (Branisavljevic & Ivetic, 2006).

q_i^c to be increased by $+\Delta q_i$ and update its value in the hydraulic model. Then run the model and let the corresponding system responses name as $\tilde{H}_{j,i}$. The sign $\sim$ indicates that the response has uncertainty. In fact, $\tilde{H}_{j,i}$ means the uncertain value for the jth system dependent response corresponding to a small positive change in the ith parameter. Repeat the same procedure for i:1 to np and $j=1$: nr where np and nr are respectively the number of network parameters and responses. The final result of this step is a $np \times nr$ matrix called the Impact Table.

(3) If $\left(\tilde{H}_{j,i} - H_j^c\right) > 0$, it means that $+\Delta q_i$ has led to an increase in H_j^c. Hence, $\bar{H}_j$ would take place when the ith parameter value takes its maximum positive uncertainty (i.e. $q_i^c + \Delta q_i$) while $\underline{H}_j$ occurs in $q_i^c - \Delta q_i$. On the contrary, for $\left(\tilde{H}_{j,i} - H_j^c\right) < 0$, $+\Delta q_i$ has caused a decrease in H_j^c. As a result, for the ith parameter, $\bar{H}_j$ and $\underline{H}_j$ would occur in $\left(q_i^c - \Delta q_i\right)$ and $\left(q_i^c + \Delta q_i\right)$, respectively. The same procedure must be done for $i=1$:np and $j=1$:nr to identify the true combinations of the parameters' extreme to find each response upper and lower bounds.

(4) By accepting the principle of superposition, the value of $\bar{H}_j$ and $\underline{H}_j$ could be explicitly calculated by introducing the true combination of extremes corresponding to $\bar{H}_j$ and $\underline{H}_j$ into the model and executing the model for the two states. If there are nr demand nodes in the network for which IA of pressure heads is going to be done, then the total number of hydraulic model runs for this step is $2 \times nr$.

Gupta and Bhave (2007) developed the ITM for UA of nodal pressures and velocities (or flow rate) in pipes as system responses against system parameters (nodal demands and roughness of pipes). Sabzkouhi and Haghighi (2018b) demonstrated that ITM is only effective for UA of nodal pressures. In fact, according to the nonlinear equations governing water distribution network hydraulics, for small uncertainties in system parameters, the nonlinear behavior of nodal pressures is almost monotonic while the variation of flow rate in pipes is non-monotonic.

7.7 EXAMPLE 7.2: INTERVAL ANALYSIS FOR THE ATM NETWORK

In this section, we are going do a UA for the ATM network by using the IA approach and accepting the ITM. Considering the deterministic information of the network according to Table 7.1, we want to obtain the interval of uncertain nodal pressure heads for a maximum uncertainty of $\pm 15\%$ in the nodal demands and HW roughness of pipes.

Note that due to existence of the storage tanks, the nodal pressures at a specific time step are affected not only by nodal demands in that time step, but are also exposed to variation in the network operating conditions at other time steps. Therefore, in order to accurately determine the interval of uncertain pressure heads, nodal demands of all time steps should be considered as parameters with uncertainty. However, since the procedure is the same, to reduce the calculations and the results to be shown, by removing the tanks from the network, we perform the analysis only for the maximum demand time step of the day (time step 12) as a sample.

7.7.1 Producing the impact table

Taking into account $\pm 15\%$ changes in nodal demands for time step 12 and the pipe roughness values mentioned in Table 7.1, with an iterative process using the 4-step procedure of the ITM illustrated in the previous section, the Impact Table was calculated. Since in the matrix of the Impact Table only the positive and negative sign of the elements (not the numeric values) are of importance, we respectively set 1 for the elements with a positive sign and –1 for the elements with a negative sign in the matrix as shown in Table 7.4. The zero-valued elements belong to pressure heads having no sensitivity with neither positive nor negative change in the corresponding parameter. Hence, for such pressure heads there is no matter which extreme of the parameter is being introduced to calculate the extreme response. In such a case, it is also possible to introduce the crisp value of the parameter to calculate the extreme pressure heads with no-sensitivity.

7.7.2 Calculating the extreme pressure heads

As explained in the last step of the ITM algorithm, to calculate the upper and lower limit of nodal pressures heads, considering a specific column of the Impact Table matrix, the appropriate upper or lower limit of each parameter must be selected. If the matrix element opposite the parameter row is 1, it means that the upper limit of the nodal pressure of the corresponding column occurs at the upper limit of that parameter, and if the element is –1, the upper limit of the node pressure is obtained at the lower limit of the parameter. As an example, according to Table 7.4, for the upper bound of pressure head in Node 30, the upper limit of roughness for Pipe 4, the lower for Pipe 30, the lower for Pipe 16, and so on must be set in the hydraulic model and the model should be once executed to calculate the numerical value of the upper pressure head of Node 30. Table 7.5 presents the value of all parameters to calculate the upper and lower bounds of pressure head in Node 30.

The results of possible intervals of pressure head variation based on UA done by ITM are illustrated in Table 7.6 and Figure 7.10. As Table 7.6 depicts, $\pm 10\%$ uncertainty in the ATM network parameters may differently affect the pressure variability in the network junctions where in Node 20 a maximum variability of $\pm 1.76\%$ is expected for pressure variation while as the most affected junction, Node 170 may experience pressure head uncertainty from -11.82 to 8.09%.

Although it was reasonably expected that due to uncertainty of the network parameters nodes that are closer to the pumping station would experience relatively less uncertainty than distant nodes in terms of pressure head variation, the UA quantified this variability for different network nodes, which could be a valuable criterion to identify the network less reliable regions for further mitigation and rehabilitation measures.

7.8 CONCLUSION

In this chapter, we reviewed the basic concepts of calibration and parameter uncertainty analysis in WDNs in simple procedures and methods. Parameter calibration is one of the most important prerequisites for using hydraulic models of WDNs for design, development and rehabilitation purposes. Typically, WDN's parameters are exposed to different types of inherent uncertainty, measurement uncertainty, and so on. Uncertainty in network modeling parameters causes variability in network responses. Quantifying uncertain network responses provides a suitable measure for planners and operators to understand the network weaknesses. Large uncertainties may result in violation of standards and minimum acceptable performance which may lead to system failure in

Table 7.4 Impact table for uncertainty analysis of nodal pressures in Example 7.2

Parameters — Roughness of Pipe #

Pipe #	Pressure Head as Response at Node #

Pipe #	20	30	110	70	60	90	100	40	50	80	150	140	170	130	160	120	55	75	115
4	0	1	1	1	1	1	1	1	1	1	1	1	1	1	1	1	1	1	1
30	0	-1	1	1	1	1	1	1	1	1	1	1	1	1	1	1	1	1	1
16	0	-1	1	1	-1	1	1	1	1	1	1	1	1	1	1	1	1	1	1
14	0	1	1	-1	1	1	1	1	1	1	1	1	1	1	1	1	1	1	1
12	0	1	1	-1	1	1	1	1	1	1	1	1	1	1	1	1	1	1	1
2	0	1	1	1	1	1	1	1	1	1	1	1	1	1	1	1	1	1	1
6	0	1	1	1	1	1	1	1	1	1	1	1	1	1	1	1	1	1	1
48	0	1	-1	1	1	1	1	1	1	1	1	1	1	1	1	-1	1	1	1
24	0	1	-1	-1	1	1	-1	1	1	1	1	1	-1	-1	-1	-1	1	1	1
10	0	1	1	-1	1	1	1	1	1	1	1	1	1	1	1	1	1	1	1
32	0	-1	1	1	1	1	1	1	1	1	1	1	1	1	1	1	1	1	1
36	0	1	-1	-1	-1	-1	-1	1	-1	-1	-1	-1	-1	-1	-1	-1	-1	-1	-1
38	0	-1	1	1	1	1	1	-1	-1	1	1	1	1	1	1	1	-1	-1	1
18	0	-1	1	1	-1	1	1	1	1	1	1	1	1	1	1	1	1	1	1
20	0	-1	-1	1	-1	1	1	-1	-1	-1	-1	-1	-1	-1	-1	-1	-1	-1	-1
66	0	-1	1	1	1	1	1	-1	-1	1	1	1	1	1	1	1	-1	1	1
40	0	-1	1	-1	-1	-1	1	1	1	-1	-1	1	1	1	1	1	1	1	1
28	0	-1	1	1	-1	-1	1	-1	-1	-1	1	1	1	1	1	1	1	1	1
22	0	0	-1	1	1	1	1	-1	-1	-1	-1	-1	-1	-1	-1	-1	-1	-1	-1
26	0	1	1	-1	1	1	-1	1	1	1	1	1	1	1	1	1	1	1	1
42	0	-1	1	-1	-1	-1	-1	1	1	-1	-1	1	1	1	1	1	1	1	1
64	0	-1	1	-1	-1	-1	-1	-1	-1	-1	-1	-1	1	1	1	1	-1	-1	-1
60	0	1	-1	1	1	1	1	1	1	1	1	1	1	-1	-1	-1	1	1	1
58	0	-1	1	-1	-1	-1	-1	-1	-1	-1	-1	-1	1	1	-1	1	-1	-1	-1
44	0	-1	1	-1	-1	-1	-1	1	1	-1	-1	1	1	1	1	1	1	1	1
50	0	1	-1	1	1	1	1	1	1	1	1	1	1	1	1	1	1	1	1
52	0	1	-1	1	1	1	1	1	1	1	1	1	1	1	1	1	1	1	1
56	0	1	-1	1	1	1	1	1	1	1	1	1	1	1	1	-1	1	1	1
62	0	-1	1	-1	-1	-1	1	-1	-1	-1	-1	-1	1	1	1	1	-1	-1	-1
46	0	1	1	-1	-1	-1	-1	1	1	1	1	1	1	1	1	1	1	1	1

Demand at Node #																		
34	0	-1	1	1	1	1	1	1	1	1	1	1	1	1	1	1	1	1
8	0	-1	1	1	1	1	1	-1	-1	1	1	1	1	1	1	1	1	1
74	0	-1	1	1	1	1	1	-1	-1	1	1	1	1	1	1	1	1	1
76	0	-1	1	1	1	1	1	-1	-1	1	1	1	1	1	1	-1	1	1
72	0	-1	1	1	1	1	1	-1	-1	1	1	1	1	1	1	-1	-1	1
68	0	1	-1	-1	-1	-1	-1	1	1	-1	-1	-1	-1	-1	-1	1	1	1
70	0	1	-1	-1	-1	-1	-1	1	1	-1	-1	-1	-1	-1	-1	1	1	1
54	0	1	-1	1	1	1	1	1	1	1	1	1	-1	1	-1	1	1	1
20	-1	-1	-1	-1	-1	-1	-1	-1	-1	-1	-1	-1	-1	-1	-1	-1	-1	-1
30	-1	-1	-1	-1	-1	-1	-1	-1	-1	-1	-1	-1	-1	-1	-1	-1	-1	-1
110	-1	-1	-1	-1	-1	-1	-1	-1	-1	-1	-1	-1	-1	-1	-1	-1	-1	-1
70	-1	-1	-1	-1	-1	-1	-1	-1	-1	-1	-1	-1	-1	-1	-1	-1	-1	-1
60	-1	-1	-1	-1	-1	-1	-1	-1	-1	-1	-1	-1	-1	-1	-1	-1	-1	-1
90	-1	-1	-1	-1	-1	-1	-1	-1	-1	-1	-1	-1	-1	-1	-1	-1	-1	-1
100	-1	-1	-1	-1	-1	-1	-1	-1	-1	-1	-1	-1	-1	-1	-1	-1	-1	-1
40	-1	-1	-1	-1	-1	-1	-1	-1	-1	-1	-1	-1	-1	-1	-1	-1	-1	-1
50	-1	-1	-1	-1	-1	-1	-1	-1	-1	-1	-1	-1	-1	-1	-1	-1	-1	-1
80	-1	-1	-1	-1	-1	-1	-1	-1	-1	-1	-1	-1	-1	-1	-1	-1	-1	-1
150	-1	-1	-1	-1	-1	-1	-1	-1	-1	-1	-1	-1	-1	-1	-1	-1	-1	-1
140	-1	-1	-1	-1	-1	-1	-1	-1	-1	-1	-1	-1	-1	-1	-1	-1	-1	-1
170	-1	-1	-1	-1	-1	-1	-1	-1	-1	-1	-1	-1	-1	-1	-1	-1	-1	-1
130	-1	-1	-1	-1	-1	-1	-1	-1	-1	-1	-1	-1	-1	-1	-1	-1	-1	-1
160	-1	-1	-1	-1	-1	-1	-1	-1	-1	-1	-1	-1	-1	-1	-1	-1	-1	-1
120	-1	-1	-1	-1	-1	-1	-1	-1	-1	-1	-1	-1	-1	-1	-1	-1	-1	-1
55	-1	-1	-1	-1	-1	-1	-1	-1	-1	-1	-1	-1	-1	-1	-1	-1	-1	-1
75	-1	-1	-1	-1	-1	-1	-1	-1	-1	-1	-1	-1	-1	-1	-1	-1	-1	-1
115	-1	-1	-1	-1	-1	-1	-1	-1	-1	-1	-1	-1	-1	-1	-1	-1	-1	-1

Table 7.5 Critical combination of parameters for the upper and lower bounds of pressure head in node 30.

Pipe ID																		
4	**30**	**16**	**14**	**12**	**2**	**6**	**48**	**24**	**10**	**32**	**36**	**38**	**18**	**20**	**66**	**40**	**28**	**22**
Pipe roughness for $\bar{H}_{30}$																		
125.6	94.6	92.5	126.6	125.9	127.5	133.5	131.8	130.9	132.5	97.7	131.3	101.3	100.9	101.1	101.6	101.9	101.4	118.5
Pipe roughness for $\underline{H}_{30}$																		
92.8	128.0	125.1	93.6	93.1	94.3	98.7	97.4	96.7	97.9	132.1	97.1	137.1	136.5	136.7	137.4	137.9	137.2	118.5

Pipe ID																		
26	**42**	**64**	**60**	**58**	**44**	**50**	**52**	**56**	**62**	**46**	**34**	**8**	**74**	**76**	**72**	**68**	**70**	**54**
Pipe roughness for $\bar{H}_{30}$																		
139.0	102.9	105.1	142.5	105.7	106.3	145.5	145.0	144.7	109.5	151.1	111.4	110.6	115.4	115.0	113.6	153.3	154.9	155.4
Pipe roughness for $\underline{H}_{30}$																		
102.8	139.3	142.1	105.3	142.9	143.9	107.5	107.2	106.9	148.1	111.7	150.8	149.6	156.2	155.6	153.8	113.3	114.5	114.8

Node ID																		
20	**30**	**110**	**70**	**60**	**90**	**100**	**40**	**50**	**80**	**150**	**140**	**170**	**130**	**160**	**120**	**55**	**75**	**115**
Nodal demand for $\bar{H}_{30}$																		
34.9	13.9	34.9	34.9	34.9	69.7	34.9	13.9	13.9	34.9	13.9	13.9	13.9	13.9	55.8	13.9	7.0	7.0	7.0
Nodal demand for $\underline{H}_{30}$																		
47.2	18.9	47.2	47.2	47.2	94.3	47.2	18.9	18.9	47.2	18.9	18.9	18.9	18.9	75.5	18.9	9.4	9.4	9.4

Table 7.6 Result of IA for the upper and lower bounds of pressure head in Example 7.2

	Node ID																		
	20	30	70	60	110	100	50	40	150	80	140	160	120	90	55	75	115	130	170
$\bar{H}_j$	84.38	73.57	73.09	72.90	72.81	72.41	72.52	72.48	50.99	72.33	63.11	50.92	50.99	72.25	63.16	63.10	63.09	50.87	50.80
Crisp H_j	82.92	70.51	69.82	69.55	69.25	68.77	68.88	68.82	47.31	68.66	59.40	47.21	47.28	68.53	59.44	59.37	59.36	47.12	47.00
$\underline{H}_j$	81.46	66.15	65.15	64.67	64.07	63.48	63.57	63.47	41.96	63.31	54.00	41.81	41.87	63.11	54.01	53.95	53.93	41.64	41.44
Upper Pressure Variation (%)	1.76	4.34	4.68	4.82	5.14	5.30	5.29	5.32	7.77	5.36	6.24	7.86	7.87	5.43	6.26	6.27	6.28	7.96	8.09
Lower Pressure Variation (%)	−1.76	−6.18	−6.69	−7.01	−7.48	−7.69	−7.70	−7.77	−11.30	−7.79	−9.09	−11.44	−11.44	−7.92	−9.13	−9.14	−9.15	−11.62	−11.82

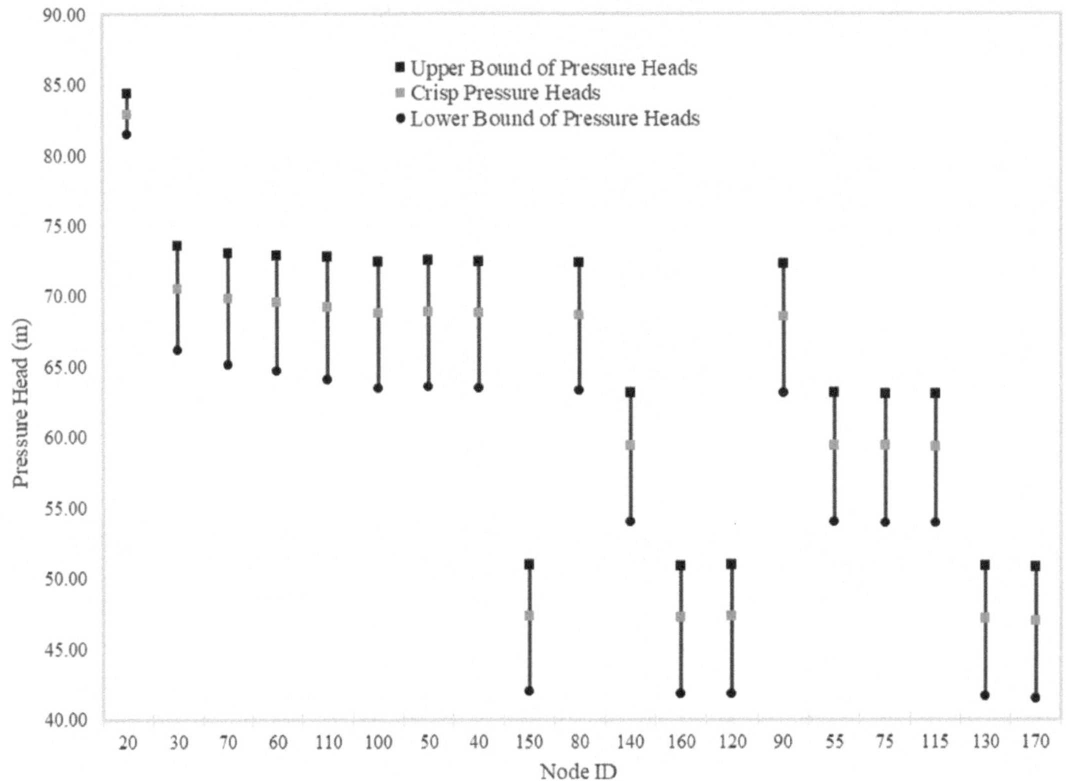

Figure 7.10 Interval for uncertain nodal pressures in Example 7.2.

terms of reliability indicators. In this regard, planners typically tend to know to what extent the system responses are exposed to uncertainties associated with information inputs the system models, so that they can more realistically promote a desirable level of system response.

What we expected in this chapter was to state the concept of uncertainty and simple methods to quantify it in applications. However, in higher level applications, indicators of uncertain network performance can be defined to handle the concepts of uncertainty in design procedures. In many cases, prudent decision makers are willing to make a more costly decision but have greater confidence that their design is more robust against uncertainties. Therefore, in recent years, development of multi-objective models to find solutions that, in addition to minimizing economic costs, have a higher resistance to network uncertainties, has been the interest of many designers and decision makers.

REFERENCES

Ahmadian S., Sabzkouhi A. M., Haghighi A. and Ranginkaman M. H.. (2019). Smart Pressure Management in Urban Water Distribution Networks for Firefighting. *Journal of Hydraulic Structures*, **5**(2), 71–89, https://dx.doi.org/10.22055/jhs.2019.30646.1119

Bao Y. and Mays L. W. (1990). Model for water distribution system reliability. *Journal of Hydraulic Engineering*, **116**(9), 1119–1137, https://doi.org/10.1061/(ASCE)0733-9429(1990)116:9(1119)

Branisavljevic N. and Ivetic M. (2006). Fuzzy approach in the uncertainty analysis of the water distribution network of Becej. *Civil Engineering and Environmental Systems*, **23**, 221–236, https://doi.org/10.1080/10286600600789425

Braun M., Piller O., Iollo A., Mortazavi I. and Deuerlein J. (2016). Uncertainty analysis toward confidence limits to hydraulic state predictions in water distribution networks. 8th International Congress on Environmental Modelling and Software, Toulouse, France.

Cimorelli L., D'Aniello A. and Cozzolino L. (2020). Boosting genetic algorithm performance in pump scheduling problems with a novel decision-variable representation. *Journal of Water Resources Planning and Management*, **146**(5), 04020023, https://doi.org/10.1061/(ASCE)WR.1943-5452.0001198

Costa L. H. M, de Athayde Prata B., Ramos H. and de Castro M. A. H. (2016). A branch-and-bound algorithm for optimal pump scheduling in water distribution networks. *Water Resources Management*, **30**(3), 1037–1052, https://doi.org/10.1007/s11269-015-1209-2

Creaco E., Franchini M. and Walski T. (2014). Accounting for phasing of construction within the design of water distribution networks. *Journal of Water Resources Planning and Management*, **140**(5), 598–606, https://doi.org/10.1061/(ASCE)WR.1943-5452.0000358

Do N. C., Simpson A. R., Deuerlein J. W. and Piller O. (2016). Calibration of water demand multipliers in water distribution systems using genetic algorithms. *Journal of Water Resources Planning and Management*, **142**(11), 04016044, https://doi.org/10.1061/(ASCE)WR.1943-5452.0000691

Duan H., Tung Y. and Ghidaoui M. (2010). Probabilistic analysis of transient design for water supply systems. *Journal of Water Resources Planning and Management*, **136**(6), 678–687, https://doi.org/10.1061/(ASCE)WR.1943-5452.0000074

Eliades D. G., Kyriakou M., Vrachimis S. and Polycarpou M. M. (2016). EPANET-MATLAB Toolkit: An open-source software for interfacing EPANET with MATLAB, *Proc. of 14th International Conference on Computing and Control for the Water Industry (CCWI)*, The Netherlands.

Ghidaoui M. (2010). Probabilistic analysis of transient design for water supply systems. Journal of Water Resources Planning and Management, **136**(6), 678–687, https://doi.org/10.1061/(ASCE)WR.1943-5452.0000074

Gupta R. and Bhave P. R. (2007). Fuzzy parameters in pipe network analysis. *Civil Engineering and Environmental Systems*, **24**(1), 33–54, https://doi.org/10.1080/10286600601024822

Haghighi A. and Asl A. Z. (2014). Uncertainty analysis of water supply networks using the fuzzy set theory and NSGA-II. *Engineering Applications of Artificial Intelligence*, **32**, 270–282, https://doi.org/10.1016/j.engappai.2014.02.010

Hwang H., Lansey K. and Jung D. (2017). Accuracy of first-order second-moment approximation for uncertainty analysis of water distribution systems. *Journal of Water Resources Planning and Management*, **144**(2), 04017087, https://doi.org/10.1061/(ASCE)WR.1943-5452.0000864

Kang D. and Lansey L. (2009). Real-time demand estimation and confidence limit analysis for water distribution systems. *Journal of Hydraulic Engineering*, **135**(10), 825–837, https://doi.org/10.1061/(ASCE)HY.1943-7900.0000086

Kang D. and Lansey K. (2011). Demand and roughness estimation in water distribution systems. *Journal of Water Resources Planning and Management*, **137**(1), 20–30, https://doi.org/10.1061/(ASCE)WR.1943-5452.0000086

Kapelan Z. S., Savic D. A. and Walters G. A. (2003). Multiobjective sampling design for water distribution model calibration. *Journal of Water Resources Planning and Management*, **129**(6), 466–479, https://doi.org/10.1061/(ASCE)0733-9496(2003)129:6(466)

Kapelan Z. S., Savic D. A. and Walters G. A. (2007). Calibration of water distribution hydraulic models using a Bayesian-type procedure. *Journal of Hydraulic Engineering*, **133**(8), 927–936, https://doi.org/10.1061/(ASCE)0733-9429(2007)133:8(927)

Lamont P. A. (1981). Common pipe flow formulas compared with the theory of roughness. *Journal – American Water Works Association*, **73**(5), 274–280, https://doi.org/10.1002/j.1551-8833.1981.tb04704.x

McEnroe B. M., Chase D. V. and Sharp W. W. (1989). Field Testing Water Mains to Determine Carrying Capacity. Miscellaneous Paper EL-89. U.S. Army Engineer Waterways Experiment Station, Vicksburg, Mississippi.

Minaei A., Sabzkouhi A. M., Haghighi A. and Creaco E. (2020). Developments in multi-objective dynamic optimization algorithm for design of water distribution mains. *Journal of Water Resources Management*, **34**(9), 2699–2716, https://doi.org/10.1007/s11269-020-02559-8

Ormsbee L. E. and Lingireddy S. (1997). Calibrating hydraulic network models. *Journal – American Water Works Association*, **89**(2), 42–50, https://doi.org/10.1002/j.1551-8833.1997.tb08177.x

Ranginkaman M. H., Ayati A. H., Bakhshipour A. E. and Haghighi A. (2019). Transient measurement site design in pipe networks using the decision table method (DTM). *Journal of Hydraulic Structures*, **5**(2), 32–48, https://doi.org/10.22055/jhs.2019.29402.1107

Rao S. S. and Berke L. (1997). Analysis of uncertain structural systems using interval analysis. *AIAA Journal*, **35**(4), 727–735, https://doi.org/10.2514/2.164

Rao Z. and Salomons E. (2007). Development of a real-time, near-optimal control process for water-distribution networks. *Journal of Hydroinformatics*, **9**(1), 25–37, https://doi.org/10.2166/hydro.2006.015

Sabzkouhi A. M. and Haghighi A. (2016). Uncertainty analysis of pipe-network hydraulics using a many-objective particle swarm optimization. *Journal of Hydraulic Engineering, ASCE*, **142**(9), 04016030, https://doi.org/1061/(ASCE)HY.1943-7900.0001148

Sabzkouhi A. M. and Haghighi A. (2018a). Closure to uncertainty analysis of pipe-network hydraulics using a many-objective particle swarm optimization. *Journal of Hydraulic Engineering*, **144**(4), 07018002, https://doi.org/10.1061/(ASCE)HY.1943-7900.0001148

Sabzkouhi A. M. and Haghighi A. (2018b). Uncertainty analysis of transient flow in water distribution networks. *Journal of Water Resources Management*, **32**(9), 1–18, https://doi.org/10.1007/s11269-018-2023-4

Sabzkouhi A. M. and Haghighi A. (2019). Analysis of hydraulic stress effects on the performance of water distribution networks using interval analysis and optimization approach. *Iranian Journal of Water and Wastewater*, **30**(3), 1–16 (in Persian), https://doi.org/10.22093/wwj.2018.97826.2487

Sabzkouhi A. M., Haghighi A. and Minaei A. (2017). Investigation of uncertainty effects on hydraulic performance of water distribution networks using the fuzzy sets theory. Proceedings of the 37th IAHR World Congress, IAHR and USAINS Holding SDN BHD, Kuala Lumpur, Malaysia, pp. 5276–5283.

Seifollahi-Aghmiuni S., Bozorg Haddad O. and Mariño M. A. (2013). Water distribution network risk analysis under simultaneous consumption and roughness uncertainties. *Water Resour Manage*, **27**(7), 2595–2610, https://doi.org/10.1007/s11269-013-0305-4

Sharp W. W. and Walski T. M. (1988). Predicting internal roughness in water mains. *Journal – American Water Works Association*, **80**(11), 34–40, https://doi.org/10.1002/j.1551-8833.1988.tb03132.x

Tsakiris G. and Splilotis M. (2017). Uncertainty in the analysis of urban water supply and distribution systems. *Journal of Hydroinformatics*, **19**(6), 823–837, https://doi.org/10.2166/hydro.2017.134

Tung Y. K. and Yen B. C. (2005). Hydro-systems Engineering Uncertainty Analysis. McGraw-Hill, New York.

Vítkovský J. P., Liggett J. A., Simpson A. R. and Lambert M. (2003). Optimal measurement site locations for inverse transient analysis in pipe networks. *Journal of Water Resources Planning and Management*, **129**(6), 480–492, https://doi.org/10.1061/(ASCE)0733-9496(2003)129:6(480)

Walski T. M., Chase D. V., Savic D. A., Grayman W., Beckwith S. and Koelle E. (2003). Advanced Water Distribution Modeling and Management. Haestad Institute Press, Connecticut.

Wood D. J. (1991). Comprehensive Computer Modeling of Pipe Distribution Networks. Civil Engineering Software Center, College of Engineering, University of Kentucky, Lexington, Kentucky.

Zanfei A., Menapace A., Santopietro S. and Righetti M. (2020). Calibration procedure for water distribution systems: comparison among hydraulic models. *Water*, **12**(5), 1421, https://doi.org/10.3390/w12051421

doi: 10.2166/9781789062380_0187

Chapter 8

Optimal pump operation

Adell Moradi Sabzkouhi[1*], Juneseok Lee[2] and Jonathan Keck[3]

[1]Department of Hydraulic Engineering, ASNRUKH, Mollasani, Iran
[2]Department of Civil and Environmental Engineering, Manhattan College, Riverdale, NY, USA
[3]Founder/Principal, Water First, LLC, Naperville, IL, USA
*Corresponding author: adellmoradi@asnrukh.ac.ir

LEARNING OBJECTIVES

At the end of this chapter, you will be able to:

(1) Review and analyze the pump characteristics and performance.
(2) Understand how to calculate flow-head characteristic curve of a pump at a new rotational speed.
(3) Know the basics for pump operation control.
(4) Understand the pump scheduling problem and how to use the simulation-optimization approach to solve it in constant and variable speed modes.
(5) Use necessary computer tools and applications to run and solve a pump scheduling problem.
(6) Understand how optimization tools can help to solve the conflict resolution of a simple Energy and Water Quality Management Systems.

8.1 INTRODUCTION

Pumps are the beating hearts of many civil and industrial projects around the world. Without pumps, proper performance of many civil infrastructures such as irrigation and drainage networks, water and wastewater treatment plants, sewer and storm water collection systems, urban and industrials water/oil/gas supply systems, and so on, could not be conceivable. Approximately 80% of municipal water processing and distribution costs relates to electrical energy, and up to 85% of this belongs to energy consumed by pumps (Güngör-Demirci *et al.*, 2020).

The structural, geometric and mechanical features of pumps are designed considering a variety of hydraulic performance expectations in operation. In addition to pump characteristics, true hydraulic performance of a pump in real-time operation also depends on features of the system within which the pump works. Although in the design stage of a pumping station taking variable demands would result in a more flexible system with more realistic insight into operation conditions, designers classically consider the most pessimistic data to size system's components. Operators, however, are generally more interested in managing the systems in a way that they have an optimum operational condition to achieve the best system performance (e.g. minimum energy consumption, improving water quality etc.).

Optimum operation could have different meanings based on objectives defined; for an aged water distribution system (WDS) suffering from a high rate of leakage, optimum system operation may be defined as maintaining pressure of the network as low as possible to minimize water loss, while meeting the minimum pressure requirements (Araujo *et al.*, 2006; Fontana *et al.*, 2018; García-Ávila *et al.*, 2019). For a network having a substantially high rate energy tariff over the peak water demand hours of the day, optimum system operation relates to setting the pumps schedule to have the minimum energy cost (Cimorelli *et al.*, 2020; Costa *et al.*, 2016; Martıinez *et al.*, 2007). Moreover, a multi-purpose approach may consider the optimum operation of network to be finding the trade-off among different conflicting objectives such as energy consumption and/or energy cost, and water quality measure (Cherchi *et al.*, 2015; Güngör-Demirci *et al.*, 2020).

Today, challenges with key resources including water shortage, limitations on energy and finance, environmental pollutions and other aspects of sustainable development have compelled decision-makers to inevitably adopt an integrated approach to make better informed decisions in practice. Hence, water companies should invest in novel **multi-objective approaches** such as Energy and Water Quality Management Systems (EWQMS) to better understand and efficiently resolve the problems, covering different concerns associated with available resources. For pumping systems operation, thinking in such an integrated way within the water-energy-environment nexus will provide and guarantee more environmentally friendly development projects with a desirable level of service to customers.

This chapter presents the framework and requirements for a water distribution network (WDN) modeling with optimal pump operations/scheduling. At the end of the chapter, an example of EWQMS is also provided, assuming that the reader has a sufficient background on water quality modeling in distribution pipe networks.

8.2 A BRIEF REVIEW ON PUMP PERFORMANCE

Centrifugal pumps are the most widely used pumps for fluid transmission and distribution in many industries, especially in WDSs (Van Zyl, 2014). Based on the direction of inflow to and outflow from the impeller, centrifugal pumps are divided into radial, axial and mixed flow types (Sanks *et al.*, 1998) among which, in drinking water supply, mainly radial flow pumps are used. The axial and mixed flow types are more common in wastewater and storm-water collection systems. Typically, radial flow impellers are more suitable for relatively low discharge and high-pressure head, whereas axial flow impellers are applied for high flow rate and low pressure head (Mackay, 2004).

For a given rotational speed N, the performance of a centrifugal pump is defined by four Characteristic Curves which relate the head (H) produced and the power (P) attracted by pump, expected pump efficiency (η) and the required Net Positive Suction Head NPSH$_r$, all as functions of the flow moving through the pump. In Figure 8.1, typical characteristic curves for a centrifugal pump with radial flow is depicted. Due to widespread use of radial flow centrifugal pumps in urban water systems, throughout this chapter, radial flow centrifugal pump is the only intended case herein, and is generally called *pump*, for brevity.

8.2.1 Head-flow characteristics

The *Head – Flow* or $H – Q$ characteristic curve is a function indicating the energy added by pump to the fluid, called *Total Dynamic Head – TDH* (approximately equals the differential pressure between the discharge and suction of the pump), against the flow. According to Figure 8.1a, for a centrifugal pump, $H – Q$ is a descending curve starting at the far left with $Q = 0$ and $H = H_{sh}$ where H_{sh} is the *shut-off head*, and ending at the far right with maximum Q with the lowest H, called *run-out point* (not shown). Note that as the flow rate becomes larger, the head values (i.e. the elevation that pump can supply) becomes smaller.

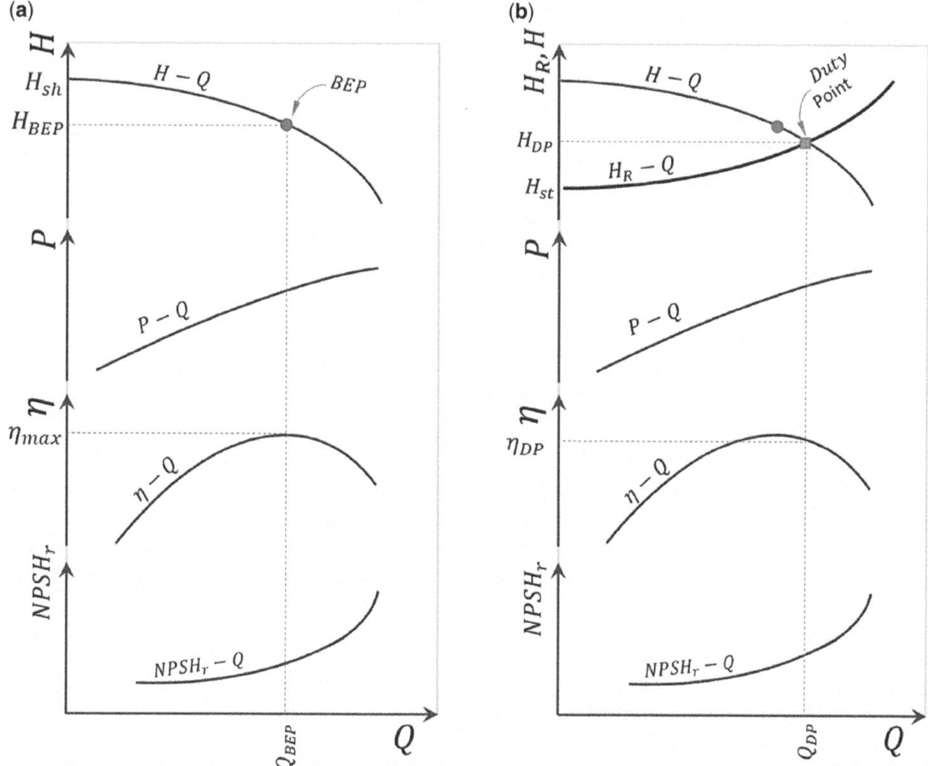

Figure 8.1 Characteristic curves for a general centrifugal pump (a) and its performance against the system curve (b).

8.2.2 Power-flow characteristics

The *Power – Flow* or $P - Q$ curve, for radial centrifugal pumps, is an ascending curve. The power in this curve refers to Brake Horse Power (P_{BH}) which means the power delivered by the electric motor to the pump or the power attracted by the pump. Since a part of P_{BH} is dissipated in the pump, the net power given to the water, called Water Horse Power P_{WH}, is less than P_{BH}. The relationships to calculate the power are as follows:

$$P_{WH} = \gamma Q H \tag{8.1}$$

$$P_{BH} = \frac{P_{WH}}{\eta} \tag{8.2}$$

where γ and η, respectively indicate the fluid specific gravity [N/m^3] and the pump efficiency.

8.2.3 Efficiency-flow characteristics

According to Equation (8.2), pump efficiency η, is defined as the ratio of water horse power to brake horse power. As shown in Figure 8.1a, the variation of η against Q follows a parabolic trend. Such variation may be also presented through ISO-Efficiency lines plotted on the $H - Q$ curve. The point corresponding to the η_{max} (see Figure 8.1a) on the $H - Q$ curve is called Best Efficiency Point (**BEP**) where the pumping flow is Q_{BEP}. Pump operation on **BEP**, in addition to optimum energy consumed by the pump, would guarantee the minimum wear and failure of the pump mechanical components (Sanks *et al.*, 1998).

8.2.4 NPSH-flow characteristics

Due to the high velocity of fluid particles inside the pump impeller, the impeller surface is prone to cavitation where relatively low pressures (below the vapor pressure) create cavitation bubbles and then high pressure in the immediate area on impeller surface force the bubbles to collapse. Such implosion can form very strong micro-jets which easily induce hundreds of *psi* on the impeller surface that make cavities on it. The required Net Positive Suction Head (NPSH$_r$) is a measure which indicates that for the flow entering the impeller with known discharge Q at temperature T, how much minimum pressure head (with respect to the vapor pressure P_v/γ) is needed to prevent the cavitation. In fact, to avoid cavitation we need the available Net Positive Suction Head (NPSH$_a$) to be equal to or greater than NPSH$_r$. In most centrifugal pumps as the discharge increases, the required net positive suction head (NPSH$_r$) goes up with increasing slope of variation for higher discharges (see Figure 8.1a). In civil engineering applications, the risk of cavitation is higher for pumps installed above water level.

8.2.5 System's curve and pump duty-point

For a known pipeline system, the System Curve or $H_R - Q$ indicates the static resistance or lift (the difference between energy level at both end boundaries) plus the dynamic resistance (energy losses) mathematically expressed as a function of discharge passing through the pipe system as follows:

$$H_R = H_{\text{st}} + H_{\text{dy}} \Rightarrow H_R = H_{\text{st}} + f(Q) \tag{8.3}$$

where H_R is the system total resistance, H_{st} is the static lift (static total head) and $f(Q)$ represents the function of friction and minor losses as a variable dependent on discharge Q. Taking the static lift fixed and the technical-operational features of the piping system (e.g. pipe diameters and status, valve openings and fittings status etc.) remain unchanged, H_R only depends on Q. Hence, similar to the pump $H - Q$ curve, $H_R - Q$ could be plotted on pressure-head/loss versus discharge coordinate plane, covering a wide range of flow rate and total resistance H_R. Assuming the Hazen–William's formula for calculating the friction losses, for a known state $S1$ of the piping system, the function $f(Q)$ may be written as (see Figure 8.1b):

$$f_1(Q) = \alpha_1 Q^{1.852} + \beta_1 Q^2 \tag{8.4}$$

where α_1 is the resistance factor for friction loss which depends on the length, diameter and Hazen–William's coefficient of the pipe, and β_1 is the minor losses factor related to valves and fittings available on the pipeline, all correspond to the state $S1$ of the piping system.

In addition to pump characteristics, the performance of a pump relies upon the system in which the pump is installed. To put it simply, while a pump with $H - Q$ curve and a pipe system having $H_R - Q$ resistance curve are connected within a pipe-pump system, according to the principle of continuity, the same discharge must flow through both segments. Additionally, the head produced by pump must compensate for the head-loss caused by system resistance (i.e. $H_R = H$). Hence, to determine the true work point of the pump and the system, called Duty Point (*DP*), it is required to intersect the $H - Q$ and $H_R = H$ curves (see Figure 8.1b). The optimal pump operation, in terms of both energy efficiency and maintenance cost, is achieved when *DP* overlies *BEP*.

As the resistance of a pipe system generally deteriorates (increases) by aging, it could be imaginable that another state $S2$ of the system in which the resistance factors for friction and minor losses have increased with α_2 and β_2 making the dynamic resistance as $f_2(Q) = \alpha_2 Q^{1.852} + \beta_2 Q^2$. This would introduce a steeper slope to the system curve causing the duty point to move to a new coordinate with a lower discharge and a higher head-loss. Using a throttling control valve, such a change in duty point may purposely be done by the operator to control the system flow and/or pressure on a desired point.

8.2.6 Affinity laws for rotational speed

Based on the Affinity Laws for rotational speed, for a centrifugal pump having speed $N1$, there are the following relationships between new speed N and the pump characteristics:

$$\frac{Q}{Q_1} = \frac{N}{N_1} \tag{8.5}$$

$$\frac{H}{H_1} = \left(\frac{N}{N_1}\right)^2 \tag{8.6}$$

$$\frac{(P_{BH})}{(P_{BH})_1} = \left(\frac{N}{N_1}\right)^3 \tag{8.7}$$

where Q_1, H_1 and $(P_{BH})_1$ are respectively the discharge, head, and brake horse power of the pump corresponding to the rotational speed N_1 (i.e. full speed), and Q, H and (P_{BH}) are the same parameters at the new rotational speed N.

According to Equations (8.5) and (8.6), for a known (Q_1, H_1) on the full speed $H - Q$ curve, the locus of discharge versus head for different reduced speeds follows a parabola function passing through the origin. If $(N \geq 0.6N_1)$, the error of Affinity Laws equations above, would rarely exceed 2 or 3% (Sanks *et al.*, 1998). Hence, in practical application, changes in pump efficiency along the parabola (expressed by Equations (8.5) and (8.6)) could be negligible. It should be noted that the *Affinity equations are used to produce the $H - Q$ curves for different speed*. However, the pump duty point follows the $H_R - Q$ curve which, compared to the parabola achieved by the Affinity Laws, has different slope and crossing point on the H_R-axis. This tip must be considered in determining the true efficiency of pump duty point for the reduced speeds which is critical in variable speed pump (VSP) operation (see Example 8.1).

8.2.6.1 Example 8.1

For a centrifugal pump the $H - Q$ and $\eta - Q$ characteristics at full rotational speed $N_1 = 1500$ rpm are available according to columns C1–C3 in Table 8.1. The pump feeds a pipeline having a system curve as $HR_{[m]} = 25 + 7 \times 10^{-5} \left(Q_{[m^3/hr]}\right)^2$, so that the duty point, based on Figure 8.2, is $DP(775.8, 67.4)$.

(A) For the case in which the pump speed would be reduced from $N = N_1$ to $N = 0.7\,N_1$, determine the $H - Q$ characteristic curves. (B) Find the duty point of the pump for $N = 0.8\,N_1$ and (C) Determine the desired speed for the system operating condition at which $Q = 653[m^3/hr]$ and $H = 55.4$ m.

Table 8.1 $H - Q$ data of the pump for full and reduced speed, and the corresponding points for the system resistance for Example 8.1 – Q and H are respectively in m³/hr and m.

C1	C2	C3	C4	C5	C6	C7	C8	C9	C10	C11	C12	C13	C14	C15	C16	C17
$N=N1=1500$ rpm			$N=0.95N1$		$N=0.9N1$		$N=0.85N1$		$N=0.8N1$		$N=0.75N1$		$N=0.7N1$		H_R-Q	
Q	H	$\eta\%$	Q	H	Q	H	Q	H	Q	H	Q	H	Q	H	Q	H_R
1000	51	77	950	46.0	900	41.3	850	36.8	800	32.6	750	28.7	700	25.0	1000	95.5
800	67	82	760	60.5	720	54.3	680	48.4	640	42.9	600	37.7	560	32.8	800	70.1
600	79.5	77	570	71.7	540	64.4	510	57.4	480	50.9	450	44.7	420	39.0	600	50.4
400	86	70	380	77.6	360	69.7	340	62.1	320	55.0	300	48.4	280	42.1	400	36.3
200	88.5	60	190	79.9	180	71.7	170	63.9	160	56.6	150	49.8	140	43.4	200	27.8
0	90	-	0	81.2	0	72.9	0	65.0	0	57.6	0	50.6	0	44.1	0	25.0

8.2.6.1.1 Solution A

For the pump discharges Q, and head H, (in column $C1$ and $C2$ of Table 8.1) using Equations (8.5) and (8.6), the $H-Q$ characteristic points are correspondingly obtained for a number of reduced speeds including: $N=0.95N_1$, $0.9N_1$, $0.85N_1$, $0.8N_1$, $0.75N_1$, $0.7N_1$; the results are listed in columns $C4$–$C15$ in Table 8.1. For instance, where $N=0.7N_1$, $Q_1=1000[\mathrm{m^3/hr}]$ and $H_1=51$ [m] the calculation will be as follows:

$$\frac{H}{H_1} = \left(\frac{N}{N_1}\right)^2 \Rightarrow H = 51\left(\frac{1050}{1500}\right)^2 \Rightarrow H = 25[\mathrm{m}]$$

$$\frac{Q}{Q_1} = \frac{N}{N_1} \Rightarrow Q = 1000\left(\frac{1050}{1500}\right) \Rightarrow Q = 700\left[\frac{\mathrm{m^3}}{\mathrm{hr}}\right]$$

The overall outcome was plotted on $H/H_R - Q$ coordinate plane in Figure 8.2 by joining the calculated points for each reduced speed separately. Consider that, in Table 8.1, for all points represented along each row from columns $C4$–$C15$ (i.e. the parabola passing through the origin) the pump efficiency remains fixed (according to Column $C3$).

8.2.6.1.2 Solution: B

The intersection point of $H-Q$ characteristic curve for $N=0.80N_1$ and the system curve in Figure 8.2 has been labeled as '(1)' whose coordinates are extracted by scaling as:

$$Q = 560\left[\frac{\mathrm{m^3}}{\mathrm{hr}}\right], \quad H = 47.3[\mathrm{m}]$$

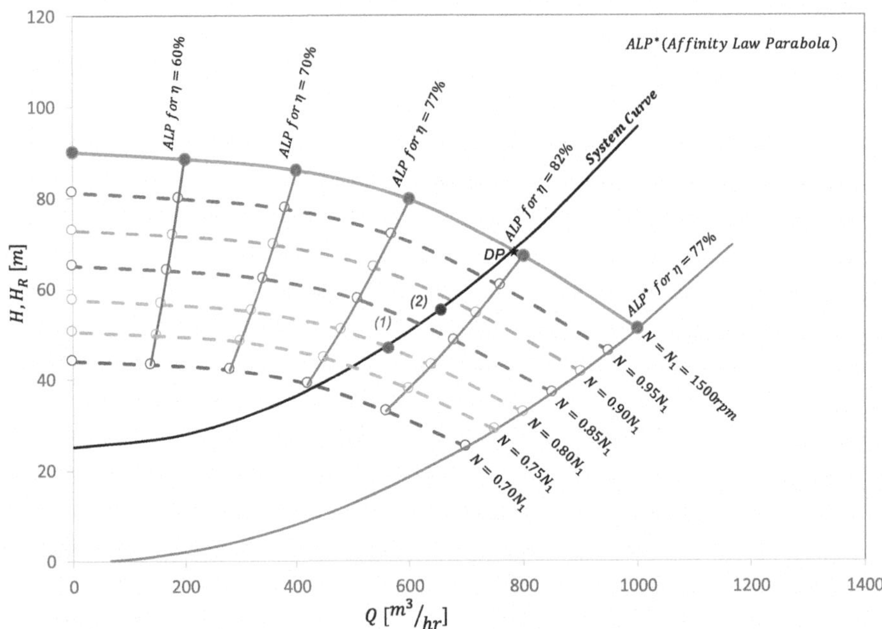

Figure 8.2 System curve and $H-Q$ characteristics curve of the pump at full and reduced speed for Example 8.1.

8.2.6.1.3 Solution: C
The new duty point (653, 55.4) labeled as '(2)' in Figure 8.2 has a corresponding unknown ISO-efficiency point on $H-Q$ curve for the full speed. To find the true coordinates of this point, simply assuming a generic linear function ($H_1 = b - c \times Q_1$) for the full speed $H-Q$ curve between (800,67) and (600, 79.5), solving a system of equation yields:

$$\begin{cases} 67 = b - c \times 800 \\ 79.5 = b - c \times 600 \end{cases} \Rightarrow b = 117, \quad c = 0.0625$$

$$\Rightarrow H_1 = 117 - 0.0625 \times Q_1$$

$$\frac{653}{Q_1} = N, \quad \frac{55.4}{H_1} = N^2 \Rightarrow \left(\frac{55.4}{N^2}\right) = 117 - 0.0625\left(\frac{653}{N}\right) \Rightarrow N = 0.884$$

8.3 MAIN CONSIDERATIONS FOR OPTIMAL PUMP OPERATION
To ensure the reliable pump operation in real conditions, a number of technical and operational considerations need to be taken into account. In this regard, the main considerations for pump scheduling are discussed below.

8.3.1 *BEP* and minimum efficiency
As long as the pump duty point exactly matches the BEP, the system works in the optimum operation condition in terms of lower energy consumption and maintenance cost. However, as a result of time-varying network demands, it is rarely possible to have a condition in which duty point exactly lies on *BEP*. Hence, the closer the duty point is to *BEP*, the better the operating conditions yields. To incorporate such consideration in practice, pump operation could be scheduled so that the pump efficiency would be limited to a minimum acceptable value η_{min}.

8.3.2 Pump discharge range
Both the minimum and maximum allowable discharge play a major role in long term successful mechanical performance of a pump, especially in issues related to different types of cavitation phenomenon (Brennen, 1994). Therefore, pump discharge preferably does not go beyond an upper and lower limit to provide a safe operating condition.

8.3.3 Pump speed
For VSP operation, it is highly recommended that the pump rotational speed should not exceed beyond a minimum and maximum threshold. This is mainly associated with cavitation, centrifugal stresses at the impeller, and issues linked to mechanical seal and bearing.

8.3.4 Pump switches and daily working hours
Increasing the cumulative working hours for a pump causes more wear and tear over time. While degradation of mechanical parts, especially the impeller, gradually increases, the $H-Q$ characteristic curve necessarily changes. The more working hours a pump has, the more degradation will occur, leading to less head produced by pump over time. These differences in the $H-Q$ curve of parallel pumps operating simultaneously results in poor performance at system pressures above the shut-off head of the weaker pump. From this point of view, it is necessary to schedule the system operation in such a way that there would be no significant difference between the cumulative working hours of identical parallel pumps in the long term. Parallel pumps also should have approximately the same number of switches for a specific duration.

8.4 PUMP OPERATION CONTROL

From both mechanical and hydraulic points of view, pump operation should be controlled in such a way that the duty point is as close to the BEP as possible. On the other hand, pump stations are generally designed and sized for the most pessimistic state regarding flow/head required by the system. Nonetheless, the time-varying nature of the system parameters, especially nodal demands, causes the pumps' duty points to regularly change during operation time. Hence, in such periodically changing conditions, pump stations need a strategy to control the duty point to increase pumping efficiency. To this end, there are mainly two approaches:

8.4.1 Change in system curve

In this approach, assuming the pump H–Q curve is fixed, by changing the status of the station or network elements (e.g. throttling valves, isolation valves, a by-pass line, etc.) the slope of system curve would intentionally alter in order to move the duty-point to a desired position (i.e. BEP). This approach mainly forces the system to have unnecessary energy dissipation in terms of excess pumping head and/or discharge. Today's strict criteria for reducing energy loss and carbon emission have led to this method being gradually abandoned over time.

8.4.2 Change in characteristic curve

Assuming no change in the system curve, this approach requires a change in the pump station characteristic curve. Two more popular approaches can be considered for this purpose. First, by keeping the rotational speed fixed at the full speed, the number of parallel working pumps changes in different time steps, and the pump station characteristic curve is regulated by On/Off pump switches based on the network requirements. This approach is called constant speed pumping (CSP). One of the main justifications for application of a parallel pumping system is to create a flexibility in operation to meet the time-varying network requirements in terms of flow and pressure. The second approach (VSP) is to let the pump rotational speed (and pump station characteristic curve as a result) vary at each time step, to adapt the desired speed for optimal pump operation. The VSP approach can save more energy as it has more flexibility than the CSP approach to match the station characteristic curve with the unchanged system curve. Hence, the problem of optimal pump operation, mostly called Pump Scheduling, is to find either the optimum number of On-switched pumps in the CSP approach, or the optimum relative rotational speed over different time steps, in the VSP approach. In other words, *in terms of hydraulic modeling, pump scheduling is defined as determining the optimum pattern for each pump.* In what follows, the simulation-optimization approach to solve the optimal pump scheduling problem is described.

8.5 OPTIMAL PUMP SCHEDULING

8.5.1 Simulation-optimization approach

Figure 8.3 depicts a general simulation-optimization framework to solve the implicit problems relevant to the design/operation of WDNs. The framework consists of two main cores: (1) a Simulation Model – EPANET (Rossman, 2000) as an example, and (2) an Optimization Engine (Genetic Algorithm or GA here) both receive 'Inputs' (the input arrows in Figure 8.3). The type of inputs to run the simulation model have known-values or predefined by the user, named out-of-control inputs (Figure 8.3). A part of these inputs belongs to known or calibrated data of the network (e.g. topology and configuration, system parameters, etc.) labeled as *Inp*1 in Figure 8.3, which is directly entered into the simulation model. Some other out-of-control inputs (*Inp*2) associated with the ranges of variables and acceptable performance expected from the network, are introduced to the optimization engine to evaluate the quality of solutions.

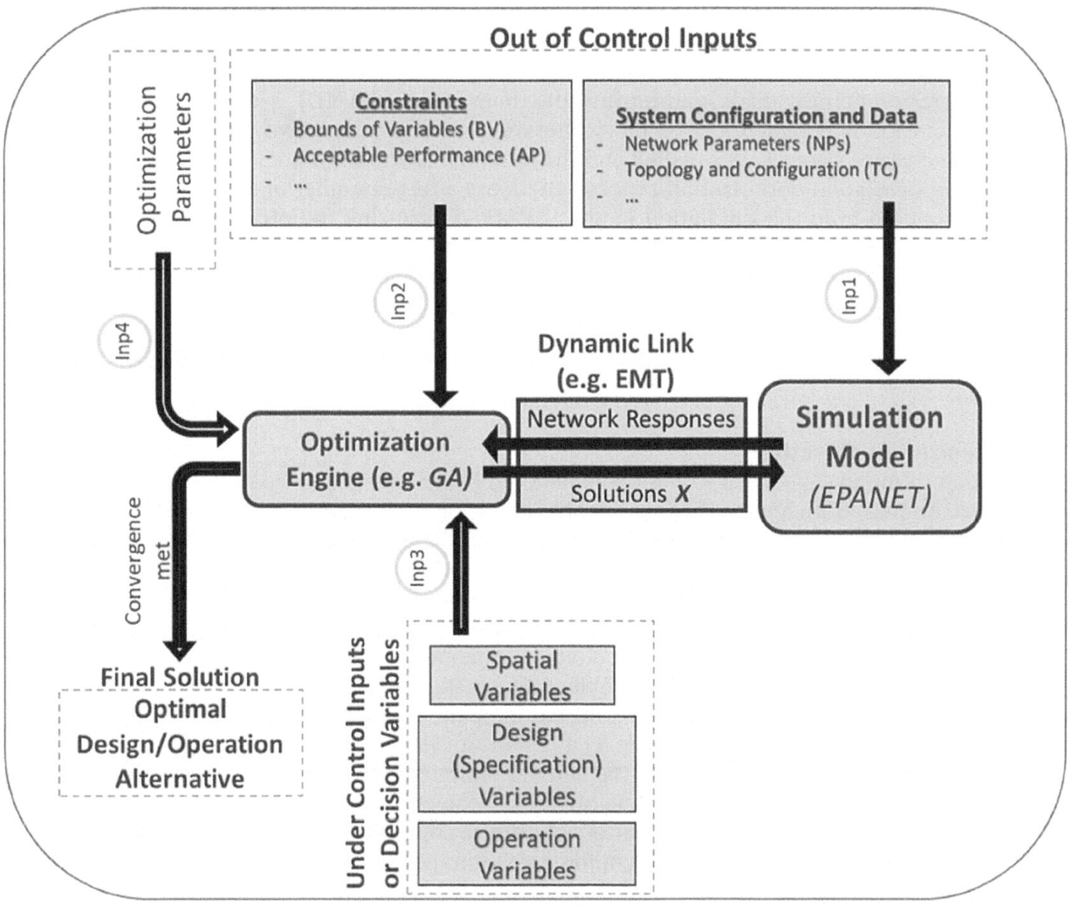

Figure 8.3 The conceptual framework of simulation-optimization model to solve the design/operation problems.

Another type of inputs (*Inp*3) called 'under control inputs' include unknowns referring to the problem's independent variables or decision variables (called *genes* in GA terminology). The GA has permission to change the value of decision variables during the optimization process. Decision variables are generally related to spatial and/or technical and/or operational features of the system elements (e.g. pipes, tanks, valves, pumping and chlorine stations, etc.). As previously stated, in pump scheduling problems, these unknowns are pump operations/patterns. An initialized set of decision variables creates a decision vector called a 'solution' or an 'alternative' or a 'chromosome' (in GA terminology). Subsequently, a series of alternatives or solutions forms the population. The aim of applying a *simulation-optimization model* is to find the optimum solution through a systematic search done by the optimization engine over successive generation of the GA.

Taking a number of tuning input parameters (*Inp*4) the GA first randomly initializes the decision variables of each individual to form the first generation of design/operation alternatives (called *initial population*). Then the alternatives are sent to the simulation model for evaluation. After solving the governing equations through the simulation model, dependent variables or the system responses (e.g. nodal pressures, power consumption, energy cost, water quality indices, etc.) return to the GA. Based on the feedback from evaluation of the solutions, and the comparison with constraints and acceptable performance (*Inp2*), the GA calculates the *objective function (or fitness function)* for each solution. Following this, by ranking the individuals and imposing the optimization operators, GA reproduces the next generation of individuals, and retransmits them to the EPANET. Then the same story of the previous generation is repeated for the new generation and the successive procedure will continue until the *convergence criteria* are satisfied and the *optimal solution* is achieved.

During the simulation-optimization process, there are a huge number of sending solutions to and receiving responses from the simulation model (EPANET), causing the process to be impossible to handle manually. Therefore, any simulation-optimization model requires a *dynamic link* to make the interaction of the simulation and optimization model automatic (see Figure 8.3). A widely popular dynamic link for WDN problems is the EPANET-MATLAB Toolkit (EMT) which is applied to solve the pump scheduling examples in this chapter later. Full details and examples about how EMT works can be found in Eliades *et al.* (2016). Additionally, a general overview of EMT is given in Chapter 7, section 7.4.2.

8.5.2 Optimization objectives

Optimization of pump scheduling could be considered based on various goals using single or multi-objective techniques. Both approaches are capable of taking performance constraints into account to achieve a minimum acceptable utility level of the network. While the final output of a single objective model is an optimum solution meeting all performance and non-performance constraints, the multi-objective approach provides a set of optimum solutions non-dominated (called *Pareto optimal front*) on the conflicting objectives coordinates plane. Moreover, it is possible to include a specific criterion as an objective in conflict with another objective (i.e. the multi-objective approach) or in the form of a constraint defined in a single-objective problem. For instance, to consider a water quality measure in scheduling for energy cost minimization, one could define the water quality measure as the second objective in a bi-objective programming, or as a constraint in a single objective problem to establish a minimum acceptable quality level. For the sake of simplicity, this chapter focuses on the second approach (i.e. constraint included in a single-objective problem).

In addition to pumping energy cost, a variety of criteria could also be considered as the objective of pump scheduling optimization. The most common criteria include pressure control and management (e.g. demand management, leakage and water loss management, crisis management), energy consumption management to minimize GHG emission, water quality management, and so on. Among the aforementioned measures, energy cost is frequently more attractive to water companies due to direct financial benefits resulting from less energy consumption over on-peak power demand time. In this chapter, energy cost is taken as the objective of pump scheduling.

The possibility to store water in storage tanks provides an ability to manage the water supply-demand in order to minimize energy cost during the day, such that the system averagely pumps water at a higher and lower flow rate over off-peak and on-peak electricity tariff pattern time steps, respectively. This basically means that when electricity cost is cheap (typically overnight), the pump is running and the pump is usually off when the electricity cost is higher.

8.5.3 Pump scheduling approaches: CSP vs. VSP

Upon the type of facilities and speed control systems available, pump scheduling could be carried out in different schemes. If pumps have no control system to regulate the pump rotational speed, the operation status of the *p*th pump in each time step *t*, $N_{p,t}$, would be either 0 (#*Off*) or 1 (#*On*). In On/Off pump switching (i.e. CSP approach), the figures 0 and 1 respectively indicate the relative rotational speed of the pump with regard to pump rated speed (or full speed). Variable speed pumps have a control system to regulate the rotational speed, $N_{p,t}$ not only on 0 or 1, but also on any continuous figure over the interval [0, 1]. Thanks to modern technologies making it possible to efficiently utilize VSP (e.g. variable frequency drives) it is attainable to cover a wide range of changing flow-head conditions required by the system. Nonetheless, due to some mechanical issues, inaccuracy related to affinity laws and the fact that the efficiency of VFDs notably decreases below $N = 0.6N_1$ the real working speed of VSPs is generally limited (Sanks *et al.*, 1998).

More complex cases of mixed CSP-VSP in a pumping station can also be considered by assuming a limited number of pumps equipped with VFDs, which may require a different representation of the problem in terms of programming, which will not be discussed here. It is also assumed here that the pumps of the pumping station are quite similar in terms of performance curves. However, in real operating conditions, for pumps with different lifespans, it is also possible to imagine cases in which, despite the same model of pumps, the performance curves would be different due to different depreciation of pumps over the past time, which needs the programming to be more complex and this is not going to be discussed here. One approach to model such a system is to simply assign a distinct decision variable to each different pump to find the best status of the pump over each time step.

The CSP approach has been traditionally implemented for many years in urban water supply pump stations having 24-hour varying demand. Notwithstanding the emergence of variable frequency drives technology with its many advantages, CSP is still being paid attention to in many cases, especially because of relatively high capital investment and the need for specialized training of operators.

In the following, the two approaches of CSP and VSP are described by two examples.

8.5.4 Example 8.2 – CSP approach

The Any-Town Modified (ATM) benchmark network has been chosen for implementing different pump scheduling approaches in this chapter. The data of ATM network was presented in Example 7.1 in Chapter 7. However, for the sake of simplicity, all roughness of pipes is taken as 120 in this example.

During a 24-hour period of time, the energy cost is generally calculated in three phases, including on-peak, mid and off-peak load demand, by multiplying a base energy price of 1 \$/Kw-hr and energy tariff multipliers illustrated in Table 7.1.

8.5.4.1 Objective function

In the CSP approach, the objective function to pump scheduling regarding energy cost could be defined as follows (Amirsardari *et al.*, 2021; Güngör-Demirci *et al.*, 2020):

$$\text{TEC} = \sum_{p=1}^{NPu} \sum_{t=1}^{T} \frac{\gamma Q_{p,t} H_{p,t}}{\eta_{p,t}} \Delta t_t \, ET_t \, x_{p,t} \tag{8.8}$$

where γ is specific gravity of water (N/m^3), TEC is total energy cost (\$) during time period T for NPu pumps. $Q_{p,t}$, $H_{p,t}$ and $\eta_{p,t}$ are respectively the flow (m³/s), total dynamic head (m) and efficiency of the pth pump at tth time step having Δt_t (hr) time interval along which the electricity tariff is ET_t[\$/ kW-hr]. $x_{p,t}$ is a binary variable representing the state of the pump p at time step t so that it equals 1(On) or 0(Off). $x_{p,t}$ is inserted as the multiplier of pump pattern in EPANET.

8.5.4.2 Decision variables and decision vector

In this example, based on the problem statement and the objective function defined, the decision variables for the CSP approach are the operating status of each pump p at each time step t, throughout the simulation period T. Given that system time-varying parameters (consumptions for instance) are periodically repeated in a 24-hour cycle, we consider the energy cost minimization within a total simulation period of $T = 24$ hours. Therefore, the generic form of the decision vector expressing the operating status of the three pumps consists of a binary string X of 24×3 digits as shown in Figure 8.4.

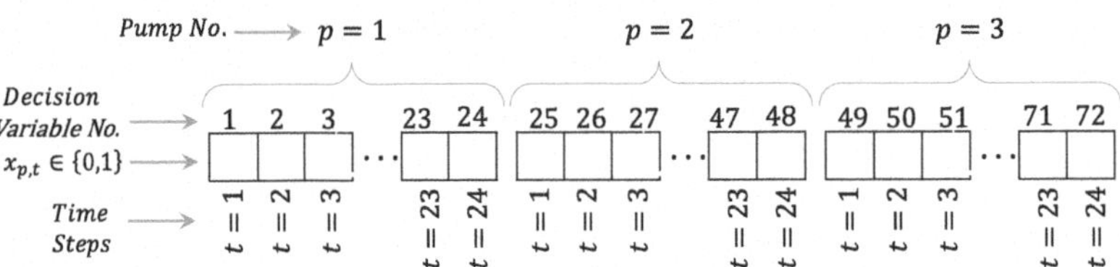

Figure 8.4 The generic form of decision vector representation in ATM example for the CSP approach.

8.5.4.3 Constraints

Constraints related to mass conservation in junctions and energy balance in closed loops are automatically handled through the simulation model, here EPANET. There are also a number of constraints for energy cost optimization in CSP scheduling problem including:

(i) The pressure constraint for each critical node j, which is mathematically stated as follows:

$$H_{j,t} \geq H_{j,t}^{\min} \quad \text{for} \quad t = 1, 2, \ldots, 24 \tag{8.9}$$

where $H_{j,t}$ and $H_{j,t}^{\min}$ are respectively the pressure head and the minimum pressure head required for the critical node j at time step t. In the ATM example the minimum pressure head required for critical nodes 55, 90 and 170 is respectively 42, 51 and 30 m. The total violation regarding the above set of constraints could be written as:

$$V_1 = \sum_{t=1}^{24} \sum_{j=1}^{NCN} \max\left(0, H_{j,t}^{\min} - H_{j,t}\right) \tag{8.10}$$

where $NCN = 3$ is the number of critical nodes.

(ii) The minimum and maximum storage tank water level constraints for each tank k, which is mathematically represented as:

$$WL_k^{\min} \leq WL_{k,t} \leq WL_{k,t}^{\max} \quad \text{for} \quad t = 1, 2, 3, \ldots, 24 \tag{8.11}$$

where $WL_k^{\min}$ and $WL_k^{\max}$ are respectively the allowable lowest and highest water levels of tank k, and $WL_{k,t}$ is the water level of tank k at time step t. In the ATM benchmark network, $WL_k^{\min}$ and $WL_k^{\max}$ respectively equal 66.53 and 71.53 m for all storage tanks. The total violation regarding the above set of constraints could be written as:

$$V_2 = \sum_{t=1}^{24} \sum_{k=1}^{NT} \left[\max\left(0, WL_k^{\min} - WL_{k,t}\right) + \max\left(0, WL_{k,t} - WL_k^{\max}\right) \right] \tag{8.12}$$

where $NT = 3$ is the number of storage tanks in the network.

(iii) The periodic repetition of initial water level in each tank k, which is mathematically represented as:

$$WL_{k,T} \geq WL_{k,0} \tag{8.13}$$

This set of constraints guarantees that the system will not encounter water deficit in terms of volume, and has sufficient energy to periodically repeat its hydraulic behavior during the next

24-hour cycles. In the ATM example, $WL_{k,0}$ is 66.93 m for all storage tanks, and the violation regarding the above set of constraints is calculated as:

$$V_3 = \sum_{k=1}^{NT} \max\left(0, WL_{k,0} - WL_{k,T}\right) \tag{8.14}$$

(iv) Violation related to the maximum number of switches ($NSW_{\max}$) for each pump p stated as follows:

$$NSW_p \leq NSW_{\max} \quad p = 1, 2 \text{ and } 3 \tag{8.15}$$

where NSW_p is the total number of pump p switches over the period T. This set of constraints avoids a high number of pump switches to provide better maintenance conditions in terms of mechanical issues. $NSW_{\max}$ for a 24-hour cycle is considered to be 3 in the ATM example. To calculate NSW_p the Heaviside Function is implemented as follows:

$$NSW_p = \sum_{t=2}^{24} \left[\left(x_{p,t} - x_{p,t-1}\right) \times Heaviside\left(x_{p,t} - x_{p,t-1}\right)\right] \quad p = 1, 2, \tag{8.16}$$

The output of Heaviside Function is 1 for positive arguments and 0 for negative arguments.

The violation of this set of constraints is now written as:

$$V4 = \sum_{p=1}^{NPu} \max\left(0, NSW_p - NSW_{\max}\right) \tag{8.17}$$

8.5.4.4 Constraint handling

Like many metaheuristic algorithms, GAs are unconstrained optimization engines, meaning that constraints cannot directly be imposed on the GA, and the violation of constraints should be indirectly introduced to the algorithm. In this regard, *penalty function* is a popular method to handle the constraints in GA. For this purpose, by scaling all types of violation through penalty factors, it is possible to sum different violations belonging to each constraint into a single positive-value violation, and to add it to the objective function which is going to be minimized. Hence, the objective function is then rewritten as follows:

$$\text{Min} \sum_{p=1}^{NPu} \sum_{t=1}^{T} \frac{\gamma Q_{p,t} H_{p,t}}{\eta_{p,t}} \Delta t_t \, ET_t \, x_{p,t} + \left(PF_1 \times V_1 + PF_2 \times V_2 + PF_3 \times V_3 + PF_4 \times V_4\right) \tag{8.18}$$

where PF_1, PF_2, PF_3 and PF_4 are scaling penalty factors related to violations, whose values should be determined by tuning. For ATM example, PF_1, PF_2, PF_3 and PF_4 were set at 500, 500, 500 and 10^5, respectively (Cimorelli *et al.*, 2020). Consider that in optimization problems, each constraint has a different degree of importance to the fitness function variation, indicating why the penalty factors should be different. To put it more clearly, different penalty factors proportion the effects of different constraints on the objective function so that by changing the decision variables, appropriate sensitivity causes the feasible search space to be more efficiently explored.

8.5.4.5 GA settings and execution

In this chapter, MATLAB version 2020a is used to perform the GA. Like many metaheuristic optimization algorithms, the GA also has a number of tuning parameters which need to be specified before the

200 — Embracing Analytics in the Drinking Water Industry

Table 8.2 The optimum pump operation schedule for CSP approach in ATM example.

Pump ID	Time steps																								TU (hr)	NS W_p	TEC ($)
	1	2	3	4	5	6	7	8	9	10	11	12	13	14	15	16	17	18	19	20	21	22	23	24			
1	1	1	1	1	1	1	1	1	0	0	1	1	1	1	1	1	1	1	0	0	0	1	1	0	18	2	3578.74
2	0	1	0	1	0	0	0	0	0	0	1	1	1	1	1	0	0	0	0	0	0	0	0	0	7	3	
3	0	0	0	0	0	0	0	0	0	0	0	0	0	0	0	0	1	0	0	0	0	1	0	0	2	2	

TU: total utilization; TEC: total energy cost; NSW: number of switches.

main execution of the algorithm. Parameters are generally adjusted either by a calibration process or by setting the values obtained by experience in similar projects. The parameters of GA in this example were considered according to Cimorelli *et al.* (2020) including: crossover probability $p_c = 0.9$, mutation probability $p_m = 0.025$, the population size of 200 and the maximum number of generations of 1000. Additionally, the values of the GA parameters not mentioned above are set to be equal to the default value considered in the MATLAB optimization toolbox. To reduce the effect of randomness of the GA operators on the optimal solution, the algorithm is run ten times and the best solution is reported.

It should be noted that in the *Optimtool* panel in *MATLAB*, the *Population Type* under the *Options→Population* must be set to *Bit string* to consider the decision variables as binary digits. A brief overview of *Optimtool* panel in *MATLAB* is explained in Chapter 7.

8.5.4.6 Results

Given the assumptions stated above, the best solution details for optimal pump operation to minimize the energy cost in CSP approach is reported in Table 8.2. According to the table, total energy cost for the best solution reaches to 3578.16 $ over a 24-hour period during which Pumps 1, 2 and 3 respectively experience 2, 3 and 2 starts, meaning that the set of constraints (iv) are completely satisfied. In addition, as seen in Figure 8.5, the best solution could provide adequate pressure head required for the nodes 55, 90 and 170 according to the set of constraints (i). Moreover, the limitations imposed by constraints (ii) and (iii) to control the water level at tanks are also met as depicted in Figure 8.6.

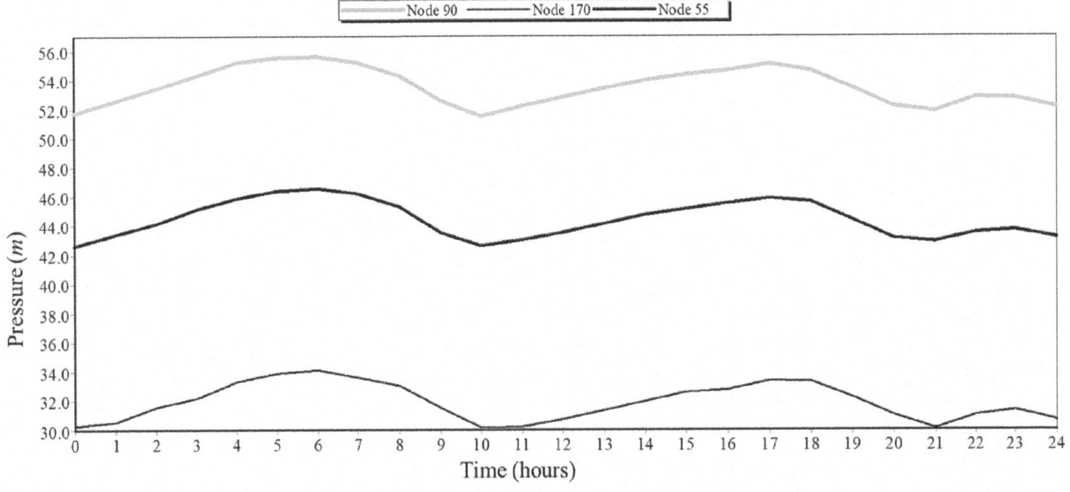

Figure 8.5 Pressure variation of the critical nodes for the best solution in ATM example for the CSP approach.

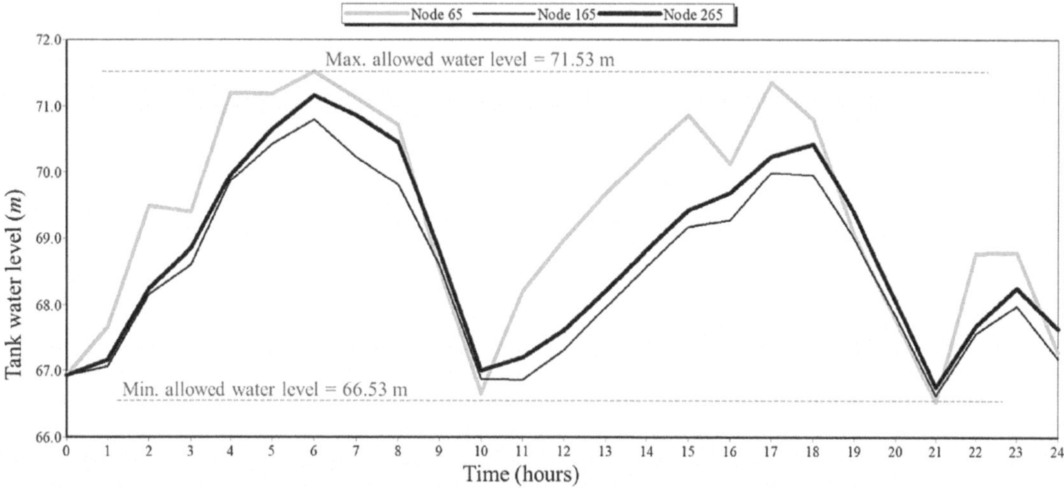

Figure 8.6 Water level variation at tanks for the best solution in ATM example for the CSP approach.

Given the logic behind the role of tanks in regulating water supply and demand, a closer simultaneous look at the pump optimum schedule (Table 8.2), the energy price tariff (Table 7.1) and the water level in tanks (Figure 8.6) reveals that the GA has tried to keep more pumps off during peak tariff hours and let the system use more water from storage in tanks. This is more evident in time steps 9, 10, 19, 20 and 21, while all three pumps are at OFF status, and the tanks release water. Obviously, this trend cannot be expected for all time steps due to the limited storage capacity available within the tanks.

8.5.4.7 Challenges and opportunities

In the above example, pump operation optimization was considered by using the pump OFF/ON status as the independent decision variables. Obviously, if operation optimization could have been combined with the network design phase, due to the possibility of having more numbers of independent variables (e.g. tank location and specifications, pipe and pump sizing, etc.) the ability of the integrated simulation-optimization model in reducing the energy cost will increase significantly. Clearly, this will require more detail and information to define the objective function in a different form in which other types of costs, including the capital investment and so on, are taken into account at the same time.

In terms of asset management and reduction of maintenance cost, operators would rather have a system of parallel pumps with approximately the same utilization time. Although this point has not been observed for the pump scheduling in Table 8.2, due to the same characteristic performance of the pumps, in the operation phase, the system can be planned in such a way that the role of each pump as Pumps 1, 2 and 3 in a periodically working program (e.g. weekly or monthly-basis) would change to balance the utilization of pumps.

As the demand pattern may vary from season to season, optimal pump operation may need updating. Additionally, due to the changes in the behavior of customers during the days of the week, especially on the weekends, it is more accurate to schedule the operation of the pumps on a weekly-basis instead of daily-basis as presented here. For the current example, in the weekly-basis scheduling each pump requires $(24 \times 7 = 168)$ digits (variables) to represent the pump pattern over the entire week, meaning that the problem totally needs $(168 \times 3 = 504)$ variables. Such a representation turns the case into a relatively *higher-dimensional problem* which increases the burden of time to run the simulation model. In such cases researchers highly tend to exploit dimension-reduction techniques to lessen the

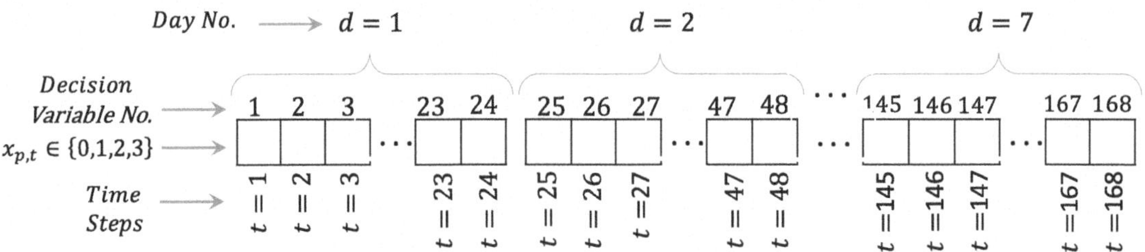

Figure 8.7 Generic form of the decision vector representation in Example 8.2 for a weekly-basis scheduling with integer programing.

simulation model run time. For instance, it is possible to formulate the weekly-basis scheduling of the ATM example in such a way that the decision vector has 168 digits, which are quantified by integer values of 0, 1, 2 or 3 representing the number of active pumps along each time step. In fact, compared to the Binary Programing, the aforementioned problem should be solved using an Integer Programing method. A generic representation of the decision vector for Integer Programing approach is shown in Figure 8.7.

8.5.5 Example 8.3 – VSP approach
In this example, the ATM problem is solved by assuming the variable frequency drives available for all three electro-pumps which make the pumps have variable speed over each time step.

8.5.5.1 Objective function
The objective function in VSP is almost similar to that of the CSP approach except that the relative rotational speed of the pump p (i.e. $x_{p,t}$) as the decision variable does not directly appear in the objective function but is indirectly concealed in $Q_{p,t}$ and $H_{p,t}$, which here are respectively the pump flow and TDH in the decreasing relative speed $x_{p,t}$. The objective function in this approach is stated as:

$$\text{TEC} = \sum_{p=1}^{NPu}\sum_{t=1}^{T} \frac{\gamma Q_{p,t} H_{p,t}}{\eta_{p,t}} \Delta t_t \, ET_t \tag{8.19}$$

Note that $\eta_{p,t}$ should also be calculated for the same parabola function from the affinity laws corresponding to the real duty point (see Example 8.1). This has not been included in EPANET, making the VSP modeling requires energy consumption calculation separately outside the EPANET environment.

8.5.5.2 Decision vector
The decision vector in this approach can be considered similar to the previous example as a string with 3×24 genes, except that the value of decision variables ($x_{p,t}$) are real numbers changing between 0 and 1, which are inserted in pump pattern and translated as pump relative rotational speed during each time step. On the other hand, in each time step, active parallel pumps should run at the same rotational speed. Hence, if the generic decision vector is defined as a 72-gene chromosome, there will be a systematic internal dependency between some genes, which makes the GA efforts not as efficient in searching the decision space. To solve this issue, the decision vector is defined as a 24-gene chromosome (corresponding to 24 time steps daily) in which genes fill with real numbers in the range [0, 1]. In addition, the real encoded value of genes could be simultaneously used for a 2-stage decoding procedure to interpret both the number and the speed of active pumps being operated, as described in Table 8.3. Simply speaking, based on the interval $x_{p,t}$ falls into, not only the number of active pumps

Table 8.3 Decoding a decision variable to both the number of active pumps and pump speed for VSP approach in Example 8.3.

Decision variables		Interval [LB, UB)				Relative pump speed	
$x_{p,t}$ (encoded in GA)		[0, 0.4)	[0.4, 0.6)	[0.6, 0.8)	[0.8, 1]	Min. acceptable	Max. acceptable
Decoded in EPANET	Number of active pumps	0	1	2	3		
	Relative pump speed ($N_{p,t}$)	0	$= \dfrac{b-a}{UB-LB}(x_{p,t}-LB)+a$			$a = 0.7$	$b = 1$

is specified but the relative rotation speed (i.e. $N_{p,t}$) of parallel active pumps is also determined at the same time. This tricky point leads to effectively decreasing the complexity of the problem in terms of dimensionality.

It should be added that in the *Optimtool* panel in *MATLAB*, the *Population Type* under the *Options* → *Population* must be set to *Double vector* and also in *Constraints* → *Bounds* the upper and lower bounds must be respectively set to *zeros*(1, 24) and *ones*(1, 24), in order to consider the decision variables as real numbers between 0 and 1.

8.5.5.3 *Constraints*

In the VSP example, the Affinity Laws (Equations (8.5) and (8.6)) play roles as constraints automatically handled through the simulation model. Additionally, all constraints defined in the CSP example are applied here except the constraint set (*iv*), instead of which the following set of constraints is replaced:

(*iv*) The relative speed for each pump p should not exceed beyond upper and lower bounds mathematically stated as follows:

$$0.7 \leq N_{p,t} \leq 1 \quad t = 1, 2, 3\ldots, 24 \tag{8.20}$$

Since the encoded decision variable $x_{p,t}$ is generated as a real value between 0 and 1, by employing the instruction stated in Table 8.3, this set of constraints will be self-adaptively imposed on the optimization model and no violation is required to handle it.

8.5.5.4 *Model execution and results*

The GA optimization model was run considering the same tuning parameters as the previous example. The result of optimal solution for VSP operation mode in ATM example is shown in Table 8.4. The optimal pump scheduling in this approach is also depicted in Figure 8.8. While all constraints applied in the problem have been satisfied, the VSP operation mode could lower the total energy cost to 3252.8 \$. In comparison with the CSP operation mode, 9.11% improvement in energy cost has been achieved by the using VSP approach. Note that EPANET does not consider variation of pump efficiency with pump speed. Therefore, for the best solution found in this example, EPANET calculates the energy cost as 3297.82 \$ that includes nearly 1.4% error with regard to the true value (i.e. 3252.8 \$).

Similar to the results obtained in the previous example, for the VSP operation mode the model has also tried to benefit the capacity of storage tanks to balance water supply and demand in such a way that for on-peak tariff time, the network is fed from the tanks as much as possible. In contrast, during off-peak hours, the system is mainly supplied with pumping and the tanks are mainly used to store water.

8.5.5.5 *Challenges and opportunities*

Notwithstanding all the advantages of the VSP alternative, due to concerns about the cost of purchase and installation of variable frequency drives, many operating companies prefer to manage their budget

Table 8.4 The optimum pump operation schedule for VSP approach in Example 8.3.

	Time steps																									TEC ($)
	1	2	3	4	5	6	7	8	9	10	11	12	13	14	15	16	17	18	19	20	21	22	23	24	
$x_{p,t}$	0.948	0.743	0.540	0.958	0.936	0.573	0.744	0.742	0.532	0.967	0.529	0.973	0.767	0.000	0.973	0.962	0.961	0.528	0.549	0.288	0.000	0.947	0.744	0.000	
$N_{1,t}$	0.922	0.914	0.908	0.937	0.904	0.956	0.916	0.914	0.898	0.951	0.893	0.960	0.950	0.000	0.960	0.944	0.941	0.892	0.924	0.000	0.000	0.921	0.911	0.000	3252.8
$N_{2,t}$	0.922	0.914	0.000	0.937	0.904	0.000	0.916	0.914	0.000	0.951	0.000	0.960	0.950	0.000	0.960	0.944	0.941	0.000	0.000	0.000	0.000	0.921	0.911	0.000	
$N_{3,t}$	0.922	0.000	0.000	0.937	0.904	0.000	0.000	0.000	0.000	0.951	0.000	0.960	0.000	0.000	0.960	0.944	0.941	0.000	0.000	0.000	0.000	0.921	0.000	0.000	

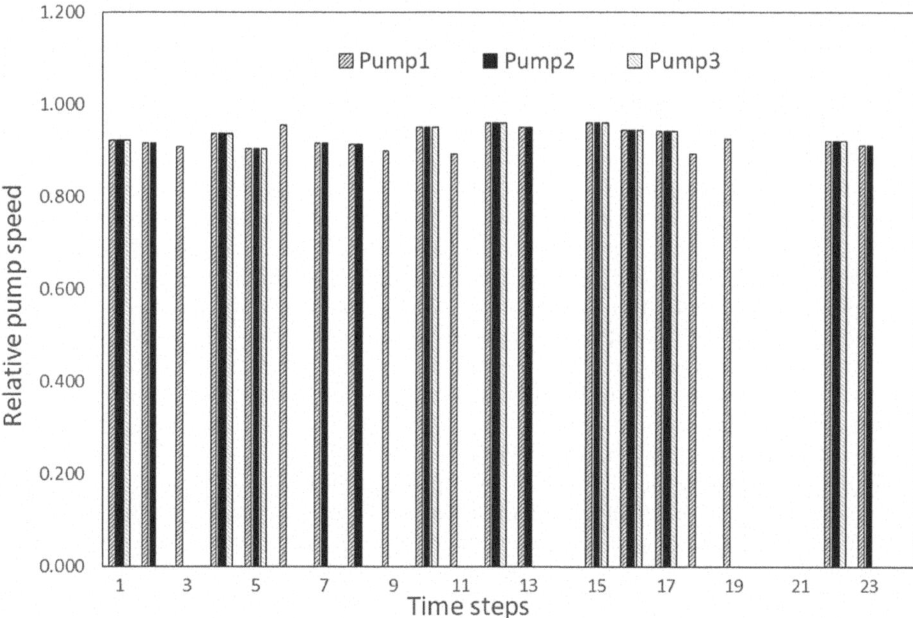

Figure 8.8 The optimum pump operation schedule for VSP approach in Example 8.3.

by implementing both approaches of constant and variable speed pumping together. Therefore, there are many pumping stations that work with a combination of constant and variable speed pumps. Such a system requires a special type of scheduling for optimal pump operation, in which the techniques used in the previous two examples should be combined.

Storage tanks play a key role in optimal pump operation when aiming to reduce energy cost. According to the principle of mass/flow continuity, without any storage tank, the pumping system must supply water equivalent to the demand at each time step. Conversely, when the system has storage capacity available, the operation could be planned in such a way that the pumped flow differs from the demand so that the total energy cost over the entire time period is reduced. For example, during the time steps with high cost of electricity, the station pumps water at a lower rate than the demand, whereas the stored water in tanks is responsible for supplying the remaining portion of demand. Conversely, during the time steps with low cost of electricity, more water is pumped than water demands and surplus was stored in tanks. In other words, to reduce pumping energy cost, the active storage capacity of tanks allows the system to manage energy consumption by shifting high rate pumping time steps from the on-peak electricity tariff to the off- peak.

From the above-mentioned perspective, although tanks have a positive role in reducing the energy cost, they prolong water stagnation in the system (possibly) resulting in poor water quality. Hence, the characteristics of the storage tanks in water distribution networks are closely related to variation of water quality indices. From this point of view, beside any optimal pump operation scheduling, the water quality analysis (see Chapter 6) should be carried out to determine associated challenges. Such a challenge is further discussed in the next section.

8.6 VSP SCHEDULING; EWQMS APPROACH

Supplying water with acceptable levels of quantity and quality for customers is an undeniable service obligation for water utilities. The term 'water quantity' mainly deals with issues related to hydraulic requirements (pressure and flow) while the term 'water quality' is related to the consideration of characteristics such as safe Residual Chlorine, Trihalomethanes (THMs), Heavy Metals, Microbial Contamination, water age, etc. Meeting such quantitative and qualitative requirements alongside energy-related limitations are among the most serious challenges faced by operating companies, so that an integrated approach of Energy and Water Quality Management Systems (EWQMS) seems essential.

The EWQMS takes all three aspects of water quantity, water quality and energy into account in WDNs planning and management. In EWQMS for WDNs, *Energy* can be significant from two perspectives: The first is the economic aspect in which the minimization of pumping 'energy cost' is considered (as emphasized in this section, too), and the second is the environmental aspect associated with the growing concern over greenhouse gas emissions, in which 'energy usage' needs to be minimized in planning and operation. Note that minimizing energy costs does not necessarily mean minimizing energy usage. If the cost of energy consumption has a single-tariff pattern, then pump scheduling to energy cost and energy usage minimization would practically lead to the same solution. However, in most cases, due to the multi-tariff pattern for energy costs, pump scheduling with the two above different approaches results in different solutions. Nonetheless, in the present section we have considered the aspect of the energy cost in EWQMS.

As already discussed in the Section 8.4, in pump scheduling with energy cost minimization as the sole optimization objective, storage tanks play an active role in balancing fluctuations in water supplied (by the pumping station) and water consumed (by customers) over a 24-hour extended period. Storage tanks, on the other hand, may possibly cause poor water quality due to the longer residence time of water within the system (called Water Age) before delivery to the customers. The degree of water quality deterioration (i.e. increasing in water age) can depend on different factors such as the characteristics of tanks, the spatiotemporal pattern of demand distribution in the system, the pipe network details, pump scheduling plan, tank flushing strategies, and so on.

In the last part of this chapter, we are going to simply bring up the idea of EWQMS in variable pump scheduling problems through the following examples in which the planning simultaneously takes both 'energy costs' and 'water quality' criteria in the planning. Moreover, we have tried to simply approach the concept of multi-objective optimization application in pump scheduling, though we are not going to review every detail of a multi-objective approach here. Rather, we express the general concept of this method in relation to Example 8.5 and leave the study of the details of the method to the reader.

8.6.1 EWQMS for ATM network – A single objective optimization approach

In Example 8.3, we optimized the pump scheduling problem of the ATM network for minimum energy cost (the optimum solution hereinafter is called $S1$ for the sake of brevity). In Example 8.4A, we will assess the status of the water age over the whole network (by a criterion defined later) for the solution $S1$. It is obvious that for $S1$, the water age criterion does not have a high level of desirability, considering that the solution was achieved based on the minimization of energy cost. As a matter of fact, such planning tries to store more water in tanks during off-peak hours of electricity tariff, and to release water from tanks during on-peak hours of electricity tariff (i.e. longer water residence time in the system). Therefore, in Examples 8.4B and 8.4C, we will show that pump scheduling driven by the energy cost minimization as the objective function conflicts with water age improvement (i.e. speeding up the water delivery to customers). Simply speaking, the better the water age index is desired, the more costly the solution of pump scheduling should be. From this point of view, energy cost and water age are considered as two opposite criteria that can be taken as planning goals in a multi-objective optimization, which in turn provides a higher level of flexibility for the decision-making process (see Example 8.5).

8.6.1.1 Example 8.4A

Setting the total duration (*T*): Assuming that the 24-hour time pattern of nodal demand remains constant during a whole week, a 24-hour simulation period is sufficient for hydraulic analysis of a network (provided that for each storage tank, the water level at the end of the simulation is equal to or greater than the water level at the beginning of the period). Conversely, a much longer simulation period is needed to attain accurate results in a normal water quality modeling. A good criterion to reliably set the 'Total Duration' (T) in EPANET is *'observation of the same repetitive periodic behavior of the water quality in network junctions over time (i.e. water quality time series)'*. A simple way to determine the appropriate T could be manual trial and error. To this end, we should execute a number of EPANET runs with increasing Ts, and check whether in 'Time Series Plot' (*Report→Graph*) the same repetitive periodic behavior for the water quality parameter (herein the water age) is observed or not. Nonetheless, since long T causes the optimization process to be very time consuming, to set T in the ATM network, it is preferred to consider a 'total time' as short as possible to ensure that a periodic trend of water age is being formed. Employing the trial and error approach for a number of random solution and checking it for the following optimal decision, $T = 336$ hr was determined as a reliable total time duration (T) for the ATM network.

Initial water quality: Water quality analysis requires calculating the initial water quality (IWQ) for each node. To calculate IWQs for a specific solution, it makes sense that under a long-enough extended period simulation, the results at the end time step could be taken as the IWQs for the same pattern being repeated in the next period. However, end time-step water qualities themselves are solution-dependent outputs since for different pump scheduling solutions, the time series of nodal heads and water levels in tanks are possibly not the same, leading inevitably to different water quality time series. However, in the optimization process concerning water quality issues, the solutions are systematically altered. Hence, to be fair, it is required to assume that we have the same set of IWQs for all solutions. This simplifying assumption avoids additional calculations to determine initial water quality for each solution separately. For this purpose, in the present example, it is assumed that for all solutions, before the beginning of simulation, a complete tank flushing has been already done while the network was being fully supplied by pumps for a long time without any limitation. In this regard, the water age at the end of this long simulation period is independent of the optimization solutions and can be considered as the IWQ for all decisions. The results of such IWQ analysis for the ATM network is presented in Table 8.5 and the time series of water age for some sample nodes are depicted in Figure 8.9.

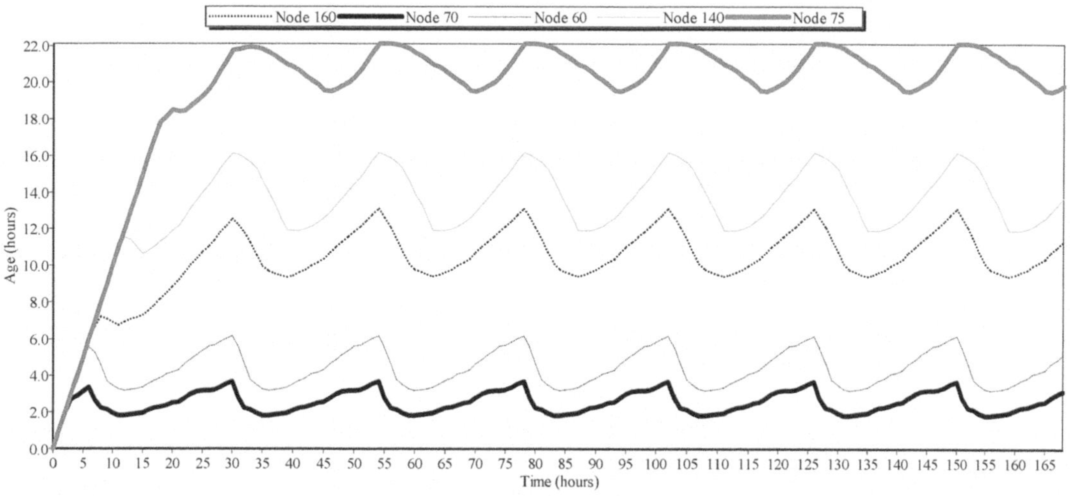

Figure 8.9 Time series for calculating IWQ in ATM network: plot of water age variation in some sample nodes for no storage tank under operation.

According to Figure 8.9, only two 24-hour cycles are sufficient to establish the same repetitive periodic behavior for water age in ATM. Nonetheless, it should be noted that such a short period of time is due to the assumption that the storage tanks are out of operation.

Water age measure: In water distribution networks, water age is the length of time that water leaves the source (or treatment plant) till the time it is delivered to the customer. According to AWWA and AWWA Research Foundation (1992), a survey showed that the average and maximum water age in 800 sample networks in the United States were 31.2 and 72 hours, respectively. In the present example, by accepting the allowable water age threshold equal to a maximum of 72 hours, we will use a water age measure (*WAM*) in the ATM network (Güngör-Demirci *et al.*, 2020) which is defined as the sum of positive deviations of the average water age (in the last 24-hour simulation period) from the allowable water age threshold (72 hours), as follows:

$$\text{WAM} = \sum_{j=1}^{N_{dn}} \max\left[0, \left(\left(\frac{\sum_{t=T-24}^{T} WA_{j,t}}{24}\right) - 72\right)\right] \tag{8.21}$$

where N_{dn} is the number of demand nodes and $WA_{j,t}$ is the water age in the *j*th demand node of the network at the *t*th time step with a 1-hour time interval.

Water age analysis for solution S1: Water quality analysis was performed for *S1* and based on a simulation period of TD = 336 hours (2 weeks), the results obtained are shown in Table 8.5. The tables shows the average water age for the last 24-hour period of the simulation (i.e. Ave$(WA)_{T-24}$) and the positive deviations of Ave$(WA)_{T-24}$ from the allowable threshold (72 hours) for each node. As seen, the WAM for this solution is 276.9 hours, which seems a high measure for this solution. According to the table, the average water age in the last 24 hours of the simulation varies from 0.2 hours in Node 20 to 164.3 hours in Node 140, while the overall average for water age in the whole network is 48.7 hours.

Results of the water age analysis for *S1*, presented in Table 8.5, along with considering location of the nodes (Figure 7.3), show that the nodes 'directly adjacent' or ' indirectly adjacent with an interface node' to the tanks have a high WAM. In addition, Tank 265, as the farthest tank from the source, has the most negative effect on deterioration of water age. In this regard, except the nodes directly adjacent to the tanks (Node 60, 140 and 160) other high-WAM nodes (Node 115, 175 and 75) are affected by Tank 265.

8.6.1.2 *Example 8.4B: VSP scheduling with WAM constraint*

In this example, due to the high WAM obtained for the solution S1 in Example 8.4A, we want to redefine the problem of VSP scheduling for ATM network (Example 8.3) by adding a new constraint in order to restrict the WAM to an upper limit. Obviously, since we have placed an additional constraint on Example 8.3, the feasible decision space becomes more limited, and inevitably the optimization objective function (energy cost) will be larger than the one achieved for *S1*. Considering this issue, the following equation is added to the mathematical model of Example 8.3:

$$WAM\left(x_{p,t}\right) \leq \left(1 - \alpha\right) WAM_{S1} \tag{8.22}$$

where $WAM(x_{p,t})$ is the water age measure as a function of the solution $x_{p,t}$, α is a ratio between 0 and 1, and WAM_{S1} is the water age measure achieved for the solution S1. Introducing the water quality constraint stated in Equation (8.22) for $\alpha_1 = 0.1$ and $\alpha_2 = 0.25$ to the VSP scheduling in Example 8.3, the model ran and the results of the optimal solutions are exhibited in Table 8.5. As can be seen, considering $\alpha_1 = 0.1$ (let us name this optimal solution S2) the solution obtained by the new form of problem statement

Table 8.5 Energy cost and water age analysis for the solution $S1$, $S2$ and $S3$.

	Node ID																			Solution ID
	20	30	110	70	60	90	100	40	50	80	150	140	170	130	160	120	55	75	115	
Initial water age (hr)	0.08	3.02	2.95	3.16	5.14	8.3	4.73	7.78	5.71	7.0	9.8	13.72	22.74	14.15	11.34	5.39	9.33	19.83	27.05	
$\text{Ave}(WA)_{T-24}$	0.2	2.4	2.3	6.5	77.0	18.2	4.2	7.0	5.0	66.7	20.5	164.3	139.7	70.5	86.1	4.6	8.4	80.6	161.3	$S1$
$\text{Max}(0.\ \text{Ave}(WA)_{T-24})$	0.0	0.0	0.0	0.0	5.0	0.0	0.0	0.0	0.0	0.0	0.0	92.3	67.7	0.0	14.1	0.0	0.0	8.6	89.3	
Criteria	**Energy cost (\$) = 45262.12**								WAM_{S1} (hr) = **276.9**										
$\text{Ave}(WA)_{T-24}$	0.2	2.4	2.3	7.3	95.1	20.7	4.3	7.0	5.0	84.1	24.3	227.0	189.1	88.4	109.3	4.6	8.5	108.2	224.0	$S2$
$\text{Max}(0.\ \text{Ave}(WA)_{T-24})$	0.0	0.0	0.0	0.0	23.1	0.0	0.0	0.0	0.0	12.1	0.0	155.0	117.1	16.4	37.3	0.0	0.0	36.2	152.0	
Criteria	**Energy cost (\$) = 45550.75**								WAM_{S2} (hr) = **248.7**										
$\text{Ave}(WA)_{T-24}$	0.2	2.4	2.3	7.3	95.1	20.7	4.3	7.0	5.0	84.1	24.3	227.0	189.1	88.4	109.3	4.6	8.5	108.2	224.0	$S3$
$\text{Max}(0.\ \text{Ave}(WA)_{T-24})$	0.0	0.0	0.0	0.0	23.1	0.0	0.0	0.0	0.0	12.1	0.0	155.0	117.1	16.4	37.3	0.0	0.0	36.2	152.0	
Criteria	**Energy cost (\$) = 48421.26**								WAM_{S3} (hr) = **223.1**										

has an energy cost of 45550.75 \$ and WAM of 248.7 hours which, in comparison with $S1$, indicates a 10.18% decrease in WAM while the energy cost has only a 0.64% increase. The solution $S2$ also has an overall average water age of 46.6 hours that shows 2.1 hours less than that of solution $S1$.

8.6.1.3 *Example 8.4C: VSP scheduling for WAM minimization*

An interesting point about the implementation of the optimization model of Example 8.4B for $\alpha_2 = 0.25$ is that the model could not achieve any feasible solution. To put it simply, despite the repetition of runs of the pump scheduling model for energy cost minimization, the GA could basically not find any solution satisfying all the constraints. This suggests that although imposing stricter constraints (i.e. WAM) leads to more costly solutions for energy cost minimization problem, it seems that the WAM constraint 'to an unknown extent' restricts the problem, so that thereinafter, even by imposing a stricter WAM on the optimization, feasible solutions are not achieved anymore. In other words, in the pump scheduling problem, same as global optimum energy cost, there exists a global minimum for WAM as well. In this regard, finding the optimal WAM may be considered as a separate optimization problem, the objective function of which can be as follows:

$$\text{Min WAM}\left(x_{p,t}\right) \tag{8.23}$$

The set of constraints for the above programming are exactly the same as for Example 8.3.

The result of VSP scheduling for WAM minimization is illustrated in Table 8.5 (named as solution $S3$). As can be seen in this table, the global optima in this run includes a solution with a WAM of 223.1 hours in which the ATM network experiences an overall average water age of 44.9 hours. The optimum solution $S3$, compared to the solution $S1$ has reduced the WAM by 19.5%. In addition, the energy cost for the optimum solution in VSP scheduling for WAM minimization reaches to 48421.26 \$ (7% larger than $S1$). Now it is clear why the optimization model in Example 8.4B, was not able to find any solution for the scenario $\alpha_2 = 0.25(!)$.

8.6.2 EWQMS for ATM network – A multi-objective optimization approach

8.6.2.1 *General concept of multi-objective optimization*

In the two recent examples (Examples 8.3 and 8.4) it was observed that pump scheduling could be modeled to minimize either the energy cost (Example 8.3, 8.4B) or the WAM (Example 8.4C). In Example 8.4B, a capability to have an upper limit of WAM (Equation (8.22)) as an additional constraint, was included as well. We could even include an upper limit of energy cost as a budget constraint, in VSP scheduling for WAM minimization in Example 8.4C, though we did not. In other words, for pump scheduling, either of energy cost or WAM criteria can be separately considered as the objective function, and the emerging problem with or without taking the other criterion as a constraint can be solved using a single optimization approach.

An important feature of the two above criteria in previous examples (i.e. energy cost and WAM) is that they are in conflict. Simply, consider the two optimal solutions $S1$ and $S3$; to improve one of the criteria (e.g., WAM), the other one (e.g., energy cost) needs to be deteriorated and vice versa. Generally speaking, in optimization-based planning problems, when two criteria (like WAM and energy cost here) are conflicting, another approach of optimization (called multi-objective) can be implemented in which both criteria are included in the planning as the objective functions.

Let us again take Example 8.3 and Example 8.4C into consideration. The result of model execution were two distinct solutions with different energy cost and WAM. If we have a look at the values of these two objectives related to $S1$ and $S3$, we realize that these solutions have no advantage over each other, since $S1$ has a better energy cost and a worse WAM against $S3$ that has a better WAM and a worse energy cost. Such optimal decisions, in the terminology of multi-objective optimization are called non-dominated solutions. The concept of domination is the logic behind a number of efficient Multi Objective Evolutionary Algorithms (MOEAs) the most well-known of which are the Strength Pareto Evolutionary Algorithm II (SPEA2) by Zitzler *et al.* (2001) and Non-dominated Sorting Genetic

Algorithm-II (NSGA-II) by Deb *et al.* (2002). Both algorithms have been successfully implemented in many researches in the field of WDNs (Creaco *et al.*, 2014; Farmani *et al.*, 2003, 2005; Güngör-Demirci *et al.*, 2020; Minaei *et al.*, 2020; Roshani & Filion, 2014; Sabzkouhi *et al.*, 2017, 2021; Zheng *et al.*, 2016).

Unlike the single-objective optimization, where only one global optima is found in each run, the multi-objective optimization model leads to a number of optimal solutions which have no predetermined distinction between each other due to the conflict between the objectives. The set of these optimal solutions is called the Pareto Optimal Front. Having the Pareto set solutions allows the best decision to always be accessible according to the priorities that may change over time. This feature does not exist in single-objective optimization where based on changes in priorities, it is required to re-run the optimization model. An up-to- date overview on multi-objective optimization in water distribution networks can be found in Sarbu *et al.* (2020).

8.6.2.2 *Example 8.5: VSP scheduling for energy cost vs WAM minimization*

According to the general description of the multi-objective optimization and its advantages, in this section the results of the implementation of multi-objective optimization model for VSP scheduling is presented for the following objective functions:

$$
\begin{cases}
\text{Min TEC} = \sum_{p=1}^{NPu}\sum_{t=1}^{T} \frac{\gamma Q_{p,t} H_{p,t}}{\eta_{p,t}} \Delta t_t \, ET_t \\
\text{Min} \quad \text{WAM}\left(x_{p,t}\right)
\end{cases}
\tag{8.24}
$$

The constraints of optimization for the above problem are exactly the same as for Example 8.3.

The above programming was solved using the NSGA-II multi-objective optimization algorithm. The algorithm settings were similar to Example 8.3. In Figure 8.10 the resulting optimal Pareto front is presented. As previously stated, the result of bi-objective optimization is a large number of non-dominated solutions that cover a range of possible values for the optimization objectives. Among the Pareto set, the solutions *S1*, *S2*, and *S3*, whose coordinates previously found in single-objective optimization in Examples 8.3 and 8.4, are also observable. Additionally, the point previously explained in Example 8.4C about an optimal value beyond which there is not any further improvement in the objective function, is well illustrated in the diagram above, where no other solution is found beyond *S1* and *S3*.

Once capturing the Pareto optimal solution set was done, a decision-making process should be carried out by decision makers to select the final solution. The Pareto frontier and how it changes in the range of objectives variation provide useful information to choose the final solution. For example, in the area adjacent to solutions *S1*, the slope of energy cost changes compared to WAM changes is very small. Obviously, if this area is desired for selecting the final decision, solutions near by the left of *S2* are in a better position. However, in the left-hand side of the *S2*, the relative cost required to improve WAM, increases dramatically due to the steeper slope of the energy cost-WAM changes.

The decision-making process should involve the views of different stakeholders (water utility companies, customers, environmentalists, etc.) who may consider the objectives with different levels of importance. Since optimization objectives are often in conflict, the final solution is usually selected using bargaining-based selection methods, the most widely-used of which is Young's bargaining method based on game theory. Interested readers are referred to Fallah-Mehdipour *et al.* (2011).

8.7 CONCLUSION

The purpose of this chapter has been to introduce the application of a simulation-optimization approach for pump scheduling and EWQMS problems. The general principles of pump scheduling in both constant speed pumping (CSP) and variable speed pumping (VSP) approaches were expressed after a brief review on pump performance basics. Since the main purpose of this chapter was mostly

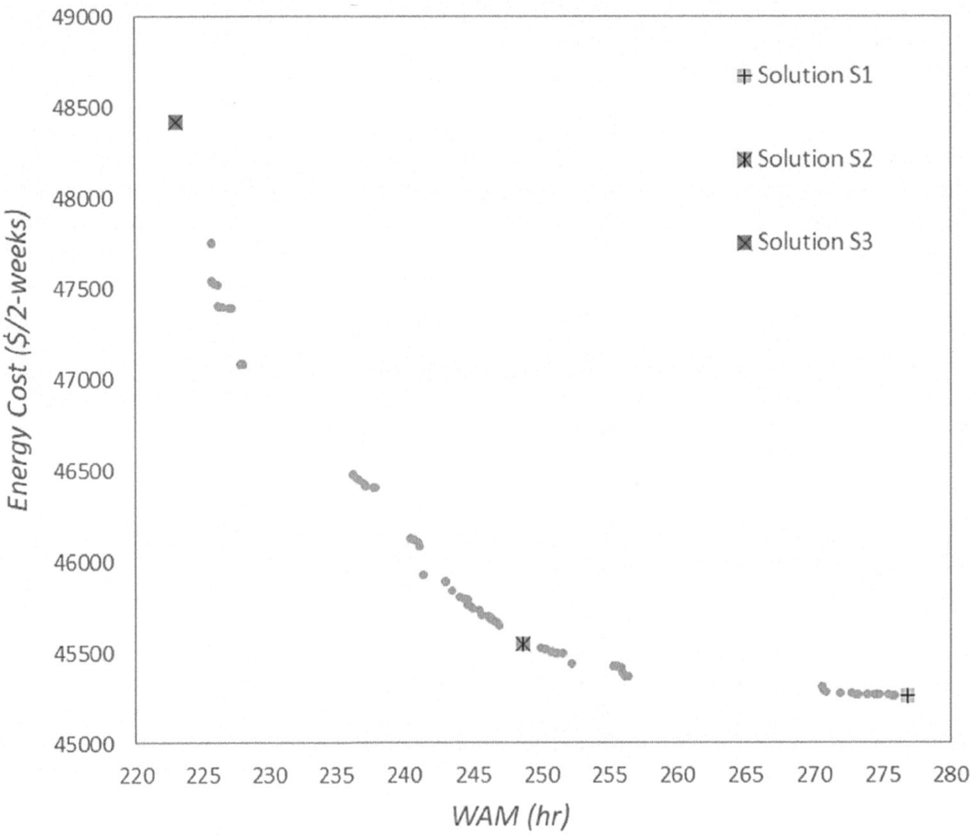

Figure 8.10 Pareto optimal frontier for Example 5.

educational aspects, we tried to simply apply the methods on the ATM network, a small benchmark pipe system from the literature. First, the classic CSP method for pump scheduling was implemented to find the minimum energy cost expressed in the form of binary programming and the resulting problem was solved using the Genetic Algorithms in the MATLAB environment. Then, the general principles of VSP scheduling was described and the model was employed in the ATM network. The result showed that the solution found by VSP approach (abbreviated as solution *S1*) decreased the energy cost by 9.11% compared to the optimum solution found by the CSP approach.

Later, expressing the Energy and Water Quality Management Systems (EWQMS) approach in water distribution networks, the capabilities of single and multi-objective simulation-optimization models to VSP scheduling was simply investigated into the ATM network. In this regard, first by calculating the water age measure (WAM) criterion to evaluate the status of water quality in different solutions, *S1* was assessed, and the reasons why *S1* had a high WAM were discussed. Since no restrictions were imposed on WAM in obtaining *S1*, the role of the tanks in increasing the water residence time in the distribution system, and accordingly the age of water delivered to the nodes, was quite obvious. Afterwards, by redefining the VSP scheduling problem to minimize the energy cost with a constraint of maximum allowable WAM, the ATM problem was again solved as a single objective optimization, which was naturally expected to result in a solution with a higher energy cost (named as solution *S2*). Finally, the EWQMS approach was implemented for VSP scheduling as a bi-objective programming to simultaneously minimize the conflicting goals of energy cost and WAM, and the importance of the Pareto optimal frontier obtained was discussed.

Nowadays, by combining multiple features (e.g. simulation packages, optimization methods, data-driven and machine learning-based models, GIS applications, etc.) alongside the facilities associated with physical assets, efficient possibilities have been provided in design, operation and management of real-world water distribution networks. Familiarity with the basic principles of these techniques and their applications can motivate engineers to learn them and effectively employ them in real problems. While the rate and cost of collecting and accessing to reliable data have been highly cost-effective, mastery of design engineers and operators of water systems with decision support systems are becoming more and more essential.

REFERENCES

Amirsardari A. R., Sabzkouhi A. M., Zahiri J. and Derakhshannia R. (2021). Improving the performance of water supply pump stations in terms of energy consumption: a comparison between the constant and variable-speed optimal pump scheduling. Proceedings of the 2nd National Conference on Agricultural Research and Environment, ASNRUKH, Mollasani, Iran (in Persian).

Araujo L., Ramos H. and Coelho S. (2006). Pressure control for leakage minimisation in water distribution systems management. *Water Resources Management*, **20**(1), 133–149, https://doi.org/10.1007/s11269-006-4635-3

AWWA and AWWARF (American Water Works Association and American Water Works Association Research Foundation) (1992). Water Industry Database, Utility Profiles. AWWA, Denver, Colorado, USA.

Brennen C. E. (1994). Hydrodynamics of Pumps. Concepts ETI, Inc. and Oxford University Press, Vermont, USA.

Cherchi C., Badruzzaman M., Oppenheimer J., Bros C. M. and Jacangelo J. G. (2015). Energy and water quality management systems for water utility's operations: a review. *Journal of Environmental Management*, **153**, 108–120, https://doi.org/10.1016/j.jenvman.2015.01.051

Cimorelli L., D'Aniello A. and Cozzolino L. (2020). Boosting genetic algorithm performance in pump scheduling problems with a novel decision-variable representation. *Journal of Water Resources Planning and Management*, **146**(5), 04020023, https://doi.org/10.1061/(ASCE)WR.1943-5452.0001198

Costa L. H. M., de Athayde Prata B., Ramos H. and de Castro M. A. H. (2016). A branch-and-bound algorithm for optimal pump scheduling in water distribution networks. *Water Resources Management*, **30**(3), 1037–1052, https://doi.org/10.1007/s11269-015-1209-2

Creaco E., Franchini M. and Walski T. (2014). Accounting for phasing of construction within the Design of Water Distribution Networks. *Journal of Water Resources Planning and Managment*, **140**(5), 598–606, https://doi.org/10.1061/(ASCE)WR.1943-5452.0000358

Deb K., Pratap A., Agarwal S. and Meyarivan T. (2002). A fast and elitist multiobjective genetic algorithm: NSGA-II. IEEE Trans. *Evolutionary Computation*, **6**(2), 182–197, https://doi.org/10.1109/4235.996017

Eliades D. G., Kyriakou M., Vrachimis S. and Polycarpou M. M. (2016). EPANET-MATLAB toolkit: an open-source software for interfacing EPANET with MATLAB. In: Proceedings of the 14th International Conference on Computing and Control for the Water Industry (CCWI 2016), Amsterdam, The Netherland.

Fallah-Mehdipour E., Bozorg Haddad O., Beygi S. and Mariño M. A. (2011). Effect of utility function curvature of Young's bargaining method on the design of WDNs. *Water Resources Management*, **25**, 2197–2218, https://doi.org/10.1007/s11269-011-9802-5

Farmani R., Savic D. A. and Walters G. A. (2003). Multi-objective optimization of water system: a comparative study. In: Cabrera *et al.* (eds.) *Pumps, Electromechanical Devices and Systems Applied to Urban Water Management*, Vol. 1, pp. 247–256.

Farmani R., Savic D. A. and Walters G. A. (2005). Evolutionary multi-objective optimization in water distribution network design. *Engineering Optimization*, **37**(2), 167–183, https://doi.org/10.1080/03052150512331303436

Fontana N., Giugni M., Glielmo L., Marini G. and Zollo R. (2018). Real-time control of pressure for leakage reduction in water distribution network: field experiments. *Journal of Water Resources Planning and Management*, **144**(3), 04017096, https://doi.org/10.1061/(ASCE)WR.1943-5452.0000887

García-Ávila F., Avilés-Añazco A., Ordoñez-Jara J., Guanuchi-Quezada Ch., Flores del Pino L. and Ramos-Fernández L. (2019). Pressure management for leakage reduction using pressure reducing valves. Case study in an Andean city. *Alexandria Engineering Journal*, **58**(4), 1313–1326, https://doi.org/10.1016/j.aej.2019.003

Güngör-Demirci G., Lee J. and Keck J. (2020). Optimizing pump operations in water distribution systems: energy cost, greenhouse gas emissions and water quality. *Water and Environment Journal*, **34**, 841–848, https://doi.org/10.1111/wej.12583

Mackay R. C. (2004). The Practical Pumping Hand Book. Elsevier, New York.

Martiinez F., Hernandez V., Alonso J. M., Rao Zh. and Alvisi S. (2007). Optimizing the operation of the Valencia water distribution network. *Journal of Hydroinformatics*, **9**(1), 65–78, https://doi.org/10.2166/hydro.2006.018

Minaei A., Sabzkouhi A. M., Haghighi A. and Creaco E. (2020). Developments in multi-objective dynamic optimization algorithm for design of water distribution mains. *Journal of Water Resources Management*, **34**(9), 2699–2716, https://doi.org/10.1007/s11269-020-02559-8

Roshani E. and Filion Y. R. (2014). Event-based approach to optimize the timing of water Main rehabilitation with asset management strategies. *Journal of Water Resources Planning and Management*, **140**(6), 04014004, https://doi.org/10.1061/(ASCE)WR.1943-5452.0000392

Rossman L. A. (2000). EPANET 2: Users Manual', Rep. No. EPA/600/ R-00/057. EPA, Cincinnati.

Sabzkouhi A. M., Haghighi A. and Minaei A. (2017). Investigation of uncertainty effects on hydraulic performance of water distribution networks using the Fuzzy Sets theory. E-proceedings. The 37th IAHR World Congress, Malaysia.

Sabzkouhi A. M., Lee J. and Keck J. (2021). Energy cost and water age management in water distribution networks via constant speed optimal pump scheduling. Proceedings of the 1st National Conference on Water Quality Management & 3rd National Conference on Water Consumption Management – Loss Reduction and Reuse, Tehran University, Iran (in Persian).

Sanks R. L., Tchobanoglous G., Bosserman B. E. and Jones G. M. (1998). Pumping Station Design, 2nd edn, Butterworth-Heinemann, Oxford, UK.

Sarbu I., Popa-Albu S. and Tokar A. (2020). Multi-objective optimization of water distribution networks: an overview. *International Journal of Advanced and Applied Sciences*, **7**(11), 74–86, https://doi.org/10.21833/ijaas.2020.11.008

Van Zyl J. E. (2014). Introduction to Operation and Maintenance of Water Distribution Systems. Water Research Commission, Pretoria, South Africa.

Zheng F., Zecchin A., Maier H. and Simpson A. (2016). Comparison of the searching behavior of NSGA-II, SAMODE, and Borg MOEAs applied to water distribution system design problems. *Journal of Water Resources Planning and Management*, **142**(7), 04016017, https://doi.org/10.1061/(ASCE)WR.1943-5452.0000650

Zitzler E., Laumanns M. and Thiele L. (2001). SPEA2: Improving the Strength Pareto Evolutionary Algorithm, TIK Report 103, Computer Engineering and Networks Laboratory (TIK), ETH Zurich, Zurich, Switzerland.

doi: 10.2166/9781789062380_0215

Chapter 9

Hydraulic transients in pipe systems

Juneseok Lee[1], Lu Xing[2] and Lina Sela[3]*

[1]*Department of Civil and Environmental Engineering, Manhattan College, Riverdale, NY*
[2]*Data Scientist, Xylem, Rye Brook, NY*
[3]*Department of Civil and Environmental Engineering, University of Texas, Austin*
**Corresponding author: juneseok.lee@manhattan.edu*

LEARNING OBJECTIVES

At the end of this chapter, you will be able to:

(1) Explain hydraulic transients in pressurized pipe systems.
(2) Use the hydraulic transients governing equations, codes and run for very simple cases.
(3) Run TSNet for simple water distribution systems.
(4) Assess and interpret modeling results.

9.1 INTRODUCTION

Many water utilities have in-house hydraulic modeling capacities to analyze their systems in terms of planning, design, operations, and management. However, many of the modeling efforts are geared toward or limited to steady state or extended period simulations, which assume that the water column is completely incompressible, and that pipe materials are not elastic. Clearly, the mass continuity and energy equations neglect to explain rapid changes that should be described by momentum equations (i.e., transient pressure waves generated due to sudden changes in flow). As is well known, the resulting pressure can result in pipe bursts and structural damage to other critical appurtenances. In addition, low flow due to transients can induce contamination intrusion in the systems (Lee, 2008; Lee *et al.*, 2012).

For transient flow analysis, Joukowski's equation is the most fundamental theory that is still used as a rough check for a head (H) change calculation:

$$\Delta H = -\frac{a}{g} \cdot \Delta V \tag{9.1}$$

where a is pressure wave speed, g is gravitational acceleration and V is the velocity. The model's assumptions are: (i) no friction in the pipe and (ii) no wave reflections in the system. In other words, there is no interaction among boundary conditions in the system. Here, we are presenting 1D classic water hammer equations (Wylie & Streeter, 1993) as follows:

Continuity equation:

$$\frac{\partial p}{\partial t} + V\frac{\partial p}{\partial x} + c^2\rho\frac{\partial V}{\partial x} = 0 \tag{9.2}$$

Momentum equation:

$$\frac{\partial V}{\partial t} + V\frac{\partial V}{\partial x} + \frac{1}{\rho}\frac{\partial p}{\partial x} + g\sin\alpha + \frac{f}{2D}V|V| = 0 \tag{9.3}$$

where p is pressure, V is velocity, c is wave speed, ρ is density, g is acceleration due to gravity, α is angle of inclination of pipe, f is friction factor, D is diameter, x is spatial dimension, and t is time.

These two equations are solved for a pipe network that incorporates suitable interior boundary conditions for appurtenances such as valves and junctions along the pipelines and external boundary conditions such as reservoirs and tanks. The solution of these equations yields the pressure/head, p or H (x, t) and velocity/flow rate, V or Q (x, t) as functions of spatial dimension x (taken along the length of the pipelines) and time t. The pressure can be highly positive and negative for hydraulic transients, and the velocity can be negative, indicating flow reversal. In the following, we provide a general overview of the mumerical schemes followed by the hydraulic transients computation tool, TSNet.

9.2 NUMERICAL METHOD CONSIDERING INITIAL AND BOUNDARY CONDITIONS

The governing equations for hydraulic transients are nonlinear hyperbolic Partial Differential Equations (PDE), so a closed-form solution is not available. Numerical methods must be used to solve these governing equations. There are several numerical methods such as, but not limited to, Methods of Characteristics (MOC), implicit Finite Difference (FD) method, and explicit FD schemes. In this article, we introduce an introductory explicit FD scheme, McCormack's method, which should be helpful for understanding why certain types of data are needed to run a specific hydraulic transients modeling package. The results obtained from McCormack's method are known to be satisfactory for many flow applications (Anderson, 1995). Please refer to Chaudhry (1987), Karney and McInnis (1990) and Wylie and Streeter (1993) for details on backgrounds as well as other numerical methods.

In McCormack's FD, the PDE is transformed into FDM (Finite Difference Method), such that the unknown conditions at a point at the end of a time step are expressed in terms of the known conditions at the beginning of the time step. Figure 9.1 shows the general schematic of the 1D explicit FD scheme. We solve unknown time level variables (H, Q) based on known time levels (H, Q). McCormack's explicit FD scheme is composed of a predictor and corrector step. One-sided FD is used for the spatial derivatives in each of these steps. First, forward FD is used in the predictor, and backward FD is used in the corrector part. Alternatively, backward FD is adopted in the predictor, and forward FD is used in the corrector part. Each alternative takes turns as the time step increases. Note that we are using simplified governing equations using (H, Q) here as follows:

$$\frac{\partial H}{\partial t} + \frac{a^2}{gA}\frac{\partial Q}{\partial x} = 0 \tag{9.4}$$

$$\frac{\partial Q}{\partial t} + gA\frac{\partial H}{\partial x} + RQ|Q| = 0 \tag{9.5}$$

where H is head, Q is flow rate, a is wave speed, g is acceleration due to gravity, f is friction factor, $R = f\Delta t/2D$, D is diameter, x is 1D spatial dimension, and t is time. See the formulations below. *: predicted values, i: space node and j: corresponding time step. H and Q's values are assumed to be known at all nodes at the time j level. We are solving for the time $(j + 1)$ level (Figure 9.1).

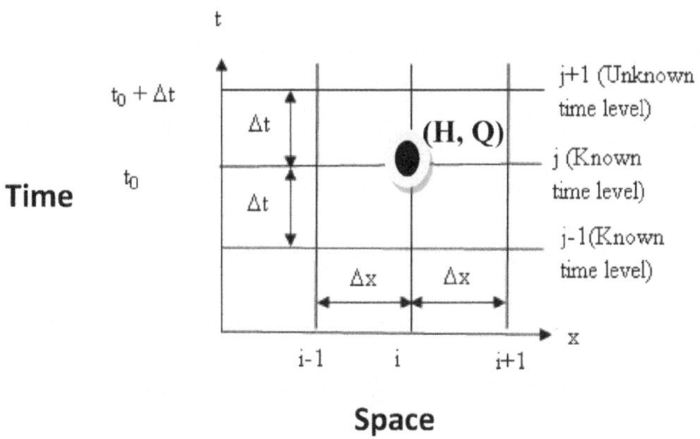

Figure 9.1 1D explicit FD scheme.

In alternative 1, the predictor part is:

$$H_i^* = H_i^j - \frac{\Delta t}{\Delta x}\frac{a^2}{gA}\left(Q_{i+1}^j - Q_i^j\right),$$

$$Q_i^* = Q_i^j - \frac{\Delta t}{\Delta x}gA\left(H_{i+1}^j - H_i^j\right) - RQ_i^j\left|Q_i^j\right|\cdot\Delta t, \quad (i = 1,\,2,...,\,n) \tag{9.6}$$

The corrector part is:

$$H_i^{j+1} = \frac{1}{2}\left[H_i^j + H_i^* - \frac{\Delta t}{\Delta x}\frac{a^2}{gA}\left(Q_i^* - Q_{i-1}^*\right)\right],$$

$$Q_i^{j+1} = \frac{1}{2}\left[Q_i^j + Q_i^* - \frac{\Delta t}{\Delta x}gA\left(H_i^* - H_{i-1}^*\right) - RQ_i^*\left|Q_i^*\right|\cdot\Delta t\right], \quad (i = 2,...,n+1) \tag{9.7}$$

In alternative 2, the predictor and corrector parts are:
  Predictor part:

$$H_i^* = H_i^j - \frac{\Delta t}{\Delta x}\frac{a^2}{gA}\left(Q_i^j - Q_{i-1}^j\right),$$

$$Q_i^* = Q_i^j - \frac{\Delta t}{\Delta x}gA\left(H_i^j - H_{i-1}^j\right) - RQ_i^j\left|Q_i^j\right|\cdot\Delta t, \quad (i = 2,...,n+1) \tag{9.8}$$

Corrector part:

$$H_i^{j+1} = \frac{1}{2}\left[H_i^j + H_i^* - \frac{\Delta t}{\Delta x}\frac{a^2}{gA}\left(Q_{i+1}^* - Q_i^*\right)\right],$$

$$Q_i^{j+1} = \frac{1}{2}\left[Q_i^j + Q_i^* - \frac{\Delta t}{\Delta x}gA\left(H_{i+1}^* - H_i^*\right) - RQ_i^*\left|Q_i^*\right|\cdot\Delta t\right], \quad (i = 1,2,...,n) \tag{9.9}$$

As mentioned, each alternative takes turns as the time step (j) increases. These formulations work for internal nodes, but we will have to consider separately for the boundary conditions. From the governing Equation (9.4), we will consider characteristic boundaries.

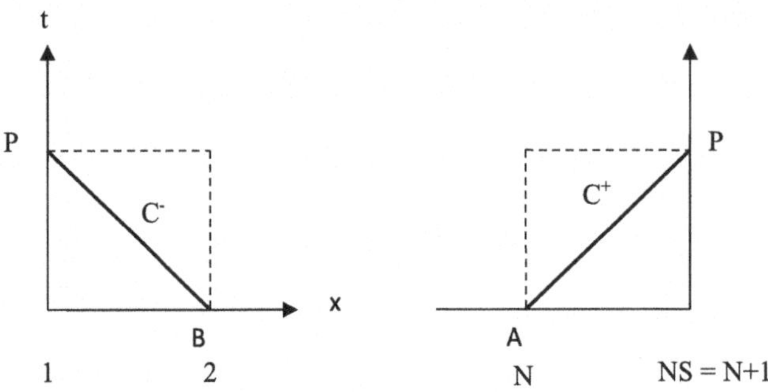

Figure 9.2 Boundary characteristic (upstream and downstream).

Multiplying Equation (9.4) by η and adding it to Equation (9.5), then we have:

$$\left(\frac{\partial H}{\partial t}+\eta gA\frac{\partial H}{\partial x}\right)+\eta\left(\frac{\partial Q}{\partial t}+\frac{a^2}{\eta gA}\frac{\partial Q}{\partial x}\right)+\eta RQ|Q|=0$$

$$(9.10)$$

$$\text{Let,}\quad \eta gA=\frac{dx}{dt}=\frac{a^2}{\eta gA},\quad \text{so}\,\eta=\pm\frac{a}{gA}$$

when

$$\lambda^+=\frac{dx}{dt}=a,\quad C+:\left(\frac{\partial H}{\partial t}+\lambda^+\frac{\partial H^+}{\partial x}\right)+\frac{a}{gA}\left(\frac{\partial Q}{\partial t}+\lambda^+\frac{\partial Q^+}{\partial x}\right)+\frac{aR}{gA}Q|Q|=0 \tag{9.11}$$

when

$$\lambda^-=\frac{dx}{dt}=-a,\quad C-:\left(\frac{\partial H}{\partial t}+\lambda^-\frac{\partial H^-}{\partial x}\right)-\frac{a}{gA}\left(\frac{\partial Q}{\partial t}+\lambda^-\frac{\partial Q^-}{\partial x}\right)-\frac{aR}{gA}Q|Q|=0, \tag{9.12}$$

At the boundary conditions, equations are solved with the condition imposed by the boundary. The characteristic of the boundary condition is shown in Figure 9.2. For the upstream boundary, the C− characteristic line is valid, and C+ is for downstream. These are used to depict the complete water phenomena. Each boundary condition is solved independently of the interior points' calculation and another end of the boundary. In this chapter, we are presenting several representative boundary conditions. Please refer to Chaudhry (1987) and Wylie and Streeter (1993) for more details.

9.2.1 Reservoir

During a short-period transient event, the upstream reservoir's hydraulic grade line elevation is assumed to be constant:

$$H_1=H_R \tag{9.13}$$

where H_R is hydraulic grade line above the reference datum, and H_1 is head value at the upstream section at point P. With $C-$ equation and $H_1=H_R$, Q can be obtained at the boundary. So, H and Q are obtained for the next time level $(j+1)$.

9.2.2 Junctions

A general junction with pipeline, non-pipeline elements (valve), and nodal inflow are shown in Figure 9.3. A continuity equation can be written at the junction. At any instant, the sum of the inflow is zero:

$$\sum Q_{in}=\sum Q_p+\sum Q_e+Q_n=0 \tag{9.14}$$

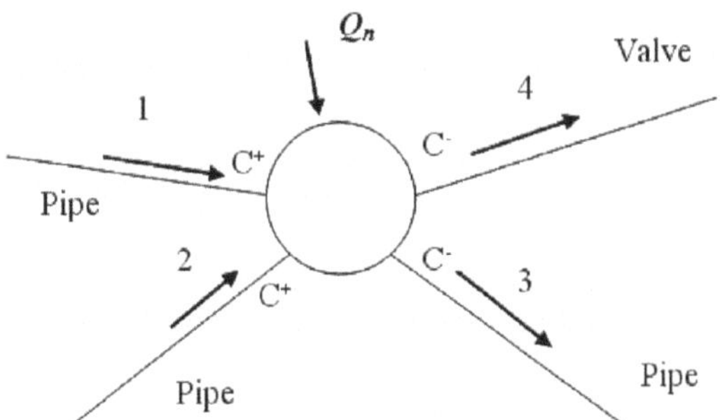

Figure 9.3 General junction.

where $\sum Q_{in}$ is the sum of the inflow (added to zero at any instant), $\sum Q_e$ is the sum of all instantaneous non-pipe flows, $\sum Q_p$ is the sum of all instantaneous pipe flows, and Q_n is a nodal flow. When minor losses are neglected at the junction, the energy equation can be written for each junction element:

$$H = H_{1,NS} = H_{2,NS} = H_{3,1} = H_{4,1} \tag{9.15}$$

Using the compatibility equations, $Q_{1,NS}$, $Q_{2,NS}$, $Q_{3,1}$, and $Q_{4,1}$ are obtained.

Figure 9.4 shows the valve located between two pipelines, where A and B show the interconnecting junctions on both sides of the valve. It is assumed that the inertia effects are neglected in the steady-state orifice equation, and the volume of fluid stored inside the valve is constant. For positive flow, $H_{1,NS} = H_A$ and $H_{2,1} = H_B$. The orifice equation is:

$$Q_{1,NS} = Q_{2,1} = Q_v = \frac{Q_0 \tau}{\sqrt{H_0}} \sqrt{H_A - H_B} \tag{9.16}$$

where H_0 is steady-state HGL drop across the valve with the flow of Q_0 $\left(t = \dfrac{C_d A_G}{(C_d A_G)_0} = \sqrt{\dfrac{K_0}{K}} = 1\right.$: dimensionless valve opening). When combined with the compatibility equation, Q_v can be obtained. Also, for reversal flows:

$$Q_{1,NS} = Q_{2,1} = Q_v = -\frac{Q_0 \tau}{\sqrt{H_0}} \sqrt{H_A - H_B} \tag{9.17}$$

Combined with the compatibility equation, Q_v can be obtained.

While working on the numerical methods or solvers, you should be aware of a few critical issues, such as convergence and stability. We are introducing fundamental concepts as follows:

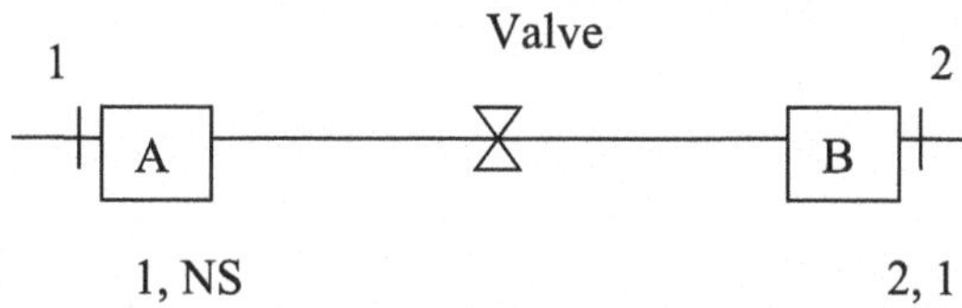

Figure 9.4 Valve located in-line.

9.2.3 Discretization error

Say, $U(x, t)$ is the exact solution of the PDE, and $u(x, t)$ is the finite difference equation's exact solution. Then, $(U-u)$ is called *discretization error*. This is introduced when replacing the PDE with finite difference approximation.

9.2.4 Truncation error

$F_i^j(u) = 0$ is the finite difference equation at grid point $i\Delta x$ and $j\Delta t$, where i and j are the number of grid points in the x and t directions. Substituting the exact solution of PDE $U(x, t)$ into the FD approximation equation, then $F_i^j(U)$ is the local truncation error at $(i\Delta x, j\Delta t)$.

9.2.5 Consistency

FDM is consistent when the truncation error tends to zero as Δx and Δt approach zero.

9.2.6 Convergence

FDM is said to be convergent as the exact solution of FDM u approaches the exact solution of PDE U, as both Δx and Δt approach zero. It is not easy to directly prove convergence. However, FDM is convergent if the scheme is proved to be *consistent and stable*.

9.2.7 Stability

When the computations are performed to an infinite number of significant figures (decimal digits), the solution $u(x, t)$ of the FDM will be exact. However, even in computers nowadays, round-off errors are introduced at each time step. So, the numerical solution we obtain is different from the exact solution. In some cases, round off errors are amplified, decrease, or stay the same. The scheme is *stable* when the amplification of the round off errors is bounded for all sections as time goes infinity. Unstable schemes result in very rapidly growing error in a few time steps. So, stability conditions must be satisfied.

9.2.8 CFL (Courant Friedrich Lewy) stability condition

$\Delta x \geq a\Delta t$, the courant number is defined as $C_N = (a/\Delta x/\Delta t) = (a\Delta t/\Delta x)$ and $C_N \leq 1$. This stability criterion applies only to linear equations (when the friction term is small). Even if the CFL condition is met, the scheme may become unstable: when the friction term is large (e.g., large friction factor, large time step, a large change in discharge, or small conduit diameter: according to the friction loss equation). The stability of the FDM may be done using von Neumann theory. In this approach, errors are expressed in a Fourier series simultaneously (for linear equations). The scheme is said to be stable if the errors decay as time increases. In the following, the pseudocode is shown for those who are interested in coding (Figure 9.5).

9.2.9 Example

For the given schematics, compare the pressure transient behavior when we shut the valve instantaneously (closing time $= 0$ at $t = 0$): Explicit scheme (McCormack's scheme) (Figure 9.6).

9.2.9.1 Given

Darcy Weisbach friction coefficient $= 0.02$; Time step $= 0.02$ sec; Distance step $= 100'$; Flow rate $= 19.5$ cfs.

9.2.9.2 Find

Pressure variation at $x = 500'$ (from the valve; at the center of the pipe).

```
# Supply initial data
old =0;
new=1;
time=0;
Time entered by the user;

Loop on m from 0 to M      % set initial data
V(old,m)=u0(x(m));
End of loop on m

Loop for time <TIME_MAX
        Time=time+k                    % time being computed
        n_time=n_time+1
        v(new,0)=beta(time)                    % set boundary conditions
        loop on m from 1 to M-1
                v(new,m)=.... % McCormack scheme
        End of loop on m
        v(new, M)=v(new,M-1)     % apply boundary condition

        old=new                    % reset for the next time step
        new=mod(n_time,2)
end of loop on time
```

Figure 9.5 Pseudocode for running hydraulic transients.

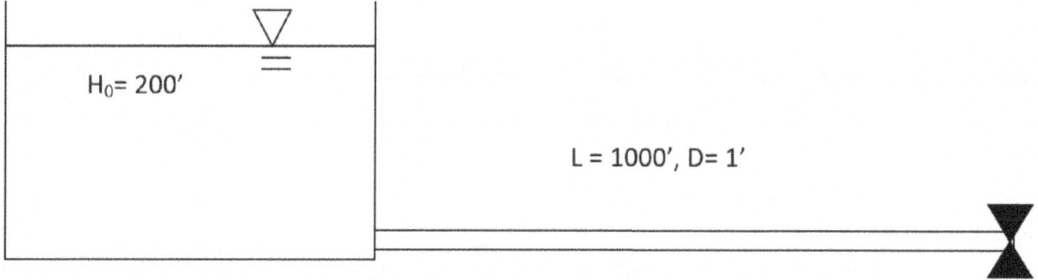

Figure 9.6 Reservoir problem.

9.2.9.3 Solution

MATLAB was used for running explicit scheme and is plotted in Excel (see Figure 9.7). It shows clear fluctuations in head values. The code is included below.

9.3 OTHER PHENOMENON OF INTERESTS | CAVITATION AND COLUMN SEPARATION

As mentioned previously, 'water hammer' is a transient flow phenomenon introduced in pipe flow systems by suddenly obstructing the flow. Consequently, there is a pressure rise and fall, and the

```
RESERVOIR PROBLEM
clear all;
delT = 0.02; %time step
h = 100;
M = 1000/h;
t_final = input('Enter end of time in seconds:');
time_step_N = t_final/delT;
a = 4000; %FT/SEC celerity
A = 0.785398; %FT2 area of pipe: PI/4*D⊥2, D = 1'
g = 32.2; %FT/sec2 gravitational acceleration
f = 0.02; %darcy-weisbach friction coeff.
D = 1;% FT; Diameter of the conduit
AA = delT*(a⊥2)/(h*g*A);%constant AA
BB = delT*g*A/h;% constant BB
CN = a*delT/h; % Courant Number should be less than 1
R = f/(2*D*A);
Hold = zeros(1,M + 1);
Qold = zeros(1,M + 1);
for m = 1:M + 1%set initial data
Hold(m) = 200-(20)*(m−1);
Qold(m) = 19.5;
end
Hstar = zeros(1,M + 1);
Qstar = zeros(1,M + 1);
Hnew = zeros(1,M + 1);
Qnew = zeros(1,M + 1);
% CP = Qold(M) + g*A/a*Hold(M)-R*delT*Qold(M)*abs(Qold(M));
% CN = Qold(2)-g*A/a*Hold(2)-R*delT*Qold(2)*abs(Qold(2));
% CA = g*A/a;
%--------------------------- McCommas Scheme-------------
for j = 1:time_step_N %time step
CP = Qold(M) + g*A/a*Hold(M)-R*delT*Qold(M)*abs(Qold(M));
CN = Qold(2)-g*A/a*Hold(2)-R*delT*Qold(2)*abs(Qold(2));
CA = g*A/a;
alterna = mod(j,2);
if alterna = =1
for i = 1:M
Hstar(i) = Hold(i)-AA*(Qold(i + 1)-Qold(i));
Qstar(i) = Qold(i)-BB*(Hold(i + 1)-Hold(i))-R*abs(Qold(i))*Qold(i)*delT;
end
```

Continued

```
for i = 2:M
Hnew(i) = 0.5*(Hold(i) + Hstar(i)-AA*(Qstar(i)-Qstar(i-1)));
Qnew(i) = 0.5*(Qold(i) + Qstar(i)-BB*(Hstar(i)-Hstar(i-1))-R*abs(Qstar(i))*Qstar(i)*delT);
end
Hnew(1) = 200; %set the left B.C.
Qnew(M + 1) = 0; %set the right B.C.
Qnew(1) = CN + CA*Hnew(1); %set the left B.C.
Hnew(M + 1) = CP/CA; %set the right B.C.
else
for i = 2:M + 1
Hstar(i) = Hold(i)-AA*(Qold(i)-Qold(i-1));
Qstar(i) = Qold(i)-BB*(Hold(i)-Hold(i-1))-R*abs(Qold(i))*Qold(i)*delT;
end
for i = 2:M
Hnew(i) = 0.5*(Hold(i) + Hstar(i)-AA*(Qstar(i + 1)-Qstar(i)));
Qnew(i) = 0.5*(Qold(i) + Qstar(i)-BB*(Hstar(i + 1)-Hstar(i))-R*abs(Qstar(i))*Qstar(i)*delT);
end
Hnew(1) = 200; %set the left B.C.
Qnew(M + 1) = 0; %set the right B.C.
Qnew(1) = CN + CA*Hnew(1); %set the left B.C.
Hnew(M + 1) = CP/CA; %set the right B.C.
end
% for mm = 1:(time_step_N) + 1
% time(mm) = mm*delT;
% end
%
% figure (1)
% plot(time,Hnew(1),'-g',time,Hnew(10),'-r', time, Hnew(21),'-r')
H1(j) = Hnew(4);
H5(j) = Hnew(5);
H10(j) = Hnew(6);
AAA = [H1;H5;H10];
for kk = 1:M + 1
Hold(kk) = Hnew(kk);
Qold(kk) = Qnew(kk);
end
end
```

pattern is repeated until the transient energy decays. Typical vapor pressures of water (10–40°C) range from 0.012 to 0.073 atmosphere; the total dissolved gas pressure of natural water is typically in the range of 0.8–1.2. When the fluid pressure drops below the constituent gases' saturation pressure, bubbles comprised of dissolved gases are formed, which is known as *gaseous cavitation* (Lee, 2008).

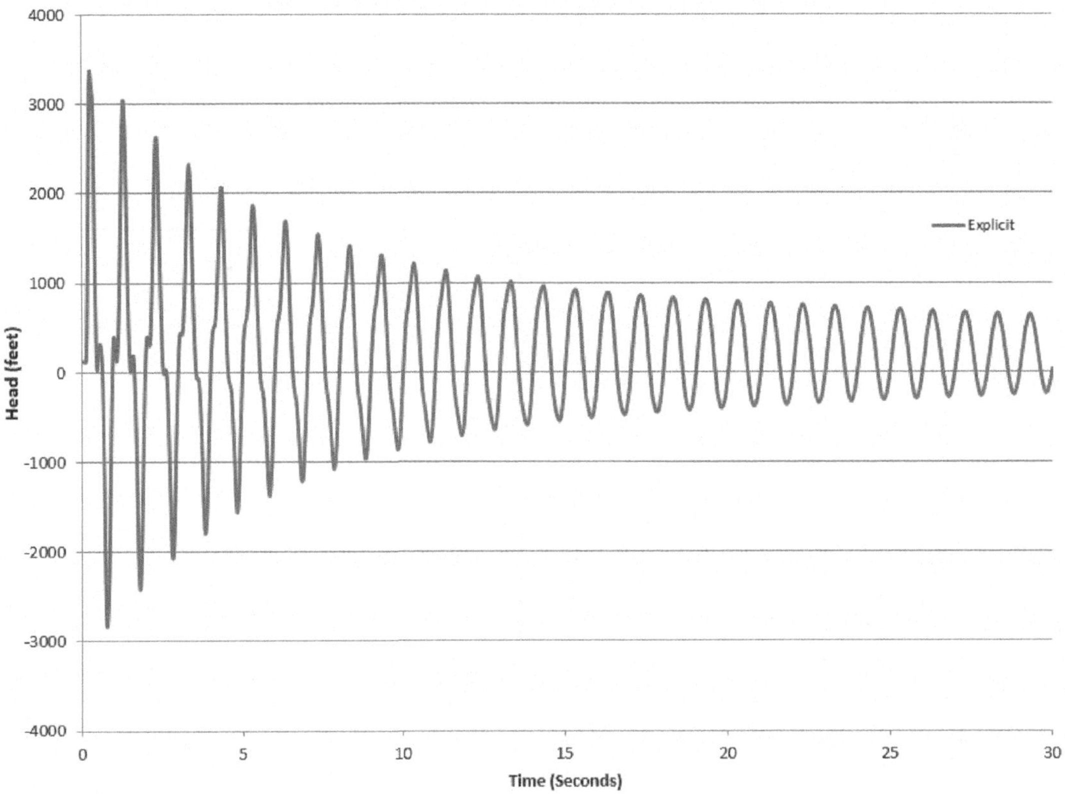

Figure 9.7 Reservoir results comparison.

When the pressure drops below the vapor pressure, vapor cavities are created in liquid by phase transformation, called vaporous cavitation.

In a liquid, gas may have two forms: dissolved and free gas. The dissolved gas is invisible in liquid and does not increase its volume and compressibility noticeably. However, free gas (or '*entrained gas*') is dispersed in the liquid as bubbles that may make the liquid look turbid. The liquid which is not entirely degassed will usually contain some entrained air in the form of microscopic or submicroscopic bubbles, either in the bulk of the liquid or near solid contaminants or near the container wall. Fluid mixtures can be categorized into five phenomena (Shu, 2003): (1) fully degassed fluid, (2) thoroughly degassed liquid with vapor, (3) liquid with dissolved gas, (4) liquid with dissolved and free gas, (5) liquid with dissolved gas, free gas, and vapor.

Vaporous cavitation can be present in cases (2) and (5). The cavity may become so large as to fill the entire or partial section of the pipe (known as *air pocket*) and divide the liquid into two columns in vertical pipes or pipes with steep slopes known as *column separation*. In horizontal pipes or mild slopes, however, a thin cavity is aggregated to the top of the pipe and extends over a long distance in the pipe (i.e. *cavitating flow*). So, the vapor cavities may be physically dispersed homogeneously or collected into a single or multiple void space, or a combination of the two phenomena. For gaseous cavitation, both liquid with dissolved and free gases can be observed. Free gas is distributed throughout the liquid in a homogeneous mix or lumped as pockets of free gas, trapped along the pipe wall, in pipe joints, in surface roughness, and crevices. In this article, we introduce modeling concepts for vaporous cavitation.

9.3.1 Discrete vapor cavity model (DVCM)

DVCM is assumed to have a vapor cavity quantity concentrated at each computational section (see Figure 9.6). This model is the most commonly used model for column separation (Bergant *et al.*, 2006). Cavities are allowed to form at any computational grid point when the computed pressure is below the vapor pressure. The pressure wave speed is assumed to be constant between the vapor. Also, the absolute pressure in the vapor is set equal to the vapor pressure:

$$p^* = p_v^* \tag{9.18}$$

The upstream and downstream discharges at a cavity are computed from *classic water hammer equations* Equations (9.2), (9.3), or (9.4), and the vapor cavity volume (V) can be obtained as below:

$$\text{Continuity equation for vapor cavity}: \frac{\partial \forall}{\partial t} = Q_2 - Q_1 \tag{9.19}$$

where $\forall$ is vapor volume, Q_1 is downstream flow rate, and Q_2 is upstream flow rate (Figure 9.8).

It is assumed that mass transfer during cavitation is ignored. Also, flow rate discontinuity is assumed at each computational node. So, there will be two predicted values of flow rates (Figure 9.6). This continuity equation for the vapor volume *V* is applied at each computing section. The liquid flow in the pipe is instantaneously and entirely separated by its vapor phase when the cavity is formed. However, in reality, when a cavity is formed in a section of the pipe, it usually expands and propagates in the direction of flow as an elongated bubble. So, this formulation's phenomenon does not necessarily occur in a horizontal or near horizontal pipe (Shu, 2003). It is known that DVCM may generate unrealistic pressure spikes with a multi-cavity collapse. However, the oscillations may be suppressed by assuming small gas volumes in each grid (Wylie & Streeter, 1993).

Column separation or cavitating flow is caused by the negative or rarefaction waves passing through the pipelines. When these waves meet a boundary such as a reservoir, they are reflected as positive waves. This raises the local pressure higher (than total dissolved gas pressure or vapor pressure). They can reduce the cavity's size during column separation and compress the bubbles in the cavitating flow region. When the cavities collapse or when the separated column rejoins, very high pressure that may burst the pipes are generated. According to Kranenburg (1974), the inclusion of gas release had no effect when only cavitating flow occurs, but the gas release effect is large when column separation occurred with the cavitating flow. The implosion of gas or air bubbles in pressurized conduits introduces extra shock waves, namely, the intense pressure wave in water produced by explosions that create violent

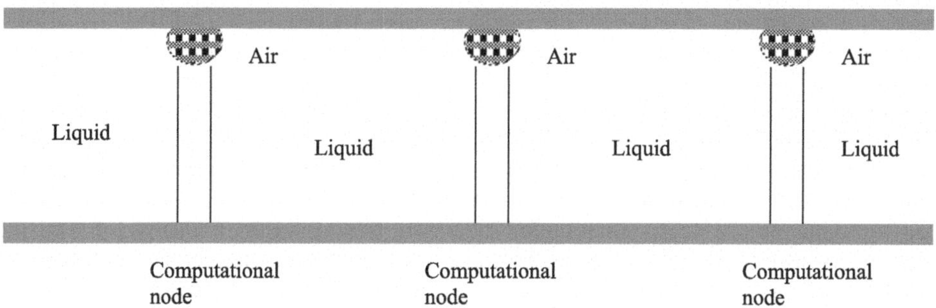

Figure 9.8 Discrete vapor cavity model sketch (modified from Bergant *et al.*, 2006).

pressure changes. Shock waves travel faster than the sound wave, and their speed increases as the amplitude is raised; however, the intensity of a shock wave also decreases faster than that of a sound wave. This is because some of the shock wave energy is dissipated due to the heat transfer in the water in which it travels. In this vein, steep waves or shocks may be generated at different boundaries due to abrupt changes in the discharge. However, damages from vaporous cavitation decrease with higher gas content as dissolved gas has a cushioning effect on implosion.

9.3.2 Short term pressure peaks following cavity collapse

According to Walsh (1964), the maximum possible pressure rise following the collapse of the first cavity at an upstream valve can be expressed as:

$$\Delta H_{\max} = \frac{a}{g}\left|V_f\right| + 2H_{RV} \tag{9.20}$$

where V_f is the velocity of the liquid column at the valve just before cavity collapse, and H_{RV} is the difference of reservoir head and vapor head at the valve. Wylie and Streeter (1993) also showed the pressure after collapsing of the first cavity for the case of instantaneous closure of upstream and downstream valves:

$$\Delta H_{\max} = \Delta H + 2\Delta H_{in} \tag{9.21}$$

Thus, the maximum pressure can be more than two times the Joukowski value.

9.4 TRANSIENT SIMULATIONS IN WATER DISTRIBUTION NETWORKS: TSNet

9.4.1 TSNet

Transient simulation in water networks (TSNet) is a Python package designed to perform hydraulic transients simulation in water distribution networks (Xing & Sela, 2020). TSNet adopts the Method of Characteristics (MOC) for solving the system of partial differential equations governing the unsteady hydraulics. The main capabilities of TSNet are: (1) allowing the user to select the computational time step and control numerical accuracy and computational complexity, (2) simulating transient system responses to the operation of valves and pumps, (3) simulating transient system response to background leakage and pipe bursts, (4) simulating open and closed surge tanks for controlling transient response, (5) simulating steady, quasi-steady, and unsteady friction models, (6) simulating instantaneous nodal demand changes using demand-pulse model, and (7) visualizing and postprocessing simulation results. In this section, we will see examples of running TSNet to simulate the transient events under different scenarios. For additional examples, see TSNet documentation (Xing & Sela, 2021a).

9.4.2 Use of Python, Spyder and Anaconda

Before looking into TSNet, it is beneficial to learn some Python basics. There are many useful resources out there, such as Python Programming and Numerical Methods: A Guide for Engineers and Scientists (Kong *et al.*, 2020).

In this section, we will be using Spyder as our Python environment. Spyder is a free and open-source integrated development environment (IDE) for scientific programming in the Python language (Spyder, 2021). Spyder is included by default in the Anaconda Python distribution (Anaconda, 2021), which comes with everything you need to get started in an all-in-one package. To download Anaconda, please visit the Anaconda website (Anaconda, 2021) and download the installer for your platform, that is Windows, Mac, or Linux. After installation, you can open Anaconda and should see something like Figure 9.9.

To run Spyder after installing it with Anaconda, you can open Anaconda Navigator, scroll to Spyder under Home, and click Launch. Then, you should see the Spyder interface as in Figure 9.10. The default layout of Sypder has three windows. *Editor* (left) is where you will be doing most of your

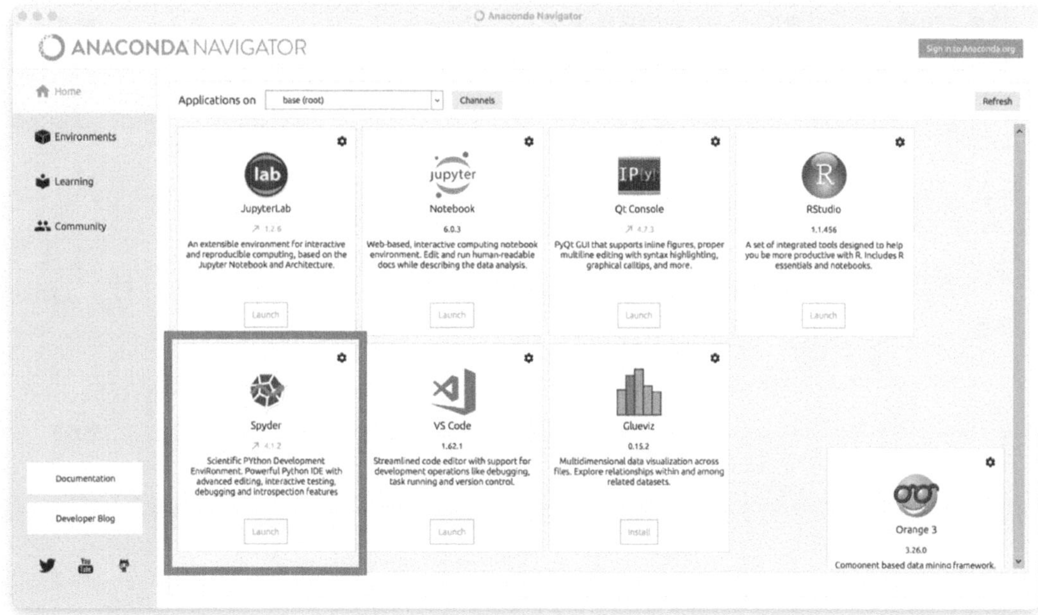

Figure 9.9 Anaconda navigator interface.

coding. In this window, you can create, edit, and save scripts and functions. You will run these scripts by selecting the green arrow at the top of the screen in the tool bar or by pressing F5. *Help* (upper right) is where you search for information on various functions within Python. For the upper right window, you can also choose to display *Variable Explorer*, where you can find all variables you create or import into Python, *Plots*, where you can view all figure outputs, and *Files*, where you can find and open various files under the current directory. On the lower right, *Console* is the main window for executing commands and viewing results. Press enter key to execute a command. Any code output and errors will be displayed in the Console. Then we can download the TSNet Python package by typing '! Pip install tsnet' in the Console as shown in Figure 9.10. This command will automatically download and install TSNet and other packages that TSNet depend on (e.g., Numpy, Matplotlib, WNTR, etc.).

9.4.3 Example application

Now that we have installed TSNet and have a basic understanding of how Python works, we can now move on to see how to use TSNet to simulate transient events. We will demonstrate how to use TSNet using an example network shown in Figure 9.11a, which is comprised of 113 pipes, 91 junctions two pumps, two reservoirs, three tanks, and one valve. The information of this network is stored in an EPANET INP file, Tnet2.inp, as shown in Figure 9.11b. The INP file and the codes that we will demonstrate here can be downloaded from Xing and Sela (2021b).

After installing TSNet, the main steps in setting up and running the transient model are: (1) read water network model input (EPANET INP) file and create the corresponding transient model; (2) set up a transient model by defining additional transient-related features, such as wave speed and time step; (3) define a transient scenario, such as operational changes in valves and pumps, and pipe bursts; (4) define if the system includes background leak conditions and obtain the initial conditions by conducting a steady-state simulation; (5) perform transient simulation using method of characteristics

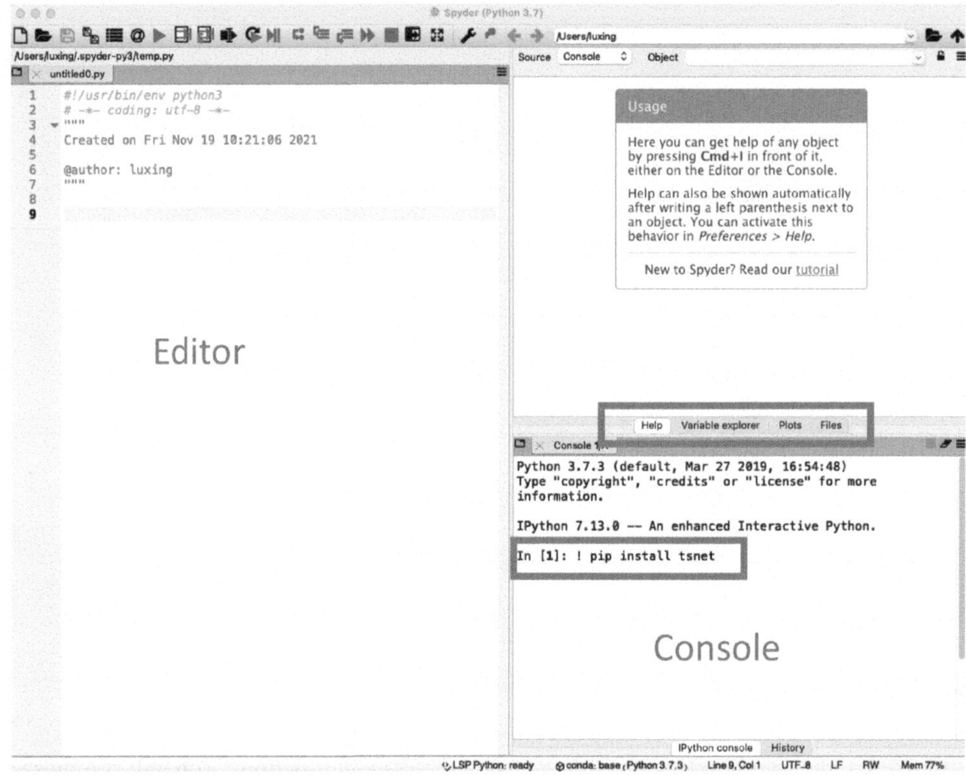

Figure 9.10 Spyder IDE interface.

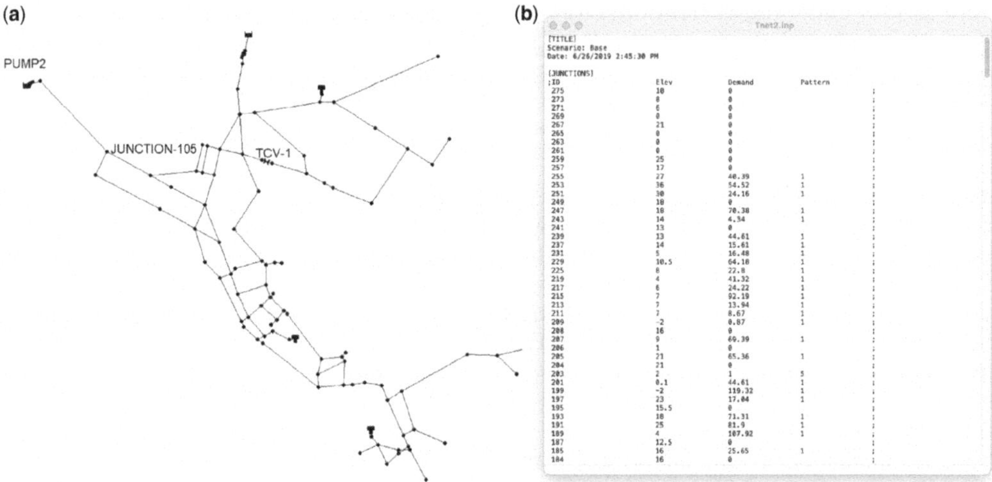

Figure 9.11 Example network: (a) network topology and (b) screenshot of Tnet2.inp.

```
1    import tsnet
2    inp_file = 'networks/Tnet2.inp'
3    tm = tsnet.network.TransientModel(inp_file)
4    # Set wavespeed and time
5    tm.set_wavespeed(1200.) # [m/s]
6    tf = 60 # simulation duration[s]
7    dt = 0.01 # time step [s]
8    tm.set_time(tf, dt)
```

Figure 9.12 Create transient model (lines 1–3) and define wave speed and time step (lines 5–8).

(MOC) (Wylie & Streeter, 1993); and (6) obtain and visualize flow and pressure results. The following sections will detail how to use TSNet to set up the transient model and simulate various transient events, including valve closure, pump shutdown, and pipe bursts.

9.4.4 Create and set up a transient model

To use TSNet for transient simulations, we first need to import the TSNet package to enable TSNet APIs (line 1), and read the EPANET INP file to import the network information and create the transient model (lines 2–3), as shown in Figure 9.12. In this example, the EPANET INP file is Tnet2.inp, which locates in the networks folder. The INP file contains all the information about network elements, for example junctions, pipes, reservoirs, tanks, pumps, and valves, as well as their characteristics, such as elevation and demands at junctions and pipe diameter, length, and roughness coefficient. More information about INP files can be found in Rossman (2000). In addition to the information included in the INP file, we also need to specify the wave speeds for each pipe (line 5), simulation duration and the time step (lines 6–8) as shown in Figure 9.12. These are required for TSNet to define the numerical grid that will be used to solve the equations that model transient hydraulics. More information about defining wave speeds and time steps in transient modeling can be found in Wylie and Streeter (1993). In this example, we assume the wave speeds for all pipes are 1200 m/s, simulation duration is 60 s, and time step is 0.01 s. It should be noted that the codes behind the # sign are comments for explanation purposes and are not executed when running the code.

Type these commands in the Editor window, save and run your script. In the Console window you should see 'Simulation time step 0.01043 s', and in the variable explorer you should see four variables (dt, inp_file, tf, tm). To test that the model was created properly, type the command tm in the Console, and you should see something like <tsnet.network.model.TransientModel at 0 × 7fb9201db4e0>. Now that we have created and set up a transient model in TSNet, we can move forward to define different scenarios for transient simulations.

9.4.5 Valve closure

Let us start with a valve closure scenario. Rapidly closing a valve in the system may cause a sudden change of flow rate, and the force resulting from the change in velocity will cause a pressure increase or decrease that may be significantly greater than the normal pressure in the pipeline. This pressure disturbance then propagates in the water network causing further pressure and velocity changes in the distribution system. We can simulate a valve closure event in TSNet by defining the valve closure start time (ts) from the beginning of the simulation, closure duration (tc), valve opening percentage when the operation is completed (se), and the operating constant (m), which characterizes the shape of the closure curve. These parameters define the valve closure curve as shown in Figure 9.13, where the solid and dashed lines represent m = 1 and m = 2, respectively.

In this example, we simulate the closure of TCV-1 (shown in Figure 9.11a), which starts at ts = 1 s, and takes tc = 1 s to completely close the valve, using the code shown in Figure 9.14.

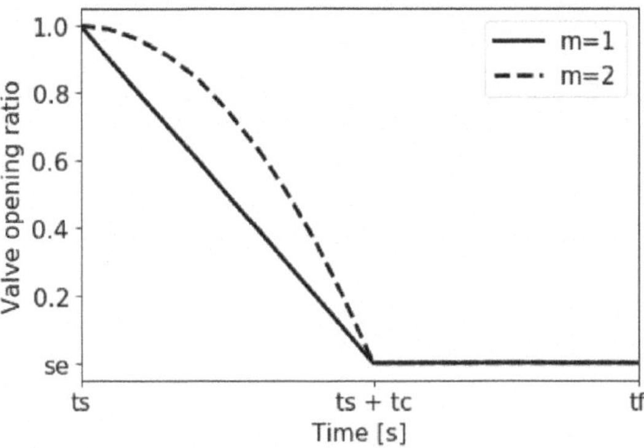

Figure 9.13 Valve closure operating curve.

```
 9    # Set valve closure
10    ts = 1 # valve closure start time [s]
11    tc = 1 # valve closure period [s]
12    se = 0 # end open percentage [dimensionless]
13    m = 1 # closure constant [dimensionless]
14    tm.valve_closure('TCV-1',[tc,ts,se,m])
```

Figure 9.14 Define valve closure.

Once the transient conditions are defined, the transient model is initialized by running a steady state simulation. TSNet uses Water Network Tool for Resilient (WNTR) (Klise *et al.*, 2018) to simulate the steady-state hydraulics, either demand or pressure driven in the event of background leaks. We initialize the hydraulic transients at t0 = 0 and use the demand driven simulator for the steady-state calculation in lines 17–19 in Figure 9.15. Moving forward, we specify the object name for saving results and run the actual transient simulation using lines 21–22 in Figure 9.15.

At the beginning of a transient simulation, TSNet will report the approximate simulation time. The computation progress will also be printed in the Console as the simulation proceeds, as shown in Figure 9.16.

Once the simulation is completed, we can then plot the pressure head at any junction, for example JUNCTION-105 (shown in Figure 9.11), using the command shown in Figure 9.17.

The pressure head at JUNCTION-105 versus time is shown in Figure 9.18. It can be observed that the transient wave induced by the valve closure arrives to JUNCTION-105 in around 7 s and causes a pressure jump of 5 m amplitude. The pressure then fluctuates and returns to the original level after approximately 35 s. The results indicate that the fast valve closure can introduce pressure transients; however, the amplitude of transient is not very significant. We will see much larger transients in the following examples.

```
16    # Initialize steady state simulation
17    t0 = 0. # initialize the simulation at 0s
18    engine = 'DD' # or PPD
19    tm = tsnet.simulation.Initializer(tm, t0, engine)
20    # Transient Simulation
21    results_obj = 'Tnet2' # name of the object for saving results
22    tm = tsnet.simulation.MOCSimulator(tm,results_obj)
```

Figure 9.15 Initialize (lines 17–19) and running transient simulation (lines 21–22).

```
Simulation time step 0.01043 s
Total Time Step in this simulation 5752
Estimated simulation time 0:06:24.526952
Transient simulation completed 9 %...
Transient simulation completed 19 %...
Transient simulation completed 29 %...
Transient simulation completed 39 %...
Transient simulation completed 49 %...
Transient simulation completed 59 %...
Transient simulation completed 69 %...
Transient simulation completed 79 %...
Transient simulation completed 89 %...
Transient simulation completed 99 %...
```

Figure 9.16 Runtime outputs – computation time and progress.

```
35      tm.plot_node_head(['JUNCTION-105'])
```

Figure 9.17 Plot head results.

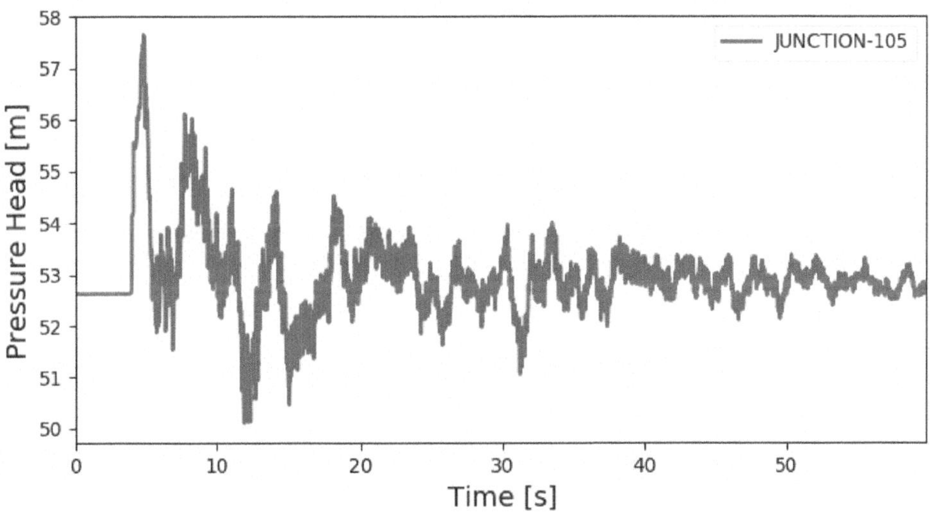

Figure 9.18 Pressure transients at JUNCTION-105 when closing valve TCV-1.

9.4.6 Pump shutdown

Now, let us move on to the pump shutdown scenario. When PUMP2 (see Figure 9.11) is being shut down, the rotating pump impeller begins to decelerate with the pressure dropping on the discharge side of the pump and rising on the suction side. The resultant transient may quickly lead to column separation with ensuing hard-to-predict consequences (Larock *et al.*, 1999). Hence, it is very important to be able to perform transient simulation to determine whether dangerous negative pressures may develop. With TSNet, we can define pump shutdown by specifying how pump rotational speed changes over time using pump shutdown start time (ts), operation duration (tc), the ratio of final pump rotational speed to the original speed (se), and the operating constant (m), which characterizes the

```
1    import tsnet
2    inp_file = 'networks/Tnet2.inp'
3    tm = tsnet.network.TransientModel(inp_file)
4    # Set wavespeed and time
5    tm.set_wavespeed(1200.) # [m/s]
6    tf = 20 # simulation duration[s]
7    dt = 0.01 # time step [s]
8    tm.set_time(tf, dt)
9    # Defien pump operation
10   tc = 1 # pump closure period
11   ts = 1 # pump closure start time
12   se = 0 # end open percentage
13   m = 1 # closure constant
14   pump_op = [tc,ts,se,m]
15   tm.pump_shut_off('PUMP2', pump_op)
16   # Initialize steady state simulation
17   t0 = 0. # initialize the simulation at 0s
18   engine = 'DD' # or PPD
19   tm = tsnet.simulation.Initializer(tm, t0, engine)
20   # Transient Simulation
21   results_obj = 'Tnet2' # name of the object for saving results
22   tm = tsnet.simulation.MOCSimulator(tm,results_obj)
23   # Visulize results
24   tm.plot_node_head(['JUNCTION-105'])
```

Figure 9.19 Transient simulation under the pump shutdown scenario.

shape of the operation curve in the same way as defined for the valve closure. Following the same simulation process as the valve scenario, we can perform the transient simulation and plot the pressure at JUNCTION-105. Figure 9.19 shows the entire code, starting with importing the TSNet package, network INP file, setting up the transient model, initializing and running the transient simulation, and plotting results. Figure 9.20 shows that the pressure wave generated by the pump shut-off reaches JUNCTION-105 after approximately 3 s and introduces a pressure drop with amplitude greater than 12 m. The pressure then fluctuates until reaching a new steady state after approximately 30 s.The

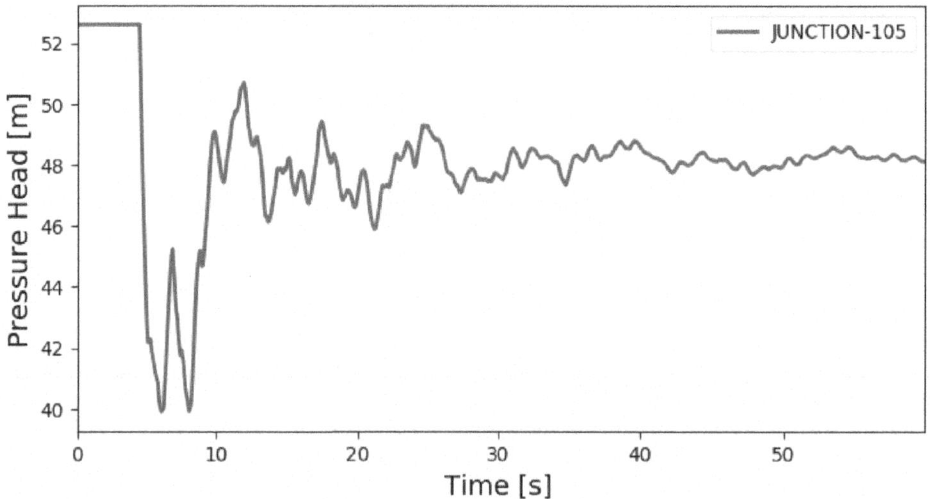

Figure 9.20 Pressure transients at JUNCTION-105 when shutting down PUMP2.

results indicate that pump shutdown, especially when operated quickly, can generate significant transients in the system. Therefore, it is essential to evaluate the impacts of pump operations on the pipelines and design an appropriate procedure to guide pump operations (Boulos *et al.*, 2005).

9.4.7 Pipe burst

Now we move to the simulation of pipe bursts. Pipe bursts, defined as sudden pipe rupture and break events, can introduce sudden and rapid hydraulic transients, which then propagate in the pipe system. TSNet simulates pipe bursts using the orifice equation, which quantifies the pressure-dependent burst discharge where $Q_b(t) = k_b(t)H_b(t)$, where t is time, H_b is the pressure head at the location of the burst, and k_b is the lumped burst coefficient, which changes with time, and aggregates the size of the leak, units, and burst coefficients (Larock *et al.*, 1999). In TSNet, pipe bursts can be specified at junctions. To model the burst occurring along a pipe, the user should introduce a new junction at the location of the burst in the INP file.

An example of simulating a pipe burst using TSNet is shown in Figure 9.21. In this example, a burst event at JUNCTION-105 is simulated by defining the burst location (line 13) and how the lumped burst coefficient (k_b) changes with time. The change in (k_b) is defined by specifying the burst start time (line 10), time for the burst to fully develop (line 11), and final burst coefficient when the burst is fully developed (line 12). In this example the burst starts at ts = 1 s, takes tc = 1 s to fully develop and reach a final burst coefficient of 0.01. The pressure head at JUNCTION-105 is shown in Figure 9.22. It can be seen that the pressure head at JUNCTION-105 decreases by more than 25 m as the burst is developing between 1 and 2 s. The pressure then recovers gradually to a pressure slightly lower than the original level. The results suggest that pipe bursts can introduce significant transient pressure changes in the system, and it is possible to detect pipe bursts by monitoring pressure signals.

9.4.8 Other applications

In addition to valve closure, pump shutdown, and pipe bursts, TSNet can also be used to simulate other transient events, such as valve opening, pump startup, and demand pulses with and without background leaks. Users can also test the effects of including open and closed surge tanks on damping pressure transients. Different friction models, that is steady, quasi-steady, and unsteady friction

```python
import tsnet
inp_file = 'networks/Tnet2.inp'
tm = tsnet.network.TransientModel(inp_file)
# Set wavespeed and time
tm.set_wavespeed(1200.) # [m/s]
tf = 20 # simulation duration[s]
dt = 0.01 # time step [s]
tm.set_time(tf, dt)
# Add burst
ts = 1 # burst start time
tc = 1 # time for burst to fully develop
final_burst_coeff = 0.01 # final burst coeff [ m^3/s/(m H20)^(1/2)]
tm.add_burst('JUNCTION-105', ts, tc, final_burst_coeff)
# Initialize steady state simulation
t0 = 0. # initialize the simulation at 0s
engine = 'DD' # or PPD
tm = tsnet.simulation.Initializer(tm, t0, engine)
# Transient Simulation
results_obj = 'Tnet2' # name of the object for saving results
tm = tsnet.simulation.MOCSimulator(tm,results_obj)
# Visulize results
tm.plot_node_head(['JUNCTION-105'])
```

Figure 9.21 Transient simulation under the pipe burst scenario.

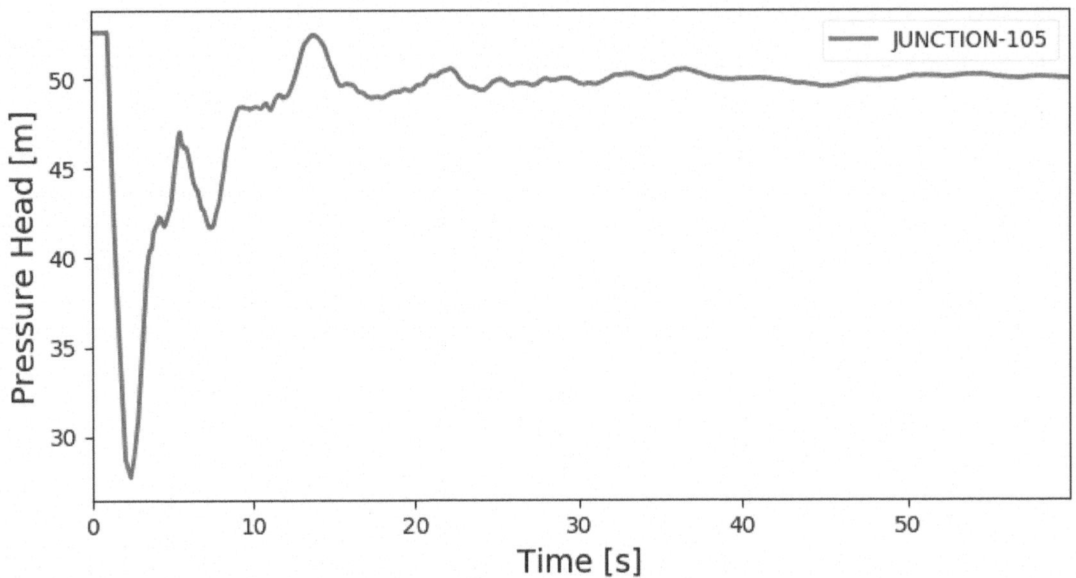

Figure 9.22 Pressure transients at JUNCTION-105 when a burst occurs at JUNCTION-105.

models, are implemented in TSNet. Additionally, more simulation results can be accessed, such as burst discharge, pipe flow rate, and pipe flow velocity. For more information about TSNet, please refer to the online documentation (Xing & Sela, 2021a).

REFERENCES

Anaconda. (2021). Anaconda: Your Data Science Toolkit. Available at: https://www.anaconda.com/products/individual (last accessed 3 March 2022)

Anderson J. D. (1995). Computational Fluid Dynamics. McGraw-Hill International Editions, Mechanical Engineering Series, New York, NY. Available at: https://www.amazon.com/Computational-Fluid-Dynamics-John-Anderson/dp/0070016852 (last accessed 10 May 2022)

Bergant A., Simpson A. R. and Tijsseling A. S. (2006). Water hammer with column separation: A historical review. *Journal of Fluids and Structures*, **22**, 135–171, https://doi.org/10.1016/j.jfluidstructs.2005.08.008

Boulos P. F., Karney B. W., Wood D. J. and Lingireddy S. (2005). Hydraulic transient guidelines for protecting water distribution systems. *Journal-American Water Works Association*, **97**(5), 111–124, https://doi.org/10.1002/j.1551-8833.2005.tb10892.x

Chaudhry M. H. (1987). Applied Hydraulic Transients. Van Nostrand Reinhold, New York.

Karney B. and McInnis D. (1990). Transient analysis of water distribution system. *Journal of American Water Resources Association*, **82**(7), 62–70.

Klise K. A., Murray R. and Haxton T. (2018). An overview of the Water Network Tool for Resilience (WNTR). WDSA/CCWI Joint Conference Proceedings. Available at: https://ojs.library.queensu.ca/index.php/wdsa-ccw/article/view/12150

Kong Q., Siauw T. and Bayen A. (2020). Python Programming and Numerical Methods: A Guide for Engineers and Scientists, pp. 480. Academic Press, Paperback ISBN: 9780128195499 eBook ISBN: 9780128195505.

Kranenburg C. (1974). Gas release during transient cavitation in pipes. *ASCE Journal of the Hydraulics Division*, **100**(HY10), 1383–1398, https://doi.org/10.1061/JYCEAJ.0004077

Larock B. E., Jeppson R. W. and Watters G. Z. (1999). Hydraulics of Pipeline Systems, pp. 552. CRC Press, Boca Raton. Available at: https://www.taylorfrancis.com/books/mono/10.1201/9780367802431/hydraulics-pipeline-systems-gary-watters-roland-jeppson-bruce-larock

Lee J. (2008). Two Issues in Premise Plumbing: Contamination Intrusion at Service Line and Choosing Alternative Plumbing Material. Doctoral dissertation, Virginia Tech., VA, USA.

Lee J., Lohani V. K., Dietrich A. M. and Loganathan G. V. (2012). Hydraulic transients in plumbing systems. *Water Science and Technology: Water Supply*, **12**(5), 619–629, https://doi.org/10.2166/ws.2012.036

Rossman L. A. (2000). *EPANET 2 User Manual. Social Studies of Science*. Available at: https://epanet22.readthedocs.io/en/latest/ (last accessed 3 March 2022)

Shu J. J. (2003). Modeling vaporous cavitation on fluid transients. *International Journal of Pressure Vessels and Piping*, **80**, 187–195, https://doi.org/10.1016/S0308-0161(03)00025-5

Spyder. (2021). Spyder: the Scientific Python Development Environment. Available at: https://docs.spyder-ide.org/current/index.html (last accessed 3 March 2022)

Walsh S. P. (1964). Pressure generated by cavitation in a pipe (Doctoral dissertation, Syracuse University).

Wylie E. B. and Streeter V. L. (1993). Fluid Transients in Systems. Prentice Hall, Upper Saddle River, NJ.

Xing L. and Sela L. (2020). Transient simulations in water distribution networks: TSNet python package. *Advances in Engineering Software*, **149**, 102884, https://doi.org/10.1016/j.advengsoft.2020.102884

Xing L. and Sela L. (2021a). TSNet Documentation. Available at: https://tsnet.readthedocs.io/en/latest/ (last accessed 3 March 2022)

Xing L. and Sela L. (2021b). Github Repositories for Transient Simulaiton in Water Networks (TSNet). Available at: https://github.com/glorialulu/TSNet/tree/master/examples (last accessed 3 March 2022)

doi: 10.2166/9781789062380_0237

Chapter 10

Innovative methods for optimal design of water network partitioning

Armando Di Nardo and Giovanni Francesco Santonastaso*

Department of Engineering, Università Della Campania 'Luigi Vanvitelli', Via Roma 29, 81031 Aversa (CE), Italy
**Corresponding author. E-mail: armando.dinardo@unicampania.it*

LEARNING OBJECTIVES

At the end of this chapter, you will be able to:

(1) Understand the main characteristics of Water Network Partitioning (WNP).
(2) Explain the advantages and drawbacks of WNP.
(3) Distinguish empirical and automatic approaches.
(4) Run a basic automatic procedure based on Python code.

10.1 INTRODUCTION

One of the most effective ways to reduce water distribution network (WDS) complexity is to apply the paradigm of 'divide and conquer' (Di Nardo *et al.*, 2014c), which exploits the property that complex systems can be better analyzed if it can be split into many sub-components.

This technique was proposed in England in the early 1980s (Water Authorities Association and Water Research Centre, 1985; Water Industry Research Ltd., 1999; Wrc/WSA/WCA Engineering and Operations Committee, 1994) and is now implemented in many countries. It consists of defining smaller water districts or sectors, defined as district meter area (DMA), obtained through the permanent insertion of boundary valves and flow meters along properly selected pipes. This can significantly improve management and maintenance, and, specifically, the water balance estimation for water leakage, pressures control, and water security from intentional contaminations (Di Nardo *et al.*, 2015a; Grayman *et al.*, 2009).

In Figure 10.1, a layout of permanent Water Network Partitioning (WNP) with three DMAs is shown, highlighting flow meters, gate valves, and district boundaries.

This technique, defined more recently in Di Nardo *et al.* (2013) as WNP, provides a series of interventions on the WDSs that require a careful economic planning by the managing authority; furthermore, it envisions the use of modern monitoring systems (remote control, etc.) which are generally becoming less expensive, and which, to be implemented, only await a new management policy. It is evident that having a network divided into smaller sub-regions makes it easier to study and manage the system (Di Nardo & Di Natale, 2011; Water Industry Research, 1999).

The definition of an optimal partitioning layout is a crucial and arduous problem. Nowadays it is possible to provide new opportunities to the traditional approach of analysis, design and management

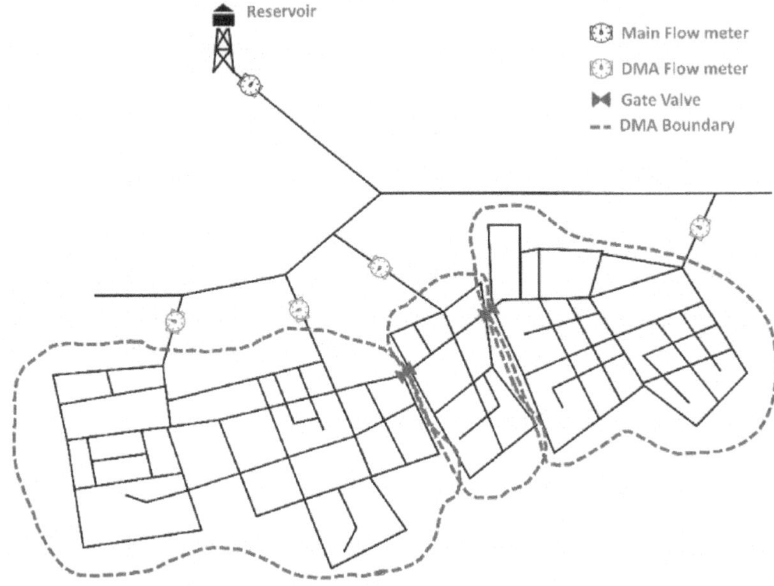

Figure 10.1 Scheme of a permanent Water Network Partitioning.

with the development of new monitoring and control technologies and with the recent growth of computational power used by simulation software. Therefore, WNP represents a crucial task not only for the technicians of the sector but also for the scientific community, because it modifies (and even challenges) some fundamental criteria followed in the design of water systems.

WNP contrasts with the traditional design criteria of the WDS with a high level of topological redundancy with many loops to have a more robust water system to face unforeseen changes in design conditions (such as pipe breaks or different distribution of water demand). Indeed, the introduction of the concept of permanent sub-districts and water sectors (Di Nardo *et al.*, 2015b) is in opposition to the traditional criterion followed in the field of the hydraulic constructions (Mays, 2000) designed with a multi-meshed network to improve its efficiency under different operating conditions. Network partitioning can indeed generate a hydraulic performance deterioration of the system (Di Nardo *et al.*, 2015b); in fact, when it is carried out in almost all cases on networks already designed and implemented using traditional design criteria, system efficiency can be partially and/or globally compromised. Indeed, the closure of some pipes with boundary valves can decrease, also significantly, the available hydraulic diameters of the whole network, with the increase of head loss and dissipated power and, consequently, worsening of the level of service for the users in terms of water pressure.

However, and conversely, the introduction of 'divide and conquer' for WDS design promotes innovation in management of water networks by introducing the concept of a Smart WAter Network (SWAN) as a key subsystem of the notion of Smart City (Di Nardo *et al.*, 2021).

Traditionally, WNP was achieved basing on empirical suggestions, such as the number of customers or parcels, length of pipes or other geometric or topological criteria; while the hydraulic alteration due to the insertion of gate, or boundary, valves is tested with hydraulic simulation based on 'trial and error' methods. These semi-empirical approaches are not effective for large water networks with thousands of nodes and links because the number of possible layouts of water districts is huge and requires heuristic optimization approaches.

In the last 10 years, many authors proposed different procedures to obtain automatically optimal water network partitioning layouts (Bui *et al.*, 2020), based on two phases, called clustering and dividing, with a systematic approach based on different innovative algorithms such as *graph algorithms, multilevel partitioning, community structure, spectral clustering*, and so on. Also, performance indices can measure the reduction of water network resilience because the reduction of network pipes availability, due to insertion of gate valves, reduces the level of service and the capacity of the water network to face different design conditions, as widely reported in Di Nardo *et al.* (2013).

The authors of this chapter developed the first automatic tool, called SWANP© (Smart Water Network Partitioning and Protection), to define the optimal layouts of water districts and sectors that is presented in this work.

10.2 ADVANTAGES OF WNP

The optimal design of DMAs simplifies monitoring and maintenance, with reference to the problems that will be explained in the following sections. Specifically, the main advantages of a permanent WNP, obtained inserting both gate valves and flow meters, can be arranged as follows:

- water balance;
- water pressure management;
- water contamination protection.

Furthermore, the data collection by monitoring of each DMA (and not of the whole network) can provide to water utilities other several detailed information related to each single district, such as demand distributions, categories of users, break frequencies, pressure levels, water quality, and so on., that can improve management, quality and cost of service.

10.3 WATER BALANCE

The most important problem of WDS management is the obsolescence of pipes and hydraulic devices (gate valves, control valves, flow meters, etc.) that generate low hydraulic performance (insufficient pressures, reduced resources during summer, poor water quality, etc.) and, above all, high values of Non-Revenue Water (NRW) both real and apparent, as reported in Lambert and Hirner (2000).

As is well known, the United Nations devoted the year 2003 (United Nation, 2003) to the problem of water in the world, and to the areas of the planet affected by water scarcity, suggesting actions to minimize waste and optimize resources. A year before, the Organization for Economic Co-operation and Development (OECD) already focused attention on the waste of water resources for the major industrialized countries, estimating that water losses in urban water networks account for around 30% (for the 30 most industrialized countries), exceeding the optimal economic level of 10 and 20% (OEAD, 2002). The more recent estimation in some industrialized countries, such as Italy, indicate water losses of about 40% (ISTAT, 2021).

Evidently, water balance estimation is crucial to evaluate the efficiency of a WDS and to help management activities reduce water leakage. The estimation of water loss is achieved as follows using a simple mass continuity statement:

$$\text{Water losses} = \text{Water Inflow} - \text{Water Consumption} \tag{10.1}$$

The practical application of the water balance is a very complex problem, from scientific and technical perspectives and for economic and management reasons. Some practical problems are: (a) water inflow depends on accuracy of flow meters; (b) water consumption depends on the ability of water utility to measure all user consumptions; (c) difficulty to identify user consumption (civil, industrial, commercial, etc.), authorized or not; (d) some water consumption is not measured (such as public fountains, schools, hospitals, etc.); (e) all measures have to be synchronized (or reported at the same time interval).

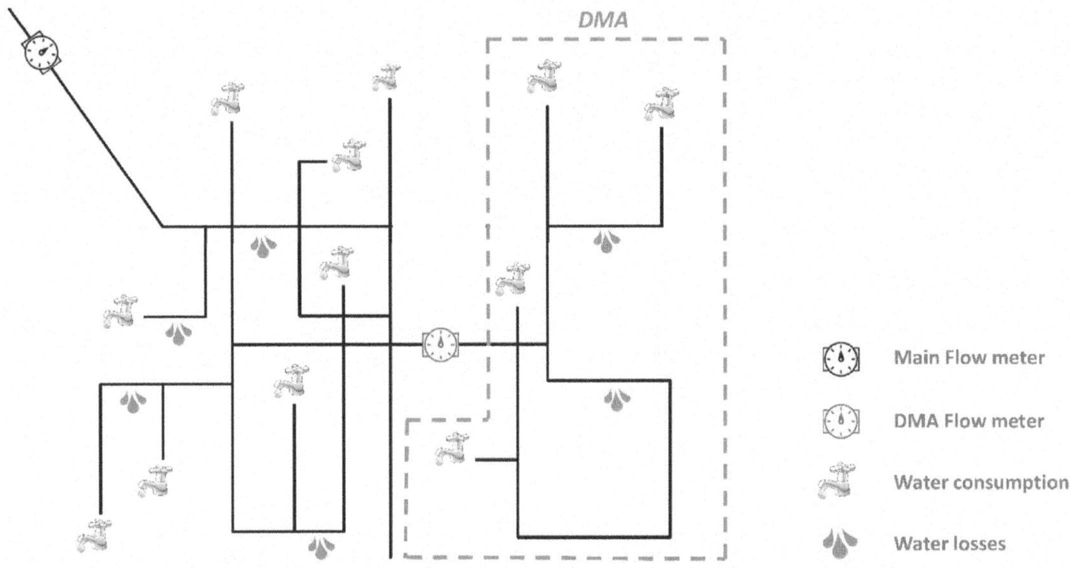

Figure 10.2 Water balance: Network vs DMA layout.

The correct application of the water balance estimation can also provide water utilities precious information about the percentage of real (or physical) losses, water really lost, apparent (or administrative) losses, and not billed water.

Furthermore, the water balance and evaluation of network integrity presupposes the exact definition of the different components of the volumes to estimate the water losses and to compare water networks of different systems in other locations. More than technical and scientific problems to correctly estimate the water balance, there have been difficulties related to the drafting of an international 'standard terminology'. So, the International Water Association (IWA) proposed a fundamental contribution (Lambert & Hirner, 2000) to define water balance components and compare the performance of the systems using evaluation indices equal for all countries (Lambert *et al.*, 1999).

Theoretically, we can carry out a water balance on the entire distribution network, but this operation is not very useful because it does not provide detailed information on which parts of the water network can be affected by higher leakage levels; so a DMA water balance is significantly better, as represented in Figure 10.2, allowing a more thorough investigation and monitoring of each district and supporting water utilities to prioritize the choice of economic investments for operations of water losses detection.

Therefore, the application of a *divide and conquer* approach with WNP optimal design allows the easier application of some methodologies for the water balance estimation developed in England (UK Water Industry, 1999; Wrc/WSA/WCA, 1994) such as minimum night flow (MNF) and minimum flow consumption (MFC).

10.4 WATER PRESSURE MANAGEMENT

Another advantage of WNP is to significantly facilitate the application of water pressure management to reduce water leakage.

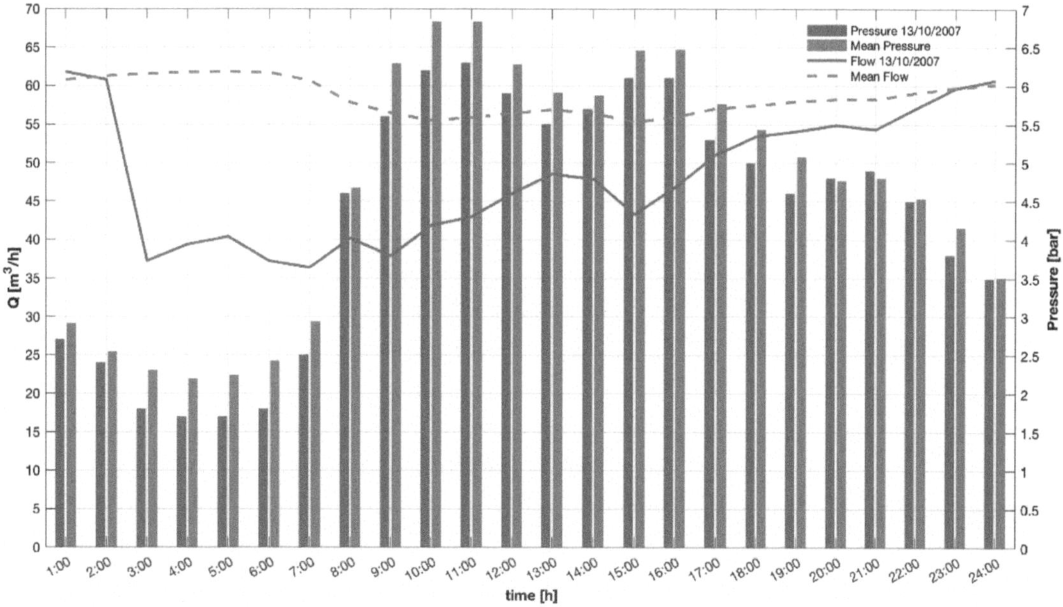

Figure 10.3 Flows reduction through pressure management.

As is known, water leakage Q^{Leakage} in the pipelines increases with increasing pressures P according to the relation (Khaled *et al.*, 1992; Lambert, 2000):

$$Q^{\text{Leakage}} = cP^{\gamma} \tag{10.2}$$

in which the values of the coefficients c and γ depend on the pipelines characteristics and the type of leak, while P (pressure or pressure head) is expressed in meters of water head.

Therefore, it is evident from Equation (10).(2) that the placement of pressure reducing valves (PRV) can bring about decreases in network water loss, as reported in Figure 10.3. The pressure reduction inevitably decreases the network hydraulic efficiency and the insertion of pressure regulation valves downstream to network reservoirs or sources can also reduce hydraulic performance of the whole water system using the same pressure control of all pipes. Therefore, a subdivision of the water network in some permanent DMAs can help the application of water pressure management inserting different PRVs upstream of each DMA and reducing water pressure for water saving. Also, it can help preserve the hydraulic performances of the system, guaranteeing the minimum level for the users in each DMA. In other terms, WNP also allows adjustment of the pressure values in each DMAs, considering the different needs of the urban areas (Alonso *et al.*, 2000).

10.5 WATER CONTAMINATION PROTECTION

Recent applications of water network partitioning have also shown interesting benefits with respect to protecting water systems from intentional contamination according to the dual-use value criteria (Di Nardo *et al.*, 2015a; Grayman *et al.*, 2009). Indeed, WNP has some primary aims ('main-use value'), related to water balance, pressure management, leakage reductions, and so on., and a secondary aim (or 'dual-use value') that consists of providing water protection from accidental or intentional contaminations. In this manner, the water distribution system protection obtained with WNP is

Figure 10.4 Simulation results of risk mitigation from terroristic attack of Matamoros water distribution network (Di Nardo *et al.*, 2015a).

capable of a likely return on investment because evidently only a small portion of the system lifetime will be spent on network protection while the majority of the system lifetime will be spent on the day-to-day management of achieving the main goals, as illustrated in Di Nardo *et al.* (2015a).

The authors investigated how WNP can reduce the risk of user contamination and limit the effects of a malicious (terroristic) act on water distribution systems. Specifically, Di Nardo *et al.* (2015a) showed that an optimal design of permanent DMAs can reduce exposures due to terrorist contamination with cyanide. This is done by closing all gate valves and quickly sectorizing the attacked district. The analysis was carried out on a real water distribution network comparing different sectorization scenarios and the simulation results showing the effectiveness of an early warning system coupled with WNP to significantly reduce the contamination risk for users.

In Figure 10.4, a simulation on the Matamoros network is reported, in which the triangle indicates the insertion point of contamination attack, light gray being the isolated DMA (i-DMA) after contamination alarm, and the dot is the exposed user without isolating district and the circle is with isolating actions. The effectiveness of WNP with isolation is clear: the number of exposed users, proportional to circle dimension, are significantly lower. More details can be found in Di Nardo *et al.* (2015a).

10.5.1 Clustering and dividing

As anticipated, the main problem of WNP is represented by the perturbation on the water distribution system due to pipe closing. Indeed, the insertion of gate valves can also significantly reduce the water network performance in terms of alternative paths of flows in case of pipe breaks (decrease of topological redundancy) and nodal water pressures (decrease of energy redundancy).

Physical (or permanent) district metering gives more opportunities than virtual districting metering, which uses only flow meters for water balance without closing the pipe (Di Nardo *et al.*, 2018). The permanent definition of DMA allows to simplify the monitoring and managing of WDS and to optimize and simplify water pressure management for leakage reduction thanks to insertion of district pressure regulation valves (PRV). In addition, physical district metering can also be used to protect water networks from accidental or intentional contamination, implementing a dual-use approach (Di Nardo *et al.*, 2015a).

On the other hand, this methodology is complicated to achieve because, by intervening in a physical way on the system (with closing pipes by gate valves), it is necessary to verify the variations of the system with respect to the initial conditions through hydraulic simulation and calibration techniques (Di Nardo & Di Natale, 2011).

The main outcomes that can be achieved through permanent WNP optimal design include (but are not limited to): (a) minimize the alteration of hydraulic performance (b) minimize the number of flow meters (the best management condition occurs when a single meter is installed for each district) in order to simplify the computation of water balance (Twort *et al.*, 2000).

The literature offers empirical suggestions for water network partitioning based on DMAs characteristics (number of users, pipes length, etc.) (Water Industry Research, 1999); or 'trial and error' approaches used with hydraulic simulation software (Di Nardo *et al.*, 2013). However, these suggestions and approaches are very difficult to apply to large water supply systems. In the last 10 years, many optimization techniques have been proposed, based on graph and network theory, that have significantly improved water network partitioning.

Several suggestions about DMA size can be found in the technical literature, that propose to include:

- 1000–3000 properties (Water Authorities Association and Water Research Centre, 1985);
- 2500–12 500 inhabitants with 5–30 km of water network (Butler, 2000);
- a number of properties up to 1000 (small DMA) and 3000 (medium DMA) and 5000 (large DMA) (as recommended by the UK Water Industry Research).

These guidelines cannot be easily extended to large water supply systems since they are based on empirical considerations, and sometimes on a small number of case studies.

Different optimization methods allow to define automatic procedures for water network partitioning (or sectorization) (Bui *et al.*, 2020). Generally, the procedures are divided into the two phases discussed below (Di Nardo *et al.*, 2016d; Perelman *et al.*, 2015).

10.5.1.1 Phase 1

Clustering is aimed at defining the shape and the dimensions of the network subsets in order to minimize the number of connections (or other characteristics like diameter, length, conductance, etc.) balancing the number of nodes (or other characteristics like flow, pressure, etc.) for each district.

As shown in Figure 10.5, with reference to a simple network clustered in two subnetworks (highlighted in red and blue colors in three different ways) shows the importance of clustering, minimizing the number N_b of boundaries and balancing the nodes. In Figure 10.5a, there are only three links (or boundaries) between two subnetworks but this solution is not well balanced with six red nodes and 12 blue nodes. Figure 10.5b shows a perfect balanced scheme with nine nodes both for blue and red nodes but a significantly higher number (seven) of boundaries. Finally, in Figure 10.5c shows the best clustering with a perfect balance of nodes (nine) and the minimum number of boundaries (three).

Therefore, the example shows that already with a very small network, different clustering layouts are possible. In a large water network, the problem to find the optimal solution in terms of minimization of elements between the clusters (links or boundaries) and of balancing of nodes or other characteristics in a way that the similarity (or the density) in each cluster is maximized (as number of nodes, length of pipes or flow delivered, etc.) is an NP-hard problem (Fortunato, 2010).

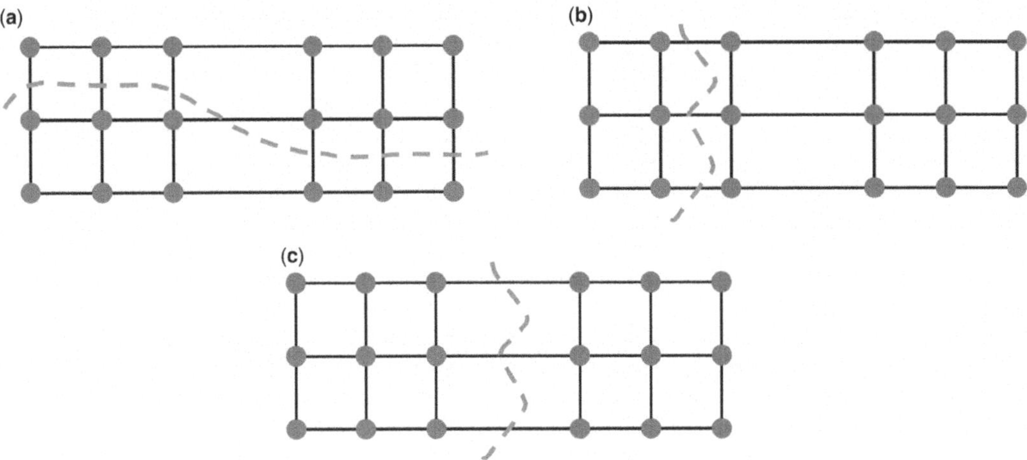

Figure 10.5 An example of possible clustering of a small network.

10.5.1.2 Phase 2
Dividing is aimed at physically partitioning the network by selecting boundaries (pipes) in which to insert flow meters or gate valves, as reported in Figure 10.6.

In the case of a small network, such as that represented in Figure 10.6, this phase, once the number N_{fm} of flow meters is fixed, can be carried out with the need of hydraulic software, permutatively, inserting the number of boundary valves $N_{bv} = (N_b - N_{fm})$, minimizing the alteration of hydraulic performance of water distribution network due to the closure of some pipes with the insertion of boundary valves between clusters. In the dividing phase, for large water networks, this problem is very complex and it is impossible to test all permutations of the possible positioning of flow meters and boundary valves in links between clusters.

This problem is an NP-hard problem (Bodlaender *et al.*, 2010) and it requires heuristic algorithms to find optimal solutions (Tindell *et al.*, 1992). In other terms, once all the N_b boundary pipes between clusters have been defined, those that can be closed must be chosen among all the possible combinations N_C of water network partitioning layouts, expressed by the following binomial coefficient:

$$N_C = \binom{N_b}{N_{fm}} = \frac{N_b!}{N_{fm}!(N_b - N_{fm})!} \tag{10.3}$$

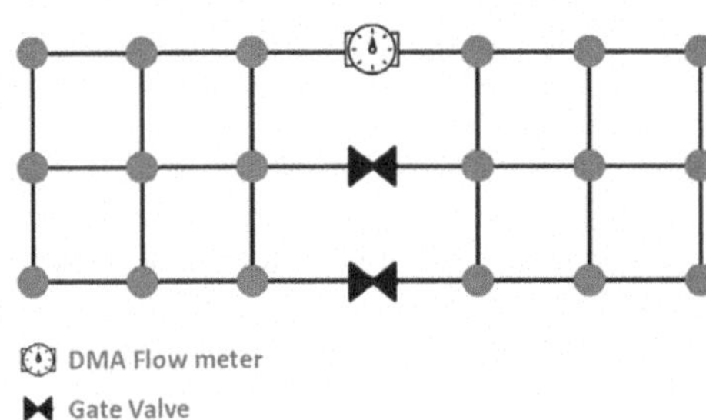

Figure 10.6 An example of possible dividing of a small network.

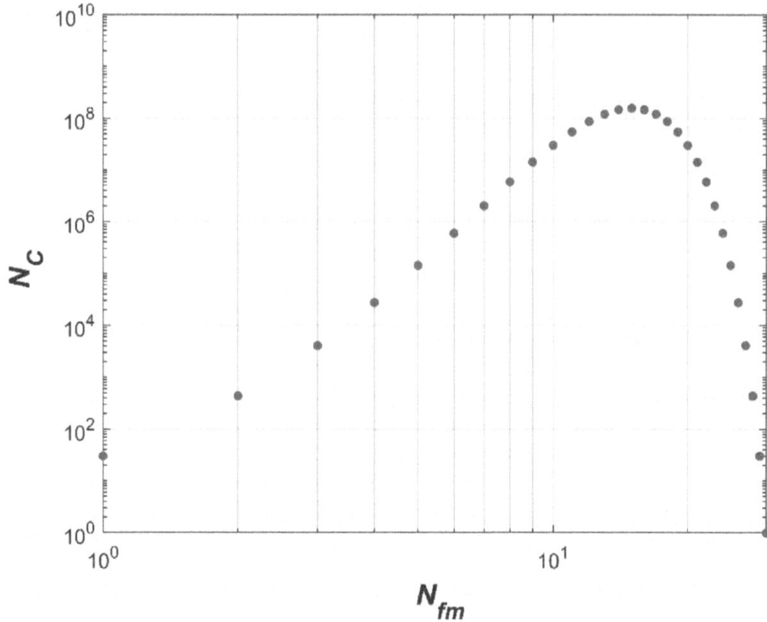

Figure 10.7 Number N_C of possible dividing layout of WNP for a small network changing N_b.

in which, already with a small network with 30 boundary pipes (N_b) and only 10 flow meters (N_{fm}), the number of all possible water partitioning layout, N_C, reachs about 3×10^7, as reported in Figure 10.7.

It is important to emphasize that even for a small water supply network and for a small number k of DMAs, N_C can be such a large number that it is often computationally impossible to investigate the entire solution space.

Therefore, it is clear that both the phases of clustering and dividing require us to define a permanent water network partitioning, and cannot be achieved using a traditional approach based on empirical suggestions or hydraulic simulation based on 'trial and error' methods if an effective optimal solution is needed. Indeed, these empirical or semi-empirical approaches are not effective for large water networks and require automatic procedures, which will be explained in the following section.

10.5.2 Innovative methods for optimal WNP design

As anticipated, traditional approaches for WNP cannot find the optimal design of DMAs for large water distribution networks. In this section, we introduce some innovative methods based on different algorithms, often developed in other disciplines for different classes of problems.

With reference to the clustering phase, the main methods proposed in the literature (Di Nardo *et al.*, 2018) to obtain a WNP are based on the following techniques:

(1) *graph algorithms* (Jacobs & Goulter, 1989; Savic & Walters, 1995; Tzatchkov *et al.*, 2006) starting from the representation of the water network as a simple weighted graph considering $G = (V, E)$, where V is the set of n vertices (or nodes) and E is the set of m edges (or pipes). Subsequently, the network is defined by a $n \times n$ connectivity matrix A and the matrix of weights W n by n (matrix of the intensity of the connections between nodes). Then, the application of different techniques of graph theory, in particular related to the search for minimum paths

(with or without the use of weights on links and nodes), allows us to obtain groupings of nodes on which it is then possible to apply the next dividing phase. Through these techniques it is possible to quickly identify the districts in the subsequent dividing phase and to guarantee a minimum service level compatible with original network reliability (Di Nardo & Di Natale, 2011; Di Nardo *et al.*, 2014a). The 'least important' or 'most redundant' sections are identified and, at the same time, the number of sections on which it is needed to insert gate valves and/ or meters is reduced.

(2) *multilevel partitioning* (Di Nardo *et al.*, 2015c) that starts from techniques implemented in informatics tools allows us to automatically obtain water network clustering, minimizing the number of links between districts. In fact, for simulations that need huge computational power like, for example, simulations based on finite element methods, parallel computation can be used. In this case, it is necessary to distribute the finite element mesh among different processors. This distribution, to improve performance, must be made according to two main rules: (1) an equal number of finite elements has to be allocated to each processor for balancing the workload; (2) a minimum number of adjacent elements between processors has to be found for reducing communication overhead. This problem can be assimilated to partitioning of a computational mesh in a k-way or in k-processors that will perform each computational process. The mesh is commonly schematized by a graph with vertices corresponding to individual computational processes (e.g., finite elements) and with links corresponding to their connections. Starting from this schematization of the mesh, partitioning techniques of a graph in k-way were developed in Computer Science for the optimal allocation of a computational mesh in parallel or distributed computing architectures. The proposed methodology is based on the similarity between a calculation mesh and a water distribution network, in particular on the analogy between the districts design criteria and those of parallel computing system, in other words: the balancing of the load of calculation to be assigned to different processors can be compared with the balancing of the number of nodes (or the flow rates) to be assigned to each water district, and the minimization of the connection elements between two processors corresponds to the minimization of the pipe closures.

(3) *community structure*, is a bottom-up hierarchical algorithm based on the measure of network density to define clusters. These algorithms identify sub graphs in an iterative manner, aggregating nodes time by time and then the groups of nodes, minimizing the density between groups and maximizing the density within each group. Density therefore becomes a measure of the quality of the clustering process, where for density it means the number of connections between nodes. Modularity and centrality of segments are generally used as metrics for measuring density (Di Nardo *et al.*, 2015c; Newman, 2004).

(4) *spectral approach*, developed in the last few years (Di Nardo *et al.*, 2016a; Herrera *et al.*, 2010) starts from considering the network as a simple graph $G = (V,E)$, where V is the set of n vertices vi (or nodes) and E is the set of m edges. Subsequently, it defined the matrix of connectivity $A\ n \times n$ and the matrix of weights $W\ n \times n$ (matrix of the intensity of the connections between nodes). In this case, methodologies and algorithms of complex networks theory are adopted (Boccaletti *et al.*, 2006), assuming water distribution networks as complex systems, constituted by thousands of elementary units (nodes and stretches), connected to form meshes (loop), and strongly geographically bound (Boccaletti *et al.*, 2006). Starting from the adjacency matrix A, it defined the diagonal matrix of the degrees $D\ n \times n$ (matrix of the degree of connection of each single node), and therefore the Laplacian matrix of the graph $L = D - A$, whose spectrum defines important characteristics of the network. In detail, if k is the number of clusters in which the network has to be divided, the first k eigenvectors of the Laplacian define a new representation of the nodes that facilitates the identification of the subsets (Fiedler, 1973). It is shown that

the obtained clustering layout minimizes the number of boundary (or infra-clusters) pipes and simultaneously balances the number of nodes for each clusters (or the sum of the weights if the graph is weighed).

With reference to the dividing phase, two different approaches are proposed in the literature:

(1) By selecting pipes for the insertion of flow meters or gate valves using recursive bisection procedure (Ferrari *et al.*, 2014);
(2) Optimization technique (Di Nardo *et al.*, 2016b) with the objective of identifying the optimal layout that minimises the economic investment and the hydraulic deterioration.

Specifically, once the number of N_b is found after the clustering phase, both methods aim to find the optimal N_C layout, which can reduce the number of flow meters N_{fm} or the number of boundary (gate) valves N_{bv}.

Usually, the optimization approaches adopted some performance indices (Di Nardo *et al.*, 2015b), both in the objective functions chosen and after the optimization process, also to compare solutions providing to operators a wide perspective of the alteration caused by the closing pipes with gate valves and, consequently, the reduction of resilience, robustness, pressure, and so on. comparing different solutions, in terms of number of flow meters and gate valves inserted in the water network for each number of cluster selected.

For this reason, often a multi-objective optimization technique is preferred in order to take into account simultaneously different performance indices and installation and maintenance costs of devices (flow meters and boundary valves).

10.5.3 WNP with SWANP© software

After more than 15 years of research work on WNP and many international experiences of case studies, the authors thought that the time was ripe to collect all knowledge, algorithms and procedures to develop an automatic software which can automatically define the optimal layout of DMAs and provide to a flexible decision support system to water utilities to find different solutions in terms of number of districts, performance indices, compliance with the physical constraints, and so on.

Therefore, the authors have developed a software in Phyton (Di Nardo *et al.*, 2014b, 2016c, 2020) in geographical information system (GIS) environment for the automatic clustering and dividing of a water distribution network. The software, called SWANP© (Smart Water Network Partitioning and Protection) and registered to Copyright Office Washington on March 10, 2019, implements different clustering algorithms and objective functions. It can carry out hydraulic simulation both in demand driven analysis (DDA) and pressure driven analysis (PDA), as well as water quality simulation to select the optimal positioning of quality detection devices to protect water systems from contamination.

SWANP© provides to the decision-maker different WNP layouts using topological, energy, hydraulic and protection performance indices.

In Figures 10.8 and 10.9, an example of the graphical user interface (GUI) of SWANP© is reported showing the results of both a clustering phase with four DMAs and a dividing phase with four flow meters and 11 gate valves for a small network in Italy.

10.5.4 Phyton code to design an optimal WNP

In this last paragraph, a Python code for students and operators to design an optimal water network partitioning is provided using a spectral method for the clustering phase and a multi-objective genetic algorithm for the dividing phase.

The code briefly gives some notes on the most important aspects (INPUT, OUTPUT, etc.) of the algorithms used. The readers can find more information in Di Nardo *et al.* (2013, 2016a).

```python
1    def cluster_phase(path,network,n_dma):

2        from epanettools import epanet2 as ep
3        import numpy as np
4        import os
5        from sklearn.cluster import SpectralClustering

6        """
7        spectral approach (Jianbo Shi, Jitendra Malik 2000) to perform
8        clustering phase
9        input:
10       network = Epanet input file of water distribution network (.inp)
11       path = directory of WDS file
12       n_dma = number of DMAs
13       output:
14       dma = labels that define cluster for each node
15       boundarypipes = pipes between two different DMAs
16       """

17       #compute the adjacency matrix of water distribution network
18       os.chdir(path)
19       err = ep.ENopen(network,'net.rpt','') #opening Epanet network file
20       err,n_node = ep.ENgetcount(ep.EN_NODECOUNT) #reading number of nodes
21       err,n_link = ep.ENgetcount(ep.EN_LINKCOUNT) #reading number of links
22       M = np.zeros((n_link,3), dtype = np.int) # array with index of link, start node and end node for each
         pipe

23
24   for i in range(0,n_link):
25       err,startnode,endnode=ep.ENgetlinknodes(i + 1)
26       M[i] [0]=i+1 # index of i-th pipe
27       M[i] [1]=startnode # start node of i-th pipe
28       M[i] [2]=endnode # end node of i-th pipe
29   ep.ENclose() #closing Epanet network file

30   A=np.zeros((n_node,n_node),dtype=np.int) # adjacency matrix of water network
31   for i in range(0,n_link):
32
33       if A[M[i][1]-1][M[i][2]-1] == 0 and A[M[i][2]-1][M[i][1]-1] == 0:
34
35       A[M[i][1]-1][M[i][2]-1]=1
36
37       A[M[i][2]-1][M[i][1]-1]=1

38   clusters=SpectralClustering(n_clusters=n_dma,affinity = 'precomputed').fit(A) # spectral clustering

39   dma=clusters.labels_
```

(*Continued*)

```
40    boundarypipes=[]

41    for k in range(0,len(M)):

42

43      cluster_node_i=dma[M[k][1]-1]
44      cluster_node_j=dma[M[k][2]-1]

45

46      if cluster_node_i != cluster_node_j:

47

48        boundarypipes.append(k + 1)

49    return dma,boundarypipes

50    def dividing_phase(path,network,boundarypipes,design_pressure):

51    import numpy as np
52    from pymoo.model.problem import Problem
53    from pymoo.factory import get_algorithm, get_sampling, get_crossover, get_mutation
54    from pymoo.optimize import minimize
55    import matplotlib.pyplot as plt
56    from epanettools import epanet2 as ep
57    import os

58    """
59    NSGAII algorithm to perform dividing phase

60    input:
61    network=Epanet input file of water distribution network (.inp)
62    path=directory of WDS file
63    boundarypipes=pipes between two different DMAs

64    output:
65    FO=values of computed objective function
66    flow_meters=array wiht optimal positioning of flow meter
67    (0 - closed pipe; 1 - opend pipe)
68    """

69    os.chdir(path)

70    n_variables=len(boundarypipes) #number of variables

71    class MyProblem(Problem):

72      def __init__(self):
73        super().__init__(n_var=n_variables, n_obj=2, n_constr=1,
74          xl=np.zeros(n_variables), xu=np.ones(n_variables),type_var=int)

75

76      def _evaluate(self, x, out, *args, **kwargs):

77
```

```
78     ep.ENopen(network,'rete.rpt','') #opening Epanet network file
79
80      err,n_node=ep.ENgetcount(ep.EN_NODECOUNT) #reading number of nodes
81
82      err,n_link=ep.ENgetcount(ep.EN_LINKCOUNT) #reading number of pipes
83
84      err,n_serb=ep.ENgetcount(ep.EN_RESERVOIR) #reading number of reservorir
85
86     dim_x=max(x.shape)
87     f1=np.zeros(dim_x)
88     f2=np.zeros(dim_x)
89     constraint=np.zeros(dim_x)
90     #chiusura dei tratti
91     for l in range(0,dim_x):
92
93     f1[l]=sum(x[l,:])
94     for k in range(0,len(boundarypipes)-1):
95       err=ep.ENsetlinkvalue(boundarypipes[k],4,np.int(x[l][k]))
96
97
98       err=ep.ENsolveH() #run hydraulic simulation
99
100      pwr_node=np.zeros(n_node-n_serb, dtype=float)
101
102      pressure=np.zeros(n_node-n_serb, dtype=float)
103
104
105      pwr_node=np.zeros(n_node, dtype=float) #compute objective function 1 (number of flow meters)
106      for k in range(0,n_node-n_serb):
107
108      err,head=ep.ENgetnodevalue(k+1,ep.EN_HEAD)
109
110      err,demand=ep.ENgetnodevalue(k+1,ep.EN_DEMAND)
111
112      err,pressure[k]=ep.ENgetnodevalue(k+1,ep.EN_PRESSURE)
113      pwr_node[k]=head*demand
114
115     f2[l]=-sum(pwr_node) #compute objective function 2 (node available power)
116
```

(Continued)

```python
117      constraint[l]=design_pressure-min(pressure)
118
119      out["F"]=np.column_stack([f1, f2])
120
121      out["G"]=constraint
122
123      #chiusura epanet
124      ep.ENclose()
125  problem=MyProblem()
126  method=get_algorithm("nsga2",
127       pop_size=100,
128       sampling=get_sampling("int_random"),
129       crossover=get_crossover("int_sbx", prob=1.0, eta=3.0),
130       mutation=get_mutation("int_pm", eta=3.0),
131       eliminate_duplicates=True,
132       )
133  res=minimize(problem,
134       method,
135       termination=('n_gen', 100),
136       seed=1,
137       save_history=True,
138       disp=False)
139  res.F[:,1]=np.abs(res.F[:,1]) #print Objective Space
140  FO=res.F
141  flow_meters=res.X
142  plt.title("Objective Space")
143  plt.scatter(FO[:, 0], FO[:, 1])
144  plt.xlabel('FO1')
145  plt.ylabel('FO2')
146  plt.grid()
147  plt.show()
148  return FO,flow_meters
149
150
151
152
153
154
155
207
```

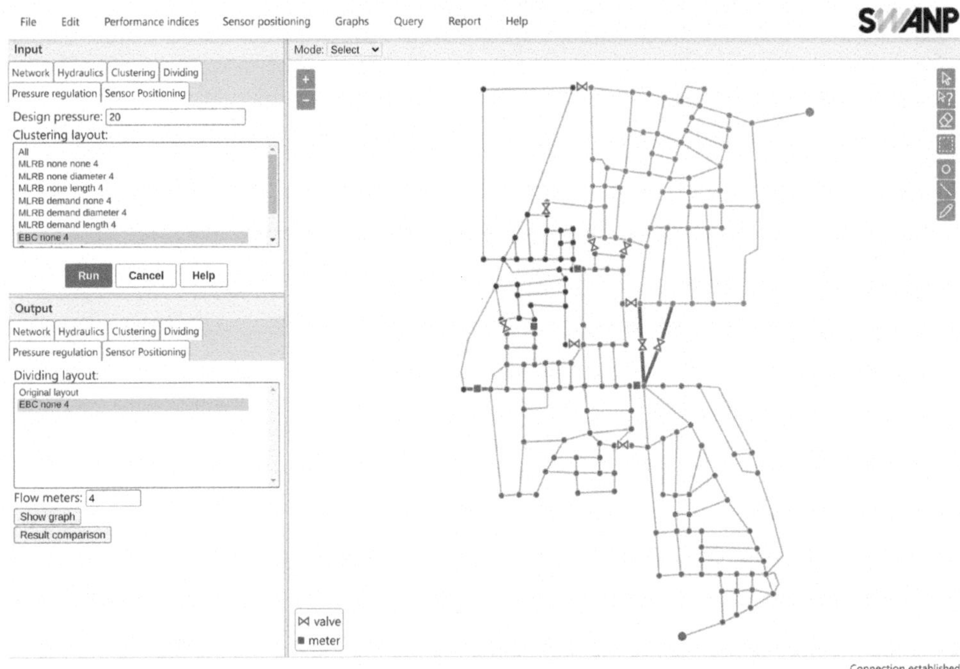

Figure 10.8 Clustering phase with SWANP©.

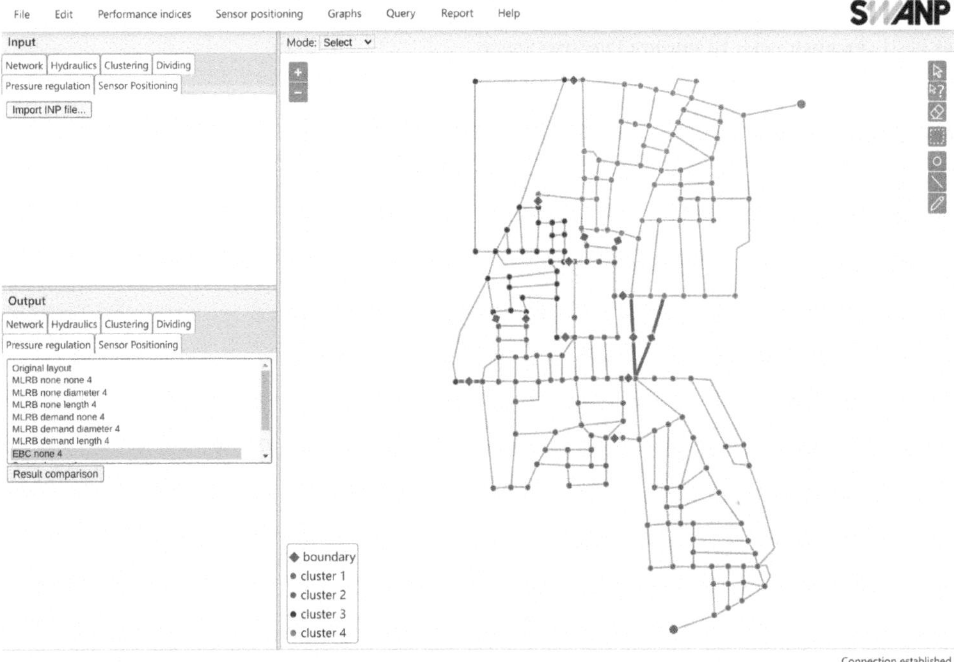

Figure 10.9 Dividing phase with SWANP©.

REFERENCES

Alonso J. M., Alvarruiz F., Guerrero D., Hernàndez V., Ruiz P. A., Vidal A. M., Martìnez F., Vercher J. and Ulanicki B. (2000). Parallel computing in water network analysis and leakage minimization. *Journal of Hydraulic Engineering ASCE*, **126**, 251–260, https://doi.org/10.1061/(ASCE)0733-9429(2000)126:3(174)

Boccaletti S., Latora V., Morenod Y., Chavezf M. and Hwanga D. U. (2006). Complex networks. Structure and dynamics. *Physics Reports*, **424**, 175–308, https://doi.org/10.1016/j.physrep.2005.10.009

Bodlaender H. L., Hendriks A., Grigoriev A. and Grigorieva N. V. (2010). The valve location problem in simple network topologies. *INFORMS Journal on Computing*, **22**(3), 433–442, https://doi.org/10.1287/ijoc.1090.0365

Bui X. K., Marlim M. S. and Kang D. (2020). Water network partitioning into district metered areas: A state-of-the-art review. *Water (Switzerland)*, **12**(4), 1–29.

Butler D. (2000). Leakage Detection and Management. Palmer Environmental, UK.

Di Nardo A. and Di Natale M. (2011). A heuristic design support methodology based on graph theory for district metering of water supply networks. *Engineering Optimization*, **43**(2), 193–211, https://doi.org/10.1080/03052151003789858

Di Nardo A., Di Natale M., Santonastaso G. F. and Venticinque S. (2013). An automated tool for smart water network partitioning. *Water Resources Management*, **27**, 4493–4508, https://doi.org/10.1007/s11269-013-0421-1

Di Nardo A., Di Natale M. and Santonasatso G. F. (2014a). A comparison between different techniques for water network sectorization. *Water Science and Technology: Water Supply*, **14**(6), 961–970, https://doi.org/10.2166/ws.2014.046

Di Nardo A., Di Natale M., Santonastaso G. F., Tuccinardi F. P. and Zaccone G. (2014b). SWANP: Software for Automatic Smart Water Network Partitioning. Proceedings of the 7th International Environmental Modelling and Software Society (iEMSs). International Congress on Environment Modelling and Software, San Diego, California.

Di Nardo A., Di Natale M., Santonastaso G. F., Tzatchkov V. and Alcocer Yamanaka V. H. (2014c). Divide and conquer partitioning techniques for smart water networks. *Procedia Engineering*, **89**, 1176–1183, https://doi.org/10.1016/j.proeng.2014.11.247

Di Nardo A., Di Natale M., Musmarra D., Santonastaso G. F., Tzatchkov V. and Alcocer-Yamanaka V. H. (2015a). Dual-use value of network partitioning for water system management and protection from malicious contamination. *Journal of Hydroinformatics*, **17**(3), 361–376, https://doi.org/10.2166/hydro.2014.014

Di Nardo A., Di Natale M., Santonastaso G. F., Tzatchkov V. G. and Alcocer-Yamanaka V. H. (2015b). Performance indices for water network partitioning and sectorization. *Water Science and Technology: Water Supply*, **15**(3), 499–509, https://doi.org/10.2166/ws.2014.132

Di Nardo A., Di Natale M., Giudicianni C., Musmarra D., Santonastaso G. F. and Simone A. (2015c). Water distribution system clustering and partitioning based on social network algorithms. *Procedia Engineering*, **119**, 196–205, https://doi.org/10.1016/j.proeng.2015.08.876

Di Nardo A., Di Natale M., Giudicianni C., Greco R. and Santonastaso G. F. (2016a). Water supply network partitioning based on weighted spectral clustering. *Studies in Computational Intelligence: Complex Networks and Their Applications*, **693**, 797–807.

Di Nardo A., Di Natale M., Giudicianni C., Santonastaso G. F., Tzatchkov V. G., Varela J. M. R. and Yamanaka V. H. A. (2016b). Water supply network partitioning based on simultaneous cost and energy optimization. *Procedia Engineering*, **162**, 238–245, https://doi.org/10.1016/j.proeng.2016.11.048

Di Nardo A., Santonastaso G. F., Giudicianni C., Di Mauro A., Di Natale M., Musmarra D. and Tuccinardi F. P. (2016c). SWANP 3.0: Software for partitioning and protection of water distribution networks. Proceedings of the H$_2$O Workshop Water Losses Management. H$_2$O, Bologna, Italy.

Di Nardo A., Di Natale M., Chianese S., Musmarra D. and Santonastaso G. F. (2016d). Combined recursive clustering and partitioning to define optimal DMAs of water distribution networks. Paper presented at the Environmental Modelling and Software for Supporting a Sustainable Future, Proceedings – 8th International Congress on Environmental Modelling and Software, iEMSs 2016. iEMSs, pp. 975–982.

Di Nardo A., Di Natale M., Di Mauro A., Santonastaso G. F. and Giudicianni C. (2018). Criteria, objectives and methodologies for water network partitioning. *Italian Journal of Engineering Geology and Environment*, **1**, 39–47.

Di Nardo A., Di Natale M., Di Mauro A., Martínez Díaz E., Blázquez Garcia J. A., Santonastaso G. F. and Tuccinardi F. P. (2020). An advanced software to design automatically permanent partitioning of a water distribution network. *Urban Water Journal*, **17**(3), 259–265, https://doi.org/10.1080/1573062X.2020.1760322

Di Nardo A., Boccelli D. L., Herrera M., Creaco E., Cominola A., Sitzenfrei R. and Taormina R. (2021). Smart urban water networks: solutions, trends and challenges. *Water (Switzerland)*, **13**(4), 1–8.

Ferrari G., Savic D. and Becciu G. (2014). Graph-theoretic approach and sound engineering principles for design of district metered areas. *Journal of Water Resources Planning and Management*, **140**, 1–13, https://doi.org/10.1061/(ASCE)WR.1943-5452.0000424

Fiedler M. (1973). Algebraic connectivity of graphs. *Czechoslovak Mathematical Journal*, **23**, 298–305, https://doi.org/10.21136/CMJ.1973.101168

Fortunato S. (2010). Community detection in graphs. *Physics Reports*, **486**(3–5), 75–174, https://doi.org/10.1016/j.physrep.2009.11.002

Grayman W. M., Murray R. and Savic D. A. (2009). Effects of redesign of water systems for security and water quality factors. *Proceedings of the World Environmental and Water Resources Congress*, **342**, 504–514.

Herrera M., Canu S., Karatzoglou A., Pérez-García R. and Izquierdo J. (2010). An approach to water supply clusters by semi-supervised learning. *Proceedings of the 5th Biennial Conference of the International Environmental Modelling and Software Society, iEMSs*, **3**, 1925–1932.

ISTAT. (2021). *ISTAT Italian Statistics on Water. 2018–2020*, Report 2021, ISTAT, Rome.

Jacobs P. and Goulter I. C. (1989). Optimization of redundancy in water distribution networks using graph theoretic principles. *Engineering Optimization*, **15**, 71–82, https://doi.org/10.1080/03052158908941143

Khaled H., Sendil U. and Al-Dhowalia. (1992). Relationship between pressure and leakage in a water distribution network. Proceedings of the AWWA Conference, June 18–22, Vancouver, BC.

Lambert A. (2000). What do we know about pressure-leakage relationships in distribution systems. Proceedings of the IWA Conference on System Approach to Leakage Control and Water Distribution System Management, IWA, Brno, Czech.

Lambert A. and Hirner W. (2000). Losses From Water Supply Systems: Standard Terminology and Recommanded Performance Measures. IWA The Blue Pages. IWA Publishing, London.

Lambert A., Brown T. G., Takizaw M. and Weimer D. (1999). A review of performance indicators for real losses from water supply systems. *Aqua*, **48**, 227–237.

Mays W. (2000). Water Distribution Systems Handbook. McGraw-Hill, New York.

Newman M. E. J. (2004). Fast algorithm for detecting community structure in networks. *Statistical, Nonlinear, and Soft Matter Physics*, **69**, 1–5.

OFWAT, ENVIRONMENT AGENCY, DEFRA. (2002). Best Practice principles in the economic level of leakage calculation, OFWAT Publication, 2002.

Perelman L. S., Allen M., Preis A., Iqbal M. and Whittle A. J. (2015). Automated sub-zoning of water distribution systems. *Environmental Modelling and Software*, **65**, 1–14, https://doi.org/10.1016/j.envsoft.2014.11.025

Savic D. A. and Walters G. A. (1995). An evolution program for optimal pressure regulation in water distribution networks. *Engineering Optimization*, **24**, 197–219, https://doi.org/10.1080/03052159508941190

Tindell K. W., Burns A. and Wellings A. J. (1992). Allocating hard real-time tasks: an NP-hard problem made easy. *Real-Time Systems*, **4**(2), 145–165, https://doi.org/10.1007/BF00365407

Twort A. C., Ratnayayaka D. D. and Brandt M. J. (2000). Water Supply, 5th edn. Butterworth-Heinemann, Oxford, UK.

Tzatchkov V. G., Alcocer-Yamanaka V. H. and Ortiz V. B. (2006). Graph theory based algorithms for water distribution network sectorization projects. Proceedings of the 8th Annual Water Distribution Systems Analysis Symposium, Cincinnati, USA.

UK WATER INDUSTRY RESEARCH Ltd. (1999). A Manual of MDA Practice, London: UK Water Industry Research.

United Nations. (2003). International Year of Freshwater. General Assembly resolution 55/196. United Nations, New York.

Water Authorities Association and Water Research Centre. (1985). Leakage Control Policy and Practice. Technical Working Group on Waste of Water, WRc Group, AQ18, National Water Council, London, UK.

Water Industry Research Ltd. (1999). A Manual of DMA Practice. WIR, London, UK.

Wrc/WSA/WCA Engineering And Operations Committee. (1994). *Managing Leakage: UK Water Industry Managing Leakage*. Report A-J, London, UK.

doi: 10.2166/9781789062380_0255

Chapter 11

Reliability analysis using optimization

*Sangamreddi Chandramouli**

Professor of Civil Engineering, Maharaj Vijayaram Gajapati, Raj College of Engineering, Vizianagaram, India
**Corresponding author: chandramouli.sangamreddi@gmail.com*

LEARNING OBJECTIVES

At the end of this chapter, you will be able to:

(1) Understand the linking of EPANET tool kit functions in MATLAB Dynamic Link Library.
(2) Work with EPANET tool kit functions.
(3) Work with Genetic Algorithm tool in MATLAB.
(4) Understand the concepts of Fuzzy logic, Optimization and Reliability.
(5) Develop a reliability-based optimization model for design of water supply pipe networks in MATLAB by combining EPANET toolkit functions.
(6) Appreciate the difference between binary logic and fuzzy logic in terms of reliability achievement for the water supply pipe networks.
(7) Work with different types of networks of water supply for their design.
(8) Analyse the results and suggest the best solution depending upon the requirement of the water users.

11.1 INTRODUCTION

11.1.1 Brief history of pipe networks

The history of water distribution systems is parallel to the history of civilization. All earlier civilizations developed on the banks of rivers. More than 2000 years ago the city of Rome had a well-developed water supply system. The means for transportation and water distribution/supply and irrigation were also developed, and artificial conduits were also constructed for the conveyance of water.

Even though water distribution systems existed earlier, modern distribution systems are of recent origin. In 1544, the British Parliament passed an act to provide clean water to the residents of London. In 1962, Boston was credited with the earliest recorded water supply in USA and used ductile iron pipes in 1968. In 1746, the first piped supply for the entire community was built in Schaeffer town, Pennsylvania. Now, modern water transmission and distribution systems throughout the world use pipes of ductile iron, steel, concrete and plastic, and so on. to meet the demands from the public.

11.1.2 Development of water supply engineering

Any community water supply/distribution systems should aim to:

- provide enough water to meet all their usage requirements;
- supply water at required pressures to draw adequate water flow;
- establish and maintain water quality integrity – supplying good quality of water which is free from disease causing bacteria and chemicals (not to mention taste and odor);
- maintain the level of service within the budgets – consistent quality, quantity, and pressure over space and time.

To meet the above requirements, every component should be designed, constructed and maintained properly. The critical tasks to be performed in the overall planning process of a water distribution network consists of three phases: *layout, design and operation.* Design plays a vital role in the water distribution system. Optimal design of a water distribution network is the aim of any agency dealing with water supply/distribution. Through a trial-and-error procedure and using a simulation model, a water distribution network may be designed. Further, the knowledge gained over experience may be utilized in choosing the design parameters. However, there is no guarantee that the design selected is optimal. Especially, when the network is large, it is difficult to reach a satisfactory solution by this method.

Hence, mathematical optimization models are preferred compared to a trial-and-error procedure. As the analysis and design process of a water distribution system is a complex and non-linear process, and with the advancements in computer technology/capacities, people nowadays use a variety of software packages or are developing their own algorithms.

11.1.3 Brief description of optimization techniques

Wide varieties of optimization tools have been applied in the past 30 years for optimal design of networks which include linear programming, different kinds of non-linear programming, heuristic methods like genetic algorithms, simulated annealing techniques, ant colony optimization, bee colony optimization, and so on. Each method has its own advantages and disadvantages in the formulation, speed of solving, handling nonlinearity, efficiency, and so on. The complexity of the optimal design of a distribution network is due to the discrete characteristics of decision variables, discrete and complicated cost function required to address the materials, labor, and overall installation/operational setting, the need for considering multiple demand loading patterns, uncertainty in demands, location of tanks, pumping station, booster pumps and valves, and so on. Selecting a network configuration with minimum pipe cost and maximum reliability is a complex process (Afshar *et al.*, 2005; Bhave and Gupta, 2006). Several works have been reported in the literature for the optimal design and some of them consider certain reliability factors also.

It may be possible to have different solutions satisfying the requirements of an engineering problem. Naturally these solutions would have different costs and the objective would be to find the least cost solution. On the other hand, in a water resource project, the objective may be to find a solution that would give maximum benefits. A solution having minimum cost or maximum benefits is termed as optimum solution (Bhave, 2003) and the concept of obtaining the optimum solution is termed *optimization.*

When a physical problem is expressed mathematically in the form of an optimization framework, the expression defining the objective (maximization or minimization) is termed as objective function, whereas different conditions which the objective has to satisfy are termed as constraints, and the entire problem consisting of the objective function and constraints is termed as optimization problem. Mathematically, such an optimization problem can be expressed as:

$$\textbf{Maximize/Minimize } Z = f\left(x_1, x_2, \ldots\ldots\ldots, x_n\right)$$

$$\textbf{Subject to } g_i\left(x_1, x_2, \ldots\ldots\ldots, x_n\right) \leq \mid \geq \mid = b_i \quad \text{where } i = 1, 2 \ldots\ldots m$$

where Z represents the objective function which involves 'n' parameters or decision variables $x_1, x_2 \ldots$ x_n. $g_i(x_1, x_2, \ldots, x_n)$ represents a set of m constraints expressed as equalities or inequalities.

Depending on the nature of the problem, there are different methods available for solving the optimization problem. Most of the problems are non-linear in nature, which are solved by search techniques. Some optimization methods that can be used for the optimal design of water distribution networks are Exhaustive Enumeration, Classical Optimization, Linear Programming, Non-Linear Programming, Dynamic Programming, Geometric Programming, Integer Programming, Stochastic Programming, Stochastic search methods, Genetic Algorithm method, Simulated Annealing Method, Goal Programming, Swarm intelligence and so on. (Bhave, 2003).

11.1.4 Brief description of reliability concept

Reliability of water distribution networks is another aspect on which considerable research has been carried out. The words 'reliable' and 'reliability' are generally used in our daily life to indicate some degree of confidence in a person/thing/system. The word reliability($=$re$+$liability) means repeated liability because of various breakdowns and failures of a device/system. Reliability of a device or system is defined as the probability that it can perform its purpose within tolerance for the period of time intended under the given operating conditions (Gurjar, 2007).

Reliability may be quantified using different methods and while assessing reliability, various factors influencing the performance are to be taken into consideration. A perfectly reliable water supply system must be able to supply desired quality of water in required quantities with desired residual heads to all consumers at their tapping points throughout the design period.

Consideration of reliability in the optimal design of water distribution networks has received increasing attention (Prasad, 2008). Reliability of a water distribution system is concerned with the ability of the network to provide an adequate supply to the consumers under both normal and abnormal operating conditions. Extensive research on reliability of water distribution systems has been performed and various measures were developed to address the reliability of the system under the failure of components or due to demand variation. However, none of these measures have been accepted universally. This is mainly due to the problem of addressing all the parameters, which affect the performance of the system as a single measure. If the network is designed with reliability alone as a prime objective, then the resulting system may also be an uneconomical one.

11.2 CONCEPT OF FUZZY SET THEORY

11.2.1 Brief description of fuzzy set theory

Zadeh (1965) first proposed fuzzy set theory in the field of system theory. Since then, it has been applied in many fields of engineering including optimization, risk analysis and resource management. The fundamental principle of fuzzy set theory is that a parameter can be 'noisy' or 'ill-defined' and can take on a range of values having a degree of membership. The incorporation of fuzzy set theory and fuzzy logic into computer models has shown a tremendous payoff in areas where intuition and judgment still play a role in the model (Ross, 1997).

Fuzzy sets in practice are often understood as fuzzy numbers and are represented through membership functions. Zadeh (1965) proposed the use of fuzzy membership function with values between 0 and 1. A fuzzy set can be represented as:

$$A = \{(x, \mu_A(x))/x \in X\}$$

where $\mu_A(x)$ *is the membership function* indicating the degree of membership or degree of belongingness with values between 0 and 1. A fuzzy set is different from a normal set. The normal set follows binary logic, that is a particular member belongs to the set or is not a member, whereas in a fuzzy set, there is gradual transition between membership and non-membership.

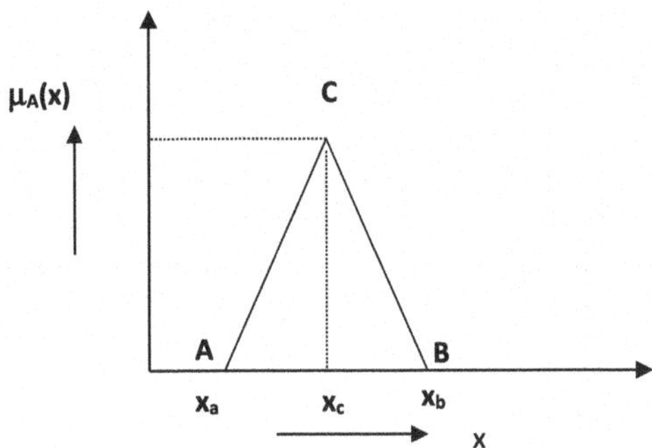

Figure 11.1 Triangular membership function.

11.2.2 Membership functions

The most common types of membership functions for fuzzy numbers are: (1) triangular; and (2) trapezoidal. A fuzzy number is any fuzzy variable defined over a real number set. A fuzzy parameter x is shown by a triangular function in Figure 11.1 and by a trapezoidal function in Figure 11.2. The value of fuzzy parameter x is shown along the x-axis and its membership value $\mu_A(x)$ is shown along the y-axis. Here, $\mu_A(x)$ gives the membership value of parameter x in fuzzy set A. In both the figures, AB represents the support; while in Figure 11.2, DE represents the core, which reduces to the most likely value, represented by points C and CB in Figure 11.1 and AD and EB in Figure 11.2 represent the boundaries.

The triangular function of Figure 11.1 corresponds to the imprecision of the type 'the parameter x is included between x_a and x_b and is most likely to be x_c'. The trapezoidal function of Figure 11.2 corresponds to the imprecision of the type 'the parameter x is certainly included between x_a and x_b and is most likely between x_d and x_e'. The membership function mathematically represents the membership of a parameter in a fuzzy set and always lies between zero and one. Thus:

$$\mu_A(x) \in [0,1]$$

in which the symbol $\mu_A(x)$ is the degree of membership of parameter x in the fuzzy set A. Therefore $\mu_A(x)$ is a value on the unit interval that signifies the degree to which parameter x belongs to fuzzy set A. Since x is a continuous variable, its membership value is also continuous between zero and one.

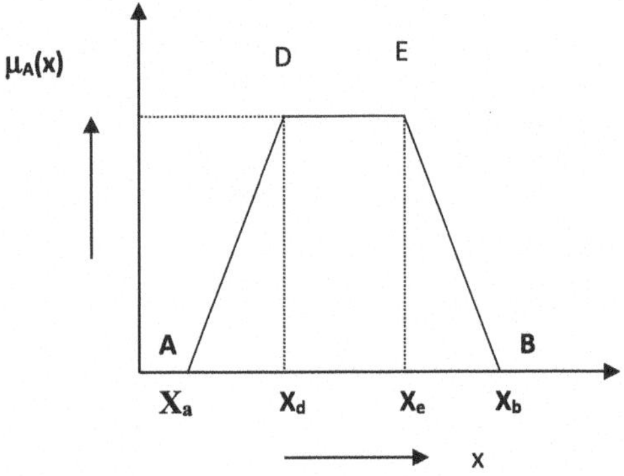

Figure 11.2 Trapezoidal membership function.

11.2.3 Types of fuzzy sets and fuzzy functions

There are two types of fuzzy sets, namely discrete fuzzy set and continuous fuzzy set. In discrete fuzzy sets, the members are associated with their degree of belongingness.

For example:

$$A = \{(2,0.1),(3,0.3),(4,0.8),(5,1),(6,0.8),(7,0.3),(8,0.1)\}$$

A continuous fuzzy set is expressed as:

$$A = \{(x,\mu_A(x))/\mu_A(x) = f(x),\ 0 \leq f(x) \leq 1\}$$

A set of elements whose degree of membership equal to 1 is called core.

$$\text{Support}(A) = \{(x/\mu_A(x) = 1\}$$

A set of elements whose degree of belongingness is greater than zero is called support:

$$\text{Core}(A) = \{(x/\mu_A(x) > 0\}$$

A fuzzy set is considered normal if core is nonempty:
α-cut (or) α-level set: α-cut of $A = \{(x/\mu_A(x) \geq \alpha\}$
strong α-cut of $A = \{(x/\mu_A(x) > \alpha\}$
Crossover points: These are the points of x, which have $\mu_A(x) = 0.5$.
Convex fuzzy set: if x_1 and x_2 have, $\mu_A(x_1)$ and $\mu_A(x_2)$, then it is said to be convex fuzzy set, if $\mu_A(\lambda x_1 + (1-\lambda x_2)) > \lambda \mu_A(x_1) + (1-\lambda)\,\mu_A(x_2)$, at an intermediate point, the membership is greater than the average.

Similar to normal set theory, fuzzy set theory also has the following functions.

11.2.3.1 Complement of fuzzy set

It indicates the non-membership of the given set. It is represented as:

$$A^1 = \{(x,\ (1 - \mu_A(x))/x \in X\}$$

11.2.3.2 Intersection

It represents the minimum degree of belongingness of a particular member in two different sets:

$$\mu_{A \cap B}(x) = \min\ (\mu A(x), \mu B(x))$$

11.2.3.3 Union

It represents the maximum degree of belongingness of a particular member in two different sets:

$$\mu_{A \cup B}(x) = \max\ (\mu A(x), \mu B(x))$$

Example: Let us consider a discrete fuzzy set $A = \{(2,0.1),(3,0.3),(4,0.8),(5,1),(6,0.8),(7,0.3),(8,0.1)\}$
Then the support, core and α-cut are extracted from the set as follows:

Support $(A) = \{2,3,4,5,6,7,8\}$
Core$(A) = \{5\}$, hence it is a normal set
Let $\alpha = 0.3$, α-cut of $A = \{3,4,5,6,7\}$
Strong α-cut of $A = \{45,6\}$

Let B = {(2,1),(3,0.7),(4,0.3),(5,0.1)}
So, from the two sets A and B, intersection and union can be obtained as follows:

$$\mu_{A \cap B}(x) = \{(2,0.1),(3,0.7),(4,0.3),(5,0.1)\}$$

$$\mu_{A \cup B}(x) = \{(2,1),(3,0.7),(4,0.8),(5,1),(6,0.8),(7,0.3),(8,0.1)\}$$

11.3 RELIABILITY ANALYSIS OF WATER SUPPLY PIPE NETWORKS

11.3.1 Definition of reliability index

Most researchers have defined reliability based on meeting consumer demands and incorporated this into optimization models. Very few researchers focused on excess residual pressures, so the satisfaction levels of these excess residual pressures have not been considered. Hence, in the present study, we defined a new parameter to assess the reliability of a water distribution network and incorporated it into the optimization model. The Network Reliability Parameter (NRP) is defined as the ratio of algebraic sum of the product of demand and satisfaction index at all the demand nodes in the network to the total demand of the network (Chandramouli, 2013).

Mathematically, it is expressed as:

$$NRP = \left(\sum Q_i * SI_i\right) / \sum Q_i \quad i = 1 \,\text{to}\, N \tag{11.1}$$

where i is the index representing the demand nodes in the network; Q is the rate of flow required in m³/s at a demand node in the network; N is the number of demand nodes in the network and SI is the Satisfaction Index at the demand node.

11.3.1.1 Satisfaction index

In finding the NRP, the satisfaction index plays a significant role. In the present study, fuzzy logic is used to obtain the satisfaction index. The satisfaction index is the membership function associated with the residual pressures. It represents the degree of belongingness of a particular value in the specified range. If x represents the residual pressure, then $\mu_A(x)$ represents the membership function corresponding to x. Here, x is the residual pressure within the range specified. The maximum value of satisfaction index is 1 whereas the minimum is 0. The trapezoidal membership function is considered, and the details are provided in the results and discussions section.

The following algorithm is adopted to find the NRP.

- Step 1: Standard benchmark network/real network are selected from the literature.
- Step 2: The networks are analysed using EPANET.
- Step 3: The heads at all the demand nodes are obtained.
- Step 4: The heads obtained are compared with the minimum heads required and then the residual heads are computed.
- Step 5: The residual heads are linked with the satisfaction level of the consumers through membership function. A trapezoidal membership function is adopted with residual heads. For example, if the residual heads are in the range of 10–15 m it adopts a satisfaction level of 100% and if the residual head is less than 0 m and greater than 25 m, it adopts a satisfaction level of 0%. In between, it adopts a straight-line variation.
- Step 6: NRP is determined using the formula mentioned above.

Table 11.1 Range-1 Residual pressures (0–15 m).

Serial No.	Range of Residual Pressures in m
1	0–5
2	5–10
3	10–15

Table 11.2 Range-2 Residual pressures (0–25 m).

Serial No.	Range of Residual Pressures in m
1	0–10
2	10–15
3	15–25

Table 11.3 Range-3 Residual pressures (0–30 m).

Serial No.	Range of Residual Pressures in m
1	0–15
2	15–25
3	25–30

11.3.2 Ranges of residual pressures and satisfaction levels based on fuzzy logic

The details of various ranges of residual pressures considered in the present study are presented below in Tables 11.1–11.3. These ranges are fixed by trial and error.

11.3.2.1 Satisfaction index based on fuzzy logic for different ranges of residual pressures

The satisfaction indices for different ranges are presented below in Tables 11.4–11.6. The trapezoidal membership functions for different ranges are shown in Figures 11.3–11.5.

11.3.2.2 Satisfaction index based on binary logic for different ranges of residual pressures

In binary logic, there are only two possibilities of a particular value belonging to the range or not. No gradual variation within the range is considered in binary logic. The satisfaction indices based on binary logic for the same ranges which are considered for fuzzy logic are given in Tables 11.7–11.9. The graphical representation of the satisfaction indices for different ranges is shown in Figures 11.6–11.8.

Table 11.4 Satisfaction index based on fuzzy logic for Range-1.

Serial No.	Range of Residual Pressures at Demand Nodes in the Network (in m)	Satisfaction Index
1	0–5	$x/5$
2	5–10	1
3	10–15	$(15-x)/5$

Table 11.5 Satisfaction index based on fuzzy logic for Range-2.

Serial No.	Range of Residual Pressures at Demand Nodes in the Network (in m)	Satisfaction Index
1	0–10	$x/10$
2	10–15	1
3	15–25	$(25\text{-}x)/10$

Table 11.6 Satisfaction index based on fuzzy logic for Range-3.

Serial No.	Range of Residual Pressures at Demand Nodes in the Network (in m)	Satisfaction Index
1	0–15	$x/15$
2	15–25	1
3	25–30	$(30\text{-}x)/5$

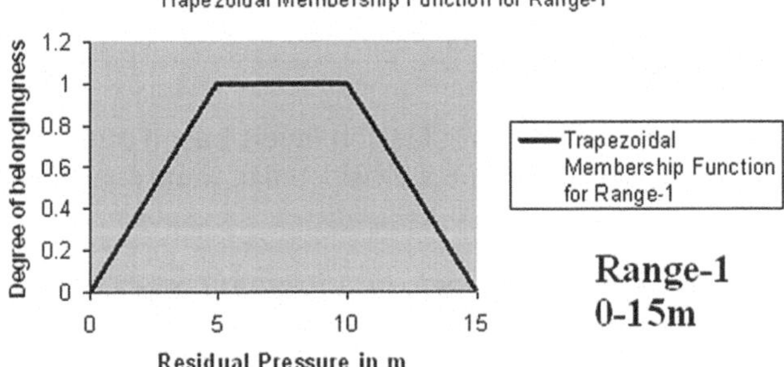

Figure 11.3 Trapezoidal membership function for Range-1.

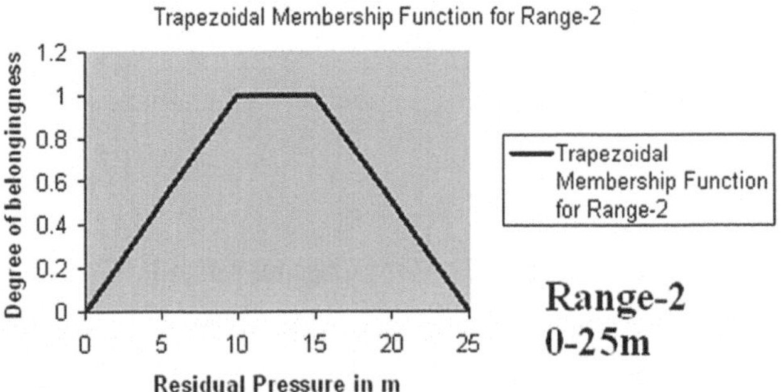

Figure 11.4 Trapezoidal membership function for Range-2.

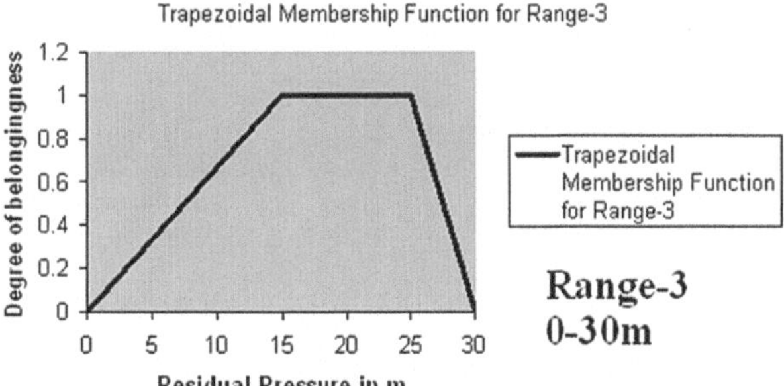

Figure 11.5 Trapezoidal membership function for Range-3.

Table 11.7 Satisfaction index based on binary logic for Range-1.

Serial No.	Range of Residual Pressures at Demand Nodes in the Network (in m)	Satisfaction Index
1	0–5	0
2	5–10	1
3	10–15	0

Table 11.8 Satisfaction index based on binary logic for Range-2.

Serial No.	Range of Residual Pressures at Demand Nodes in the Network (in m)	Satisfaction Index
1	0–10	0
2	10–15	1
3	15–25	0

Table 11.9 Satisfaction Index based on Binary Logic for Range-3.

Serial No.	Range of Residual Pressures at Demand Nodes in the Network (in m)	Satisfaction Index
1	0–15	0
2	15–25	1
3	25–30	0

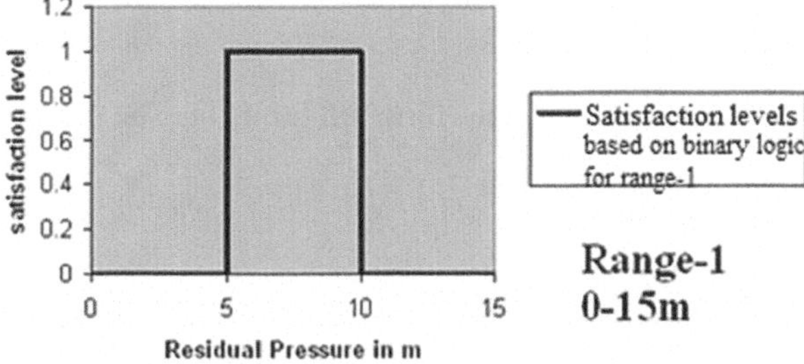

Figure 11.6 Satisfaction index based on binary logic for Range-1.

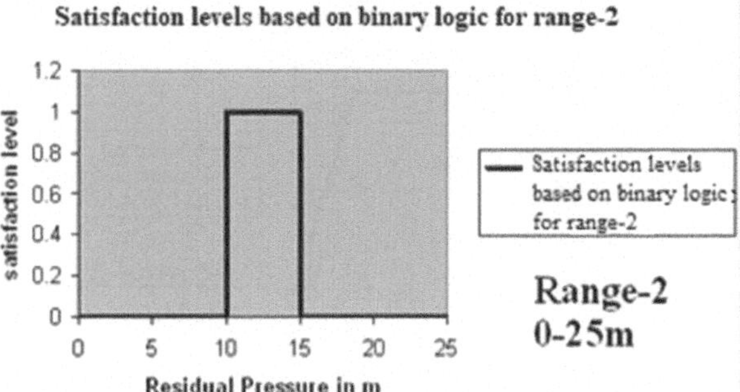

Figure 11.7 Satisfaction index based on binary logic for Range-2.

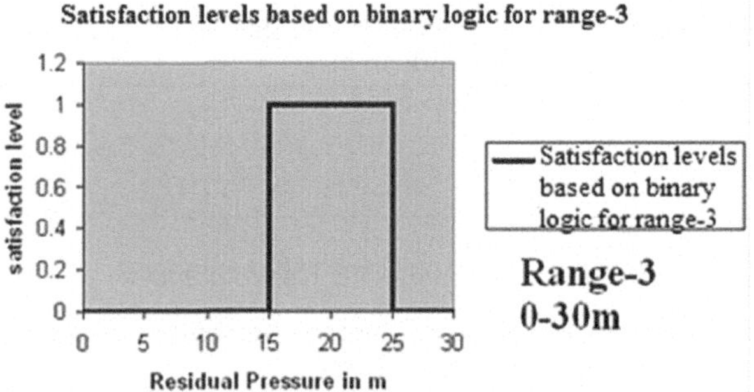

Figure 11.8 Satisfaction index based on binary logic for Range-3.

11.4 RELIABILITY BASED OPTIMIZATION OF PIPE NETWORKS

11.4.1 Description of objective function

Minimization of the total cost of the network and maximization of network reliability are considered as two objectives – hence the problem is multi-objective. The aim is to obtain several optimal solutions with different costs and having different reliabilities to obtain a Pareto-Optimal front.

Minimize C_T

where:

C_T = total cost of the network = $C_T = \sum L_j \times D_j \times C(D_j)$

L_j = Length of the link 'j'

D_j = Diameter of the link 'j'

$C(D_j)$ = Unit cost for the diameter of the link 'j'

(or)

$C_T = \sum c_j x L_j$

$c_j{}^*$ = Unit cost according to continuous function (units may be the currency of that particular country)

= 1.2654 $D_j^{1.327}$ (*Bhave, 2003)

11.4.2 Description of constraints

The objective function stated above is subject to satisfaction of the following constraints

$$(TH)_i \geq (TH_{\min})_i \quad i = 1, 2, 3 \ldots\ldots\ldots\ldots N \tag{11.2}$$

$$V_{\min} \leq V_j \leq V_{\max} \quad j = 1, 2, 3 \ldots \ldots \ldots \ldots M \tag{11.3}$$

$$\sum Q_k = 0 \quad k = 1, 2, 3 \ldots \ldots \ldots \ldots P \tag{11.4}$$

$$\sum h_x = 0 \quad x = 1, 2, 3 \ldots \ldots \ldots \ldots R \tag{11.5}$$

$$D_j \geq 0 \tag{11.6}$$

where: i is the index used to represent demand node in the network; j is the index used to represent the link in the network; k is the index used to represent any node in the network; x is the index used to represent any link in the loop of network; N is the number of demand nodes in the network; M is the number of links in the network; P is the number of links joined at the node; R is the number of links in the loop; TH is the total head at the demand node in m; $TH_{\min}$ is the minimum HGL required at the demand node; $V_{\min}$ is the minimum velocity of flow in the link in m/s; $V_{\max}$ is the maximum velocity of flow in the link in m/s; V is the actual velocity of flow in the link in m/s; h_L is $(10.68 \text{ L } Q^{1.852})/(C^{1.852} D^{4.871})$; h_L is the head loss in the pipe in m; C is the Hazen–Williams roughness coefficient; D is the pipe diameter (m); L is the pipe length (m); Q is the flow rate (m³/s).

Equations (11).(4) and (11).(5) are used to determine the hydraulic balance of the network and they are basically continuity and energy equations respectively. The continuity equation is applied at each node with Q_k being the flow rate (in and out of the node). The energy equation is applied to each loop in the network with h_x being the head loss in each pipe. The head loss is the sum of minor losses and losses due to friction, that is major losses. The friction head loss can be computed by various formulae, in the present study the Hazen–Williams head loss equation is used.

11.4.3 Incorporation of reliability index into optimization
In the present study, the maximization of reliability is considered as a constraint to the main objective function, which is a typical practice in multi-objective optimization. Hence, some penalty in the form of cost is added to the objective function if it violates the minimum value of network reliability parameter that is specified.

$$\mathrm{NRP} \geq \mathrm{NRP}_{\min} \tag{11.7}$$

where NRP = Network Reliability Parameter; $\mathrm{NRP}_{\min}$ = minimum required Network Reliability Parameter.

The minimum required Network Reliability Parameter is considered to be between 0.5 to 1 in the present study.

11.5 DESCRIPTION OF EPANET TOOLKIT FUNCTIONS
11.5.1 Introduction
EPANET is a program that analyses the hydraulic and water quality behavior of water distribution systems. EPANET performs extended period simulation of hydraulic and water quality behavior within pressurized pipe networks. A network can consist of pipes, nodes (pipe junctions), pumps, valves and storage tanks or reservoirs. EPANET tracks the flow of water in each pipe, the pressure at each node, the height of water (HGL or piezometric head) in each tank, and the concentration of chemical species throughout the network during a multi-time period simulation. In addition to chemical species, water age and source tracing can also be simulated (United States Environmental Protection Agency, 2012).

The Programmer's Toolkit is an extension of the EPANET simulation package. The EPANET Programmer's Toolkit is a dynamic link library (DLL) of functions that allows developers to customize EPANET's computational engine for their own specific needs. The functions can be incorporated into

32-bit Windows applications written in C/C++, Delphi Pascal, Visual Basic, or any other language that can call functions within a Windows DLL. The Toolkit DLL file is named EPANET2.DLL and is distributed with EPANET. The Toolkit comes with several different header files, function definition files, and.lib files that simplify the task of interfacing it with C/C++, Delphi, and Visual Basic code. The Toolkit provides a series of functions that allow programmers to customize the use of EPANET's hydraulic and water quality solution engine to their own applications. Before using the Toolkit, one should become familiar with the way that EPANET represents a pipe network and the design and operating information it requires to perform a simulation.

11.5.2 A typical usage of the toolkit functions to analyse a distribution system

(1) Use the **ENopen** function to open the Toolkit system, along with an **EPANET Input file.**
(2) Use the **ENsetxxx** series of functions to change selected system characteristics.
(3) Run a full hydraulic simulation using the **ENsolveH** function (which automatically saves results to a Hydraulics file) or use the **ENopenH – ENinitH – ENrunH – ENnextH – ENcloseH** series of functions to step through a hydraulic simulation, accessing results along the way with the **ENgetxxx** series of functions.
(4) Run a full water quality simulation using **ENsolveQ** (which automatically saves hydraulic and water quality results to an Output file) or use the **ENopenQ – ENinitQ – ENrunQ – ENnextQ (or ENstepQ) – ENcloseQ** series of functions to step through a water quality simulation, accessing results along the way with the **ENgetxxx** series of functions.
(5) Return to Step 2 to run additional analyses or use the **ENreport** function to write a formatted report to the Report file.
(6) Call the **ENclose** function to close all files and release system memory.

11.5.3 Input file format with examples

An input file (in notepad with.dat or.txt format) according to the format specified in the EPANET tool kit is to be prepared. A sample of the input file is given below.

> **[Title]**
> (Here, the title can be mentioned as 'Hydraulic Analysis of a Standard benchmark network')
> **[Junctions]**
> (Here the details of the demand nodes need to be provided as per the format given below)
> ID Elevation demand pattern
> **[Reservoirs]**
> (here the details of the source node such as reservoirs/tanks need to be provided as per the format given below)
> ID Head Pattern
> **[PIPES]**
> (here, the details of the pipes in the network need to be provided as per the following format)
> ID Node1 Node2 Length Diam. Roughness Mloss Status
> **[OPTIONS]**
> (here, we can provide the details of analysis options)
> **[UNITS]**
> (here, the units to be followed can be mentioned)
> **[HEADLOSS]**
> (there are three different formulae available in EPANET, so we can mention the formula i.e. adopted, H-W, or D-W or C-M formula)

11.5.4 Linking of EPANET tool kit functions in MATLAB (input, output, opening and closing)

The EPANET tool kit is a shared library which is a collection of functions. On Windows systems, the library is precompiled into a dynamic link library (.dll) file named EPANET2.dll. At run-time,

the library is loaded into memory and made accessible to all applications. The MATLAB Interface to Generic DLLs enables interaction with functions in dynamic link libraries. This interface can load an external library into MATLAB memory space and then access any of the functions defined therein.
To load and unload EPANET library into MATLAB, the following functions are used.

```
loadlibrary('epanet2','epanet2')
unloadlibrary('epanet2')
```

Generally, at the beginning of the program, the library is to be loaded into memory and at the end of the program, it is to be unloaded. To invoke library functions, the *calllib* function is used.
For loading an input file and creating a report file, the following function is used.

```
calllib('epanet2','ENopen','input2.inp','report2.rpt',' ')
```

11.5.4.1 *Description of functions for assigning parameters*
To assign a value to a link in the network, the following function is used.
```
calllib('epanet2','ENsetlinkvalue',1,0100)
```

11.5.4.2 *Description of functions for performing analysis of the network*
To analyse the network, the following functions are used:

```
calllib('epanet2','ENsolveH')
calllib('epanet2','ENsolveQ')
calllib('epanet2','ENreport')
```

11.5.4.3 *Description of functions for extracting values of parameters*
To extract any value from the node/link, the following functions are used:

```
calllib('epanet2','ENgetnodevalue',1,11,0)
calllib('epanet2','ENgetlinkvalue',19,0)
```

11.6 PROCESS OF OPTIMIZATION USING GENETIC ALGORITHMS IN MATLAB USING GA TOOL KIT FUNCTIONS

11.6.1 Genetic algorithms
Genetic algorithms (GAs) are one of the most popular methods used for optimization of water distribution networks. GAs are inspired by Darwin's theory about evolution. GA is a search algorithm based on natural selection and the mechanism of population genetics (Goldberg, 2000). GA simulates mechanisms of population generation and normal rules of survival. It relies on the collective learning process within a population of individuals, each of which represents a point in space of feasible or infeasible solutions. In GA, an initial population is generated randomly. The population consists of number of individuals and each individual is a point on the solution space. The algorithm is started with a **set of solutions** (represented by **chromosomes**) called a **population**. Solutions from one population are taken and used to form a new population. This is motivated by a hope that the new population will be better than the old one. Solutions which are selected to form new solutions (**offspring**) are selected according to their fitness – the more suitable they are the more chances they have to reproduce. This is repeated until some condition (e.g. number of populations or improvement of the best solution) is satisfied. In the case of pipe network optimization, giving random values for diameters of all links will generate one individual. The values given to the individuals may be real or binary numbers.

11.6.2 Description of GA tool kit functions
In MATLAB, (Prasad *et al.*, 2003) the commands *ga* and *gatool* are used to implement the genetic algorithm to minimize an objective function. *ga* implements the genetic algorithm at the command

line to minimize an objective function. *gatool* opens the Genetic Algorithm Tool, a graphical user interface (GUI) to the genetic algorithm. In the present study *ga* is used to implement the genetic algorithm. The description of *ga* is presented in detail below:

Syntax of 'ga':

x = ga(fitnessfcn, nvars, options) applies the genetic algorithm to an optimization problem, using the parameters in the options structure.
options = gaoptimset;
gaoptimset('PopulationType','DoubleVector','PopulationSize',4,'InitialPopulation',[50;60;70;80],'PopInitrange',[50;100],'Generations',50)
ga(@fitnessfcn,4)

fitnessfcn -- Fitness function, nvars -- Number of independent variables for the fitness function

Please refer to the Matlab's GA options for more detailed descriptions.

11.7 IMPLEMENTATION OF RELIABILITY BASED OPTIMIZATION FOR PIPE NETWORK DESIGN USING MATLAB AND EPANET TOOLKIT FUNCTIONS

11.7.1 Development of coding in MATLAB – step by step process

The methodology adopted in the present study of reliability based optimal design of water distribution networks is explained in the following steps. *The combination of genetic algorithms toolbox in Matlab for optimal design and EPANET toolbox for analysis of the network is used.* The coding is developed in the MATLAB editor file.

(1) Standard benchmark networks/a real network are selected.
(2) Input files are prepared for all the networks according to the format specified in the EPANET tool kit and are stored in the computer directory.
(3) Through dynamic link libraries, EPANET2.DLL is linked within the MATLAB to access EPANET tool kit functions.
(4) EPANET Tool kit functions are used within the objective function in the MATLAB editor file and coding is developed in M-file with the same name as that of the objective function. For example:
function f = myfun(x,arg1,…..)
path(path,'c:\Matlab\R2014\extern\examples\shrlib')
loadlibrary('epanet2','epanet2');
calllib('epanet2','ENopen','input2.inp','report2.rpt','');
d1 = x(1); d2 = x(2); d3 = x(3); d4 = x(4);
calllib('epanet2','ENsetlinkvalue',1,0,d1);
calllib('epanet2','ENsetlinkvalue',2,0,d2);
calllib('epanet2','ENsolveH');
calllib('epanet2','ENsolveQ');
calllib('epanet2','ENreport');
calllib('epanet2','ENclose');
f = (632*x(1)∧1.3 + 506*x(2)∧1.3 + 759*x(3)∧1.3 + 2538*x(4)∧1.3) + 10⁴*(arg1 + arg2 + ….)
end
(here, myfun is the name of the objective function and also name of the M-file)
(5) The objective function developed in the M-file mentioned in step 4 is linked with Genetic Algorithms functions in another M-file. For example:
options = gaoptimset;

gaoptimset('PopulationType','DoubleVector','PopulationSize',4,'InitialPopulation',[50;60;70; 80],'PopInitrange',[50;100],'Generations',50)
ga(@myfun,4)

(6) The second M-file is executed to get the optimal solution.

The internal process of optimization is explained below in the following steps.

(1) Initially, GA optimization tool randomly generates the decision variables, that is pipe diameters, of equal number of links in the network within the given range of diameters.
(2) The generated link diameters are assigned to all the links in the network.
(3) The network is analyzed using the EPANET tool kit functions which are linked in the MATLAB through DLLs based on the input file supplied.
(4) A report file is generated and is stored in the default directory of the computer.
(5) The required information such as total heads, pressure heads, and demands at all demand nodes is extracted and velocities of all the links of the network also extracted using the specified functions in EPANET tool kit.
(6) The obtained total heads at the demand nodes are compared with the minimum required heads, and also the velocities with the maximum permissible velocities. If the obtained values are violating the specified conditions, penalties are assigned in the form of cost which is added to the objective function. The *penalty cost* is calculated as the product of the unit cost of violation of constraint and the amount of deviation.
(7) The pressure heads obtained are compared with the minimum pressures and the residual pressure heads are computed.
(8) Network Reliability Parameter (NRP) is determined.
(9) The penalty is added in the form of cost if NRP is less than the minimum value. The penalty cost is calculated as the product of the unit cost of violation of NRP and the amount of deviation of NRP.
(10) The total cost of the network is computed.
(11) The optimization algorithm is repeated till it gets the optimal cost of the network by assigning different values of diameters randomly at each generation.
(12) The optimal diameters obtained are converted into commercially available pipe sizes and then the network is analyzed until all the required parameters are obtained.
(13) The total heads are compared with the minimum required pressures, if they are satisfied, the optimal results are noted down, otherwise the procedure is repeated.
(14) This procedure is repeated for a number of times to obtain more optimal solutions. While selecting the next solution, the sizes of several pipes around nodes with residual pressures near zero are increased. Similarly, for a low-lying area where residual pressures are high, several pipe sizes are reduced. This helps in improving reliability and keeping the solution near the Pareto-optimal front. This will reduce the total number of solutions required to be generated to obtain the Pareto-optimal front.
(15) A graph of network reliability parameter versus network cost is plotted.
(16) A Pareto-optimal front is obtained.
(17) All solutions lying on the Pareto-optimal front are Pareto-optimal solutions, and the decision maker can choose the proper one depending upon the available funds.

11.8 STUDY ON A STANDARD BENCHMARK NETWORK

11.8.1 Description of network

The network shown in Figure 11.9 was first used by Alperovits and Shamir (1977) for optimal design using Linear Programming. Subsequently the same network has been used by several researchers

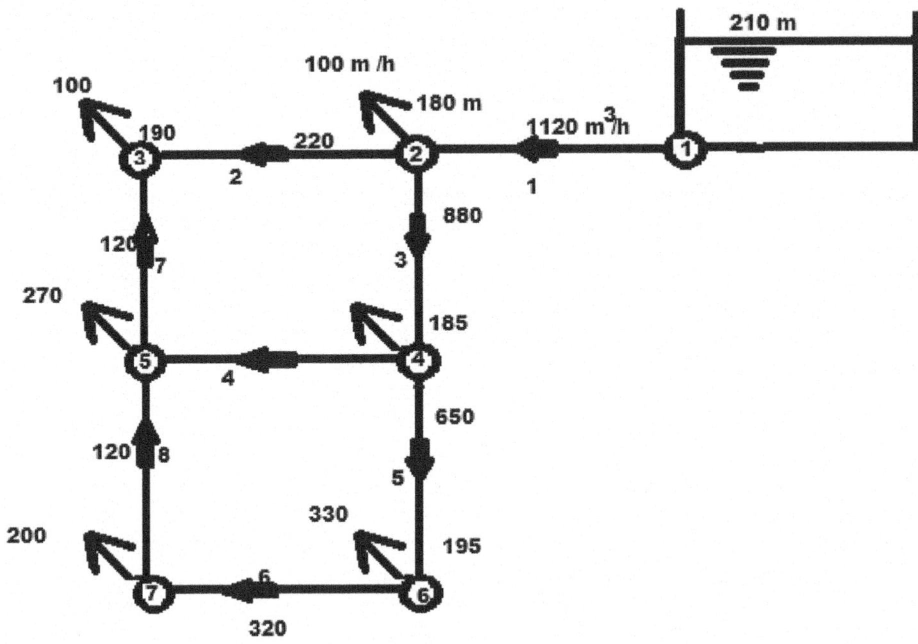

Figure 11.9 Two loop gravity network.

Table 11.10 Node details for two loop gravity network.

Node	Elevation (m)	Min. HGL (m)	Demand (m³/hr)
2	150	180	100
3	160	190	100
4	155	185	120
5	150	180	270
6	165	195	330
7	160	190	200

for optimal as well as reliability study. Some of them are Quidry *et al.* (1979), Goulter *et al.* (1986), Fujiwara *et al.* (1987), Kessler and Shamir (1989), Savic and Walters (1997), Abebe and Solomatine (1998), Cunha and Sousa (1999), Eusuf and Lansey (2003), Shie-Yui Liong *et al.* (2004), Keedwell and Khu (2005), Samani and Mottaghi (2006), Suribabu *et al.* (2006), Prassad *et al.* (2008), Van Dijk *et al.* (2008) and Afshar (2009).

The two loop gravity network consists of eight links, six demand nodes and one reservoir. The nodal information for this network is given in Table 11.10. Node 1 is a source node with HGL of 210 m and a demand of –1120 m³/hr. All the links in the network have a length of 1000 m and the Hazen–Williams coefficient (C_{HW}) is taken to be 130. The minimum required HGL values at demand nodes are given in Table 11.10. Table 11.11 shows the cost of pipes which can be used in the network.

Table 11.11 Cost data for pipes of two loop gravity network.

Diameter (inches)	Diameter (mm)	Cost (units)
1	25.4	2
2	50.8	5
3	76.2	8
4	101.6	11
6	152.4	16
8	203.2	23
10	254.0	32
12	304.8	50
14	355.6	60
16	406.4	90
18	457.2	130
20	508.0	170
22	558.8	300
24	609.6	550

11.8.2 Input file preparation for the standard benchmark network
[Title]

Analysis of a test network (Two loop gravity network)
[Junctions]
;ID   Elevation   demand   pattern
;---
2   150   1668   default
3   160   1668   default
4   155   2000   default
5   150   4500   default
6   165   5500   default
7   160   3334   default
[Reservoirs]
     ;ID   Head   Pattern
     ;------------------------
     1   210   default
[PIPES]
     ;ID Node1 Node2 Length Diam. Roughness Mloss Status
     ;---
     1   1   2   1000   558.8 130   0   open
     2   2   3   1000   609.6 130   0   open
     3   2   4   1000   406.9 130   0   open
     4   4   5   1000   254.0 130   0   open
     5   4   6   1000   355.6 130   0   open
     6   6   7   1000   203.2 130   0   open
     7   3   5   1000   355.6 130   0   open
     8   5   7   1000   152.4 130   open

```
[OPTIONS]
UNITS LPM
HEADLOSS H-W
[TIMES]
DURATION 0
[REPORT]
NODES ALL
LINKS ALL
HEAD YES
FLOW YES
LENGTH YES
DIAMETER YES
STATUS FULL
SUMMARY YES
VELOCITY PRECISION 2
```

11.8.3 Developing the code in matlab
11.8.3.1 First M-file

```
function f=fristMfile(x)
%path(path,'C:\Program Files\MATLAB\R2014a\extern\examples\shrlib')
%loading library into Matlab
loadlibrary('epanet2','epanet2');
%Calling EPANET input file prepared for the network
calllib('epanet2','ENopen','input.inp','report.rpt','');
% Setting the diameters for the pipes in the network
calllib('epanet2','ENsetlinkvalue',1,0,x(1));
calllib('epanet2','ENsetlinkvalue',2,0,x(2));
calllib('epanet2','ENsetlinkvalue',3,0,x(3));
calllib('epanet2','ENsetlinkvalue',4,0,x(4));
calllib('epanet2','ENsetlinkvalue',5,0,x(5));
calllib('epanet2','ENsetlinkvalue',6,0,x(6));
calllib('epanet2','ENsetlinkvalue',7,0,x(7));
calllib('epanet2','ENsetlinkvalue',8,0,x(8));
% Performing analysis of network
calllib('epanet2','ENsolveH');
calllib('epanet2','ENsolveQ');
calllib('epanet2','ENreport');
%Extracting hydraulic heads at various demand nodes of the network
[a h2]=calllib('epanet2','ENgetnodevalue',1,11,0);
[a h3]=calllib('epanet2','ENgetnodevalue',2,11,0);
[a h4]=calllib('epanet2','ENgetnodevalue',3,11,0);
[a h5]=calllib('epanet2','ENgetnodevalue',4,11,0);
[a h6]=calllib('epanet2','ENgetnodevalue',5,11,0);
[a h7]=calllib('epanet2','ENgetnodevalue',6,11,0);
%calculation of reliability index based on range-1 residual pressures
if h2>25 | h2<0          SI2=0;
elseif h2>=10 &h2<=15 SI2=1;
elseif h2>15 & h2<=25   SI2=(25-h2)/10;
elseif h2>0 & h2<10      SI2=h2/10;
```

```
end
if h3>25 | h3<0            SI3 = 0;
elseif h3> = 10 &h3< = 15 SI3 = 1;
elseif h3>15 & h3< = 25   SI3 = (25-h3)/10;
elseif h3>0 & h3<10       SI3 = h3/10;
end
if h4>25 | h4<0            SI4 = 0;
elseif h4> = 10 &h4< = 15 SI4 = 1;
elseif h4>15 & h4< = 25   SI4 = (25-h4)/10;
elseif h4>0 & h4<10       SI4 = h4/10;
end
if h5>25 | h5<0            SI5 = 0;
elseif h5> = 10 &h5< = 15 SI5 = 1;
elseif h5>15 & h5< = 25   SI5 = (25-h5)/10;
elseif h5>0 & h5<10       SI5 = h5/10;
end
if h6>25 | h6<0            SI6 = 0;
elseif h6> = 10 &h6< = 15 SI6 = 1;
elseif h6>15 & h6< = 25   SI6 = (25-h6)/10;
elseif h6>0 & h6<10       SI6 = h6/10;
end
if h7>25 | h7<0            SI7 = 0;
elseif h7> = 10 &h7< = 15 SI7 = 1;
elseif h7>15 & h7< = 25   SI7 = (25-h7)/10;
elseif h7>0 & h7<10       SI7 = h7/10;
end
%extraction of demands at various demand nodes of the network
[a de2] = calllib('epanet2','ENgetnodevalue',1,1,0);
[a de3] = calllib('epanet2','ENgetnodevalue',2,1,0);
[a de4] = calllib('epanet2','ENgetnodevalue',3,1,0);
[a de5] = calllib('epanet2','ENgetnodevalue',4,1,0);
[a de6] = calllib('epanet2','ENgetnodevalue',5,1,0);
[a de7] = calllib('epanet2','ENgetnodevalue',6,1,0);

sum1 = de2*SI2 + de3*SI3 + de4*SI4 + de5*SI5 + de6*SI6 + de7*SI7;
sum2 = de2 + de3 + de4 + de5 + de6 + de7;
NRP = sum1/sum2
%Pressure constraint violation
if ge(h2,30) flag2 = 0;
else flag2 = (30-h2); end
if ge(h3,30) flag3 = 0;
else flag3 = (30-h3); end
if ge(h4,30) flag4 = 0;
else flag4 = (30-h4);
end
if ge(h5,30) flag5 = 0;
else flag5 = (30-h5);
end
if ge(h6,30) flag6 = 0;
else flag6 = (30-h6);
```

```
end
if ge(h7,30) flag7=0;
else flag7=(30-h7);
end
%Extracting velocities at various pipes in the network
[a ve1]=calllib('epanet2','ENgetlinkvalue',1,9,0);
[a ve2]=calllib('epanet2','ENgetlinkvalue',2,9,0);
[a ve3]=calllib('epanet2','ENgetlinkvalue',3,9,0);
[a ve4]=calllib('epanet2','ENgetlinkvalue',4,9,0);
[a ve5]=calllib('epanet2','ENgetlinkvalue',5,9,0);
[a ve6]=calllib('epanet2','ENgetlinkvalue',6,9,0);
[a ve7]=calllib('epanet2','ENgetlinkvalue',7,9,0);
[a ve8]=calllib('epanet2','ENgetlinkvalue',8,9,0);
v1=double(ve1);
v2=double(ve2);
v3=double(ve3);
v4=double(ve4);
v5=double(ve5);
v6=double(ve6);
v7=double(ve7);
v8=double(ve8);
%velocity constraint violation
if ge(v1,6) flagv1=0;
else flagv1=(6-v1); end
if ge(v2,6) flagv2=0;
else flagv2=(6-v2); end
if ge(v3,6) flagv3=0;
else flagv3=(6-v3);
end
if ge(v4,6) flagv4=0;
else flagv4=(6-v5);
end
if ge(v5,6) flagv5=0;
else flagv5=(6-v5);
end
if ge(v6,6) flagv6=0;
else flagv6=(6-v6);
end
if ge(v7,6) flagv7=0;
else flagv7=(6-v7);
end
if ge(v8,6) flagv8=0;
else flagv8=(6-v8);
end
%NRP violation
if ge(NRP,0.5) flagNRP=0;
else flagNRP=(1-NRP);
end
%cost function including penalities
```

$$f = 1265.4*(x(1)\wedge1.327 + x(2)\wedge1.327 + x(3)\wedge1.327 + x(4)\wedge1.327 + x(5)\wedge1.327 + x(6)\wedge1.327 + x(7)\wedge1.327 +$$
$$x(8)\wedge1.327) + 25000*(flag2 + flag3 + flag4 + flag5 + flag6 + flag7) + 5000\ x(flagv1 +$$
$$flagv2 + flagv3 + flagv4 + flagv5 + flagv6 + flagv7 + flagv8) + 3000*flagNRP;$$
%Closing
calllib('epanet2','ENclose');

11.8.3.2 Second M-file
options = gaoptimset('PopulationSize',[8],'Generations',1500,'PopInitRange',[25.4;610],'StallGenLimit',
1000,'StallTimeLimit',1000,'PlotFcns',[@gaplotbestf],'EliteCount',4);
[x fval 'reason' output population scores] = ga(@firstMfile,8,options)

11.8.4 Screenshots for the program implemented in matlab
For better understanding, the various screenshots are shown in Figures 11.10–11.16.

11.8.4.1 Screenshot-1: storing EPANET toolkit functions in Matlab shared library
Note: (i) the content highlighted within the top rectangular box shows the path for the DLL file and Header File stored in Matlab; (ii) the content highlighted within the bottom rectangular box shows the files.

11.8.4.2 Screenshot-2: code developed in Matlab editor for first M-file
Note: (i) the content highlighted within the rectangular boxes show that the function name and file name are same.

11.8.4.3 Screenshot-3: Run file (second M-file) showing options and ga function
Note: (i) the content highlighted within the first rectangular box shows that the options set to ga function; ii) the content highlighted within the second rectangular box shows the ga function which involves first M-file, number of decision variables and options.

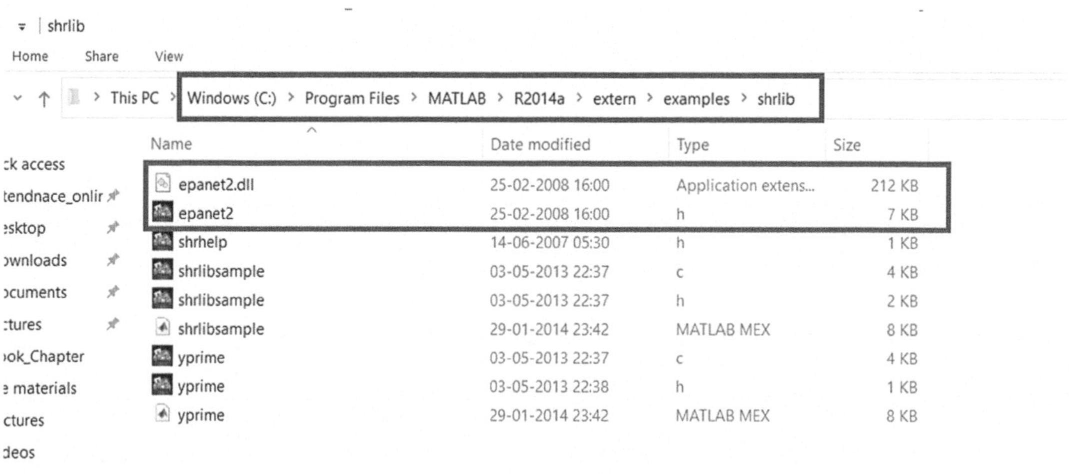

Figure 11.10 Storing EPANET toolkit functions in Matlab shared library.

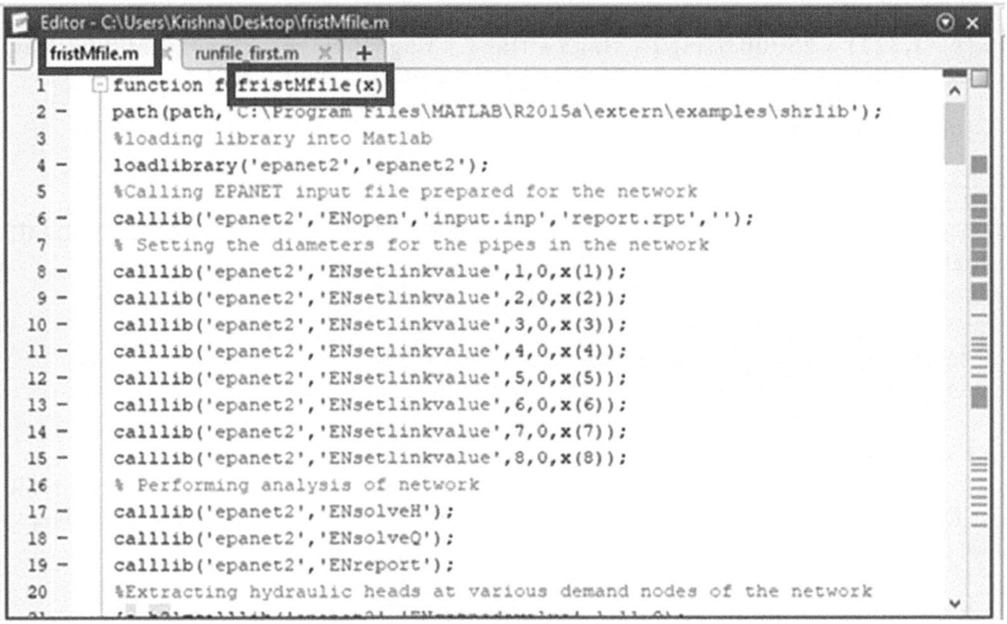

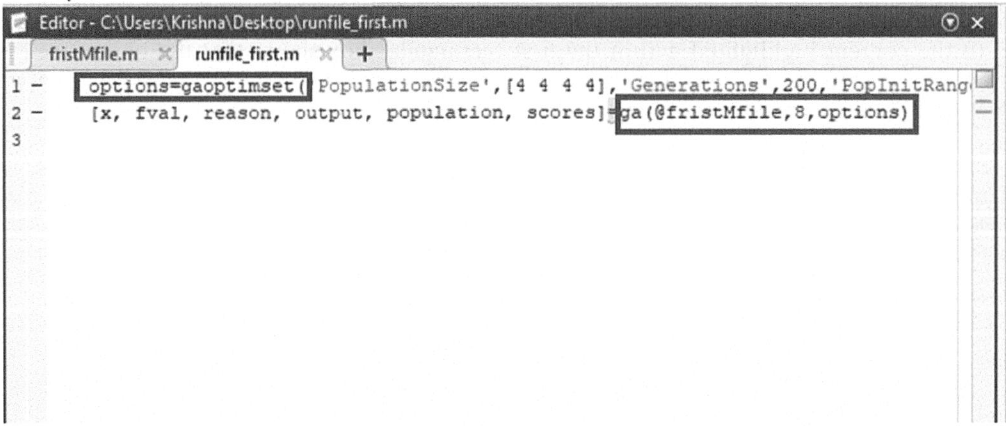

Figure 11.11 Code developed in Matlab editor for first M-file.

Figure 11.12 Run file (Second M-file) showing options and ga function.

11.8.4.4 Screenshot-4: input file created in notepad showing the details of the network

Note: the content highlighted within the rectangular box shows the name of the input file which contains all the details of pipe network.

11.8.4.5 Screenshot-5: report file generated in notepad showing the analysis results of the network

Note: the content highlighted within the rectangular box shows the name of the report file generated which contains all the pressures and velocities and so on.

```
[Title]
Analysis of a test network (Two loop gravity network)
[Junctions]
; ID     Elevation       demand      pattern
;--------------------------------------------------
   2      150                 1668            default
   3      160         1668        default
   4      155                 2000        default
   5      150         4500        default
   6      165         5500        default
   7      160         3334        default
[Reservoirs]
;ID   Head   Pattern
;------------------------
1    210     default
[PIPES]
;ID  Node1 Node2 Length  Diam.  Roughness Mloss  Status
;------------------------------------------------------------------
   1       1     2    1000    558.8     130         0     open
   2       2     3    1000    609.6     130         0     open
   3       2     4    1000    406.9     130         0     open
   4       4     5    1000    254.0     130         0     open
```

Figure 11.13 Input file created in notepad showing the details of the network.

```
*report  - Notepad
File  Edit  Format  View  Help
 Page 1                          Sun Aug 15 17:08:15 2021

  ******************************************************************
  *                 E P A N E T              *
  *           Hydraulic and Water Quality           *
  *           Analysis for Pipe Networks         *
  *                 Version 2.0              *
  ******************************************************************

  Analysis of a test network (Two loop gravity network)

       Input Data File .................... input.inp
       Number of Junctions................ 6
       Number of Reservoirs............... 1
       Number of Tanks ................... 0
       Number of Pipes ................... 8
       Number of Pumps ................... 0
       Number of Valves .................. 0
       Headloss Formula .................. Hazen-Williams
       Hydraulic Timestep ................ 1.00 hrs
       Hydraulic Accuracy ................ 0.001000
       Maximum Trials .................... 40
       Quality Analysis .................. None
```

Figure 11.14 Report file generated in notepad showing the analysis results of the network.

11.8.4.6 Screenshot-6: population generated in Matlab

Note: the content highlighted within the rectangular box shows the population generated a function in Matlab.

11.8.4.7 Screenshot-7: final output in terms of decision variables

Note: the content highlighted within the rectangular box shows the variables -x in Matlab (final output in terms of decision variables).

	1	2	3	4	5	6	7	8	9
1	147.5434	365.7829	153.2384	332.5514	427.4856	475.0243	326.1320	136.3047	
2	440.0461	341.5162	87.2497	335.4392	235.4048	257.3643	542.3507	314.7804	
3	163.5004	533.9675	89.5291	528.8223	455.8644	498.0573	369.1600	111.6918	
4	517.5918	343.3557	137.6598	51.3879	505.7782	474.8722	90.7032	25.7054	
5	380.4299	211.3461	261.9175	255.4146	424.9249	246.0254	142.2398	522.7266	
6	288.5505	95.0928	287.5188	417.9186	436.9861	151.6847	263.3058	353.1031	
7	293.5709	574.8243	239.2561	458.7394	283.9717	487.4721	463.0934	568.8493	
8	147.5434	365.7829	153.2384	332.5514	427.4856	475.0243	326.1320	136.3047	
9	475.7089	305.6942	392.4682	228.6728	218.8195	216.8948	487.2124	366.0996	
10	230.1375	399.1447	476.6997	113.0884	273.4513	417.8212	211.6093	502.0812	
11	412.4108	343.8410	570.7462	368.0294	183.4001	281.8319	337.6139	539.2715	
12	380.4299	211.3461	261.9175	255.4146	424.9249	246.0254	142.2398	522.7266	
13	517.5918	343.3557	137.6598	51.3879	505.7782	474.8722	90.7032	25.7054	
14	512.3232	446.9239	106.5859	466.7340	276.7321	123.1764	105.0766	531.3354	
15	175.3154	330.8508	432.4373	167.3323	544.3909	529.3138	422.1401	383.5064	
16	475.7089	305.6942	392.4682	228.6728	218.8195	216.8948	487.2124	366.0996	

Figure 11.15 Population generated in Matlab.

	1	2	3	4	5	6	7	8	9
1	147.5434	365.7829	153.2384	332.5514	427.4856	475.0243	326.1320	136.3047	
2									
3									
4									
5									

Figure 11.16 Final output in terms of decision variables.

11.8.5 Analysis of the results
11.8.5.1 Best optimal solutions
The advantage of the coding developed is that it gives different optimal solutions for every run, so that feature is captured and the best optimal solutions obtained are presented below. The full details such as diameters, nodal pressure heads and total heads at demand nodes and velocity and flows in the links of the network and Total Cost (TC), Network Reliability Parameter (NRP), Cost Reliability Ratio (CRR) and Cost per Unit Reliability and Unit Length (CURUL) are presented in Table 11.12.

Table 11.12 Full details of the best solutions obtained for the two loop gravity network.

Solution/Parameter	1	2	3	4	5	6	7
D1 in mm	508	508	457.2	457.2	457.2	508.00	457.2
D2 in mm	304.8	355.6	355.6	355.6	355.6	254.00	304.8
D3 in mm	406.4	406.4	406.4	406.4	406.4	406.40	406.4
D4 in mm	254	76.2	76.2	254	254	50.80	203.2
D5 in mm	355.6	355.6	355.6	355.6	355.6	355.60	406.4
D6 in mm	254	50.8	50.8	254	254	304.80	203.2
D7 in mm	254	406.4	406.4	304.8	304.8	254.00	254
D8 in mm	254	355.6	355.6	304.8	254	101.60	203.2
H2 in m	55.96	55.96	53.24	53.24	53.24	55.96	53.24
H3 in m	41.74	39.64	36.93	40.18	40.28	32.74	38.38
H4 in m	45.65	48.63	45.91	43.96	43.87	46.24	43.26
H5 in m	46.91	47.36	44.65	46.5	46.78	35.76	42.42
H6 in m	31.79	36.2	33.49	30.8	30.42	30.05	31.01
H7 in m	35.50	36.44	33.73	35.38	34.67	32.63	30.03
Th2 in m	205.96	205.96	203.4	203.24	203.24	205.96	203.24
Th3 in m	201.74	199.64	196.93	200.18	200.28	192.74	198.38
Th4 in m	200.65	203.63	200.91	198.96	198.87	201.24	198.26
Th5 in m	196.91	197.36	194.65	196.5	196.78	185.76	192.42
Th6 in m	196.79	201.2	198.49	195.8	195.42	195.76	196.01
Th7 in m	195.50	196.44	193.73	195.38	194.67	192.63	190.03
V1 in m/s	1.54	1.54	1.9	1.9	1.9	1.54	1.90
V2 in m/s	1.14	1.56	1.56	1.06	1.04	1.88	1.23
V3 in m/s	1.54	0.99	0.99	1.38	1.39	1.45	1.49
V4 in m/s	0.95	0.59	0.59	0.76	0.7	0.74	1.05
V5 in m/s	1.20	0.93	0.93	1.07	1.13	1.54	0.97
V6 in m/s	0.53	0.39	0.39	0.29	0.4	0.84	1.06
V7 in m/s	1.09	0.98	0.98	1.06	1.03	1.33	1.22
V8 in m/s	0.56	0.55	0.55	0.56	0.7	0.74	0.65
Q1 in lpm	18 670.00	18 670	18 670	18 670	18 670	18 670.00	18 670.00
Q2 in lpm	4985.00	9293	9293	6296	6176	5717.00	5384.00
Q3 in lpm	12 017.00	7708	7709	10 706	10 826	11 285.00	11 617.00
Q4 in lpm	2891.00	161	161	2309	2114	90.00	2046.00
Q5 in lpm	7126.00	5548	5548	6397	6712	9195.00	7571.00
Q6 in lpm	1626.00	48	48	897	1212	3695.00	2071.00
Q7 in lpm	3316.00	7625	7625	4628	4508	4049.00	3716.00
Q8 in lpm	1708.00	3286	3286	2437	2122	−361.00	1263.00
TC in Units	498 000	543 000	503 000	504 000	486 000	450 000	461 000
NRP	0.591	0.6762	0.586	0.5705	0.5545	0.321	0.454
CRR	842 782	803 017	858 362	883 436	876 465	1 404 056	1 016 090
CURUL	105	100	107	110	110	176	127

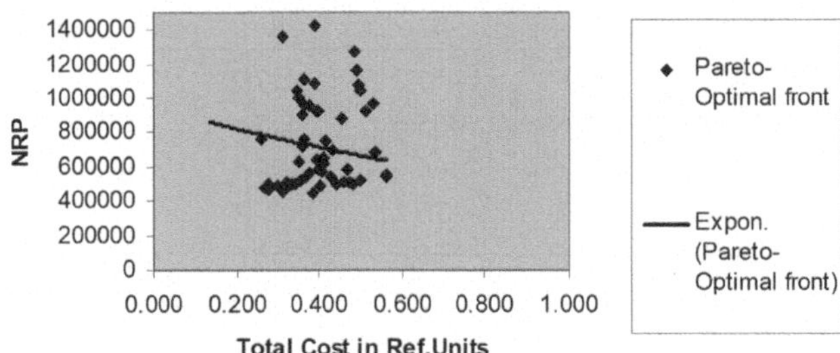

Figure 11.17 Pareto-optimal front for Range-1.

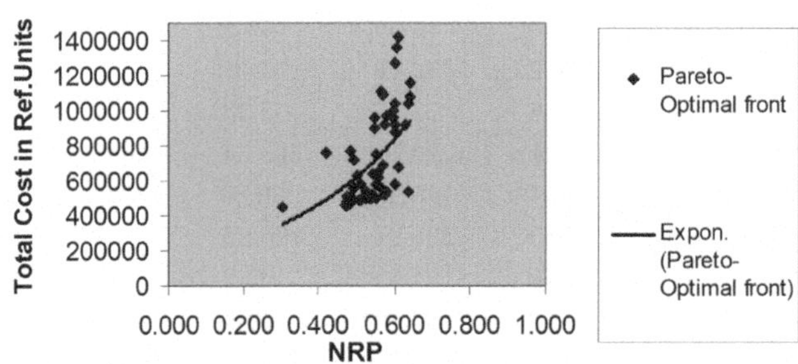

Figure 11.18 Pareto-optimal front for Range-2.

In the present study, the following parameters are used to know the overall network performance.

(1) Total Cost of the network (TC)
(2) Reliability Parameter (NRP)
(3) Cost Reliability Ratio (CRR)
(4) Cost per Unit Reliability and Unit Length (CURUL)

The above parameters are also used to compare the results with the previous literature.

11.8.5.2 Pareto-optimal solutions

The Pareto-optimal solutions are determined for different ranges of residual pressures using the TC and NRP obtained for different solutions and are shown in Figures 11.17–11.19.

The exponential trend line of pareto-optimal solution of Range-1 is showing differently to that of Range-2 and Range-3. Actually, with the increase in reliability, the cost has to increase. Range-3 is following this fact. Hence in the present study it is proposed to use pareto-optimal front developed based on Range-3 which is having reliability of more than 0.8 for the network under consideration.

11.8.5.3 Comparison of results with previous researchers

The optimal solutions obtained by different researchers for the two loop network are compared with the present research and the details are given in the Table 11.13.

The complete sets of optimal values of the parameters obtained on the network by previous researchers are compared with the present work and are presented in Figures 11.20–11.25.

Based on the performance indicators (NRP, CRR, CURUL), it is concluded that the values obtained for these parameters in the present study are giving better results when compared to other researchers.

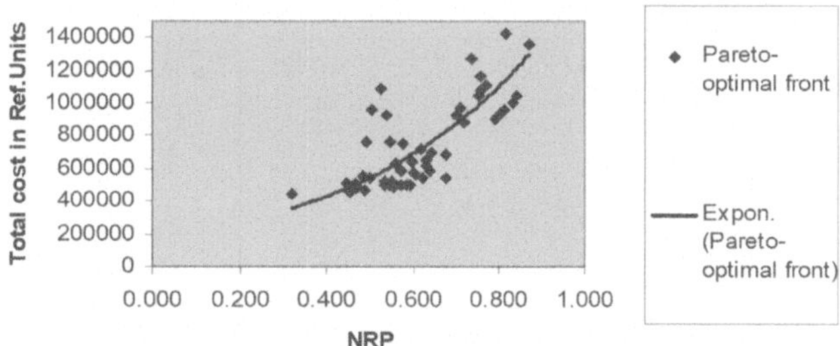

Figure 11.19 Pareto-optimal front for Range-3.

11.8.6 Comparison of reliability Index of a simple pipe network based on binary logic and fuzzy logic

11.8.6.1 NRP results obtained on two loop gravity network (network 1) based on fuzzy logic and binary logic

The proposed methodology is applied to a standard two loop gravity network. This network is taken from the literature which is used by most of the researchers. Several optimal solutions are obtained in which 54 optimal solutions are identified as the best solutions for comparison.

The range wise comparison is made and is presented graphically in Figures 11.26–11.28.

Based on binary logic, range-3 shows partially higher values of NRP for some solutions, range-2 for some other solutions and range-1 for remaining solutions so there is no clear cut demarcation for the ranges based on this logic. However, all the NRP values obtained are lesser than those obtained based on fuzzy logic. Based on fuzzy logic, range-3 gives the highest NRP values in almost all the optimal solutions.

In all the three ranges, almost all the values of NRP based on fuzzy logic are higher than those obtained based on binary logic. Hence, it is concluded that range-3 is the best range based on fuzzy logic for incorporating in the optimal design of water distribution networks for maximum reliability.

Table 11.13 Optimal solutions for two loop network obtained by different researchers.

Reference	D_1	D_2	D_3	D_4	D_5	D_6	D_7	D_8	C_T (10^3)	FEN[a] (10^3)
Savic and Walters (1997) (GA1)	457.2	254	406.4	101.6	406.4	254	254	25.4	419	65
Savic and Walters (1997) (GA2)	508	254	406.4	25.4	355.6	254	254	25.4	420	65
Abebe and Solomatine (1998)	457.2	254	406.4	101.6	406.4	254	254	25.4	419	1.373
Cunha and Sousa (1999)	457.2	254	406.4	101.6	406.4	254	254	25.4	419	25
Wu and Simpson (2001)	457.2	254	406.4	101.6	406.4	254	254	25.4	419	7.467
Eusuf and Lansey (2003)	457.2	254	406.4	101.6	406.4	254	254	25.4	419	11.323
Prassad *et al.* (2003)	450	250	400	100	400	250	250	25	419	-
Liong and Atiquazzaman (2004)	457.2	254	406.4	101.6	406.4	254	254	25.4	419	1.091
Van Dijk *et al.* (2008)	457.2	254	406.4	101.6	406.4	254	254	25.4	419	100
Afshar (2009)	457.2	254	406.4	101.6	406.4	254	254	25.4	419	3
Present Study	457.2	254	406.4	101.6	406.4	254	254	25.4	419	0.8

[a]*FEN:* Function Evaluation Number.

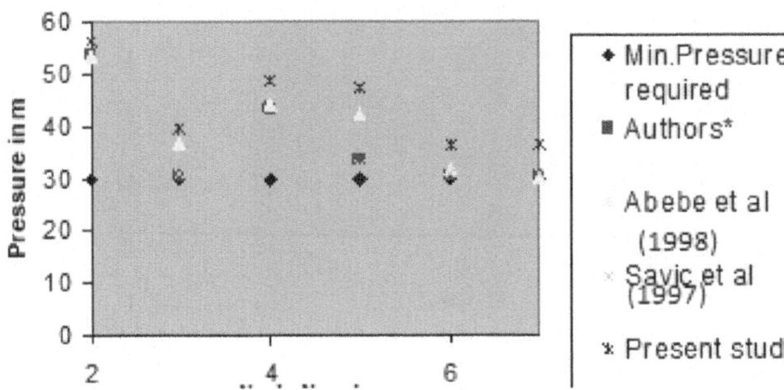

Figure 11.20 Comparison of pressure heads at demand nodes of network 1 with previous researchers.

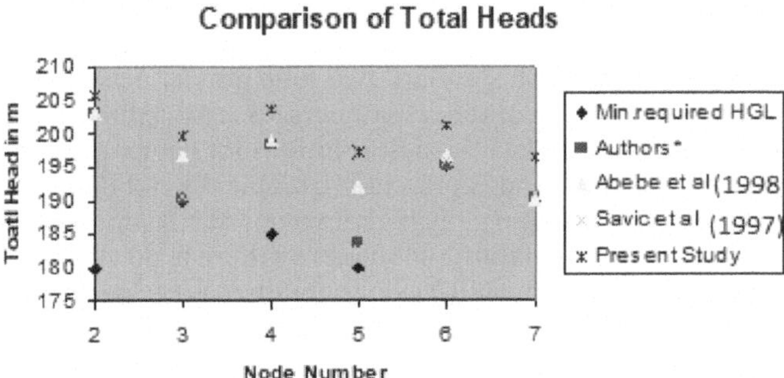

Figure 11.21 Comparison of total heads at demand nodes of network 1 with previous researchers.

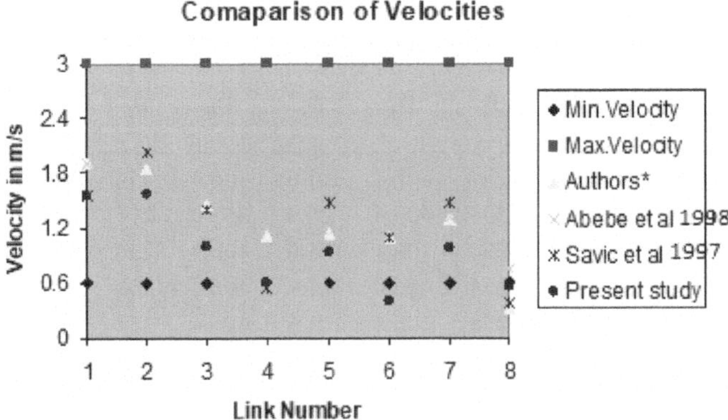

Figure 11.22 Comparison of velocities in the links of network 1 with previous researchers.

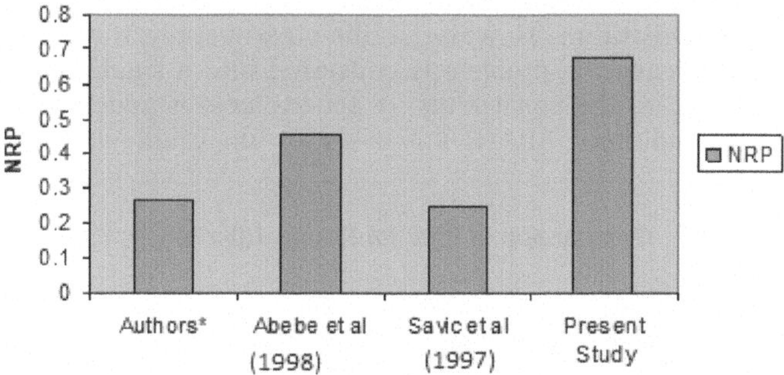

Figure 11.23 Comparison of NRP of network 1 with previous researchers.

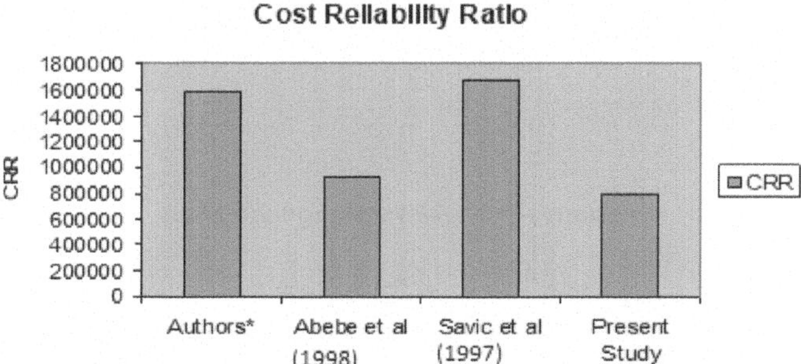

Figure 11.24 Comparison of CRR of network 1 with previous researchers.

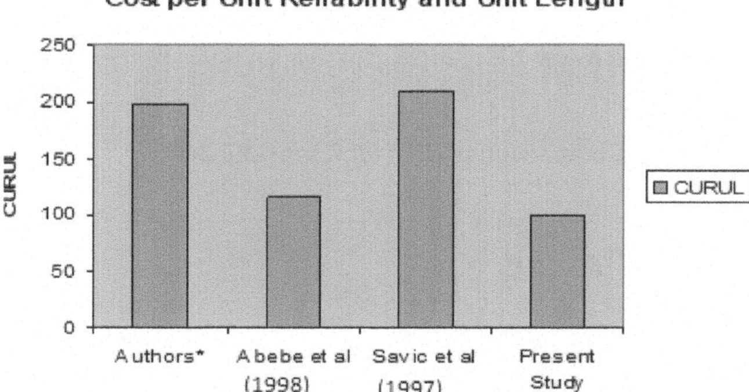

Figure 11.25 Comparison of CURUL of network 1 with previous researchers.

11.8.7 Summary of results and conclusions

GA optimization techniques are very effective in minimizing an objective function if it has only a single objective function without constraints. However, in the present study, the problem is a two-objective problem with constraints. Hence the problem is converted into a single objective optimization by considering reliability as one of the constraints. A penalty cost is added to the network cost if it is violating the specified conditions. EPANET is based on the gradient method of analysis which

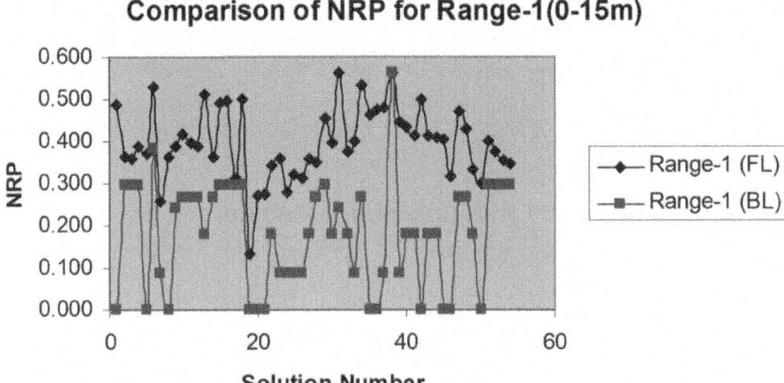

Figure 11.26 Comparison of NRP for range-1.

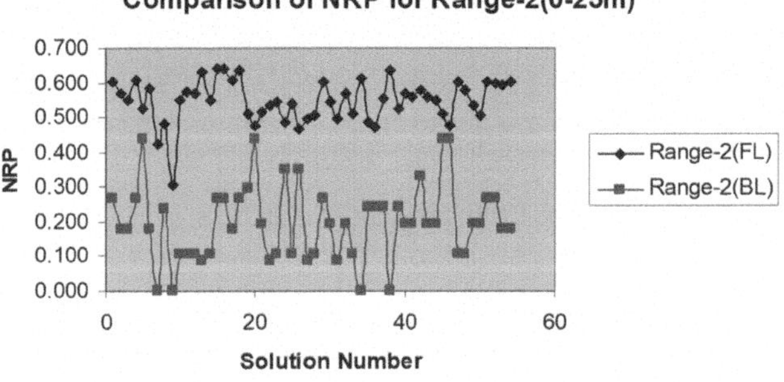

Figure 11.27 Comparison of NRP for range-2.

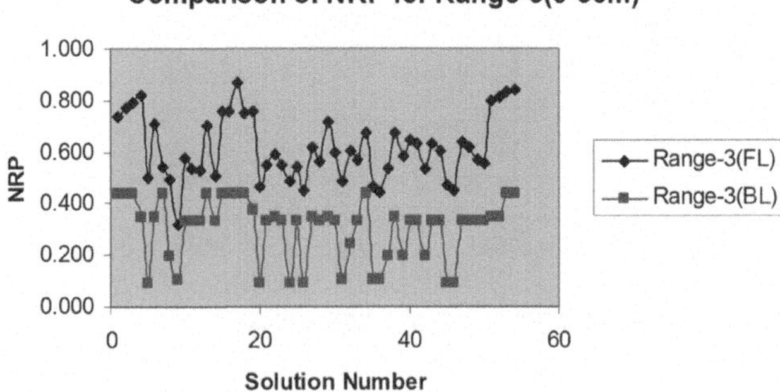

Figure 11.28 Comparison of NRP for range-3.

is basically a demand driven network analysis method. Hence in the present study, the developed algorithm for reliability-based optimal design of water distribution networks has been conducted numerous times until it obtains the optimal solutions satisfying the pressure constraints at the demand nodes and for the maximum reliability so that the optimal solutions obtained are satisfying both the demands and the minimum pressure requirements at the demand nodes in the network.

In the two-loop gravity network, to know the variation of the NRP, three different ranges of residual pressures which are in excess of the minimum pressure requirements at the demand nodes are considered in the present study for fixing the best range of residual pressures such that the reliability of the network is maximum. A comparison is also made between fuzzy logic and binary logic in estimating the network reliability parameter for three different ranges of residual pressures. The range-3 (0–30 m) residual pressures are fixed as the best range for the network. The optimal results obtained are compared with those obtained by previous researchers.

Following are the major conclusions arrived at from the present study:

- For the two loop gravity network, three different ranges of residual pressures that is 0–15, 0–25 and 0–30 m at the demand nodes are considered for fixing the best range of residual pressures such that the reliability of the network is maximum. The ranges are fixed based on the elevation difference between the source node and demand node of the network. Fuzzy logic and binary logic are used separately to link the residual pressures with the satisfaction index. Based on the results obtained, it is concluded that the range-3 pressures based on fuzzy logic are the best to incorporate in the optimal design for achieving maximum reliability of the network considered.
- The Function Evaluation Number (FEN) is an indication of how an optimal solution is obtained quickly, so in the present study it is obtained as 800 for the two loop network using the proposed approach, which is the least when compared to the number obtained by previous researchers. Also, the runtime to reach the optimal solution is less. So it is concluded that the developed method gives the solution quickly with less number of function counts.
- All the optimal output parameters of the two loop network obtained, such as diameters, pressures and so on., are comparable with those obtained by previous researchers.
- The best NRP, CRR and CURUL obtained among the 54 optimal solutions for the two loop network are 0.6762, 803017 and 100 which are also better than those obtained by the previous researchers.
- The pareto-optimal front obtained based on range-3 is the best one when compared with the other two ranges since it gives the highest reliabilities for least cost for the two loop network.

11.9 LIMITATIONS OF THE PRESENT STUDY AND SCOPE FOR FUTURE RESEARCH

The following are the limitations of the present research work.

- The optimization cost considered in the present study is *only* the cost of pipes in the network.
- In the present study the network reliability parameter is based on excess residual pressures and the demands at demand nodes of the network.
- Only gravity looped water distribution networks are considered in the present study.
- The best range of residual pressures obtained is applicable only if the elevation difference between the reservoir and the demand nodes is more than or equal to 60 m.
- The algorithm searches the optimal solution by assigning the decision variables to the optimization function randomly.

The scope for continuing the present work in future is described below.

- In the present study the optimization cost considered is the cost of links in the network only, so one may also include installation cost of other components of the water distribution network such as reservoir, pumps and so on., and can also include the operational cost of the network.

- In the network reliability parameter, one may also consider quality of water as one of the criteria in satisfying the consumer needs and cost of treatment may also be taken into consideration.
- Failure probability of components could also be included in assessing the reliability. In this study it is assumed that all the pipes are functioning well.

11.10 PRACTICE PROBLEMS

11.10.1 Practice problem-1

This test network is a three loop water distribution network of Hanoi city water distribution system, which consists of 32 nodes, 34 pipes and a single reservoir and is shown in Figure 11.29. The input data for this problem are given in the Tables 11.14 and 11.15.

Perform reliability-based optimization for the Hanoi City Water Distribution Network by using the above data. Assume any data, if required suitably.

11.10.1.1 Solution

The optimal diameters obtained for the Hanoi network and FEN (Function Evaluation Number) and run time are compared with the previous researchers and the data are presented in Table 11.16.

The performance indicators of the network TC, NRP, CRR and CURUL obtained for the Hanoi network are compared with that of the previous researchers and the data are presented in Table 11.17 and in Figures 11.30–11.33.

Based on the above values, it is clearly understood that the present proposed methodology shows better results when compared with other researchers' results.

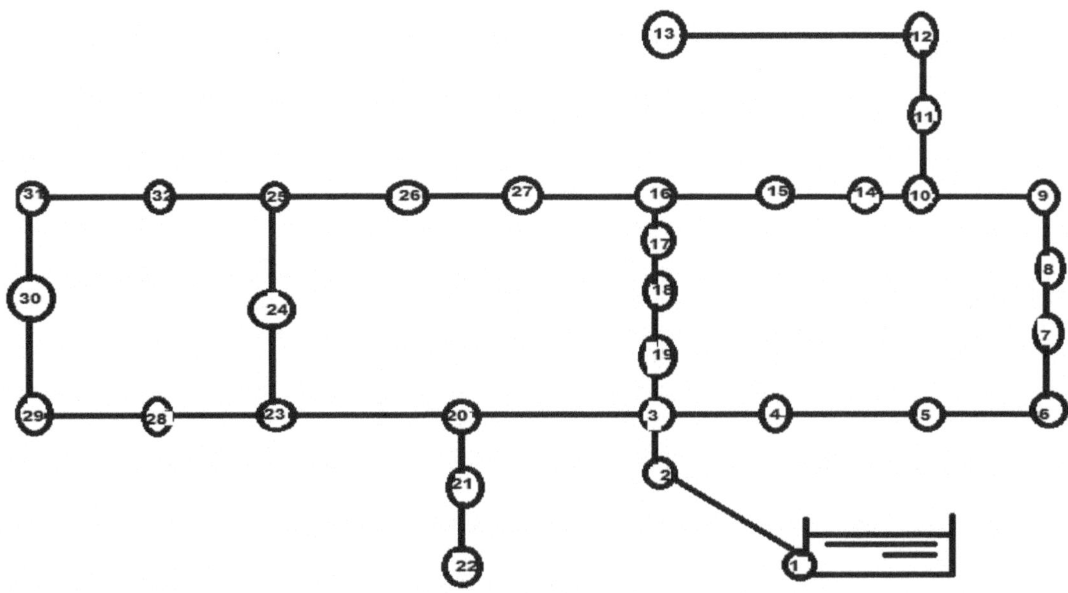

Figure 11.29 Hanoi City Water Distribution Network (three loop network).

Table 11.14 Pipe cost for Hanoi network.

Diameter (inches)	Diameter (mm)	Cost (units)
12	304.8	45.73
16	406.4	70.40
20	508.0	98.38
24	609.6	129.33
30	762.0	180.8
40	1016.0	278.3

Table 11.15 Node and link data for Hanoi network.

Node Number	Demand (m³/hr)	Link Index	Arc	Length (m)
1	−19 940	1	(1,2)	100
2	890	2	(2,3)	1350
3	850	3	(3,4)	900
4	130	4	(4,5)	1150
5	725	5	(5,6)	1450
6	1005	6	(6,7)	450
7	1350	7	(7,8)	850
8	550	8	(8,9)	850
9	525	9	(9,10)	800
10	525	10	(10,11)	950
11	500	11	(11,12)	1200
12	560	12	(12,13)	3500
13	940	13	(10,14)	800
14	615	14	(14,15)	500
15	280	15	(15,16)	550
16	310	16	(16,17)	2730
17	865	17	(17,18)	1750
18	1345	18	(18,19)	800
19	60	19	(19,3)	400
20	1275	20	(3,20)	2200
21	930	21	(20,21)	1500
22	485	22	(21,22)	500
23	1045	23	(20,23)	2650
24	820	24	(23,24)	1230
25	170	25	(24,25)	1300
26	900	26	(25,26)	850
27	370	27	(26,27)	300
28	290	28	(27,16)	750
29	360	29	(23,28)	1500
30	360	30	(28,29)	2000
31	105	31	(29,30)	1600
32	805	32	(30,31)	150
		33	(31,32)	860
		34	(32,25)	950

Table 11.16 Comparison of optimal diameter and other parameters of Hanoi network with previous research.

Pipe Number	Pipe Diameter (inches)						
	Savic and Walters (1997)		Abebe and Solomatine (1998)		Cunha and Sousa (1999)	Liong and Atiquazzaman (2004)	Present Study
	GA1	GA2	GA	ACCOL			
1	40	40	40	40	40	40	40
2	40	40	40	40	40	40	40
3	40	40	40	40	40	40	30
4	40	40	40	40	40	40	30
5	40	40	30	40	40	40	30
6	40	40	40	30	40	40	30
7	40	40	30	40	40	40	24
8	40	40	30	40	40	30	30
9	40	30	30	24	40	30	20
10	30	30	30	40	30	30	40
11	24	30	30	30	24	30	30
12	24	24	39	40	24	24	24
13	20	16	16	16	20	16	30
14	16	16	24	16	16	12	40
15	12	12	30	30	12	12	24
16	12	16	30	12	12	24	20
17	16	20	30	20	16	30	24
18	20	24	40	24	20	30	20
19	20	24	40	30	20	30	40
20	40	40	40	40	40	40	40
21	20	20	20	30	20	20	40
22	12	12	20	30	12	12	24
23	40	40	30	40	40	30	40
24	30	30	16	40	30	30	40
25	30	30	20	40	30	24	20
26	20	20	12	24	20	12	40
27	12	12	24	30	12	20	30
28	12	12	20	12	12	24	24
29	16	16	24	16	16	16	30
30	16	16	30	40	12	16	40
31	12	12	30	16	12	12	30
32	12	12	30	20	16	16	40
33	16	16	30	30	16	20	40
34	20	20	12	24	24	24	40
FEN	–	–	16 910	3055	53 000	25 402	1800
Run time	3 hr	3 hr	1 hr 15 min	15 min	2 hr	11 min	8 min

11.10.2 Practice problem 2

This network is a two reservoir seven loop network consisting of two source nodes and 13 demand nodes and 21 links. The network is shown below in Figure 11.33. The details of the network are given below in Tables 11.18 and 11.19.

Perform reliability-based optimization for the above two reservoir Water Distribution Network by using the above data. Assume any data if required, suitably.

Table 11.17 Comparison of NRP, CRR and CURUL obtained for Hanoi network with previous researchers.

Researcher	NRP	Cost (millions)	CRR (Rel/millions)	CURUL (rel/millions/km)
Abebe *et al.* (1998) (GA)	0.4222	7.01	16.60	0.422
Abebe *et al.* (1998) (ACCOL)	0.5908	7.84	13.27	0.337
Liong and Atiquazzaman (2004)	0.5754	6.22	10.81	0.275
Zecchin *et al.* (2006)	0.5754	6.14	10.67	0.271
Present study	0.7323	7.78	10.62	0.269

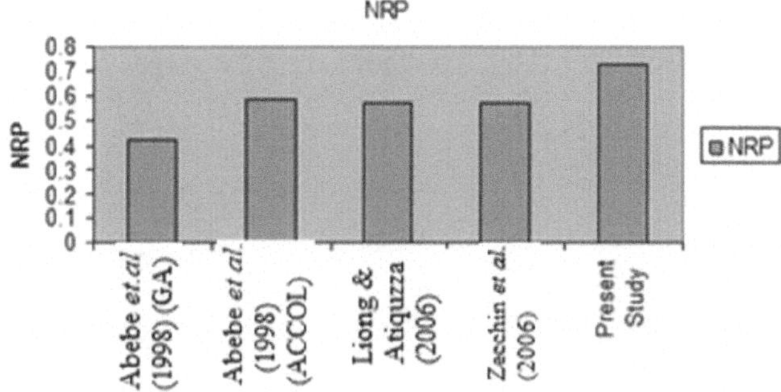

Figure 11.30 Comparison of NRP with previous researchers of Hanoi Network.

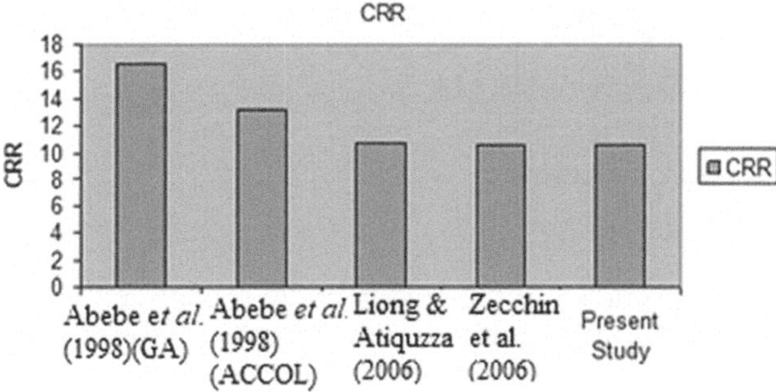

Figure 11.31 Comparison of CRR with previous researchers of Hanoi Network.

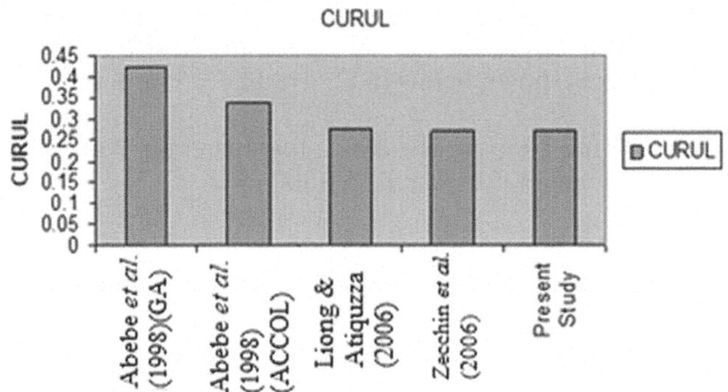

Figure 11.32 Comparison of CURUL with previous researchers of Hanoi Network.

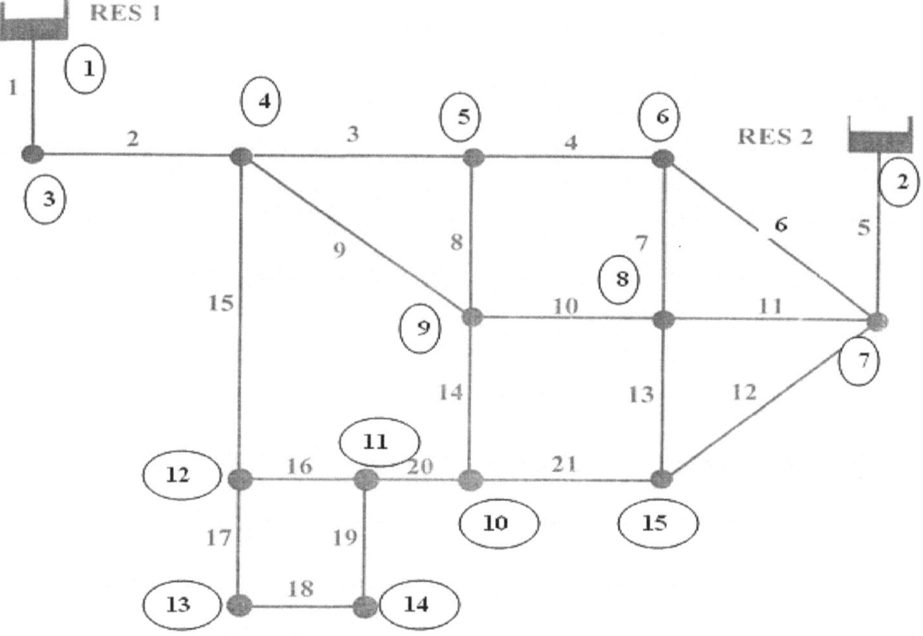

Figure 11.33 Two reservoir seven loop network.

11.10.2.1 Solution

The two reservoir network considered in the present study is analyzed by previous researchers for three different cases (cases 1–3). In the present study, the proposed methodology is applied on the network and the pressures obtained are compared with those three cases. The observed pressure heads and the obtained pressure heads are shown in Table 11.20 and in Figure 11.34.

From Figure 11.34, the pressures obtained in the present study show higher values than the observed values and also with the values obtained by the previous researchers. Hence, one can rely on the present proposed methodology.

Table 11.18 Node characteristics of network 3.

Node Id	Elevation (m)	Demand (lpm)
3	27.43	0
4	33.53	3540
5	28.96	3540
6	32	10 680
7	30.48	3540
8	31.39	11 400
9	29.56	10 680
10	31.39	5460
11	32.61	0
12	34.14	0
13	35.05	1800
14	36.58	1800
15	33.53	0
RES–1	60.96	N/A
RES–2	60.96	N/A

Table 11.19 Pipe characteristics of network 3.

Pipe ID	Start Node	End Node	Length (m)	Hazen–William Coefficient (CHW)
1	1	2	609.60	130
2	3	4	243.80	128
3	4	5	15 240.00	126
4	5	6	1127.76	124
5	2	7	1188.72	122
6	6	7	640.08	120
7	6	8	762.00	118
8	5	9	944.88	116
9	4	9	1676.40	114
10	8	9	883.92	112
11	7	8	883.92	110
12	7	15	1371.60	108
13	8	15	762.00	106
14	9	10	822.96	104
15	4	12	944.88	102
16	11	12	579.00	100
17	12	13	487.68	98
18	13	14	457.20	96
19	11	14	502.92	94
20	10	11	883.92	92
21	10	15	944.88	90

Table 11.20 Pressure heads obtained for network 3.

Node	Observed Pressure Head (m)	Pressure Head (m) for Case 1	Pressure Head (m) for Case 2	Pressure Head (m) for Case 3	Present Pressure Head (m) for Present Study
3	32.28	31.96	31.91	32.25	33.84
4	25.67	25.07	25.21	25.63	27.74
5	27.11	25.92	26.98	27.08	28.78
6	22.99	22.08	22.85	23.08	25.65
7	24.6	24.24	24.65	24.67	26.89
8	18.46	17.05	18.8	18.49	22.98
9	20.39	19.53	20.72	20.38	21.98
10	17.56	17.13	17.99	17.64	20.07
11	19.62	19.57	19.87	19.8	23.45
12	19.4	19.3	19.39	19.57	22.87
13	13.93	13.74	13.55	13.64	19.64
14	12.17	12.15	11.93	11.96	17.54
15	18.61	17.08	18.91	18.58	22.3

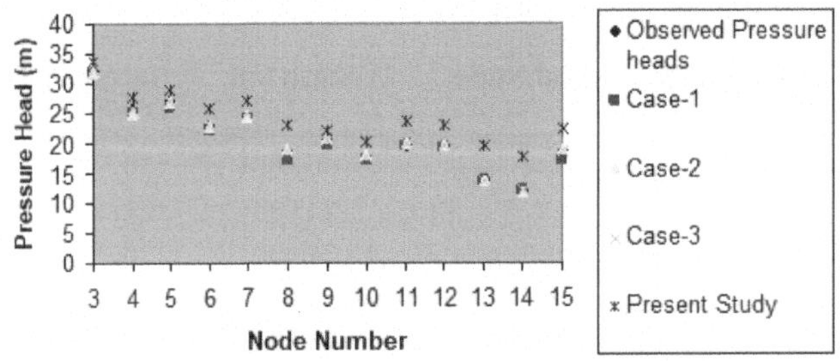

Figure 11.34 Comparison of pressure heads of two reservoir network.

REFERENCES

Abebe A. J. and Solomatine D. P. (1998). Application of global optimization to the design of pipe networks. Proceedings of 3rd International Conference on Hydroinformatics, Copenhagen, Balkema, Rotterdam, pp. 1–8.

Abebe A.J. and Solomatine D.P. (1998). "Application of global optimization to the design of pipe networks." 3rd International Conferences on Hydroinformatics, Copenhagen, Denmark, pp. 989–996.

Afshar M. H., Akbari M. and Marino M. A. (2005). Simultaneous layout and size optimization of water distribution networks: engineering approach. *Journal of Infrastructure Systems, ASCE*, **11**(4), 221–230, https://doi.org/10.1061/(ASCE)1076-0342(2005)11:4(221)

Afshar M.H.(2009). "Application of a Compact Genetic Algorithm to pipe network optimization problems" *Sceitia Iranica, Transaction A: Civil Engineering, Sharif University of Technology*, **16**(3), 264–271.

Alperovits E. and Shamir U. (1977). Design of optimal water distribution systems. *Water Resources Research*, **13**(6), 885–900. https://doi.org/10.1029/wr013i006p00885

Bhave P. R. (2003). Optimal Design of Water Distribution Networks. Narosa Publishing House, New Delhi.

Bhave P. R. and Gupta R. (2006). Analysis of Water Distribution Networks. Narosa Publishing House, New Delhi.

Chandramouli S. (2013). Reliability Based Optimal Design of Water Distribution Networks for Municipal Water Supply. PhD thesis, Submitted to Andhra University, Visakhapatnam, India.

Cunha M. and Sousa J. (1999). Water distribution network design optimization: simulated annealing approach. *ASCE Journal of Water Resources Planning and Management*, **125**(4), 215–221, https://doi.org/10.1061/(ASCE)0733-9496(1999)125:4(215)

Devi Prasad T., Hong S.-H. and Park N. (2003). Reliability based design of water distribution networks using multi-objective genetic algorithms. *KSCE Journal of Civil Engineering*, **7**(3), 351–361.

Eusuf M. M. and Lansey K. E. (2003). Optimization of water distribution network design using the shuffled frog leaping algorithm. *ASCE Journal of Water Resources Planning and Management*, **129**(3), 210–225, https://doi.org/10.1061/(ASCE)0733-9496(2003)129:3(210)

Fujiwara, O., Jenchaimahakoon B. and Edirisinghe N. C. P. (1987). A modified linear programming gradient method for optimal design of looped water distribution networks. *Water Resource Research*, **23**(6), 977–982.

Gilat A. (2004). MATLAB: An Introduction with Applications, 2nd edn. John Wiley and Sons, Hoboken, NJ, USA.

Goldberg D. E. (2000). Genetic Algorithms in Search, Optimization and Machine Learning. Addison-Wesley Publishing Co Inc., Reading, Massachusetts, USA.

Goulter, I. C., Lussier B. M. and Morgan D. R. (1986). Implications of head loss path choice in the optimization of water distribution network. *Water Resource Research*, **22**(5), 819–822.

Gurjar J. S. (2007). Reliability Technology – Theory and Applications. I.K. International Publishing House Pvt. Ltd., New Delhi.

Keedwell E. and Khu S. T. (2005). A hybrid genetic algorithm for the design of water distribution networks. *Engineering Applications of Artificial Intelligence*, **18**(4), 461–472.

Kessler A. and Shamir U. (1989). Analysis of the linear programming gradient method for optimal design of water supply networks. *Water Resources Research*, **25**, https://doi.org/10.1029/89WR00428. ISSN: 0043–1397.

Liong S. Y. and Atiquazzaman M. (2004). Optimal design of water distribution network using shuffled complex evolution. *Journal of Institution of Engineers, Singapore*, **44**(1), 93–107.

Prasad G. V. K. S. V. (2008). Optimal Design of Water Distribution Network for Uniform Supply in Intermittent System. PhD thesis, Submitted to National Institute of Technology, Warangal, India.

Prasad T. D. and Park N. S. (2004). Multiobjective genetic algorithms for design of water distribution networks. *Journal of Water Resource Planning Management*, **130** (1), 73–82. https://doi.org/10.1061/(ASCE)0733-9496

Quindry, G. E., Brill E. D., Liberman J. C. and Robinson A. R. (1979). Comment on "Design of optimal water distribution system" by E. Alperovits and U. Shamir. *Water Resource Research*, **15**(6), 1651–1654.

Ross T. J. (1997). Fuzzy Logic with Engineering Applications. Mc-Graw-Hill, Inc., New Delhi.

Samani H. and Mottaghi A. (2006). Optimization of Water Distribution Networks Using Integer Linear Programming. *Journal of Hydraulic Engineering*, **132**, 501–509.

Savic D. A. and Walters G. A. (1997). Genetic algorithms for least-cost design of water distribution networks. *ASCE Journal of Water Resources Planning and Management*, **123**(2), 67–77, https://doi.org/10.1061/(ASCE)0733-9496(1997)123:2(67)

Shie-Yui Liong, Md Atiquzzaman (2004). Optimal design of water distribution network using shuffled complex evolution. *Journal of the Institution of Engineers, Singapore*, **44**(1), 93–107.

Suribabu C. R. and Neelakantan T. R. (2006). Design of water distribution networks using particle swarm optimization. *Urban Water Journal*, **3**(2), 111–120, https://doi.org/10.1080/15730620600855928

Todini E. and Pilati S. (1987). A gradient method for the analysis of pipe networks. International Conference on Computer Application for Water Supply and Distribution, Leicester Polytechnic, UK.

United States Environmental Protection Agency. (2012). *Programmer's Toolkit*. Available at: https://www.epa.gov/. Updated 5 January 2012. Retrieved on 21 January 2012

van Dijk M., van Vuuren S. J. and van Zyl J. E. (2008). Optimising water distribution systems using a weighted penalty in a genetic algorithm. *Journal of Water SA*, **34**(5), 537–548, https://doi.org/10.4314/wsa.v34i5.180651

Wu Z. Y. and Simpson A. R. (2001). Competent genetic-evolutionary optimization of water distribution systems. *Journal of Computing in Civil Engineering*, ASCE, **15**(2), 89–101.

Zadeh L. A. (1965). Fuzzy sets. *Information and Control*, **8**(3), 338–353, https://doi.org/10.1016/S0019-9958(65)90241-X

Zecchin A. C., Simpson A. R. and Maier H. R. (2006). Application of two ant colony optimization algorithms to water distribution system optimization. *J. Math. Comput. Model.*, **44**, 451–668, https://doi.org/10.1016/j.mcm.2006. 01.005

doi: 10.2166/9781789062380_0295

Chapter 12

Water network tool for resilience

*Lucinda-Joi Chu-Ketterer[1], Jonathan Burkhardt[2], Katherine Klise[3] and Terranna Haxton[2]**

[1]*ORISE Fellow at United States Environmental Protection Agency, Oak Ridge Institute of Science and Education, 26 W. Martin Luther King Dr., Cincinnati, OH 45268, USA*
[2]*Environmental Engineer, United States Environmental Protection Agency, Office of Research and Development, 26 W. Martin Luther King Dr., Cincinnati, OH 45268, USA*
[3]*Sandia National Laboratories, P.O. Box 5800, Albuquerque, NM 87185, USA*
**Corresponding author: haxton.terra@epa.gov*

LEARNING OBJECTIVES

This chapter introduces the Water Network Tool for Resilience (WNTR) and how it can be used to evaluate drinking water distribution system (WDS) resilience. At the end of this chapter, you will be able to:

(1) Install and run WNTR.
(2) Set up and run various disaster scenario simulations.
(3) Calculate resilience metrics.
(4) Create simple network plots.

12.1 INTRODUCTION

Resilience can be defined as the capability of an object to recover or adjust after a source of strain or change. In the context of drinking water distribution systems (WDSs), resilience is the ability of the system to continue delivering sufficient water to users in a damaged state while working to return the system to regular service as quickly as possible (EPA, 2015). Predicting and measuring resilience in WDSs is helpful to prioritize strategies to improve resilience, perform cost-benefit analyses, measure progress, and identify critical components within a WDS (NAS, 2012). Tools that can quantify system resilience are important and help improve system security and general operations even when confronted with natural or human induced disruptions.

Models simulate the dynamic relationships between components of a system and can help identify how different components affect each other during a disaster and subsequent response. Modeling highlights interactions, side effects, and consequences of each action that are used to improve resilience of a WDS (Fiksel, 2006). Modeling tools can provide quantitative information, which provides more clarity when performing benefits analysis for different corrective or response actions related to building system resilience. Metrics are needed to assess resilience quantitatively. Several metrics have been developed to help assess resilience for WDSs (EPA, 2015).

Modeling tools can be used to understand system resilience, but such tools need to be: (1) capable of providing simulation results even during damaged states; and (2) able to dynamically change models to reflect damage and recovery for various types of scenarios. The United States Environmental Protection Agency (EPA) and Sandia National Laboratories developed the Water Network Tool for Resilience (WNTR) (pronounced 'winter') to address these modeling needs. WNTR extended the capabilities of EPANET (Rossman, 2000) for applications to resilience-oriented problems. EPANET 2.00.12 (Rossman, 2000) provided only a demand-driven hydraulic solver, which did not realistically reflect water demands if pressures were low due to the system being damaged or stressed. WNTR included a pressure dependent hydraulic solver engine based on the work of Wagner (Wagner *et al.*, 1988). A pressure dependent hydraulic solver engine is now available in EPANET 2.2 (Rossman *et al.*, 2020).

The second needed component of a resilience modeling tool is flexibility to dynamically change the water distribution model to simulate an incident. How damage manifests within a WDS (generation of a damage state) and how a system responds (response action) can both be affected by the type of disaster or system disruption under consideration. The first disaster scenario implemented in WNTR was an earthquake (Klise *et al.*, 2017). WNTR has the capability to: (1) simulate how an earthquake might damage pipes (i.e. identifies which pipes could be damaged given an earthquake of a certain magnitude and location); and (2) modify the network to reflect that damage (e.g. break pipes or add leaks). WNTR has also been used to simulate other disaster scenarios such as pipe breaks (Logan *et al.*, 2021; Mazumder *et al.*, 2020; Tomar *et al.*, 2020), power loss or source isolation (Abdel-Mottaleb *et al.*, 2019), and cyber-security related incidences (Moraitis *et al.*, 2020; Nikolopoulos *et al.*, 2021). The results from these types of applications can be used to identify important system components that help improve system resilience. Analysis using WNTR can be used to evaluate and potentially improve response actions through failure planning exercises and to develop more effective mitigation strategies for the future. WNTR can also be used to model more routine exercises such as fire flow analysis to assess WDSs ability to respond to everyday incidents.

This chapter discusses: (1) the challenges disasters pose on WDS infrastructure and the process to apply WNTR to assess these challenges; (2) the steps to install WNTR; (3) the types of disasters that can be currently modeled; (4) the available resilience metrics; and (5) tutorials. While not explicitly discussed here, interested researchers and developers can support WNTR through EPA's GitHub repository (https://github.com/USEPA/WNTR). WNTR is actively being used and extended within the Water Distribution Systems Analysis community for a variety of topic areas.

12.2 RESILIENCE OF DRINKING WATER SYSTEMS

WDSs are critical infrastructure that provide residential and commercial consumers with safe drinking water. A WDS also supplies water used in fire-fighting activities, healthcare facilities, electrical sectors, and others. For this reason, it is imperative to maintain all components of a WDS.

12.2.1 Disasters

Potential threats that can disrupt water service include natural disasters, releases of hazardous materials, or intentional attacks. Disasters can lead to water loss, water quality issues, power outages, and/or fires. In the event of disasters, multiple consequences are possible. For example, the 2020 Texas winter storm observed power outages and pipe breaks across the whole state (NPR, 2021). Similarly, the 2014 earthquake near Napa, California caused pipe breaks and fires throughout the city (USGS, 2014). Fires create unique challenges by dramatically increasing the water demand at a specific site. Additionally, wildfires can destroy and damage large parts of the system. During the Camp Fire in 2018 across Butte County, California, Paradise had 85% of the town's buildings and infrastructure, including water pipes, destroyed (NIST, 2021). Water contamination can also occur and prevent the water utility from providing safe drinking water. For example, the 2014 Elk River chemical spill in West Virginia impacted over 300 000 residents in Charleston and left them without access to clean

Potential Disasters	Consequences	Response Actions	Mitigation Strategies
Natural Disasters	Human Health	Public Health Advisories	Backup Power/Fuel Storage
Drought	Pipe Breaks	Repairing Pipe Breaks	
Earthquakes	Other Infrastructure Damage	Fixing Infrastructure Damage	Earthquake resistant pipes
Floods	Power Outage		Securing facilities/assets
Hurricanes	Service Disruption (source treatment, distribution, storage)	Restoring Power	Water Quality Monitoring
Tornados		Treating Water	Increased Redundancy
Tsunamis	Loss of Access to Facilities/Supplies	Repairing Roads/Access	Practiced Emergency Response Plans
Wildfires		Fighting fires	
Winter Storms	Loss of Pressure/Leaks	Communication with Customer	
Terrorist Attacks	Change in Water Quality		
Cyber Attacks	Environmental/Financial/Social	Conservation	
Transportation/industrial Accidents and Spills			

Figure 12.1 Summary of potential disasters, their consequences on a WDS, response actions and mitigation strategies.

drinking water due to crude methylcyclohexane methanol (MCHM), a chemical used in washing coal, entering the drinking water system (USGS, 2015). As the MCHM contamination moved downstream, numerous utilities along the Ohio River, which use the river as their source water, had to close their intakes as the chemical plume passed them. Sometimes disaster consequences may cause long term impacts to a system which makes the system more susceptible to additional consequences in the future. For example, in 2021, Hong Kong, China had pipes burst within their system which led to a landslide (SCMP, 2021). When modeling disasters, it is also important to identify potential cascading impacts. The modeling framework in WNTR allows the user to develop effective response actions or mitigation strategies. These actions and strategies can be modeled within WNTR using component attributes and controls. Examples on using controls are provided in the tutorials (Section 12.6) and additional information be found in the WNTR user manual (Klise *et al.*, 2020, https://wntr.readthedocs.io).

Figure 12.1 lists potential hazards that WDSs can face and their potential consequences. The additional columns highlight response actions and ways to reduce risk in the future. Select topics will be discussed in later sections of this chapter.

12.2.2 Measuring resilience

Understanding the consequences that a disaster can have to a water utility is important for identifying ways to mitigate severe impacts in the future. Similarly, understanding how to measure the resilience of a WDS provides a baseline for how prepared a water utility may be against disasters. System modeling tools, like WNTR, can be used to measure resilience by calculating the predicted impact due to certain hazards on a given system. Modeling tools can also help test the effectiveness of response actions and mitigation strategies related to the hazard. Metrics are needed to assess resilience quantitatively. WNTR includes hydraulic, topographic, water quality, and cost metrics to help quantify a system's resilience. Section 12.5 provides more information on these metrics.

12.2.3 Challenges with modeling system resilience

Modeling a WDS to determine resilience is not without its challenges. Damaged or impaired WDSs are likely to experience low or no pressure conditions, which need to be accurately simulated. Simulating extreme events and the potential associated damage (i.e. multiple pipe breaks, change in flow demand) can cause the hydraulic equations to not converge and result in failed simulations. Modeling tools for resilience applications need to be able to handle more complex conditions than would typically be encountered in a regular extended period simulation. Furthermore, tools should be able to stop and restart simulations to capture operational or network changes (e.g. repairing a pipe) mid simulation. Tools should be able to build and manage a variety of network states to capture the disaster/recovery during a simulated incident and include probabilistic analyses. Once simulated, the results need to be analyzed across a range of metrics and presented effectively to utility decision makers.

Specific models for the disaster or response action may not be available or easily translated into a model. Similarly, there are limited data at the appropriate scales needed to develop or validate models for resilience applications (e.g. costs, weather data, or impacts from previous incidents). WNTR was designed to be a flexible modeling tool, and as additional disaster models or increased data become available its capabilities can be expanded.

12.3 WATER NETWORK TOOL FOR RESILIENCE

Researchers from the EPA and Sandia developed WNTR to help drinking water utilities assess a WDS's resilience to disasters. WNTR is an open-source Python package made available through EPA's public GitHub repository (https://github.com/USEPA/WNTR). At the time of writing this chapter, the WNTR release version was 0.4.2. Specific releases can be found at https://github.com/USEPA/WNTR/releases.

12.3.1 Overview

WNTR is based on EPANET and integrates hydraulic and water quality simulations, damage estimates and response options, and resilience metrics into a single platform. Since WNTR is a Python package, this allows for customized modeling and analysis of more complex network states than previously available within EPANET. Users are encouraged to be familiar with Python and EPANET or understand hydraulics and pressurized pipe network modeling before using WNTR. WNTR can simulate and analyze resilience of water distribution systems using EPANET network input files (i.e. INP). Network files represent a collection of pipes, pumps, valves, junctions, tanks, and reservoirs. Given WNTR's flexible application programming interface (API), changes to the network can easily be made to account for structural and operational changes associated with disaster scenarios and simulate recovery actions. Figure 12.2 outlines the general process for analyzing a WDS. WNTR allows users to evaluate the resilience of the system under different disaster scenarios. This information provides the utility with options to determine the most effective actions to improve resilience.

12.3.2 Installation and requirements

WNTR requires 64-bit Python (tested on versions 3.6, 3.7, 3.8, and 3.9) along with several Python package dependencies. For the latest requirements and other installation information see https://wntr.readthedocs.io.

WNTR can be installed into the user's Python environment using PyPI, as:

```
pip install wntr
```

or from Conda, as:

```
conda install -c conda-forge wntr
```

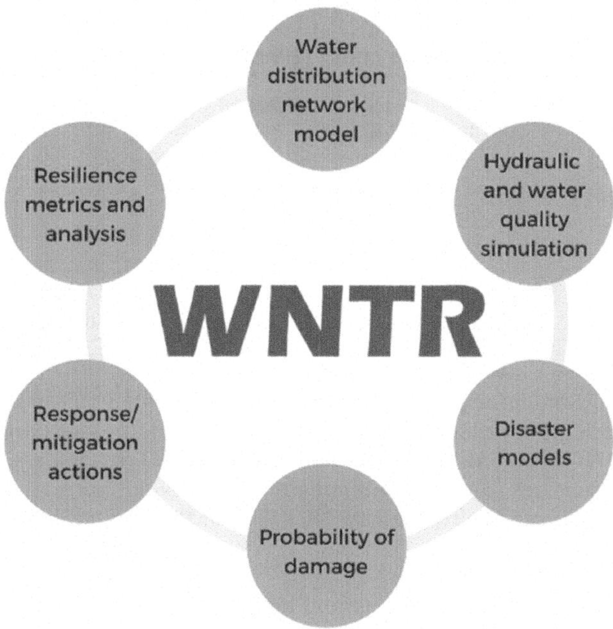

Figure 12.2 Diagram of WNTR process and capabilities.

Those interested in supporting development can also use the developer approach from the command prompt, as:

```
git clone https://github.com/USEPA/WNTR
cd WNTR
python setup.py develop
```

12.3.3 Units
WNTR uses SI (International System of Units) units (Newell & Tiesinga, 2019). When importing an EPANET INP file into WNTR, units are automatically converted to SI units. Users must convert their preferred unit system to SI when developing simulation scripts or functions. Table 12.1 highlights common parameters and their associated default base unit for EPANET and WNTR. The full table of default units can be found at https://wntr.readthedocs.io.

12.3.4 Available solvers
WNTR has two available simulators: (1) EpanetSimulator and (2) WNTRSimulator. The EpanetSimulator by default uses the EPANET 2.2 toolkit, but the EPANET 2.0 toolkit can also be selected when initializing the simulator. The EPANET 2.2 toolkit can be used to simulate both demand-driven analyses (DD) and pressure dependent demand analyses (PDD). Only DD is available when using the EPANET 2.0 toolkit. Since the EPANET engine/toolkit is written in C, the runtimes of the EpanetSimulator are typically faster than the WNTRSimulator, which is written in Python.

Table 12.1 EPANET and WNTR hydraulic unit conventions.

Parameter	US Customary Units (EPANET)	SI Units (WNTR)
Time	min	s
Demand or flow	gal/min (GPM)	m^3/s
Diameter (for pipes)	in	m
Diameter (for tanks)	ft	m
Length	ft	m
Elevation	ft	m
Pressure	psi	mH_2O (assuming a fluid density of 1000 kg/m^3)

While slower, the WNTRSimulator provides more flexibility in how PDD simulations are configured compared to the EpanetSimulator. Specifically, the WNTRSimulator can handle node-specific PDD parameters, where EPANET can only handle globally defined PDD parameters. Since Python can also handle complex data structures, the WNTRSimulator can provide more advanced analysis options that would not be easily achieved using the EPANET GUI or the EpanetSimulator.

12.3.5 Examples and demos

The WNTR package includes example scripts and demos to help new users familiarize themselves with WNTR's capabilities. These were developed to capture the current capabilities and will evolve with continued WNTR development. When WNTR is installed using PyPI or Conda, the examples folder is not included. Examples and demos can be found within the WNTR repository at https://github.com/USEPA/WNTR/tree/main/examples.

Python scripts can be run in a variety of integrated development environments (IDEs), but common software programs are Spyder, Jupyter Notebook, PyCharm and Atom. The WNTR examples are Python scripts, and the demos are Jupyter Notebooks. A Jupyter Notebook requires additional software installation (https://jupyter.org/), or it can be accessed through Anaconda. Jupyter Notebook is an open-sourced web-based application that acts as both an IDE and a presentation tool. Tutorials of the Jupyter Notebooks are provided in Section 12.6 and interested readers can check the WNTR examples repository folder for available Jupyter Notebook examples. Tables 12.2 and 12.3 provide a brief description of each example and demo file.

Table 12.2 Script examples.

File	Purpose
getting_started.py	Demonstrates how to import WNTR, generate network model from an INP file, simulate hydraulics, and plot node pressures 5 hours into the simulation
fire_flow.py	Runs hydraulic simulations with and without fire flow demand to a single fire node. Plots and compares node pressures 24 hours into simulation
pipe_criticality.py	Runs multiple hydraulic simulations to compute the impact that different individual pipe closures have on water pressure. Plots junction pressure impact from a single pipe break
sensor_placement.py	Uses WNTR with Chama (https://chama.readthedocs.io) to optimize the placement of sensors that minimize detection time of contamination incidents. Plots location and detection time for each sensor
stochastic_simulation.py	Runs multiple realizations of pipe leaks where each pipe is assigned a probability failure related to pipe diameter. Calculates and plots water service availability and tank water levels for each realization

Table 12.3 Jupyter notebook demos.

File	Purpose
pipe_break_demo.ipynb	Runs multiple hydraulic simulations to compute the impact that different individual pipe breaks have on network pressure. Plots pressure and population impacts for all junctions impacted by pipe breaks
segment_break_demo.ipynb	Runs multiple hydraulic simulations to compute the impact that different segment breaks have on network pressure. Plots pressure and population impacts for all junctions impacted by segment breaks
fire_flow_demo.ipynb	Runs multiple hydraulic simulations with and without fire flow demand to multiple fire hydrant nodes. Plots pressure and population impacts for junctions impacted by fire demand nodes
earthquake_demo.ipynb	Runs hydraulic simulations of earthquake damage with and without repair efforts. Plots fragility curves, peak ground acceleration, peak ground velocity, repair rate, leak probability, and damage states. Compares junction pressure 24 hours into the simulation, and tank and junction pressure over time. Also plots water service availability and population impacted by low pressure conditions

12.3.5.1 Basic example with getting_started.py

The basics of using WNTR can be found in the *getting_started.py* example. The full example can be found in the GitHub repository. To start using WNTR, open a Python console and import the package.

```
import wntr
import matplotlib.pyplot as plt
```

Next an EPANET compatible INP file needs to be supplied – shown here as providing a file location (using referential or explicit file path and file name) and supplying that to the WNTR `WaterNetworkModel` function. To avoid errors, make sure the file path to the network file is correct.

```
inp_file = "networks/Net3.inp"
wn = wntr.network.WaterNetworkModel(inp_file)
```

To ensure the correct INP file was imported, a simple network map can be created.

```
wntr.graphics.plot_network(wn, title=wn.name)
plt.show()
```

After all the steps, the user should see a network map like Figure 12.3.

All examples and demos use the Net3.inp file, which consists of 92 junctions, two reservoirs, three tanks, 117 pipes, and two pumps. Additional example networks can be found at https://github.com/USEPA/WNTR/tree/main/examples/networks and range from a 9-junction network to a 3000-junction network. The University of Kentucky also provides additional networks at https://uknowledge.uky.edu/wdsrd/ for testing purposes.

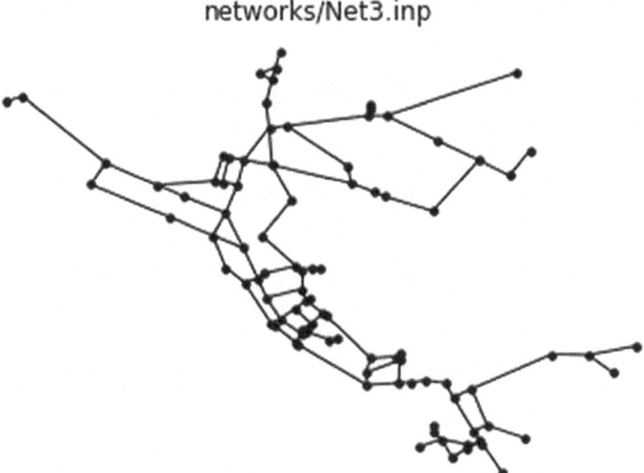

Figure 12.3 Network map of Net3.

12.4 DISASTER SCENARIOS

WDSs can experience disruptions in their ability to deliver water to their customers in a variety of ways. Disruptions caused by a disaster could result in short- or long-term issues. Persistent pipe leaks, population fluctuations (e.g. after loss of housing), or changes to supply and demand are some of the impacts that could be experienced. This section describes common disaster scenarios that can be modeled using WNTR.

12.4.1 Pipe breaks

A pipe break is the simplest form of damage that is considered. Pipe breaks, or leaks, prevent water from being delivered to consumers within a network or reduce available downstream pressure. In WNTR, users can identify pipes of interest by diameter or other attributes and simulate a break or a leak. Pipe breaks can be simulated three ways within WNTR: (1) using the controls to simulate pipe closure; (2) using the `split_pipe` method to add a leak to the pipe; or (3) using the `break_pipe` method to create a break in the system. By closing a pipe, water is no longer able to flow through that pipe during a simulation. Controls and rules can be used to change the status of a pipe, pump, or valve between 'open' or 'closed'. The `split_pipe` method splits the pipe of interest into two new pipes and junction. The new junction has a base demand of 0 and the default demand pattern. The new junction can then be used to simulate a leak. The `break_pipe` method breaks the pipe of interest into two separate pipes by adding two new junctions (with the same coordinates). The junctions have a base demand of 0 and the default demand pattern. The `break_pipe` method is ideal for simulating a break at a specific location on the pipe and to stop water flow through the pipe. The example and demo files use the pipe closure method to simulate the pipe breaks. The results identify which junctions in the system experience a drop in pressure below a specified threshold. The simplest scenario is a single pipe break; however, it is possible that multiple pipes, or segments, may break at once.

12.4.2 Segment isolation

The basic purpose of an isolation valve is to control water flow within a WDS. Since in some cases isolation valves are not available for individual pipes – because they are paved over, broken, or otherwise unavailable – segment break analyses can provide a more realistic picture of the impact of damage to a pipe segment. In segment break analysis, isolation valves are placed throughout the network where pipes between two valves create a segment. In the event valve locations are not known within the network, WNTR can be used to identify potential valve locations to generate a valve layer. Using the `generate_valve_layer` method, users can choose between (1) strategic placement and (2) random placement to add isolation valves within the network. For the strategic placement method, a variable, n, is used to determine the number of pipes from each node that do not include a valve. The random placement method randomly places n valves throughout the network. During analysis, all pipes within the segment are closed using pipe attributes or controls. Like the pipe break analysis, the results identify which junctions in the system experience a drop in pressure below a specified threshold.

12.4.3 Earthquakes

Earthquakes can cause significant damage to a WDS that could take weeks, or months, to repair. Damage could be to pipes, tanks, pumps, and other infrastructure. Additionally, earthquakes can cause power outages and source loss leaving a majority, if not all, of the users without safe drinking water. WNTR can simulate these consequences to understand the severity of the impact on the system. For instance, pipe leaks, tanks leaks, and pump closures can be changed in the model prior to running a simulation. WNTR includes the ability to calculate peak ground acceleration (PGA) using the `pga_attenuation_model` method, peak ground velocity (PGV) using the `pgv_attenuation_model` method, and repair rates (RR) using the `repair_rate_model` method. Fragility curves can be used to define the probability of damage to a component with respect to PGA, PGV, and/or RR (Klise *et al.*, 2017, 2020).

12.4.4 Fires

Fires affect water distribution systems by increasing water demand at the hydrant location to support firefighting activity. While the minimum required fire flow and duration vary by state and building type, most small residential fires may require 0.095 m³/s (1500 GPM) for 2 hours while large commercial spaces may require 0.505 m³/s (8000 GPM) for 4 hours (International Code Council, 2011). The size and spread of a fire can have a large effect on the rest of the system since the additional water demand near the fire location may cause users to lose water for an extended period of time. WNTR can be used to simulate firefighting conditions to understand these impacts. Demand, location, time, and duration of firefighting are all parameters that can be specified within WNTR.

12.4.5 Loss of source water

The loss of source water within the system can hinder the ability to deliver safe drinking water to users. This could be a result of pump failure or power loss at water treatment facilities or source water contamination. During an incident, a water utility may advise consumers to conserve water or enact a water boil notice to prolong the availability of water in storage tanks. Water conservation efforts can be simulated within WNTR by reducing junction demands to understand how they impact a utility's ability to continue delivering water during an incident. Changes to source water availability can be simulated within WNTR as pump closures or adjustments to the water demands.

12.4.6 Power outage

Power outages can have both short- and long-term effects on a system depending on the length of the outage. Power is required to operate many WDS components and power outages can disrupt the ability to operate the WDS normally. Of particular concern, power outages can cause pump stations to shut down, resulting in reduced water pressure. Reduced water pressures may prevent certain portions of a WDS from receiving their expected demands – in some cases, receiving no water at all. This can lead to depressurization in the system, which in turn can cause long-term damage such as pipe breaks. WNTR has a method `add_outage` that allows users to quickly add time controls to pumps and specify when to start and stop a power outage.

12.4.7 Other scenarios

The previous scenarios are just a few examples of how a disaster may affect a system, but there are many others. For example, environmental changes, such as drought, may lead to long term water shortages. Chemical, microbial, or radiological contamination could also render the water unsafe to drink even with boil water alerts. Moreover, disasters can have multiple consequences to a system as seen with floods, hurricanes, tornadoes, and winter storms to name a few. Given its flexibility, WNTR can be extended to model a range of these disasters if appropriate damage-state estimation techniques and response actions are available.

12.5 WNTR RESILIENCE METRICS

Metrics seek to provide a standardized quantitative measure of system resilience. Resilience can refer to design, maintenance, or operations of a system and are all vital for providing safe drinking water to the community. The following section describes key hydraulic metrics (i.e. water service availability, Todini index, and modified resilience index) and briefly highlights topographic, water quality, water security, and economic metrics that are included in WNTR.

12.5.1 Water service availability

Water service availability (WSA) is the ratio of delivered consumer demand to expected (requested) demand (Ostfeld *et al.*, 2002). This metric captures the amount of water a user will actually receive during an incident relative to the amount they would normally receive. In WNTR, the `water_service_availability` method can be used to calculate WSA as a function of time or space (Equation (12.1)):

$$\frac{D_i}{D_{\exp}} \tag{12.1}$$

where D_i is the actual demand at junction i and $D_{\exp}$ is the expected demand at junction i. Expected demand can be calculated using the `expected_demand` method.

12.5.2 Todini index

While WSA only considers a system's ability to deliver water, this may not capture the full breadth of a system's resilience. The Todini index (Todini, 2000) quantifies a system's capability to continue meeting consumer demands and pressures at junctions, reservoirs, and pumps while overcoming failures within the system. The Todini index reports resilience as a ratio of surplus internal power to the maximum power for a given time while satisfying junction demands and head. The Todini index is best suited for networks with a single water source. In WNTR, the `todini_index` method can be used to compute the Todini index (Equation (12.2)):

$$\frac{\Sigma_{i=1}^{N}D_i\left(H_i\right)-\Sigma_{i=1}^{N}D_i(P_{\text{star}}+Elev_i)}{\Sigma_{j=1}^{M}D_j\left(H_j\right)+\Sigma_{k=1}^{Z}F_k\left(HL_k\right)-\Sigma_{i=1}^{N}D_i(P_{\text{star}}+Elev_i)} \tag{12.2}$$

where D_i is the actual demand at junction i, H_i is the head at junction i, P_{star} is the pressure threshold (i.e. required pressure), $Elev_i$ is the elevation of junction i, N is the total number of junctions, D_j is the actual demand at reservoir j, H_j is the head at reservoir j, M is the total number of reservoirs, F_k is the flowrate at pump k, HL_k is the headloss at pump k, and Z is the total number of pumps.

12.5.3 Modified resilience index

Like the Todini index, the modified resilience index (MRI) (Jayaram & Srinivasan, 2008) quantifies surplus energy within the network but has the additional ability to calculate surplus power at each junction for a given specific time or as a system average for each timestep. MRI is more versatile than the Todini index since it calculates a system's excess energy more accurately for networks with multiple water sources. In WNTR, the `modified_resilience_index` method can be used to calculate MRI per junction (Equation (12.3)) or over all junctions (Equation (12.4)):

$$\frac{(P_i + Elev_i) - \left(P_{star} + Elev_i\right)}{Elev_i + P_{star}} \tag{12.3}$$

where P_i is the pressure at junction i, $Elev_i$ is the elevation of junction i, and P_{star} is the pressure threshold (i.e. required pressure),

$$\frac{\Sigma_{i=1}^{N}D_i(P_i + Elev_i) - \Sigma_{i=1}^{N}D_i(P_{star} + Elev_i)}{\Sigma_{i=1}^{N}D_i(P_{star} + Elev_i)} \tag{12.4}$$

where D_i is the actual demand at junction i and N is the total number of junctions.

12.5.4 Additional metrics

While the hydraulic-based metrics discussed above can provide useful information, they are unlikely to fully describe the impacts for all aspects of a system's ability to deliver water. Additional resilience metrics available in WNTR are categorized as topographic, water quality, water security, and economic metrics (Klise *et al.*, 2020). Topographic metrics are based on graph theory and are used to assess the strength of connectivity within a WDS (e.g. how many pipes serve a given node). Water quality metrics quantify contaminant concentrations or water age, which provides valuable information about water quality but at the expense of more complex simulations. Water security metrics measure the effect contaminated water may have on the consumers (e.g. estimate exposure), extending the use of water quality simulations. Economic metrics measure costs associated with operating the WDS as well as greenhouse gas emissions.

12.6 TUTORIALS

This section provides step by step tutorials of resilience analysis demos using four simple disaster scenarios: (1) pipe break, (2) segment isolation, (3) fire flow, and (4) earthquake. All tutorials use default WNTR SI units (refer to Section 12.3.3). Figure 12.4 outlines the basic steps for each analysis. The code provided below follows these steps. The snippets of code provided assume readers have a moderate understanding of Python and its structure. More guidance, detailed comments, and additional visualization options for each scenario can be found in the Jupyter Notebook demo files located at https://github.com/USEPA/WNTR/tree/main/examples/demos. Demo files are not automatically downloaded with the PyPI or Conda installations. The Jupyter Notebook demo files are in color, which highlights the subtleties and provides more details of the graphs included in this section.

12.6.1 Pipe break

Step 1: Import Python packages and create a water network. Numpy is required to support data handling. Once the network model has been created, continue to the sample code for the disaster scenario of interest. The file path in this example assumes that the user is working from the demos folder directory.

```python
import numpy as np
import wntr

inp_file = "../networks/Net3.inp"
wn = wntr.network.WaterNetworkModel(inp_file)
```

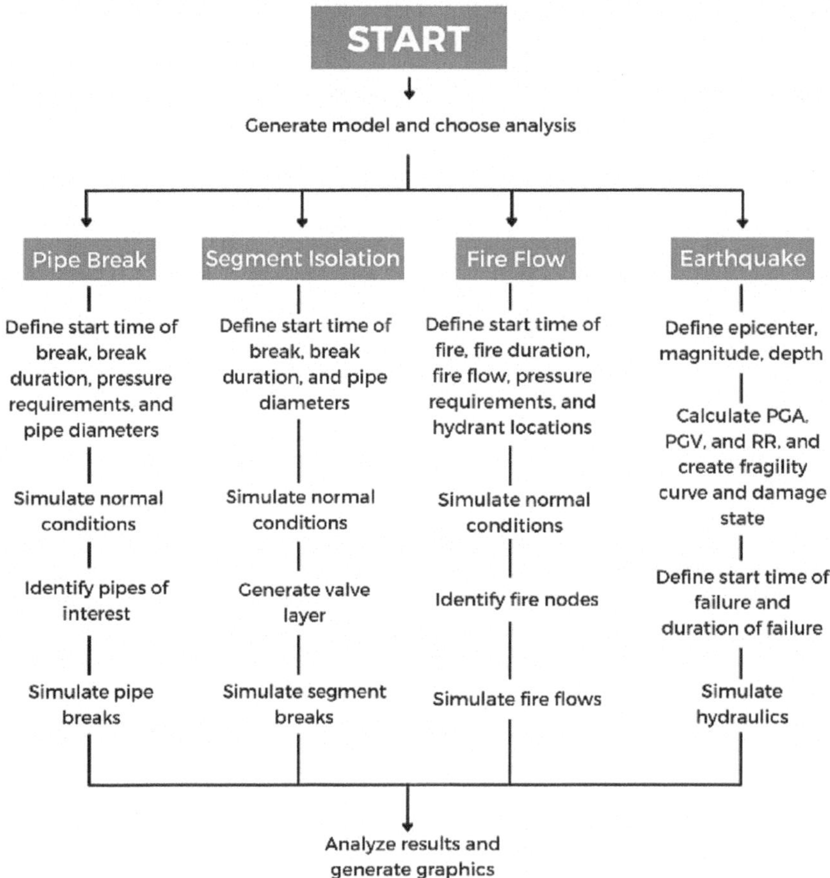

Figure 12.4 Flow chart demonstrating basic steps for resilience analysis of various disaster scenario simulations.

Step 2: Define the start time of the break (s), the break duration (s), the system pressures requirements (mH$_2$O), and the pipe diameters of interest (m). The parameters *minimum_pressure* and *required_pressure* are used for PDD simulations. Nodes with pressures below minimum pressure will not receive any water, and node pressures need to be at least the required pressure to receive all of the requested demand. The parameter *min_pipe_diam* defines the lower limit of pipe diameters to include in analysis.

```
start_time = 2*3600 # 2 hours
break_duration = 12*3600 # 12 hours
total_duration = start_time + break_duration # 14 hours

minimum_pressure = 3.52 # 5 psi
required_pressure = 14.06 # 20 psi

min_pipe_diam = 0.3048 # 12 inch
```

Step 3: Identify non-zero demand (NZD) junctions by calculating average expected demand (AED) for each junction and selecting those with a value greater than zero. Simulate normal conditions. Identifying junctions that experience pressures below *minimum_pressure* during normal conditions helps identify junctions that experience low pressures during the disaster simulations as a direct result of the disaster. Note, it is standard practice to set the *report_timestep* and *hydraulic_timestep* to an hour or less.

```
AED = wntr.metrics.average_expected_demand(wn)
nzd_junct = AED[AED > 0].index

wn.options.hydraulic.demand_model = 'PDD'
wn.options.time.duration = total_duration
wn.options.hydraulic.minimum_pressure = minimum_pressure
wn.options.hydraulic.required_pressure = required_pressure
wn.options.time.report_timestep = 3600 # 1 hour
wn.options.time.hydraulic_timestep = 3600 # 1 hour

sim = wntr.sim.WNTRSimulator(wn)
results = sim.run_sim()

pressure = results.node['pressure'].loc[start_time::, nzd_junct]
normal_pressure_below_pmin = pressure.columns[(pressure <
minimum_pressure).any()]
```

Step 4: Identify pipes of interest. The parameter *pipes_of_interest* include all pipes in the network with diameters greater than *min_pipe_diam* defined in *Step 2*.

```
pipes_of_interest = wn.query_link_attribute('diameter',
np.greater_equal, min_pipe_diam)
```

Step 5: Simulate pipe breaks for each of the identified pipes of interest. Criticality is determined by the impact of a pipe break on the system. Iterate through the list of pipes and simulate a break. For each pipe break, save out results for impacted junctions. A junction is considered impacted if the pressure drops below *minimum_pressure* during the disaster time interval. This list does not include junctions with pressure drops below *minimum_pressure* during normal operations. A *try/except/ finally* approach is taken to ensure the script can finish running and still catch any convergence issues a single pipe break might cause. The user is shown which simulations failed to complete, and all successfully run simulations are saved to *analysis_results*. With large numbers of simulations, it is sometimes necessary to use such an approach to prevent one error from causing a user to have to repeat the entire set of simulations. A user can revisit nodes with failed simulations individually to determine the cause of failure, if desired.

```python
analysis_results = {}
for pipe_name in pipes_of_interest.index:
    wn = wntr.network.WaterNetworkModel(inp_file)
    wn.options.hydraulic.demand_model = 'PDD'
    wn.options.time.duration = total_duration
    wn.options.hydraulic.minimum_pressure = minimum_pressure
    wn.options.hydraulic.required_pressure = required_pressure
    wn.options.time.report_timestep = 3600 #1 hour
    wn.options.time.hydraulic_timestep = 3600 #1 hour

    pipe = wn.get_link(pipe_name)
    act = wntr.network.controls.ControlAction(pipe, 'status', 0)
    cond = wntr.network.controls.SimTimeCondition(wn, 'Above',
    start_time)
    ctrl = wntr.network.controls.Control(cond, act)
    wn.add_control('close pipe ' + pipe_name, ctrl)

    try:
        sim = wntr.sim.WNTRSimulator(wn)
        sim_results = sim.run_sim()

        sim_pressure = sim_results.node['pressure'].loc[start_time::,
        nzd_junct]
        sim_pressure_below_pmin = sim_pressure.columns[(sim_pressure
        < minimum_pressure).any()]
        impacted_junctions = set(sim_pressure_below_pmin)
        - set(normal_pressure_below_pmin)

except Exception as e:
    impacted_junctions = None
    print(pipe_name, ' Failed:', e)

finally:
    analysis_results[pipe_name] = impacted_junctions
```

Step 6: Calculate pressure and population impacts for each pipe break and visualize results in the network. The parameters *node_range* and *link_range* in the `plot_network` method can be changed to best fit the analysis results. If users are working through the code in Jupyter Notebooks, the resulting graphs will automatically generate. However, if the code is run in a different IDE and graphs do not appear, `import matplotlib.pyplot as plt` and `plt.show()` should be added to the code as shown in Section 12.3.5.1.

```python
Population = wntr.metrics.population(wn)

num_junctions_impacted = {}
num_people_impacted = {}
for pipe_name, impacted_junctions in analysis_results.items():
    if impacted_junctions is not None:
```

```
        num_junctions_impacted[pipe_name] = len(impacted_junctions)
        num_people_impacted[pipe_name] = population[impacted_
        junctions].sum()

wntr.graphics.plot_network(wn, link_attribute=num_junctions_impacted,
node_size=0,
                        link_width=2,
                        link_range=[0,10], link_colorbar_
                        label='Junctions
                        Impacted',
                        title='Number of junctions impacted by each
                        pipe closure')
wntr.graphics.plot_network(wn, link_attribute=num_people_impacted,
node_size=0, link_width=2,
                        link_range=[0,5000], link_colorbar_label=
                        'Population',
                        title='Number of people impacted by each pipe
                        closure')
```

The resulting graphs are shown in Figure 12.5. The left side shows the pipe breaks that impacted junctions within the network where the color and pipe thickness corresponds to the number of junctions impacted. The right side shows the pipe breaks that impacted consumers where the color and pick thickness corresponds to the number of consumers impacted. Pipe thickness increases with junction and population impacts. The graphs indicate the pipe with the most impacts affected over ten junctions and over 5000 consumers.

12.6.2 Segment isolation

Step 1: Import Python packages and create a water network. Numpy and Matpotlib are required to support data handling and graphics/plotting. Once the network model has been created, continue to the sample code for the disaster scenario of interest. The file path in this example assumes that the user is working from the demos folder directory.

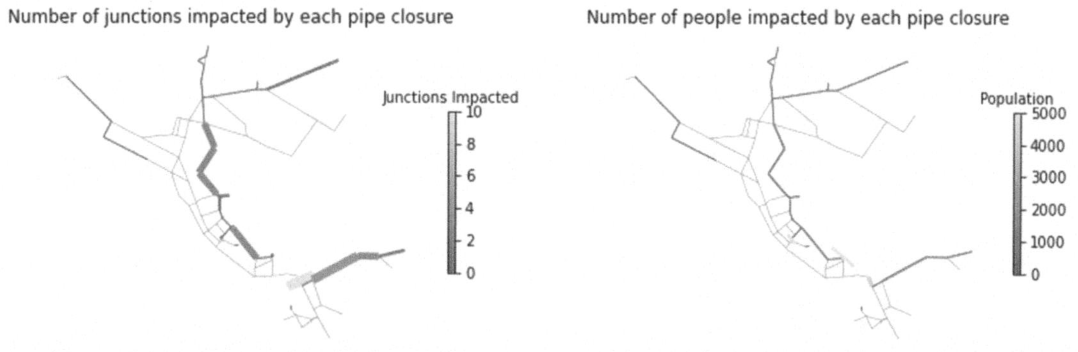

Figure 12.5 Net3 results from pipe break analysis.

```python
import numpy as np
import matplotlib.pylab as plt
import wntr

inp_file = "../networks/Net3.inp"
wn = wntr.network.WaterNetworkModel(inp_file)
```

Step 2: Define the start time of break (s), the break duration (s), and the system pressures (mH$_2$O). The parameters *minimum_pressure* and *required_pressure* are used for PDD simulations. Nodes with pressures below the minimum pressure will not receive any water, and node pressures need to be at least the required pressure to receive all of the requested demand.

```python
start_time = 2*3600 # 2 hours
break_duration = 12*3600 # 12 hours
total_duration = start_time + break_duration # 14 hours

minimum_pressure = 3.52 # 5 psi
required_pressure = 14.06 # 20 psi
```

Step 3: Identify non-zero demand (NZD) junctions by calculating average expected demand (AED) for each junction and selecting those with a value greater than zero. Simulate normal conditions. Identifying junctions that experience pressures below the *minimum_pressure* during normal conditions helps identify junctions that experience low pressures during the disaster simulation are as a direct result of the disaster. Note, it is standard practice to set the *report_timestep* and *hydraulic_timestep* to an hour or less.

```python
AED = wntr.metrics.average_expected_demand(wn)
nzd_junct = AED[AED > 0].index

wn.options.hydraulic.demand_model = 'PDD'
wn.options.time.duration = total_duration
wn.options.hydraulic.minimum_pressure = minimum_pressure
wn.options.hydraulic.required_pressure = required_pressure
wn.options.time.report_timestep = 3600 #1 hour
wn.options.time.hydraulic_timestep = 3600 #1 hour

sim = wntr.sim.WNTRSimulator(wn)
results = sim.run_sim()

pressure = results.node['pressure'].loc[start_time::, nzd_junct]
normal_pressure_below_pmin = pressure.columns[(pressure < minimum_pressure).
any()]
```

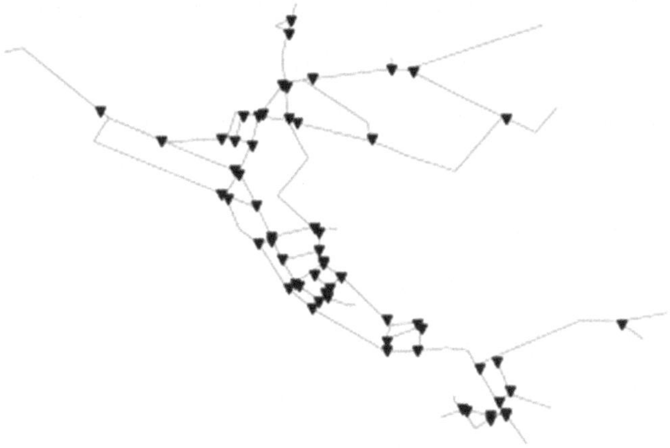

Figure 12.6 Net3 isolation valve locations using 'strategic' valve placement.

Step 4: Generate, save, and visualize the valve layer. A valve layer represents valve placement within a network, where pipes between two valves are considered a segment. While the pipe break analysis is helpful in identifying impacts of specific, individual pipes within the network (since it assumes isolation valves at the end of each pipe), the segment isolation analysis is more realistic since it reflects the position of isolation valves across the network. The example provided uses $n = 2$ *strategic* valve placement, which indicates that for every node, two pipes connected to that node do not have a valve. The resulting graph is shown in Figure 12.6 where triangles represent isolation valve placement.

```
valve_layer = wntr.network.generate_valve_layer(wn, placement_type=
'strategic', n=2, seed=123)
G = wn.get_graph()
node_segments, link_segments, seg_sizes = wntr.metrics.valve_segments
(G, valve_layer)
```

Step 5: Simulate segment breaks for each segment. Iterate through the list of segments and simulate each break. For each segment break, save out results for impacted junctions. A junction is considered impacted if the pressure drops below the *minimum_pressure* during the disaster time interval. This list does not include junctions with pressure drops below *minimum_pressure* during normal operations. Like the pipe break analysis, the *try/except/finally* approach is taken to ensure the script can finish running and still catch any failures a single pipe break may cause.

```
analysis_results = {}
for segment in link_segments.unique():
    wn = wntr.network.WaterNetworkModel(inp_file)
    wn.options.hydraulic.demand_model = 'PDD'
```

```python
    wn.options.time.duration = total_duration
    wn.options.hydraulic.minimum_pressure = minimum_pressure
    wn.options.hydraulic.required_pressure = required_pressure
    wn.options.time.report_timestep = 3600 #1 hour
    wn.options.time.hydraulic_timestep = 3600 #1 hour

    pipes_in_seg = link_segments[link_segments == segment]

    for pipe_name in pipes_in_seg.index:
    pipe = wn.get_link(pipe_name)

    act = wntr.network.controls.ControlAction(pipe, 'status', 0)
    cond = wntr.network.controls.SimTimeCondition(wn, 'Above',
    start_time)
    ctrl = wntr.network.controls.Control(cond, act)
    wn.add_control('close pipe ' + pipe_name, ctrl)

    try:
    sim = wntr.sim.WNTRSimulator(wn)
    sim_results = sim.run_sim()

    sim_pressure = sim_results.node['pressure'].loc[start_time::,
    nzd_junct]
    sim_pressure_below_pmin = sim_pressure.columns[(sim_pressure <
    minimum_pressure).any()]
    impacted_junctions = set(sim_pressure_below_pmin) - set(normal
    _pressure_below_pmin)
    except Exception as e:
    impacted_junctions=None
    print(segment, ' Failed:', e)

finally:
    analysis_results[segment]=impacted_junctions
```

Step 6: Calculate the pressure and population impacts for each segment and visualize results onto a network. The impacts for each segment are calculated and then mapped to each pipe within that segment. This allows for the impacts to be plotted onto a network map correctly. The parameters *node_range* and *link_range* in the `plot_network` method can be changed to best fit the analysis results. If users are working through the code in Jupyter Notebooks, the resulting graphs will automatically generate. However, if the code is run in a different IDE and graphs do not appear, `import matplotlib.pyplot as plt` and `plt.show()` should be added to the code as shown in Section 12.3.5.1.

```python
population=wntr.metrics.population(wn)

num_junctions_impacted_per_segment={}
num_people_impacted_per_segment={}
```

```
for segment, impacted_junctions in analysis_results.items():
    if impacted_junctions is not None:
        num_junctions_impacted_per_segment[segment] =
        len(impacted_junctions)
        num_people_impacted_per_segment[segment] =
        population[impacted_junctions].sum()

num_junctions_impacted=link_segments.
map(num_junctions_impacted_per_segment)
num_people_impacted=link_segments.
map(num_people_impacted_per_segment)

wntr.graphics.plot_network(wn, link_attribute=num_junctions_impacted,
node_size=0,
                           link_width=2,
                           link_range=[0,10], link_colorbar_label=
                           'Junctions Impacted',
                           title='Number of junctions impacted by each
                           segment closure')

ax=plt.gca()
wntr.graphics.plot_valve_layer(wn, valve_layer, add_colorbar=False,
include_network=False,ax=ax)

wntr.graphics.plot_network(wn, link_attribute=num_people_impacted,
node_size=0, link_width=2,
                           link_range=[0,5000], link_colorbar_label=
                           'Population Impacted',
                           title='Number of people impacted by each
                           segment closure')

ax=plt.gca()
wntr.graphics.plot_valve_layer(wn, valve_layer, add_colorbar=False,
include_network=False,ax=ax)
```

The resulting graphs are shown in Figure 12.7. The left side shows the number of junctions each segment impacted represented with color and segment thickness. The right side shows the number of consumers each segment impacted represented with color and segment thickness. Segment thickness increases with junction and population impacts. Unlike the pipe break analysis which only simulated breaks in pipes of specified diameters, the segment break analysis simulated the impact for all pipes. The pipe break analysis results show the most impactful pipe is located near the bottom of the network. However, when taking into consideration isolation valve placement and grouping pipes into segments, the most impactful segments are located both near the top of the network and the bottom.

Number of junctions impacted by each segment closure

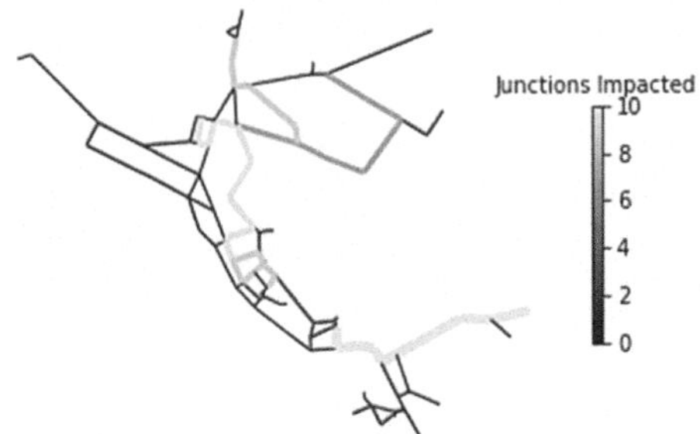

Number of people impacted by each segment closure

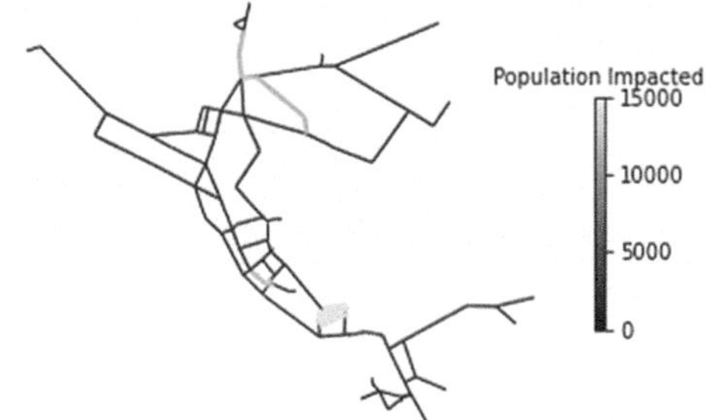

Figure 12.7 Net3 results from segment break analysis.

12.6.3 Fire flow

Step 1: Import Python packages and create a water network. Numpy and Matplotlib are required to support data handling and graphics/plotting. Once the network model has been created, continue to the sample code for the disaster scenario of interest. The file path in this example assumes that the user is working from the demos folder directory.

```
import numpy as np
import matplotlib.pylab as plt
import wntr

inp_file = "../networks/Net3.inp"
wn = wntr.network.WaterNetworkModel(inp_file)
```

Step 2: Define the start time of fire (s), the fire duration (s), the fire flow (m³/s), the system pressures (mH$_2$O), and the pipe diameters (m). The parameters *minimum_pressure* and *required_pressure* are used for PDD. Nodes with pressures below minimum pressure will not receive any water, and node pressures need to be at least the required pressure to receive all of the requested demand. Hydrants are typically attached to pipes with diameters between 0.15 m (6 inches) and 0.20 m (8 inches). For this reason, pipes selected for this analysis are within that range.

```
start_time = 2*3600 # 2 hours
fire_duration = 4*3600 # 4 hours
total_duration = start_time + fire_duration

fire_demand = 0.5047 # 8000 GPM

minimum_pressure = 3.52 # 5 psi
required_pressure = 14.06 # 20 psi

min_pipe_diam = 0.1524 # 6 inch
max_pipe_diam = 0.2032 # 8 inch
```

Step 3: Identify non-zero demand (NZD) junctions by calculating average expected demand (AED) for each junction and selecting those with a value greater than zero. Simulate normal conditions. Identifying junctions that experience pressures below *minimum_pressure* during normal operations help determine which junctions that experience low pressures during the disaster are as a direct result of the disaster. Note, it is standard practice to set the *report_timestep* and *hydraulic_timestep* to an hour or less.

```
AED=wntr.metrics.average_expected_demand(wn)
nzd_junct=AED[AED > 0].index

wn.options.hydraulic.demand_model='PDD'
wn.options.time.duration=total_duration
wn.options.hydraulic.minimum_pressure=minimum_pressure
wn.options.hydraulic.required_pressure=required_pressure
wn.options.time.report_timestep=3600 #1 hour
wn.options.time.hydraulic_timestep=3600 #1 hour

sim=wntr.sim.WNTRSimulator(wn)
results=sim.run_sim()

pressure=results.node['pressure'].loc[start_time::, nzd_junct]
normal_pressure_below_pmin=pressure.columns[(pressure < minimum_pressure).
any()]
```

Step 4: Identify hydrant locations. Assuming that there are hydrants at every junction in the network model, hydrants of interest are identified as nodes connected to pipe diameters of interest. This snippet shows how to select only certain size pipes and a unique set of nodes that are connected to that selection of pipes. In this tutorial, hydrant locations are referred to as *junct_of_interest* or fire nodes.

```
pipe_diameter=wn.query_link_attribute('diameter')
pipes_of_interest=pipe_diameter[(pipe_diameter <=max_pipe_diam) &
                                (pipe_diameter >=min_pipe_diam)]

junct_of_interest=set()
for pipe_name in pipes_of_interest.index:
    pipe=wn.get_link(pipe_name)
    if pipe.start_node_name in wn.junction_name_list:
        junct_of_interest.add(pipe.start_node_name)
    if pipe.end_node_name in wn.junction_name_list:
        junct_of_interest.add(pipe.end_node_name)
```

Step 5: With the list of identified fire nodes, simulate fire flows for each. Iterate through the list of fire nodes and simulate the increased fire demand flow for each. For each fire node, save out results for impacted junctions. A junction is considered impacted if its pressure drops below *minimum_pressure* during the disaster time interval. If the junction experiences pressures below *minimum_pressure* during normal conditions, it is not included in the list of impacted junctions. Like pipe and segment break analyses, a *try/except/finally* approach is taken to ensure the script can finish running and still catch any failures a single pipe break may cause.

```
analysis_results = {}
for junct in junct_of_interest:
    wn = wntr.network.WaterNetworkModel(inp_file)
    wn.options.hydraulic.demand_model = 'PDD'
    wn.options.time.duration = total_duration
    wn.options.hydraulic.minimum_pressure = minimum_pressure
    wn.options.hydraulic.required_pressure = required_pressure
    wn.options.time.report_timestep = 3600 #1 hour
    wn.options.time.hydraulic_timestep = 3600 #1 hour

    fire_flow_pattern = wntr.network.elements.Pattern.binary_pattern(
        'fire_flow',
        start_time = start_time,
        end_time = total_duration,
        step_size = wn.options.time.pattern_timestep,
        duration = wn.options.time.duration
        )
    wn.add_pattern('fire_flow', fire_flow_pattern)

    fire_junct = wn.get_node(junct)
    fire_junct.demand_timeseries_list.append((fire_demand, fire_flow_pattern,
    'Fire flow'))

    try:
        sim = wntr.sim.WNTRSimulator(wn)
        sim_results = sim.run_sim()

        sim_pressure = sim_results.node['pressure'].loc[start_time::,
        nzd_junct]
        sim_pressure_below_pmin = sim_pressure.columns[(sim_pressure
        < minimum_pressure).any()]
        impacted_junctions = set(sim_pressure_below_pmin) -
        set(normal_
        pressure_below_pmin)

    except Exception as e:
        impacted_junctions = None
        print(junct, ' Failed:', e)
    finally:
        analysis_results[junct] = impacted_junctions
```

Step 6: Calculate pressure and population impacts for each fire node and visualize results onto a network. The parameters *node_range* and *link_range* in the `plot_network` method can be changed to best fit the analysis results. If users are working through the code in Jupyter Notebooks, the resulting graphs will automatically generate. However, if the code is run in a different IDE and graphs do not appear, `import matplotlib.pyplot as plt` and `plt.show()` should be added to the code as shown in Section 12.3.5.1.

```
population=wntr.metrics.population(wn)

num_junctions_impacted={}
num_people_impacted={}
for pipe_name, impacted_junctions in analysis_results.items():
    if impacted_junctions is not None:
        num_junctions_impacted[pipe_name]=len(impacted_junctions)
        num_people_impacted[pipe_name]=population[impacted_junctions].
    sum()

wntr.graphics.plot_network(wn, node_attribute=num_junctions_impacted,
node_size=20,
                           link_width=0,
                           node_range=[0,3], node_colorbar_label=
                           'Junctions Impacted',
                           title='Number of junctions impacted by each
                           fire demand')

wntr.graphics.plot_network(wn, node_attribute=num_people_impacted,
node_size=20, link_width=0,
                           node_range=[0,2000], node_colorbar_label-
                           'Population',
                           title='Number of people impacted by each fire
                           demand')
```

The resulting graph is shown in Figure 12.8. The left side shows which fire nodes impacted junctions where the color and circle size corresponds to the number of junctions impacted. The right side shows which fire nodes impacted consumers where color and circle size corresponds to the number of consumers impacted. Circle size increases with junction and population impacts. The graphs indicate the fire nodes with the greatest junction impact are at the bottom of the network and each impacted three other junctions. However, the fires nodes with the greatest population impact are at the top of the network and each impacted 1902 consumers.

12.6.4 Earthquake

Step 1: Import Python packages and create a water network. Numpy, Pandas, Matplotlib, and SciPy are required to support data handling and graphics/plotting. Once the network model has been created, continue to the sample code for the disaster scenario of interest. The file path in this example assumes that the user is working from the demos folder directory.

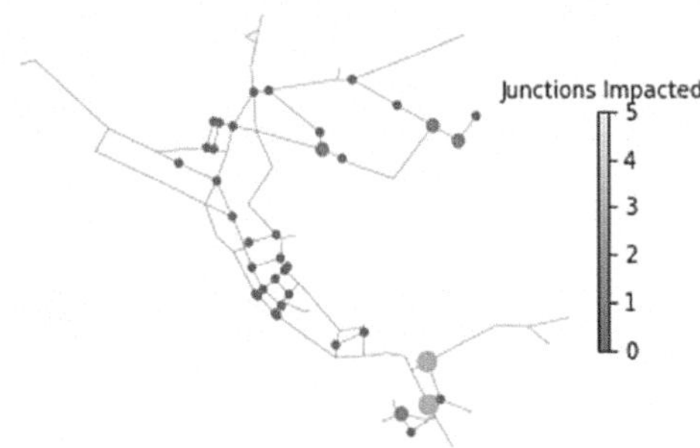

Figure 12.8 Net3 results from fire flow analysis.

```python
import numpy as np
import pandas as pd
import matplotlib.pylab as plt
from scipy.stats import expon
import wntr


inp_file = "../networks/Net3.inp"
wn = wntr.network.WaterNetworkModel(inp_file)
```

Step 2: Define all parameters. For the earthquake analysis, this includes the epicenter (*x,y*-coordinates: same used by EPANET), the magnitude (Richter scale), and the depth (m).

```python
epicenter = (32000,15000) # m (x,y)
magnitude = 6.5 # Richter magnitude
depth = 10000 # m


total_duration = 24*3600 # 24 hours


minimum_pressure = 3.52 # 5 psi
required_pressure = 14.06 # 20 psi


leak_start_time = 5*3600 # 5 hours
leak_repair_time = 15*3600 # 15 hours
```

Step 3: Create the earthquake object and calculate PGA, PGV, and RR. Create fragility curve and damage states. Coordinate morphing is optional.

```
wn=wntr.morph.scale_node_coordinates(wn, 1000)
earthquake=wntr.scenario.Earthquake(epicenter, magnitude, depth)

R=earthquake.distance_to_epicenter(wn, element_type=wntr.network.Pipe)
pga=earthquake.pga_attenuation_model(R)
pgv=earthquake.pgv_attenuation_model(R)
RR=earthquake.repair_rate_model(pgv)

L=pd.Series(wn.query_link_attribute('length', link_type=wntr.network.Pipe))

pipe_FC=wntr.scenario.FragilityCurve()
pipe_FC.add_state('Minor leak', 1, {'Default': expon(scale=0.2)})
pipe_FC.add_state('Major leak', 2, {'Default': expon()})
pipe_Pr=pipe_FC.cdf_probability(RR*L)
pipe_damage_state=pipe_FC.sample_damage_state(pipe_Pr, seed=123)
```

Step 4: Define the hydraulic parameters, the start time of failure, and the duration of failure. In this demo, failure is simulated as pipe leaks. Note, it is standard practice to set the *report_timestep* and *hydraulic_timestep* to an hour or less.

```
wn.options.hydraulic.demand_model='PDD'
wn.options.time.duration=total_duration
wn.options.hydraulic.minimum_pressure=minimum_pressure
wn.options.hydraulic.required_pressure=required_pressure
wn.options.time.report_timestep=3600 #1 hour
wn.options.time.hydraulic_timestep=3600 #1 hour
```

Step 5: Add all pipe leaks to the network and simulate hydraulics. First, the area of the pipe leak needs to be defined, where this code classifies a leak as 'Major leak' or 'Minor leak' to specify leak area. In this example, a 'Major Leak' assumes a 25% leak relative to the pipe's diameter and a 'Minor Leak' assumes a 10% leak relative to the pipe's diameter. To add the leak to a pipe, the `split_pipe` method is used to split a pipe and add a node while retaining the characteristics of the original pipe. The leak is then applied to the node just created to simulate a leak in the pipe of interest.

```
for pipe_name, damage_state in pipe_damage_state.items():
    pipe_diameter=wn.get_link(pipe_name).diameter
    if damage_state is not None:
        if damage_state== 'Major leak':
            leak_diameter=0.25*pipe_diameter
            leak_area=np.pi/4.0*leak_diameter**2
        elif damage_state== 'Minor leak':
            leak_diameter=0.1*pipe_diameter
            leak_area=np.pi/4.0*leak_diameter**2
        else:
            leak_area=0
```

```
        wn=wntr.morph.split_pipe(wn,pipe_name, pipe_name+'A', 'Leak'
        +pipe_name)
        n=wn.get_node('Leak'+pipe_name)
        n.add_leak(wn, area=leak_area, start_time=leak_start_time)

sim=wntr.sim.WNTRSimulator(wn)
results=sim.run_sim()
```

Step 6: Simulate hydraulics with repair efforts included. Below simulates a partial repair of the leaks for each pipe of interest. Remember, leaks are applied to nodes within WNTR to simulate pipe leaks. The code below specifies four pipes will be repaired.

```
wn.reset_initial_values()

leaked_demand=results.node['leak_demand']
leaked_sum=leaked_demand.sum()
leaked_sum.sort_values(ascending=False, inplace=True)

number_of_pipes_to_repair=4
leaks_to_fix=leaked_sum[0:number_of_pipes_to_repair]

for leak_name in leaks_to_fix.index:
    node=wn.get_node(leak_name)
    leak_area=node.leak_area
    node.remove_leak(wn)
    node.add_leak(wn, area=leak_area, start_time=leak_start_time,
    end_time=leak_repair_time)

results_wrepair=sim.run_sim()
```

Step 7: Visualize and compare earthquake scenarios with and without repair efforts. The time index is first converted from seconds to hours. The code below provides figures for: (1) network map without repair, (2) network map with repair, and (3) average system pressure with and without repair. The parameters *node_range* and *link_range* in the `plot_network` method can be changed to best fit the analysis results. If users are working through the code in Jupyter Notebooks, the resulting graphs will automatically generate. However, if the code is run in a different IDE and graphs do not appear, `import matplotlib.pyplot as plt` and `plt.show()` should be added to the code as shown in Section 12.3.5.1.

```
pressure=results.node['pressure']
pressure_wrepair=results_wrepair.node['pressure']
pressure.index=pressure.index/3600
pressure_wrepair.index=pressure_wrepair.index/3600
```

```
pressure_at_24hr=pressure.loc[24,wn.junction_name_list]
wntr.graphics.plot_network(wn, node_attribute=pressure_at_24hr, node_
size=20,
                          node_range=[0,90], node_colorbar_label=
                          'Pressure (m)',
                          title='Pressure at 24 hours, without repair')

pressure_at_24hr_wrepair=pressure_wrepair.loc[24,wn.junction_name_list]
wntr.graphics.plot_network(wn, node_attribute=pressure_at_24hr_wrepair,
node_size=20,
                          node_range=[0,90], node_colorbar_label=
                          'Pressure (m)',
                          title='Pressure at 24 hours, with repair')

plt.figure()
ax=plt.gca()
pressure.loc[:,wn.junction_name_list].mean(axis=1).plot(label='Without
repair', ax=ax)
pressure_wrepair.loc[:,wn.junction_name_list].mean(axis=1).
plot(label='With repair', ax=ax)
ax.set_xlabel('Time (hr)')
ax.set_ylabel('Average system pressure (m)')
ax.legend()
```

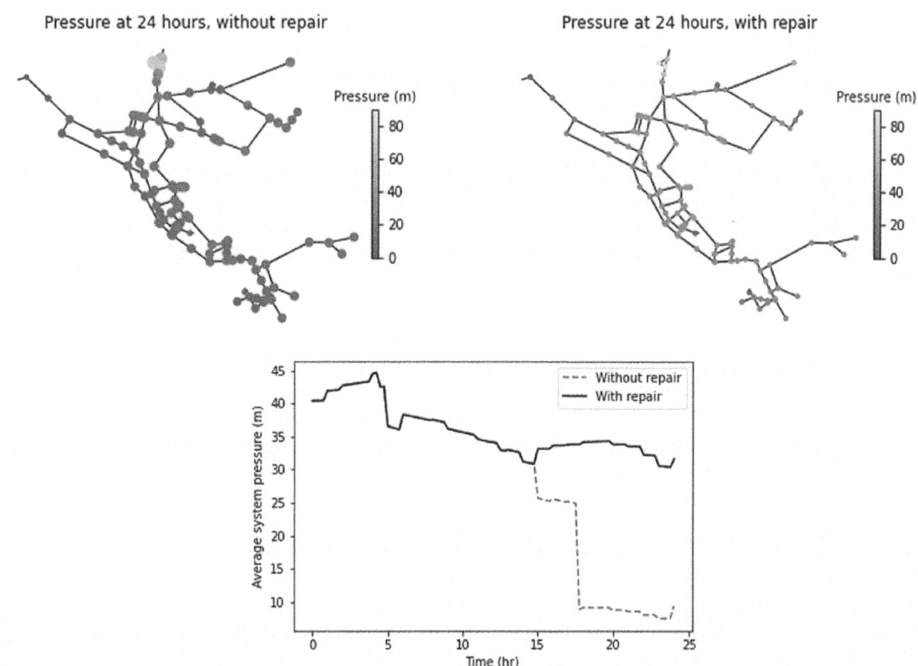

Figure 12.9 Net3 results from earthquake analysis.

The resulting graphs are shown in Figure 12.9. The network graphs and pressure graphs show that the partial repair restored junction pressure to most of the junctions. This is supported by the graph showing the average system pressures represented with color and circle size. Circle size increased with average system pressure.

12.7 CONCLUSIONS

WNTR was developed to be a flexible and extendable framework for modeling resilience of water distribution systems. This chapter highlights numerous disaster scenarios that can be modeled by WNTR and some associated tutorials.

Estimating how a disaster impacts a WDS and modeling the simulated disaster and associated response actions can help a water utility prepare for such disasters. WNTR is a tool capable of dealing with such complex problems. Furthermore, the integration of the data structures available within Python and the ability to deal with complex inputs can provide necessary flexibility for addressing a range and combination of disaster scenarios. Analysis results can help utilities identify which components are critical to their system resilience and take steps to ensure that they: (1) use the best material/design configuration to maximize resilience (e.g. strengthened to a certain type of failure); (2) have available backup options to ensure rapid replacement if damaged; or (3) have emergency response plans in place to manage failures to continue operations and minimize impact to customers. Fire flow analyses provide utilities with information about the impact of firefighting activities to overall system pressure and ability to meet system demands. The ability to model disasters in a realistic manner and analyze impacts and responses with standard metrics provides utilities with a more quantitative concept of their resilience to that disaster. Working through such exercises can help utilities to prepare and prioritize mitigation strategies to help build resilience and ensure security in their WDS.

12.8 DISCLAIMER

The U.S. Environmental Protection Agency (EPA), through its Office of Research and Development, funded and managed the research described herein under Interagency Agreement (IA # DW08992524701) with Department of Energy's Oak Ridge Associated Universities (ORAU) and Interagency Agreement (IA #DW08992513801) with the Department of Energy's Sandia National Laboratories. It has been subjected to review by the Office of Research and Development and approved for publication. Any mention of trade names, manufacturers or products does not imply an endorsement by the United States Government or the U.S. Environmental Protection Agency. EPA and its employees do not endorse any commercial products, services, or enterprises.

REFERENCES

Abdel-Mottaleb N., Ghasemi Saghand P., Charkhgard H. and Zhang Q. (2019). An exact multiobjective optimization approach for evaluating water distribution infrastructure criticality and geospatial interdependence. *Water Resources Research*, **55**(7), 5255–5276, https://doi.org/10.1029/2018WR024063

EPA (U.S. Environmental Protection Agency) (2015). Systems Measures of Water Distribution System Resilience, EPA/600/R-14/383, U.S. Environmental Protection Agency, Washington, DC.

Fiksel J. (2006). Sustainability and resilience: toward a systems approach. *Sustainability: Science, Practice, and Policy*, **2**(2), 14–21, https://doi.org/10.1080/15487733.2006.11907980

International Code Council (2011). 2012 International Fire Code, Appendix B – Fire-Flow Requirements for Buildings. International Code Council, Country Club Hills, IL.

Jayaram N. and Srinivasan K. (2008). Performance-based optimal design and rehabilitation of water distribution networks using life cycle costing. *Water Resources Research*, **44**(1), W01417, https://doi.org/10.1029/2006WR005316

Klise K., Bynum M., Moriarty D. and Murray R. (2017). A software framework for assessing the resilience of drinking water systems to disasters with an example earthquake case study. *Environmental Modeling & Software*, **95**, 420–431, https://doi.org/10.1016/j.envsoft.2017.06.022

Klise K., Hart D., Bynum M., Hogge J., Haxton T., Murray R. and Burkhardt J. (2020). Water Network Tool for Resilience (WNTR) User Manual: Version 0.2.3, EPA/600/R-20/185, U.S. EPA Office of Research and Development, Washington, DC.

Logan K. T., Leštáková M., Thiessen N., Engels J. I. and Pelz P. F. (2021). Water distribution in a socio-technical system: resilience assessment for critical events causing demand relocation. *Water*, **13**(15), 2062, https://doi.org/10.3390/w13152062

Mazumder R. K., Salman A. M. and Li Y. (2020). Post-disaster sequential recovery planning for water distribution systems using topological and hydraulic metrics. *Structure and Infrastructure Engineering*, **8**(5), 728–743, https://doi.org/10.1080/15732479.2020.1864415

Moraitis G., Nikolopoulos D., Bouziotas D., Lykou A., Karavokiros G. and Makropoulos C. (2020). Quantifying failure for critical water infrastructures under cyber-physical threats. *Journal of Environmental Engineering*, **146**(9), 04020108, https://doi.org/10.1061/(ASCE)EE.1943-7870.0001765

NAS (National Academy of Sciences) (2012). Disaster Resilience: A National Imperative. Prepared by the NAS Committee on Science, Engineering, and Public Policy. The National Academies Press, Washington, DC.

Nikolopoulos D., Ostfeld A., Salomons E. and Makropoulos C. (2021). Resilience assessment of water quality sensor designs under cyber-physical attacks. *Water*, **13**(5), 647, https://doi.org/10.3390/w13050647

Newell D. and Tiesinga E. (2019). The International System of Units (SI). National Institute of Standards and Technology, Gaithersburg, MD, https://www.nist.gov/pml/special-publication-330 (last accessed 16 November 2021)

NIST (National Institute of Standards and Technology (2021). *New Timeline of Deadliest California Wildfire Could Guide Lifesaving Research and Action*. Available at: https://www.nist.gov/news-events/news/2021/02/new-timeline-deadliest-california-wildfire-could-guide-lifesaving-research (last accessed 16 November 2021)

NPR (National Public Radio) (2021). *Winter Storm Leaves Many in Texas Without Power and Water*. Available at: https://www.npr.org/2021/02/17/968665266/millions-still-without-power-as-winter-storm-wallops-texas (last accessed 16 November 2021)

Ostfeld A., Kogan D. and Shamir U. (2002). Reliability simulation of water distribution systems – single and multiquality. *Urban Water*, **4**(1), 53–61, https://doi.org/10.1016/S1462-0758(01)00055-3

Rossman L. (2000). EPANET 2.0 User Manual, EPA/600/R-00/057, U.S. Environmental Protection Agency, Washington, DC. https://nepis.epa.gov/Adobe/PDF/P1007WWU.pdf (last accessed 16 November 2021)

Rossman L., Woo H., Tryby M., Shang F., Janke R. and Haxton T. (2020). EPANET 2.2 User Manual, EPA/600/R-20/133, U.S. Environmental Protection Agency, Washington, DC, https://cfpub.epa.gov/si/si_public_record_report.cfm?Lab=CESER&dirEntryId=348882 (last accessed 16 November 2021)

South China Morning Post (2021). *Underground Water Pipe Bursts, Triggers Landslide in Hong Kong's Upscale The Peak District*. Available at: https://www.scmp.com/news/hong-kong/transport/article/3155127/underground-water-pipe-bursts-triggers-landslide-hong (last accessed 30 November 2021)

Todini E. (2000). Looped water distribution networks design using a resilience index based heuristic approach. *Urban Water*, **2**(2), 115–122, https://doi.org/10.1016/S1462-0758(00)00049-2

Tomar A., Burton H. V., Mosleh A. and Lee J. Y. (2020). Hindcasting the functional loss and restoration of the Napa water system following the 2014 earthquake using discrete-event simulation. *Journal of Infrastructure Systems*, **26**(4), 04020035, https://doi.org/10.1061/(ASCE)IS.1943-555X.0000574

USGS (United States Geological Survey) (2014). *M6.0 South Napa, California Earthquake – August 24, 2014*. https://www.usgs.gov/natural-hazards/earthquake-hazards/science/m60-south-napa-california-earthquake-august-24-2014?qt-science_center_objects=0#qt-science_center_objects (last accessed 16 November 2021)

USGS (2015). *Chemicals in Elk River Spill Lingered Longer, Travelled Farther*. https://www.usgs.gov/news/chemicals-elk-river-spill-lingered-longer-traveled-farther (last accessed 16 November 2021)

Wagner J. M., Shamir U. and Marks D. H. (1988). Water distribution reliability: simulation methods. *Journal of Water Resources Planning and Management*, **114**(3), 276–294, https://doi.org/10.1061/(ASCE)0733-9496(1988)114:3(276)

Part III
Management

doi: 10.2166/9781789062380_0327

Chapter 13

Optimal replacement time of water mains

*Juneseok Lee**

Department of Civil and Environmental Engineering, Manhattan College, Riverdale, NY
**Corresponding author: Juneseok.Lee@manhattan.edu*

LEARNING OBJECTIVES

At the end of this chapter, you will be able to:

(1) Explain asset management and the concepts of pipe replacement program.
(2) Calculate the threshold break rate.
(3) Compute optimal replacement time given failure datasets.

13.1 INTRODUCTION

According to the USEPA (2022), Asset Management (AM) is defined as 'maintaining a desired level of service for what you want your assets to provide at the lowest life cycle cost. Lowest life cycle cost refers to the best appropriate cost for rehabilitating, repairing or replacing an asset'. Collectively, interpretation and implementation of AM definitions and programs depend on the water utility, but typical water mains replacement program is composed of several, interconnected parts (Figure 13.1):

- Performance Management;
- Failure Mode Analysis;
- Operations and Maintenance (O&M);
- Risk Analysis;
- Prioritization and Capital Specification.

Many water utilities define their pipes with an individual ID; a pipe ID is assigned in GIS for every piece with the same diameter, material, install date, and project code based on the assumption that all pieces are degrading at the same rate and a leak will spring up at a random location along the pipe. More details on this will be covered in Chapter 18 about GIS applications. Below are brief descriptions on the different phases in water mains replacement programs.

13.1.1 Performance management

Performance management governs the entire asset management program and tracks the company's overall service level.

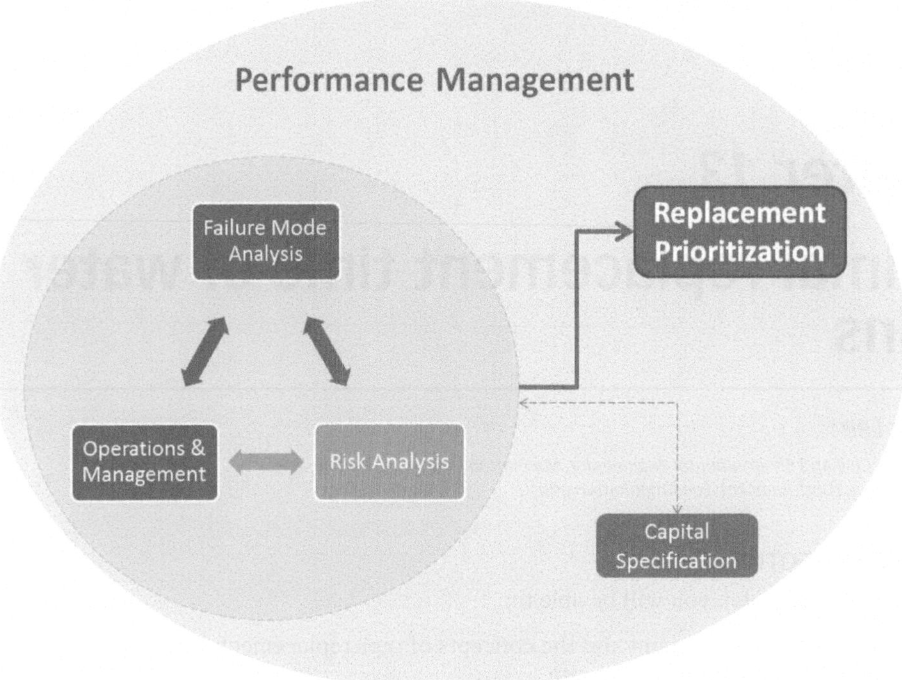

Figure 13.1 Overall structure of water mains replacement program.

13.1.2 Failure mode analysis

Every asset is vulnerable to four modes of failure: (i) Mortality, (ii) Capacity, (iii) Efficiency, and (iv) Level of Service (LOS). Mortality refers to the end of life of the asset, specifically the financial end of life. The concept of *threshold break rate* to determine end of life can be considered here (and will be covered later in this chapter). Prior to this threshold, a pipe is replaced if three or more leaks have occurred or have operated based upon an engineering judgment/rule of thumb basis. Capacity of the asset refers to its performance in terms of hydraulic parameters such as flow and pressure. Water utilities can consider replacement with increased diameter if a pipe does not provide adequate flow or pressure due to insufficient diameter or heavy tuberculation. In addition, if a small diameter pipeline results in high water velocities which can carry sediment to the customers resulting in complaints, then the utility can propose replacement with a larger diameter. Efficiency is often interpreted as an output-to-input ratio, with perhaps the clearest water utility example being that of pump and motor efficiency.

LOS is defined by quality, quantity, reliability, environmental standards, and associated system performance goals, both short- and long-term. Information about customer demand, as well as data from utility commissions or boards and other stakeholders, can be utilized to develop LOS requirements. Therefore, developing/defining LOS concepts is critical for the water utilities. The AWWA benchmark recommends comparing the number of leaks per 100 miles of pipe, while the Partnership for Safe Water sets their optimal distribution system standard at 15 breaks per 100 miles of pipe per year. The five-year running average is expected to be trending down as a result of effective asset management.

13.1.3 Operations and maintenance

O&M is composed of pipeline inspection, condition assessment, valve maintenance, and so on. Pipeline inspection is done by individual districts and based upon visual assessment (i.e., with the basis being if it looks bad, then it is bad) and only occurs following a report of an existing leak. Many utilities currently may not have a standardized process for determining the optimum course of action based upon visual inspection. Also, leak detection may not be a part of O&M, but many utilities are showing interest in incorporating this in the replacement program.

13.1.4 Risk analysis

Risk Analysis involves Business Risk Exposure, which is comprised of the likelihood of failure (LOF) and the consequences of failure (COF). Each asset can be categorized by low, medium or high COF and low, medium, or high LOF. More details on this will be covered in Chapter 18 with respect to GIS Analytics. Any pipes that are categorized as 'high' for both categories are assigned a high priority for replacement. A GIS leak database where the leak locations are geocoded are typically available for water utilities. LOF is typically based upon observed pipe failure (leak/breaks) histories. COF levels are determined by their proximity to roads, schools, and sensitive environmental areas such as waterways, marine protected areas, and locations listed in the national wetland inventory using buffer zones in GIS. COF also takes into account situations in which an additional leak would incur additional costs for the utilities and/or pose a potential danger to the area.

13.1.5 Capital specification

Capital specification refers to the material and design of the assets. These specifications are dependent upon the generally accepted practice at the time of installation.

13.1.6 Prioritization

Prioritization of pipe replacement considers both affordability and risk management. This is discussed in more detail in the following.

13.2 OPTIMAL REPLACEMENT

In a water distribution system, the repair/replacement cost and possible water damage cost must be balanced by the water utility when deciding at the time of a leak/break whether to repair or replace the system. Accelerated replacement refers to replacing the system well in advance of the optimal replacement time, while delaying replacement beyond the optimal replacement time will lead to consequences through neglecting repairs, which may effectively amount to the utility paying a penalty to compensate for the high replacement cost. To manage the integrity of water main infrastructure through its entire life cycle, we introduce a replacement program for water utilities in this section. This program is expected to ensure affordability, manage risk, and support a high level of confidence in the decisions reached.

The following construct is utilized to assess the contribution of costs towards present worth (see Loganathan and Lee (2005)). At the time of the nth leak, a decision has to be made whether to replace the system at a cost of F_n or to repair it at a cost of C_n. The scenario also implies that for the previous $(n-1)$ leaks only repairs have been performed. If we assume that the system will be replaced (the value of C_n included in the sum should be adjusted for F_n when necessary) at the time of the nth leak, t_n, we can write the present worth of the total cost of the pipe as:

$$T_n = \sum_{i=1}^{n} \left\{ \frac{C_i}{(1+R)^{t_i}} \right\} + \frac{F_n}{(1+R)^{t_n}} \tag{13.1}$$

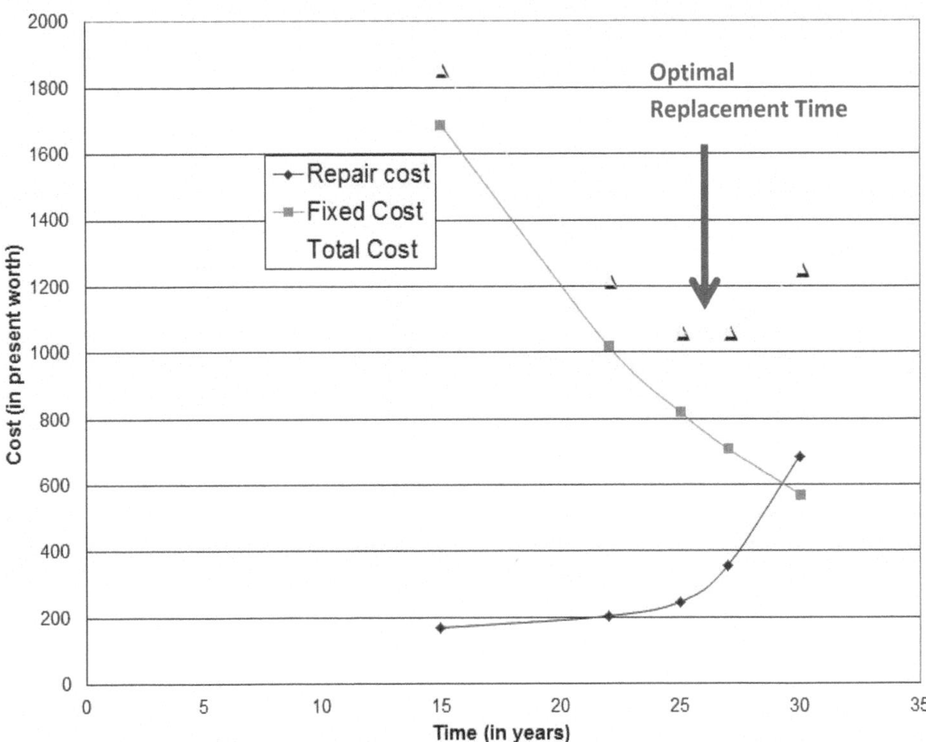

Figure 13.2 Present worth of total cost curve over time.

where R = the discount rate, t_i = the time of the ith leak measured from the installation year (in years), C_i = the repair cost of the ith leak, F_n = the replacement cost at time t_n, T_n = the total cost at time '0' (present worth).

When the system is new, it tends to experience very few leaks, while an old system experiences more leaks under the same conditions. Therefore, the combination of varying time interval between leaks (accelerated leak incidences towards the end), relatively smaller repair costs, and a generally large replacement cost (fixed cost) generates a 'U' shaped present worth of the total cost curve over time (Figure 13.2). The optimal replacement refers to the point at which the total cost is minimized.

Following Loganathan *et al.* (2002), the point of minimum total cost in Equation (13.1) occurs at the time when the inequality is satisfied for the first time:

$$t_{n+1} - t_n < \frac{\ln((C_{n+1} / F_n) + (F_{n+1} / F_n))}{\ln(1 + R)} \tag{13.2}$$

where C_n = the repair cost at the nth leak and F_n = the replacement cost, R = the discount rate, and t_n = the leak occurrence time. Furthermore, the threshold break is defined as follows:

$$\text{Threshold break rate} = \frac{\ln(1 + R)}{\ln(1 + (C_n / F_n))} \tag{13.3}$$

For a water system using the data from data sets, we can calculate the time that minimizes the total cost. That is, when the inter-arrival time becomes smaller than the threshold year, the system should be replaced. Clearly, these values will change based on the available data.

The above methods can be applied to develop their prioritization model. In theory, it is expected that the time between leaks should become smaller as the pipe ages. A majority of pipes do follow this pattern, but some pipes do not. For those pipelines where the assumption of increasing break rate is not satisfied, utilities have analysts consider the shape of the curve; if the curve shape was trending up, flat, or nearly flat, the utility can propose replacement as the pipeline's annualized costs were near or at a minimum value. It is noted that repair and replacement costs may vary considerably due to site specific conditions or accounting procedures. Therefore, utilities can implement a risk-based annualized cost curve analysis for the prioritization of replacement decisions. To this end, utilities should consider: (i) the risk score based upon LOF and COF, (ii) the life cycle cost curve and (iii) the hydraulic performance. Based upon the level of confidence of the analyst, each parameter can be weight summed to assign a replacement priority.

13.3 PRACTICAL EXAMPLES

13.3.1 Example 1
Table 13.1 shows the break time and the number of pipe breaks for a water system. Calculate and plot the repair, replacement, and total cost.

13.3.2 Example 2
Answer the following. Pipe repair cost is $1000, replacement cost is $150 K. use discount rate of 7.5% (adopted from Khambhammettu, 2001; see Figure 13.3 and Table 13.2 for pipe break data):

(1) Compute the cumulative costs when the pipes were replaced in year 5.
(2) Compute the present worth cost in year 0 to cover the cumulative costs up to year 3.
(3) Compute the optimal replacement year when the present worth cost is the minimum.

13.3.3 Solution

(1) Total repair cost = $1000 * (3 + 5 + 7 + 9 + 11) = $34 000

   Replacement cost = $150 000. So, the Cumulative cost = $34 000 + $150 000 = $184 000

(2) The present worth is obtained by discounting the yearly repair costs and the replacement cost to the origin and calculate the sum. The Present worth would be obtained by summing up the repair and replacement components. So, the repair components are:

$$\text{Repair component} = \frac{\$3000}{1 + 0.075} + \frac{\$5000}{(1 + 0.075)^2} + \frac{\$7000}{(1 + 0.075)^3} = \$12\,752$$

$$\text{Replacement component} = \frac{\$150\,000}{(1 + 0.075)^3} = \$120\,744$$

Present Worth = $12 752 + $120 744 = $133 496

Table 13.1 Pipe break data and calculated costs.

Break Time (years)	#brks	Repair Cost ($)	Fixed Cost ($)	Total Cost ($)
15	1	169	1690	1859
22	2	204	1019	1222
25	3	246	820	1066
27	5	355	709	1064
30	12	685	571	1256

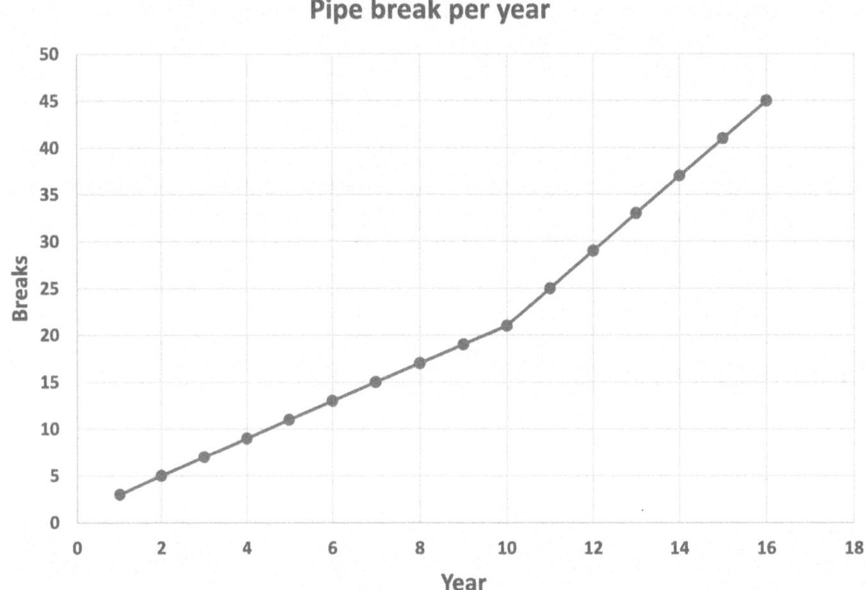

Figure 13.3 Pipe break for each year.

Table 13.2 Pipe break data.

Year	Breaks	Year	Breaks
0	3	16	49
1	5	17	53
2	7	18	57
3	9	19	61
4	11	20	65
5	13	21	69
6	15	22	73
7	17	23	77
8	19	24	81
9	21	25	85
10	25	26	89
11	29	27	93
12	33	28	97
13	37	29	101
14	41	30	105
15	45		

Table 13.3 Present worth costs.

Time	Breaks	Repair	Replacement	Present Worth
0	3	$2791	$139 535	$142 326
1	5	$7117	$129 800	$136 917
2	7	$12 752	$120 744	$133 496
3	9	$19 491	$112 320	$131 811
4	11	$27 153	$104 484	$131 637
5	13	$35 577	$97 194	$132 771
6	15	$44 618	$90 413	$135 031
7	17	$54 150	$84 105	$138 256
8	19	$64 060	$78 238	$142 298
9	21	$74 249	$72 779	$147 028
10	25	$85 533	$67 701	$153 234
11	29	$97 709	$62 978	$160 687
12	33	$110 597	$58 584	$169 182
13	37	$124 040	$54 497	$178 537
14	41	$137 896	$50 695	$188 591
15	45	$152 044	$47 158	$199 202
16	49	$166 374	$43 868	$210 242
17	53	$180 793	$40 807	$221 600
18	57	$195 218	$37 960	$233 178
19	61	$209 578	$35 312	$244 890
20	65	$223 812	$32 848	$256 660
21	69	$237 868	$30 557	$268 425
22	73	$251 702	$28 425	$280 126
23	77	$265 275	$26 442	$291 717
24	81	$278 557	$24 597	$303 154
25	85	$291 523	$22 881	$314 404
26	89	$304 152	$21 284	$325 436
27	93	$316 427	$19 800	$336 227
28	97	$328 338	$18 418	$346 756
29	101	$339 874	$17 133	$357 007
30	105	$351 031	$15 938	$366 968

(3) Optimal replacement time can be determined from Table 13.3. The fifth year has the lowest Present Worth of $131,637. So, the threshold break rate is calculated as follows:

$$\text{Threshold break rate} = \frac{\ln(1+R)}{\ln(1+(C/F))}$$

$$\text{Threshold break rate} = \frac{\ln(1+0.075)}{\ln(1+(1000/150,000))} = 10.88 \text{ breaks.}$$ From Table 13.3, it is shown that that the optimum indeed occurs for a break rate of 11 breaks/year. So, the optimal replacement time is the fifth year!

13.4 CONCLUSIONS

This chapter presents the general concepts of asset management and replacement program. The optimal replacement program and concepts can lay the foundation for a standardized platform of sustainable life cycle assessments for individual elements of the water infrastructure. Given the pace of infrastructure aging and deterioration, combined with workforce retirements and aggressive technology changes/adoption, it is thus imperative that this type of systematic decision support system be adopted to address the challenges that this will entail for the water industry.

REFERENCES

Khambhammettu P. (2001). A Comprehensive Decision Support System (CDSS) for Optimal Pipe Renewal using Trenchless Technologies. MS thesis, Virginia Tech, VA, USA.

Loganathan G. V. and Lee J. (2005). Decision tool for optimal replacement of plumbing systems. *Civil Engineering and Environmental Systems*, **22**(4), 189–204, https://doi.org/10.1080/10286600500279964

Loganathan G. V., Park S. and Sherali H. D. (2002). Threshold break rate for pipeline replacement in water distribution systems. *Journal of Water Resources Planning and Management*, **128**(4), 271–279.

USEPA. (2022). Avaialble on https://www.epa.gov/dwcapacity/about-asset-management (last accessed date 10 May 2022)

doi: 10.2166/9781789062380_0335

Chapter 14

Water mains replacement decision using GIS analytics

Diego Martinez Garcia

Associate Engineer, Public Works Department, City of Palo Alto, USA
Corresponding author: diegomartinezgarcia@outlook.com

LEARNING OBJECTIVES

At the end of this chapter, you will be able to:
(1) Perform water main and water main failure data preprocessing.
(2) Run a basic multilinear regression model using R.
(3) Visualize findings through GIS.
(4) Optimize the model parameters to improve performance of a GIS.
(5) Assess results based on model accuracy and perform interpretation of GIS results.

14.1 INTRODUCTION

Water mains are typically the largest and most significant asset for any water utility. Depending on the number of served customers, large water utilities can manage hundreds of miles of water mains made of different materials and diameters. When water mains fail, utilities are affected by the loss of treated and energized water (Güngör-Demirci *et al.*, 2018). To be delivered to customers, potable water is treated and then pressurized so it can be distributed to meet EPA water quality standards. Potable water lost between treatment and customers is defined as nonrevenue water (AWWA, 2017). Additionally, rising failure rates in distribution systems increase the capital improvement and maintenance budgets which likely lead to higher bills to their customers and a negative public perception (Folkman, 2018; Giustolisi *et al.*, 2006; Martínez García *et al.*, 2020; Shi *et al.*, 2013). Water main failures could cause lower pressure and flow in premise plumbing and possibly deteriorating water quality (Lee & Tanverakul, 2015; Lee *et al.*, 2012). The general public can be also affected by traffic and construction disruptions when water mains are being repaired as these processes can take several days.

Recent reports issued by the American Society of Civil Engineers (ASCE, 2013) and the American Water Works Association (AWWA, 2017), indicate that the aging of water distribution assets represents one of the most critical technical and financial challenges in the United States. Both estimate that the cumulative gap between needs and likely (feasible) investments from water utilities was $84 billion by 2020 and $144 billon by 2040 (ASCE, 2013; AWWA, 2017).

Water mains can fail due to multiple reasons and it is generally thought that failures occur due to a combination of multiple factors. Internal factors such as pipe length, material and diameter often interact in different ways with external factors such as soil types, climatic conditions and external loads (e.g. traffic). In addition, operational parameters like internal water pressure and quality often play an important role (Christodoulou *et al.*, 2008; de Oliveira *et al.*, 2011; Giustolisi *et al.*, 2006; Goulter & Kazemi, 1988; Kettler & Goulter, 1985; Shi *et al.*, 2013; Tabesh *et al.*, 2010; Wang *et al.*, 2009).

Although an aggressive capital program to repair or replace all affected water mains will reduce the amount of revenue loss, economic and financial constraints make it impossible to replace all failed water mains at the same time. Therefore, supporting water utilities to make informed decisions about the time and location to perform water mains repairs or replacements has attracted attention from researchers in the water industry. In recent years more research has been influenced by technological resources available and advanced computational capacities (de Oliveira *et al.*, 2011; Giustolisi *et al.*, 2006; Rajani & Kleiner, 2001; Wang *et al.*, 2009).

Monitoring (including condition assessment) and identifying water mains that need rehabilitation and replacement is important. As mentioned, unnoticed aging water mains and failures could lead to disruption in the service and water contamination. As introduced in previous chapters, several computational and machine learning-based techniques have been developed to support water utilities in allocating resources for an optimal replacement/rehabilitation of pipes (e.g. artificial neural networks, clustering, etc.)

Clustering analysis involves spatial and temporal grouping of attributes under consideration, such as crime rates, motor vehicle accidents and pipeline failures, to identify areas of high-risk zones in future for planning and replacement/rehabilitation purposes. Clustering analysis considers the spatial and temporal aspects of the pipeline failures to identify areas in high risk of failure. In this chapter, we present different approaches that can assist in decision making for optimal replacement of water mains by using clustering techniques coupled with GIS analysis.

A cluster in the context of water mains is a geographical area with an anomalously higher number of pipe leaks or breaks when compared to surrounding areas. The extent of the cluster will depend on the pipe density, for example in New York City where water mains are located in every block, the definition of a cluster will be different than in a rural city in Iowa, for instance. In the late 1980s, Goulter and Kazemi (1988) utilized clustering techniques that were available to a dataset of water mains and their failures. By utilizing these techniques, they found a strong spatial and temporal clustering in pipe failures. In other words, most water main failures (58%) occurred within 20 m (65 ft) of the original one and around the same time (within same week or month).

The tools presented in this chapter can provide valuable information about the spatiotemporal trend of water main failures. By applying these techniques (along with a more comprehensive enterprise asset management program that focus on maintaining water main integrity), water utilities can save economic resources in avoided failures, reduced water loss and energy savings. In addition, an asset management program (or water mains integrity program) can help select improved materials and sizing can provide other benefits to customers such as improvement in water supply reliability, system resilience, and level of service.

14.2 DATA

The analytical techniques described in this chapter require datasets that the water utilities should have in their database. In general, large water utilities collect reported water main failure data and have datasets that cover all their water assets.

14.2.1 Water main failures

In performing these techniques, a water main failure is defined as a leak and/or break incident that requires attention from the utility's field crews and is available in the utility's database (e.g. GIS database or asset management records). It must be recognized that not all water main failures are reported and some of them remain out of sight. While preparing the data, it is important to note that

Figure 14.1 Example of a water district distribution system data. White lines represent water mains and dots reported water main failures.

some water utilities keep records of water main failures for those that have been already replaced. For example, a cast iron (CI) pipe segment was installed in 1950, then it failed three times between 1950 and 1990 and was replaced in 1995 with an asbestos cement (AC) pipe. Some utilities will still show the three failures assigned to the AC pipe segment without specifying that these failures occurred in the old CI pipe that is not currently in place. Therefore, it is important to verify that all water main failures in the dataset you will use are actual attributes of the currently installed water mains.

Depending on the water utility capability to collect information, water main failure databases may include the following data: reported date, location, type of failure and repair cost. For the purposes of these techniques, reported date and location are the minimum parameters to be collected.

14.2.2 Water mains

In addition to collecting water main failure data, it is common practice for water utilities to have a database of their assets such as reservoirs, tanks, valves, pump stations and water mains (as covered in previous chapters). The water main databases could have various degrees of information, depending on the water utility, for example diameter, material, pressure rating, date of installation, bedding, depth of installation, joint details, and so on. Our experience tells us that material, diameter and installation date are the minimum parameters to be collected for a reliable spatiotemporal analysis.

14.2.3 Base map

A base map or reference map contains local geographic features and is used to overlay the data we are analyzing to facilitate visualization. Depending on the software used, base maps are freely available or can be obtained from the software server (e.g. GIS). Base maps can contain streets, highways, parcels, neighbors, rivers, and so on. For this application, it is recommended to have a street and county/city base map to identify location of water mains and failures (Figure 14.1).

14.3 MULTILINEAR REGRESSION MODEL

14.3.1 Description of linear model

In Chapter 2, we covered the concept of regression, and we know that multiple curves can be fitted to create a model that represents the relationship between an independent variable 'x' and a dependent variable 'y'. Regression models can have multiple independent variables that all combined influence the dependent variable. We described earlier that water mains can fail due to multiple reasons and those failures are usually a combination of factors such as age, material, diameter, type of soil, and so on. In this section, we create a Multi-linear Regression (MLR) model to predict pipe longevity based on historical failure data.

14.3.2 Age based linear model

The age based linear model that we are going to set up is based on pipe longevity (i.e. pipe installation date till the first failure). This means that we will be able to make a prediction of how long a pipe will last under given conditions.

In the model, pipe longevity is selected as the dependent variable 'y'. In this context, pipe longevity can be defined as the difference in years between the installation date of the pipe and the reported date of the first water main failure. For example, water main 'A' was installed in May 1957 and there is a failure record in December 2008. The longevity of this pipe would be calculated using the difference between the failure year (2008) and the installation date (1957), a longevity of 51 years.

One of the disadvantages of this type of modelling approach is how to manage water mains that have failed on multiple occasions. As mentioned earlier, one of the principles of water main failure clustering is the fact that failures tend to occur close to earlier failures and in many cases caused by them. However, in this type of modeling we are trying to predict the first failure which is not related to previous failures but caused by the combination of internal and external factors discussed in the previous section(s). Therefore, to be consistent with one of the fundamental assumptions of regression modeling (independent samples), dependent failures within each pipe need to be removed from the dataset (Martínez García *et al.*, 2018) and the MLR model will only contain the first failure of each pipe segment. Dependent failures in this context refer to the second or subsequent failure that occurred in the same water main following the definition of Jacobs and Karney (1994). They defined dependent failures as those that occur within 20 m and 90 days from previous failures in the same pipe segment.

Depending on the amount of data available, the regression model can utilize independent variables. The general structure of this model is as shown in Equation (14.1). Table 14.1 contains a generic description of water main variables that can be included in the model:

$$\text{Pipe longevity} = \alpha_i + \beta_1\,\text{material} + \beta_2 * \text{diameter} + \beta_3 * \text{pressure}$$
$$+ \beta_4 * \text{season} + \beta_5 * \text{soil} + \beta_6 * \text{length} \tag{14.1}$$
$$+ \beta_7 * \text{air temperature} + \beta_8 * \text{water content} + \varepsilon$$

From your math classes, you remember there are three types of variables: *continuous, discrete and categorical.* A continuous variable can take any value, for example the height of your classmates. For

Table 14.1 Description of variables for modelling water main failures.

Pipe material	Diameters (mm)		Pressure (kPa)	Season	Soil Type	Pipe Age (years)	Pipe Length (m)
Asbestos	25	300	Varies from	Spring	Entisols	Varies from 1 to	Varies from
cement	50	350	0 to 1400	Summer	Alfisols	about 100 years	1 m to 500 m
Cast iron	75	400		Fall	Inceptisols		(typically)
Ductile iron	100	450		Winter	Vertisols		
PVC	150	500			Mollisols		
Steel	200						
W. iron	250						
Categorical variables						Continuous variables	

this model, some of the continuous variables will be pipe length and pressure. On the other hand, a categorical variable can only take certain values and is not necessarily arranged in a logical order. These variables are identified with levels or categories. For example, pipe diameter where each pipe size would constitute a 'level' of the variable 'diameter': 2 in (50 mm), 4 in (100 m), and so on. Another example of categorical variable is pipe material where a level will be each different material: asbestos cement, cast iron, ductile iron, PVC, and so on. For each variable and each level, a binary set of dummy regressors has to be included by assigning a 0 value or 1. Refer to Example 17.1 to understand how these dummy regressors help to set up the model.

One characteristic of categorical variables is that the model results will be based on the intercept values. In a linear equation $y = mx + b$, 'b' is called the 'y' intercept because it is the value where the line (represented by the equation $y = mx + b$) crosses the 'y' axis when the independent variable 'x' takes a value of zero. In an equation with categorical variables, since the independent variable cannot take a value of zero (there is no pipe with a diameter of 0 in (0 mm) and there is no material called 0), we need to assign certain values to these variables so the model can utilize those as the baseline scenario for intercept. If we set up the model with *the reference level being an asbestos cement pipe with a diameter of 6 in (150 mm)*, the model will calculate the intercept coefficient as the pipe longevity of a water main with this material/diameter combination. All other materials and diameters will have a coefficient which is based on the intercept value.

For example, in Table 14.2 there are partial results of a model that predict pipe longevity using two categorical variables: pipe material and pipe diameter. Because these variables are categorical, they cannot take continuous values as seen in Table 14.2. *The intercept coefficient is 31.3 years, which means that a pipe segment made of asbestos cement (AC), with a diameter of 6 in (150 mm) have an average longevity of 31 years.* The rest of the coefficients represent the difference in years between the intercept and other variables. For example, a water main with a diameter of 2 in (50 mm) has a coefficient of −5.31 years, which means that its estimated longevity is 5.3 years less than the intercept (31.32 − 5.31 = 26 years). In the case of material, the coefficient compares the longevity of other materials with asbestos cement pipes. Cast iron is estimated to last 27.1 years more for example. For any given water main, the model should be like:

- Pipe longevity $= \alpha_i + \beta_1\text{material} + \beta_2\text{diameter}$
- Pipe longevity $= 31.33 + \beta_1\text{material} + \beta_2\text{diameter}$

Table 14.2 Variable coefficient in years.

Variable	Coefficient
Intercept – AC	31.33
Intercept – 6 in (150 mm)	
Diameter 2 in (50 mm)	−5.31
Diameter 4 in (100 mm)	0.91
Diameter 6 in (150 mm)	Intercept
Diameter 8 in (200 mm)	−3.03
Diameter 10 in (250 mm)	8.26
Diameter 12 in (300 mm)	1.50
Asbestos cement	Intercept
Cast iron	27.10
S. Steel	15.45
Steel	20.02

Example 18.1

Given the following set of water mains, fill the following table using binary dummy regressors to develop the model

Pipe A: 6 in (150 mm), PVC, 3500 ft (1067 m), pressure rate 150 psi
Pipe B: 3 in (80 mm), Cast iron, 4500 ft (1372 m), pressure rate 200 psi
Pipe C: 8 in (200 mm), Ductile iron, 2000 ft (610 m), pressure rate 150 psi

Step 1: Identify the variables from the information provided: diameter, material, length and pressure rate.

Step 2: Identify the levels for each variable

Diameters – 3 levels: 3 in (80 mm), 6 in (150 mm) and 8 in (200 mm).
Materials – 3 levels: cast iron (CI), ductile iron (DI) and PVC
Length – no levels because it is a continuous variable
Pressure rate – no levels because it is a continuous variable

Step 3: Create a table adding the four variables and their respective levels.
Step 4: Add the dummy regressors for each categorical variable.

Variable	Level	Pipe A	Pipe B	Pipe C
Diameter	3 in (80 mm)	0	1	0
	6 in (150 mm)	1	0	0
	8 in (200 mm)	0	0	1
Material	Cast iron	0	1	0
	Ductile iron	0	0	1
	PVC	1	0	0
Length (ft)	No level	3500 (1067 m)	4500 (1372 m)	2000 (610 m)
Pressure rate (psi)	No Level	150	200	150

14.3.3 R linear regression model

As we covered in a few previous chapters, 'R' is a language and environment for statistical computing and graphics (The R Foundation, 2016). R is an open source software and provides a wide variety of statistical techniques with hundreds of extensions for more advanced processes. R can be used for linear and nonlinear modeling, statistical tests, classification, graphing, and so on. R is available as free software and runs on different operating systems such as MacOS and Windows, hence, it can be used in most computers. In this section, we are going to learn how to setup a linear model in R to predict water main longevity given a set of data. The full example with explanation is available in the online repository.

The following data contains data from a hypothetical water main failure database. Table 14.3 contains data from 180 water main failures, but we only show the selected rows as an example. Table 14.3 contains the reported date of failure and the installation date of the water main. With that information, the pipe longevity was calculated. Table 14.3 also contains water main's information such as diameter, material and length. In this particular dataset, the season of the failure is also included because we wanted to determine what the impact of air and soil temperature is on the different pipe materials that could possibly cause a failure.

Using R language and a basic code, a linear regression is generated and shown below. The water main longevity is the dependent variable and the data is stored with the name RegExampledata which is a name given by the user:

Table 14.3 Pipe failure dataset.

Longevity	Reported Date	Install Date	Diameter	Material	Season	Length
44	02/03/2001 00:00	14/11/1957 00:00	6	AC	Spring	105.514
41	15/10/1993 00:00	01/01/1952 00:00	4	STL	Fall	106.0233
59	05/01/2007 00:00	01/01/1948 00:00	2	CI	Winter	108.9022
62	22/09/2010 00:00	01/01/1948 00:00	2	CI	Fall	108.9022
59	05/01/2007 00:00	01/01/1948 00:00	2	CI	Winter	108.9022
49	06/12/2001 00:00	01/01/1952 00:00	2	CI	Winter	110.1174
62	28/10/2014 00:00	01/01/1952 00:00	2	CI	Fall	110.1174
55	21/12/2001 00:00	01/01/1946 00:00	2	CI	Winter	110.1796
54	24/05/2000 00:00	01/01/1946 00:00	2	CI	Spring	110.1796
50	11/09/2013 00:00	01/01/1963 00:00	6	AC	Fall	110.5255
34	27/09/1993 00:00	16/07/1959 00:00	6	AC	Fall	111.8343
52	30/03/1998 00:00	01/01/1946 00:00	6	CI	Spring	112.4503
50	06/03/1996 00:00	01/01/1946 00:00	6	CI	Spring	112.4503

$$RegExampledata < lm(Exampledata\$Age \sim Exampledata\$Diameter$$
$$+ Exampledata\$Material + Exampledata\$Season + Exampledata\$Length)$$

After R generates the regression model, the next step is to call the model results with the 'summary' function by typing *summary(RegExampledata)*. The end result is the following table (Figure 14.2). An explanation of each section is included below.

The first line includes the regression model with the variables we included:

```
Call:
lm(formula = Exampledata$Age ~ Exampledata$Diameter + Exampledata$Material +
Exampledata$Season + Exampledata$Length)
```

The residuals describe the distribution of longevity compared to the intercept.

```
Residuals:
Min     1Q Median 3Q  Max
-36.559 −6.144 1.161 8.518 23.441
```

The third section includes the coefficients. *Each coefficient measures the pipe longevity compared with the reference levels (in years)*, the first column depicts the variables and their respective levels. For diameter, there were four levels (2 in (50 mm), 4 in (100 mm), 6 in (150 mm) and 8 in (200 mm)). You can see diameter 6 in is not listed, this is because it is already included in the intercept. Take a look at Problem 14.1 to learn how to use the function "relevel" to set the intercept in R. In this example, the intercept contains a water main made of asbestos cement (AC) and 6 in (150 mm) diameter that failed during the spring. The coefficient indicated that the model predicts a longevity of 40.2 years before failure for a water main with these characteristics.

Let us start comparing the other diameters, the coefficient for a 2 in (50 mm) is −0.67 years. This indicates that an asbestos cement, 2 in (50 mm) in diameter will last 40.2 years minus 0.67 years = 39.53 years. With regard to the material, the cast iron pipe coefficient is 17.35 years, this means that CI water mains last more than AC. For example, a 6 in (150 mm) CI pipe will last (40.2 years + 17.5 years = 67.7 years). The length variable is the only one that is not categorical. This means it can take any value. The coefficient is −0.000028 and is multiplied by the actual pipe length (m or

```
Call:
lm(formula = Exampledata$Age ~ Exampledata$Diameter + Exampledata$Material +
    Exampledata$Season + Exampledata$Length)

Residuals:
    Min      1Q  Median      3Q     Max
-36.559  -6.144   1.161   8.518  23.441

Coefficients:
                            Estimate  Std. Error  t value              Pr(>|t|)
(Intercept)              40.21827868  4.83741163    8.314  0.000000000000027 ***
Exampledata$Diameter2    -0.66994340  3.01171795   -0.222              0.824
Exampledata$Diameter4    -2.49985208  2.63576691   -0.948              0.344
Exampledata$Diameter8    -0.19240419  3.08450507   -0.062              0.950
Exampledata$MaterialCI   17.35294206  2.68585703    6.461  0.000000001030988 ***
Exampledata$MaterialSTL  -0.15298068  2.93443689   -0.052              0.958
Exampledata$SeasonFall   -0.06785262  2.86827320   -0.024              0.981
Exampledata$SeasonSummer -0.05799121  2.98571686   -0.019              0.985
Exampledata$SeasonWinter  3.76128040  3.15713977    1.191              0.235
Exampledata$Length       -0.00002768  0.02097316   -0.001              0.999
---
Signif. codes:  0 '***' 0.001 '**' 0.01 '*' 0.05 '.' 0.1 ' ' 1

Residual standard error: 12.77 on 172 degrees of freedom
Multiple R-squared:  0.3358,    Adjusted R-squared:  0.301
F-statistic: 9.662 on 9 and 172 DF,  p-value: 0.000000000006901
```

Figure 14.2 Results from regression model.

ft depending on the data that you used). Therefore, the coefficient for a pipe with 1000 ft (305 m) of length would be ($-0.000028 \times 1000 = -0.28$ years). In general, the longer the pipe the more prone to water main failures and that is what the negative sign of the coefficient is indicating.

Another interesting data from this summary is the adjusted R-square which indicates how well the model fits the data.

Finally, another item to pay attention to is the *p*-value located in the last column on the right. The *p*-value for each term tests the null hypothesis that the coefficient is equal to zero (no effect). A low *p*-value (<0.05) indicates that you can reject the null hypothesis. Therefore, a variable with a low *p*-value is likely a meaningful addition to the model because *a small variation in this independent variable will produce a significant increase or decrease in pipe longevity*. On the other hand, a larger *p*-value suggests that a change in a variable is not associated with a pipe longevity change. In the results we can see that the cast iron *p*-value is low so water that mains made of this material are likely to have a larger pipe longevity based on the sign of the coefficient.

14.4 HOT SPOT ANALYSIS OF WATER MAIN FAILURES

Now that you have a good understanding of what a cluster of water main failures is (a region with an anomalously higher number of water main failures compared to its surroundings!), we are going to present a methodology to analyze clusters by utilizing the 'Hot Spot Analysis' tool in ArcGIS.

14.4.1 Hot spot analysis tool

The Hot Spot Analysis Tool was developed by the Environmental Science Research Institute (ESRI) to be utilized under their ArcGIS platform. The tool calculates the Getis-Ord Gi* statistic for each feature in a dataset. The Getis-Ord-Gi* is a statistic that measures the intensity of spatial clustering.

The Hot Spot Analysis Tool produces a *z*-score with its associated *p*-value, which indicates if features with either a high or low value are grouped together by looking at a feature within the context of neighboring features (ESRI, 2021). For example, in Figure 14.3 there are two distributions of events (red circles), which could represent recorded events such as crimes or water main failures. If we place a grid over the distribution and get a count of the number of events per square, we can see that the cell in the center has five events. In the context of the Hot Spot Analysis tool, a feature with a high value such as the red cell could be interesting to analyze but may not be detected as a statistically significant

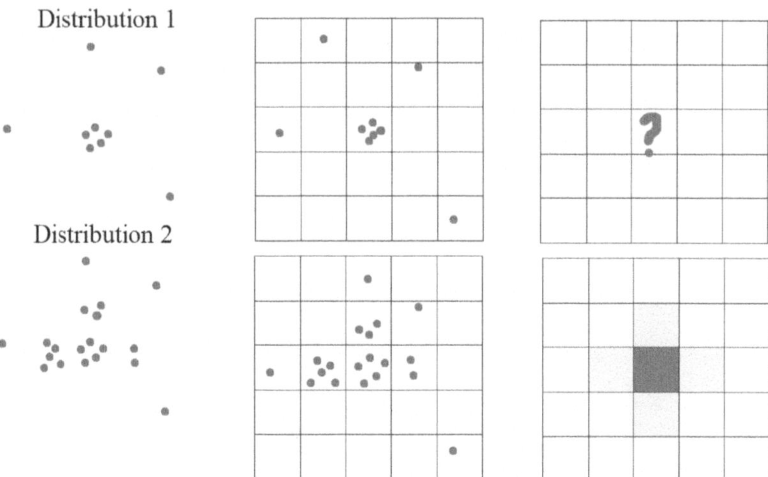

Figure 14.3 Two-point distributions and how the hot spot tool classifies clustering.

hot spot. The tool will only identify a feature as a hot spot if it is surrounded by other features with high values as well. Beyond assessing the density of points in a given area, hotspot techniques also measure the extent of point event interaction to understand spatial patterns (Baddeley, 2010). The local sum of a feature and its neighbors (light gray cells) is compared proportionally to the sum of all features (ESRI, 2021), when the sum is very different from the average local sum and the difference is too large to be the result of random chance, a *hot spot* is identified. The same occurs with features with a low count. When a feature with a low count is surrounded by others with low counts as well, and the difference of the local area compared to the average is too low to be the result of random chance, a *cold spot* is identified.

Because the data analysis to identify hot spots is based on quantifying the value of a feature (number of events occurring within an area), it is necessary to aggregate data spatially. This means that we must create a polygon where all the events are aggregated. There are multiple ways to achieve this. An option could be to create a circular area around each failure (buffer) and count the number of events. The end result in this case would be a hot spot definition per event. Another approach is to create a square grid and have each cell to be a polygon where the number of events will be counted. In the practice problems, the creation of grid to aggregate data is included to visualize the process. The result is shown in Figure 14.4.

Once the features are aggregated, the Hot Spot Analysis tool calculates the Getis-Ord Gi* statistic in each bin. The Gi* statistic returns a z-score for each cell (Figure 14.5). A z-score is a standard deviation and the p-value is the probability that the observed pattern is result of a random process (ESRI, 2021). Both the z-score and p-value are related with the normal distribution.

Very high or very low (negative) z-scores, associated with very small p-values, are found in the tails of the normal distribution (ESRI, 2021). When we run this tool and it yields small p-values and either a very high or a very low z-score, *this indicates it is unlikely that the observed spatial pattern is caused by random pattern* (ESRI, 2021). If the z-score is very large, clustering of high values is occurring in that cell, which means that a cell in the grid and its neighbors have a higher number of water main failures than the average in the whole study area (i.e. hot spots).

Likewise, small negative z-scores are called cold spots and indicate clustering of low values, which means the number of water main failures is not high. The z-scores and associated p-values for each water main failure indicate whether the null hypothesis (i.e. that water main failures are following a random process) should be rejected based on given confidence levels (Martínez García *et al.*, 2018). A combination of high z-scores and low p-values indicate that the water main failures are exhibiting significant clustering rather than a random pattern. The detection of clustering can be a sign of phenomena that is causing a higher number of water main failures in a region.

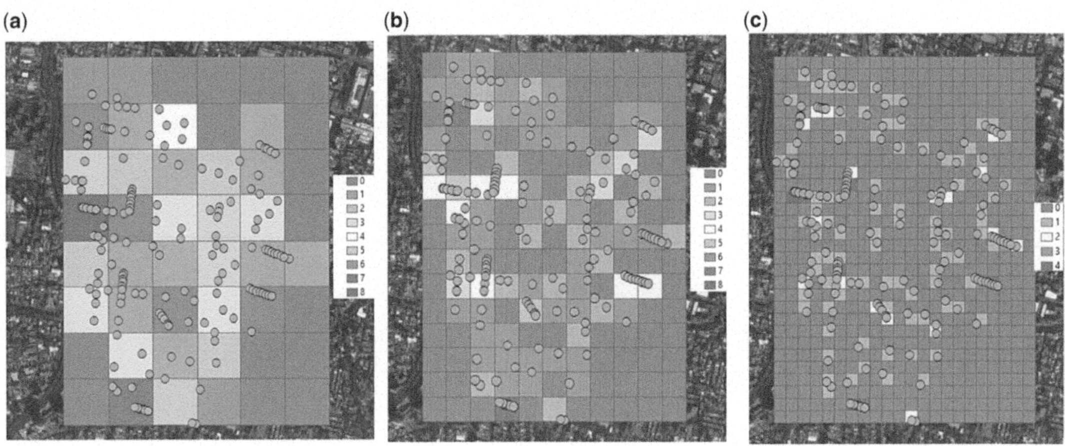

Figure 14.4 Water main failure count per bin with different grid sizes (A) 200 m, (B) 100 m and (C) 50 m.

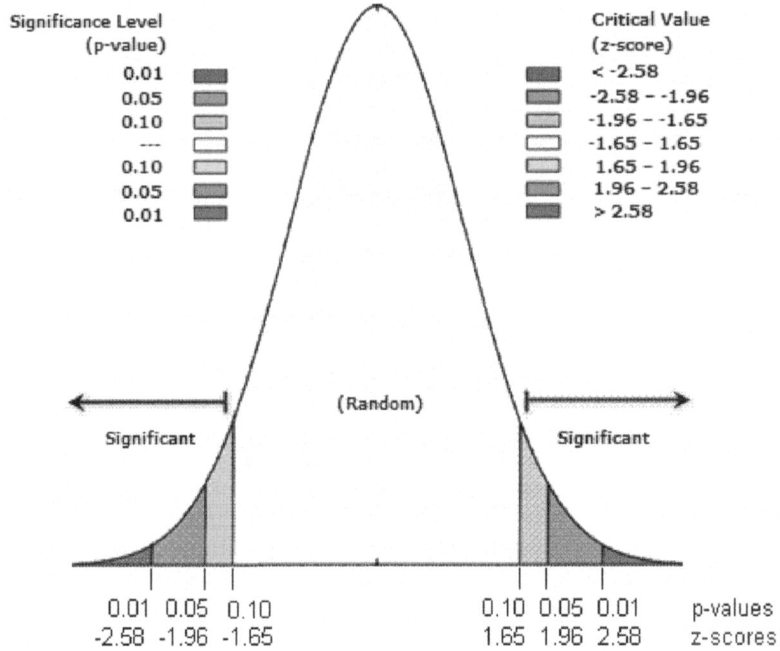

Figure 14.5 Normal distribution *p*-values and associated *z*-scores (image credit ESRI, 2021).

14.4.2 Interpretation of results

The result of this analysis will be a table (Figure 14.6) with the Hot Spot Analysis results and a new shapefile depicting the grid with the results from the tool. The table will have as many rows as bins were created. The attributes of the table are:

(A) FID and Source ID: ID of each bin.
(B) Shape: Type of bin object, it will always be a polygon.

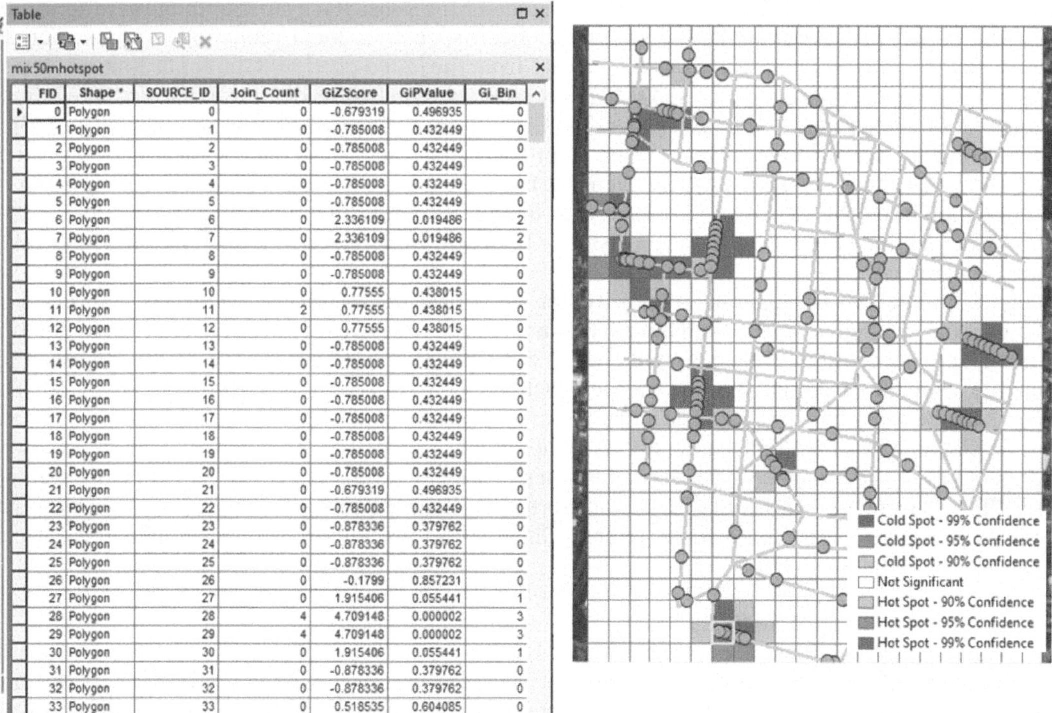

FID	Shape *	SOURCE_ID	Join_Count	GiZScore	GiPValue	Gi_Bin
0	Polygon	0	0	-0.679319	0.496935	0
1	Polygon	1	0	-0.785008	0.432449	0
2	Polygon	2	0	-0.785008	0.432449	0
3	Polygon	3	0	-0.785008	0.432449	0
4	Polygon	4	0	-0.785008	0.432449	0
5	Polygon	5	0	-0.785008	0.432449	0
6	Polygon	6	0	2.336109	0.019486	2
7	Polygon	7	0	2.336109	0.019486	2
8	Polygon	8	0	-0.785008	0.432449	0
9	Polygon	9	0	-0.785008	0.432449	0
10	Polygon	10	0	0.77555	0.438015	0
11	Polygon	11	2	0.77555	0.438015	0
12	Polygon	12	0	0.77555	0.438015	0
13	Polygon	13	0	-0.785008	0.432449	0
14	Polygon	14	0	-0.785008	0.432449	0
15	Polygon	15	0	-0.785008	0.432449	0
16	Polygon	16	0	-0.785008	0.432449	0
17	Polygon	17	0	-0.785008	0.432449	0
18	Polygon	18	0	-0.785008	0.432449	0
19	Polygon	19	0	-0.785008	0.432449	0
20	Polygon	20	0	-0.785008	0.432449	0
21	Polygon	21	0	-0.679319	0.496935	0
22	Polygon	22	0	-0.785008	0.432449	0
23	Polygon	23	0	-0.878336	0.379762	0
24	Polygon	24	0	-0.878336	0.379762	0
25	Polygon	25	0	-0.878336	0.379762	0
26	Polygon	26	0	-0.1799	0.857231	0
27	Polygon	27	0	1.915406	0.055441	1
28	Polygon	28	4	4.709148	0.000002	3
29	Polygon	29	4	4.709148	0.000002	3
30	Polygon	30	0	1.915406	0.055441	1
31	Polygon	31	0	-0.878336	0.379762	0
32	Polygon	32	0	-0.878336	0.379762	0
33	Polygon	33	0	0.518535	0.604085	0

Figure 14.6 Hot spot analysis example results.

(C) Join_Count: Number of events per bin. In our case the number of water main failures per bin.

(D) GiZscore: Calculated z-score for each bin. If the z-score is very large, clustering of high values is occurring in that cell, which in our context means that a cell in the grid and its neighbors has a higher number of water main failures than the average in the whole study area (hot spots).

(E) GiPValue: Calculated p-value for each bin.

(F) Gi_Bin: This attribute can take values from -3 to $+3$. These numbers do not represent a magnitude but instead represent the confidence levels based on each combination of z-score and p-value (see the legend included in Figure 14.6). For confidence levels above 99% to be a cold spot, bin value is -3, confidence levels between 95 and 99%, bin value is -2 and so forth.

14.5 SPATIOTEMPORAL ASSESSMENT OF WATER MAIN FAILURES

14.5.1 Introduction

Research of water main failures has determined that the probability of a water main that has already failed increases substantially after the first event (de Oliveira *et al.*, 2011; Goulter & Kazemi, 1988). In other words, spatial and temporal failure clustering could be caused by previous failures. The first failure in each cluster is considered to be an independent one, and subsequent adjacent ones are then treated as dependent failures (Berardi *et al.*, 2008; Bogárdi & Fülöp, 2011; Ganesan *et al.*, 2017; Jacobs & Karney, 1994; Martínez García *et al.*, 2018).

A history of failure could be a proper forecaster of future water main failures, that is, a water main that has broken one or more times will be a likely candidate to fail again, and the time intervals between failures are likely to become smaller (Goulter & Kazemi, 1988; see Chapter 13 for more details). This may be due to the inevitable disturbance of the subsurface surrounding the water main due to the repair process combined with other concerns, such as local reduction in the structural integrity of the water main, as well as water leakage from the failed portion (Goulter & Kazemi, 1988).

By using the space time cube tool in ArcMap, we can identify clusters of failures within a predefined distance '*d*' of each other, as well as a specified time interval '*t*' between one failure and the other. *Now we consider time-dimension as well!* By identifying space and time clusters, these clusters can be used to infer the effect of distance '*d*' between two failures on the time interval of occurrence between one failure and the other. The result of the spatio-temporal clustering will be identifying areas with a high risk of pipeline failure based on the distance between the failures, and also the time interval of failure occurrence

This technique can be adapted to identify hot spots of water main failures to assist in decision making for optimal replacement of pipelines. For instance, if there are intensifying hot spots of pipeline failures at a location, then those are spots which need immediate attention and can be assessed for repair/replacement based on cost and lifecycle factors (see Chapter 13 for cost-benefit analysis). The outcome of emerging hot spot analysis is the identification of areas which are in high risk of pipeline failure which may warrant replacement. Once emerging hot spots are identified, further steps can be taken to analyze the repair cost versus replacement cost. For example, a water utility can establish that if the replacement cost is 50% higher or more than the repair cost, then repair is a best option when compared to replacement. Again, refer to Chapter 13 for the various logics in pipeline optimal replacement strategies.

14.5.2 Emerging hot spot analysis tool

This spatiotemporal analytics tool is also provided in the catalog of ArcMap. The first step to use this tool is to create a space time cube. This cube is a 3D representation of geographical events through time. To analyze water main failures, a space time cube can be established by water district. In the cube, the (x, y) location of each failure is represented by the X and Y dimension of the cube, similar to a map. The vertical axis of the cube (Z) represents time. The newest events are located on top of the cube (Figure 14.7).

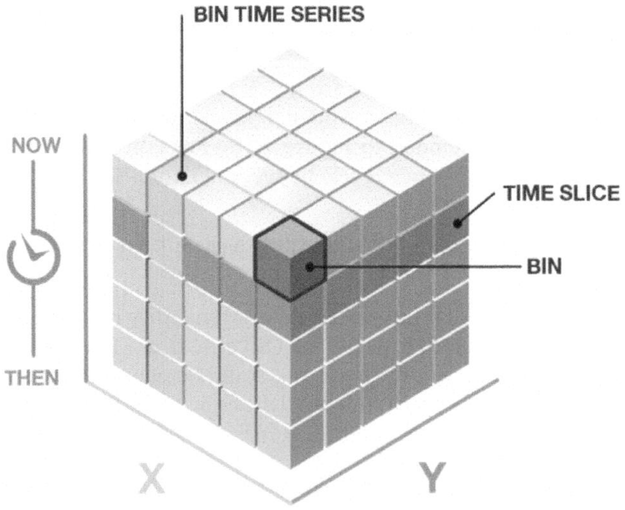

Figure 14.7 Parameters of the space time cube (image credit: ESRI, 2021).

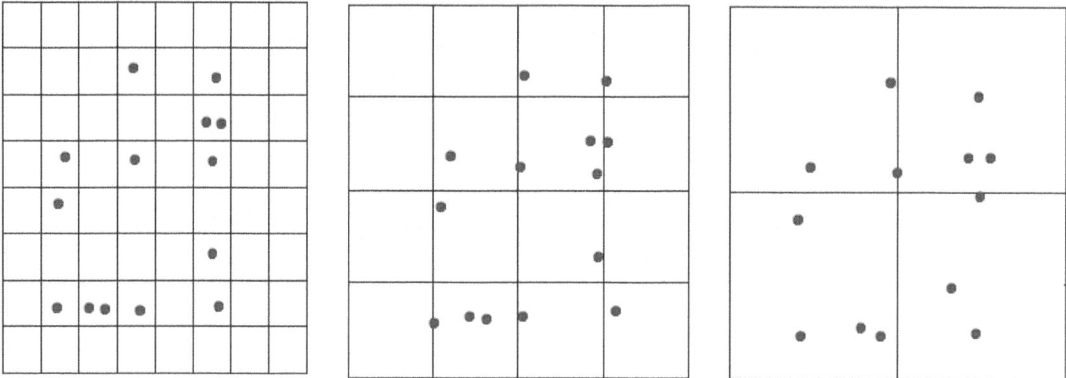

Figure 14.8 Examples of different bin size for the same distribution of point events.

For a given study area (city, neighborhood) the cube is divided horizontally and vertically to create bins. The vertical size of the bins represent time. This parameter can be adjusted depending on the variable we are studying. In the case of studying water main failures, one year intervals could be used to start the analysis. This interval is called *time step*. The tool will compare trends in the number of failures per time step. The horizontal size of the bins (x,y) represent a geographical area. It is recommended to test different sizes to select the optimum clustering intensity. If the bin dimensions are too small, no clusters would be detected because the count of failures within each bin would be low (Figure 14.8 left). If the dimensions are too large, no spatial variance would be detected as most of the bins would have a high number of failures (Figure 14.8 right). There is not a recommended bin size and different sizes should be tested depending on the type of water district (urban, suburban, rural), water main density and district size. Based on analysis previously carried out, we recommend starting with dimensions 2000×2000 ft (610×610 m) for the districts with low density of water mains (suburban areas) and 1000×1000 ft (305×305 m) districts with higher density of water mains (urban areas, city centers).

Once the space time cube is created, the tool calculates the number of failures within each bin and the Getis-Ord-Gi* statistic as described in the previous section. The only difference is that the Gi* statistic is calculated for each time step. To categorize a bin as either a hot spot or a cold spot, the tool also evaluates the failure count of neighboring bins using two parameters: *neighborhood distance and neighborhood time*. These two parameters define the extension of each bin's neighborhood both in space (x,y dimensions) and time (z dimension).

Again, there are no specific recommended values for these parameters, but we advise starting with an additional bin in each direction. Also, a good starting point is to let ArcMap decide an automatic configuration and then explore increasing or decreasing both parameters. *Caution should be taken if the neighborhood distance parameter is too large because the tool will evaluate a very similar area to the entire area of study, and it may not detect any hot spots.*

For example, given a water main failure dataset, it was decided to create space time cube where spatial bin dimensions are 400×400 m. Then, it was decided to divide the cube temporally in intervals of 1 day (Figure 14.9 left). The total extension of the cube would be 1600×1600 m and 4 days in the vertical dimension. For the middle bin, if you set the neighborhood distance to 801 m, the spatial neighbors (gray) will extend two bins to each direction and one bin diagonally as shown in Figure 14.9. In addition, there are temporal neighbors (dark gray). If we set up the neighborhood time step to 2, the temporal neighbors will be all bins in the same location as the target bin and its spatial neighbors (gray bins on top of the cube) for the two preceding time periods as shown below (Figure 14.9 right, dark gray bins).

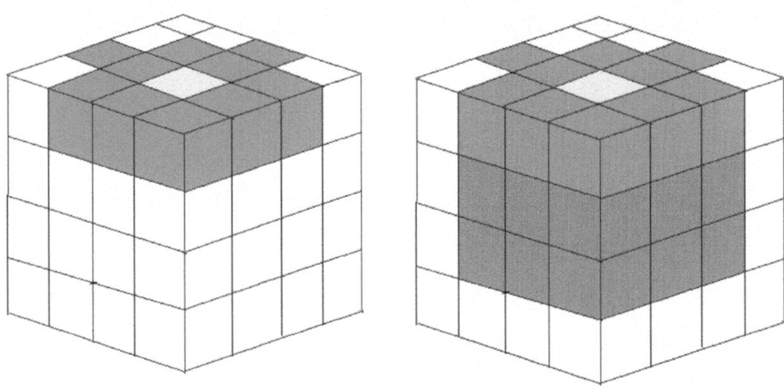

Figure 14.9 Extent of neighborhoods based on a neighborhood distance of 801 m (gray bins) and a neighborhood time of 2 (dark gray bins).

After the space time cube is created, within each bin (column), the events or points are counted and the trend for bins across time at each location is measured using the *Mann–Kendall* statistic. The *Mann–Kendall* trend test is performed on every location with data as an independent bin time-series test (ESRI, 2021). This test compares the number of events per bin for the first time period against the second one. If the second period count is larger, the result is a +1 (increasing water main failures), if the second period is smaller, the result is a −1 (decreasing water main failures). If the counts are similar between time periods, the result is zero. The hypothesis for this test is that the expected sum is zero, meaning that there is no trend in the count over time. The observed sum is compared to the expected sum to determine if the difference is statistically significant or not (ESRI, 2021). The trend for each bin is recorded as a z-score and a p-value (ESRI, 2021). A small p-value indicates the trend is statistically significant. The sign associated with the z-score determines if the trend is an increase in bin counts (positive z-score) or a decrease in bin counts (negative z-score) (ESRI, 2021).

With the resultant trend z-score and p-value for each location with data, and with the hot spot z-score and p-value for each bin, the Emerging Hot Spot Analysis tool categorizes each study area location in 16 different groups (ESRI, 2021). The description of the groups is available on the ArcGIS website; https://pro.arcgis.com/en/pro-app/2.7/tool-reference/space-time-pattern-mining/learnmoreemerging.htm. Depending on the type of variable that you are analyzing each of these categories may be helpful to study. Figure 14.10 shows a typical example of the results of the Emerging Hot Spot Analysis tool. The purple dots represent water main failures and the black lines represent water mains. The cells created using the 'Create Space Time Cube' can be seen as well. Depending on the temporal pattern of cold and hot spots, ArcGIS automatically categorized the bins in one of the 16 groups mentioned earlier. Let us take a look at the categories in the following section

14.5.3 Interpretation of results

Among all the categories that the tool offers, the following categories are the most useful to analyze water main failures because they provide information about increasing failure patterns:

- **New Hot Spot**: A location that was identified as a statistically significant hot spot for the final time step but has never been identified as a statistically significant hot spot before. This category would allow identification of an area with recent water main failures.
- **Consecutive Hot Spot**: A location where the last bins in the time series are identified as statistically significant hot spots but historically has not been a hot spot.

Figure 14.10 Emerging hot spot analysis results for a water distribution district.

- **Intensifying Hot Spot:** A location that has been a statistically significant hot spot for 90% of the time-step intervals (ESRI, 2021), including the final time step. In addition, the intensity of clustering of high counts in each time step is increasing overall and that increase is statistically significant (ESRI, 2021).
- **Persistent Hot Spot:** A location that has been a statistically significant hot spot for 90% of the time-step intervals with no discernible trend indicating an increase or decrease in the intensity of clustering over time (ESRI, 2021).

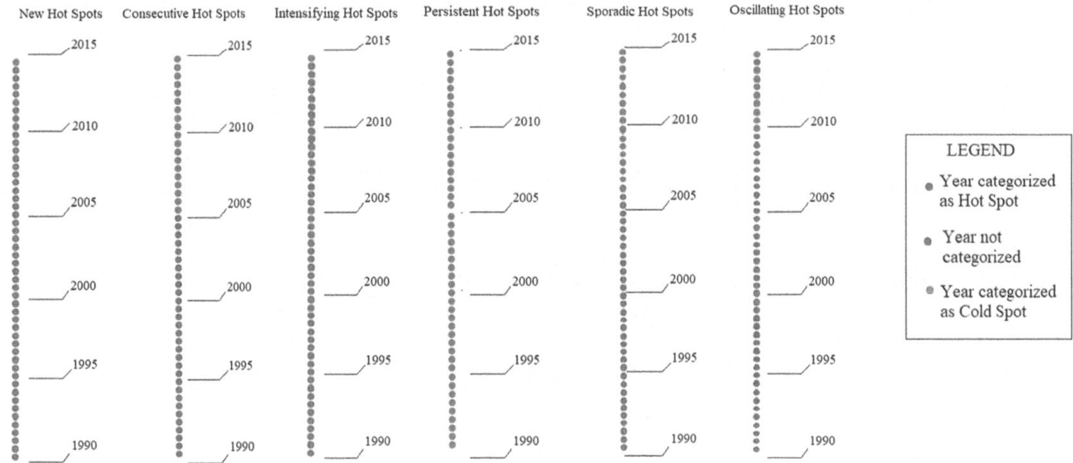

Figure 14.11 Visualization of six categories of patterns detected by the emerging hot spot analysis tool.

To help you to understand these definitions, Figure 14.11 (Martínez García *et al.*, 2019) explains six different bin locations that were classified using the definitions above. The time series was 25 years (1990–2015) and the time-step was one year.

On most occasions, the definitions above may not provide enough information to study spatiotemporal trends in the area. The Emerging Hot Spot Analysis tool provides a summary table with attributes related to each location. The results are compiled for all time steps for each of the bins in space, therefore each row in the results table provides a summary of all the time series for a location (Figure 14.12). The attributes of the table are:

(a) FID: Identification of bin.
(b) Shape: Type of object, for this tool it will be a polygon representing a bin.
(c) Shape length: Size of each bin.
(d) Shape area: Area of each bin.
(e) Category and pattern: Classification made by ArcMap based on the definitions above.
(f) Trend_Z: Z-score of the trend for each bin. If positive means that the number of failures is increasing through time for that bin. Depending on the magnitude it will indicate if the trend is statistically significant.
(g) Trend_P: Probability associated with the trend *p*-value based on the normal distribution.

A final step to help understand what these patterns mean is to utilize the Visualize Cube in the 3D tool. This tool allows the user to see a three-dimensional model of the results from the emerging hot spot analysis. The user can see the bins created along the area of study extruded to vertical (*z*-axis) to represent time. Each vertical partition represents a time step. For each individual location the tool will indicate if for a time step the bin was classified as hot spot or cold spot. The cube can be visualized in ArcScene, where the size of each bin can be adapted a user-defined figure, size and color. In Figure 14.13, the left image shows an example of the Emerging Hot Spot Analysis for a city. The image on the right shows the same layer along the extruded time series with its respective bins. Each column represents a location. The Space Time Cube can be represented as a series of extruded columns to identify each location separately. ArcScene allows the user to rotate the view to better understand clustering patterns (Figure 14.13).

Figure 14.14 shows an isometric view of the 'Space Time Cube' viewed as a series of extruded columns that represent time. For example, in the map, the red cells represent Consecutive Hot Spots,

FID	Shape	OBJECTID	Shape_Leng	Shape_Area	LOCATION	CATEGORY	PATTERN	PERC_HOT	PERC_COLD	TREND_Z	TREND_P
37	Polygon	910	799.9984	39999.84	17506	2	Consecutive Hot Spot	23.076923	0	1.046278	0.295433
38	Polygon	920	799.9984	39999.84	17636	2	Consecutive Hot Spot	23.076923	0	1.105046	0.26914
39	Polygon	925	799.9984	39999.84	17660	2	Consecutive Hot Spot	30.769231	0	3.247886	0.001163
40	Polygon	926	799.9984	39999.84	17661	2	Consecutive Hot Spot	30.769231	0	3.38075	0.000723
41	Polygon	928	799.9984	39999.84	17667	2	Consecutive Hot Spot	23.076923	0	2.89265	0.00382
42	Polygon	944	799.9984	39999.84	17828	2	Consecutive Hot Spot	15.384615	0	2.589831	0.009602
43	Polygon	981	799.9984	39999.84	18163	2	Consecutive Hot Spot	23.076923	0	1.105046	0.26914
44	Polygon	998	799.9984	39999.839999	18442	2	Consecutive Hot Spot	23.076923	0	0.190757	0.848716
45	Polygon	1048	799.9984	39999.84	19085	2	Consecutive Hot Spot	30.769231	0	2.112657	0.03463
46	Polygon	1049	799.9984	39999.839999	19086	2	Consecutive Hot Spot	30.769231	0	2.179005	0.029331
47	Polygon	1053	799.9984	39999.84	19104	2	Consecutive Hot Spot	30.769231	0	2.466506	0.013644
48	Polygon	1059	799.9984	39999.84	19232	2	Consecutive Hot Spot	15.384615	0	3.106849	0.001891
49	Polygon	1060	799.9984	39999.84	19233	2	Consecutive Hot Spot	23.076923	0	3.038851	0.002375
50	Polygon	1066	799.9984	39999.84	19265	2	Consecutive Hot Spot	30.769231	0	2.236931	0.025291
51	Polygon	1067	799.9984	39999.84	19393	2	Consecutive Hot Spot	15.384615	0	2.709208	0.006744
52	Polygon	1134	799.9984	39999.84	20338	2	Consecutive Hot Spot	15.384615	0	0.468405	0.639495
53	Polygon	1209	799.9984	39999.84	21334	2	Consecutive Hot Spot	15.384615	0	3.946409	0.000079
54	Polygon	1252	799.9984	39999.84	21990	2	Consecutive Hot Spot	15.384615	0	0.468405	0.639495
55	Polygon	1258	799.9984	39999.84	22143	2	Consecutive Hot Spot	30.769231	0	3.78961	0.000151
56	Polygon	1260	799.9984	39999.84	22146	2	Consecutive Hot Spot	15.384615	0	-0.281207	0.778552
57	Polygon	1276	799.9984	39999.84	22302	2	Consecutive Hot Spot	23.076923	0	2.485479	0.012938
58	Polygon	1277	799.9984	39999.84	22304	2	Consecutive Hot Spot	38.461538	0	3.435734	0.000591
59	Polygon	1288	799.9984	39999.84	22463	2	Consecutive Hot Spot	23.076923	0	2.485479	0.012938
60	Polygon	1339	799.9984	39999.839999	23133	2	Consecutive Hot Spot	23.076923	0	2.170608	0.029961
61	Polygon	1340	799.9984	39999.84	23134	2	Consecutive Hot Spot	23.076923	0	2.170608	0.029961
62	Polygon	1341	799.9984	39999.84	23135	2	Consecutive Hot Spot	30.769231	0	2.112657	0.03463
63	Polygon	1412	799.9984	39999.84	24076	2	Consecutive Hot Spot	23.076923	0	3.206457	0.001344
64	Polygon	1422	799.9984	39999.84	24221	2	Consecutive Hot Spot	15.384615	0	2.979651	0.002886
65	Polygon	1451	799.9984	39999.839999	24560	2	Consecutive Hot Spot	46.153846	0	3.300628	0.000965
66	Polygon	1464	799.9984	39999.84	24722	2	Consecutive Hot Spot	23.076923	0	2.236931	0.025291

Figure 14.12 Typical result table of the emerging hot spot analysis tool.

which if you remember the definition referred to areas where the last bins were hot spots (red), meaning that in recent time steps the location has seen hot spots, in our case a significant number of water main failures. If you take a look at the red bins, the extruded columns are gray in the bottom, because at the beginning the location was not identified as a hot spot but as we move in time (vertically) we see a consecutive run of hot spots at the end (top of the column).

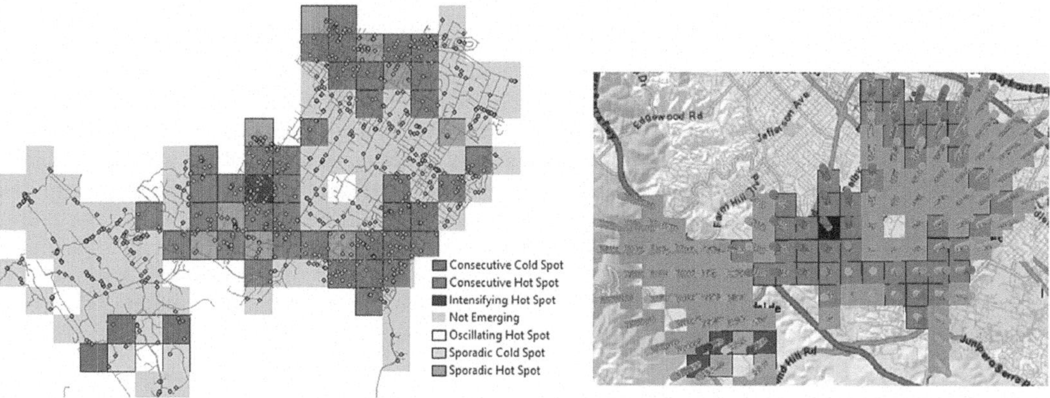

Figure 14.13 Example of the emerging hot spot analysis tool and 3D visualization of the space time cube (refer to online version of the book to visualize colors).

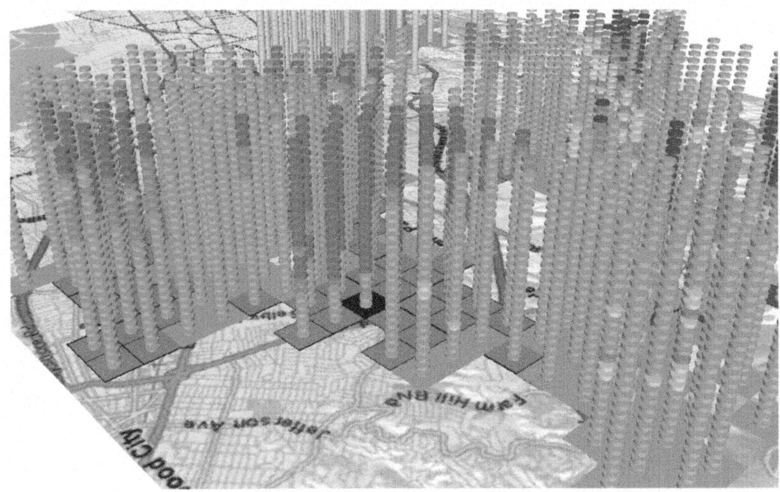

Figure 14.14 Isometric view of the 'Space Time Cube' for a water district (refer to online version of the book to visualize colors).

14.6 SPATIOTEMPORAL BASED BUSINESS RISK EXPOSURE ANALYSIS

A business risk exposure (BRE) is an approach that can be adopted to generate a risk prioritization ranking for individual water main segments. This method can be used in any water distribution system that applies a BRE based integrity program.

The BRE concept is the product of the probability or likelihood of failure (LOF) and the consequence of failure (COF). In this case, applied to water main failures. The likelihood of failure can be estimated based on historical data using one of the methodologies presented in previous sections. The consequence of failure can be estimated using a variety of factors depending of the water utility needs. Some examples of consequence of failure include, but are not limited to, environmental concerns, proximity to critical infrastructure, number of customers affected, failure to meet hydraulic requirements, water quality performance concerns, and so on.

14.6.1 Likelihood of failure

To calculate the Likelihood of Failure (LOF) we can utilize the Space-Time Pattern Mining tool in ArcGIS. The outcome of this step is to identify consecutive/increasing hot spot areas which are zones with a high concentration of failures during the analyzed times, which are likely to fail again.

The hypothesis of this analysis is that in an area identified as a consecutive or increasing hot spot, there may be an underlying phenomenon that is negatively affecting the integrity of water mains. Since the external factors that caused these failures (type of soil, weather, pressure, operations) are likely to remain the same, it is probable that as water mains age, more failures could appear in these areas. Figure 14.15 shows an example of the results of the Emerging Hot Spot Analysis for a water distribution district. The colors indicate the z-score for the entire time series. We can identify two areas with hot spots. The bin size for this example is 1000 ft (305 m). Then, each water main is assigned the z-score of the bin where it is located.

14.6.2 Consequence of failure

To estimate the consequence of failure, we are going to utilize a set of critical facilities in the study area. In more refined analysis other parameters could be used to evaluate the consequence of failure

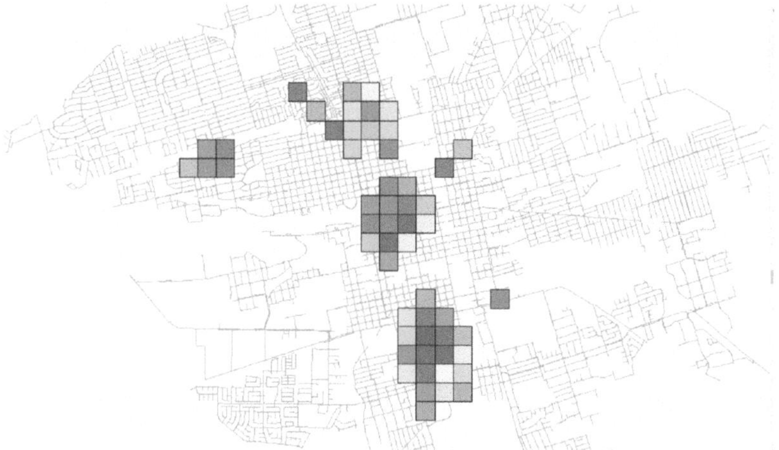

Figure 14.15 Emerging Hot spot analysis z-scores to be used as LOF. Blue bins represent lower z-scores and red bins higher z-scores (refer to online version of the book to visualize colors).

such as number of impacted customers, asset value, pressure or water quality. As a starting point we will work with critical facilities that could be affected by water service loss such as airport boundaries, roads, highways, railroads, backyard easement, emergency centers, fire stations, hospitals, police stations, schools, and so on. Another group of critical infrastructure could be marine protection areas and water bodies such as lakes, reservoirs and groundwater contaminated areas to prevent water quality degradation from surface or groundwater leaching into potable water mains.

These data can be gathered as shapefiles (points, lines or polygons) from public databases. Figure 14.16 shows example of these datasets. The point features refer to critical facilities such as hospitals, schools and fire stations. The line features represent highways, important streets and railroads which can be negatively affected by a water main failure.

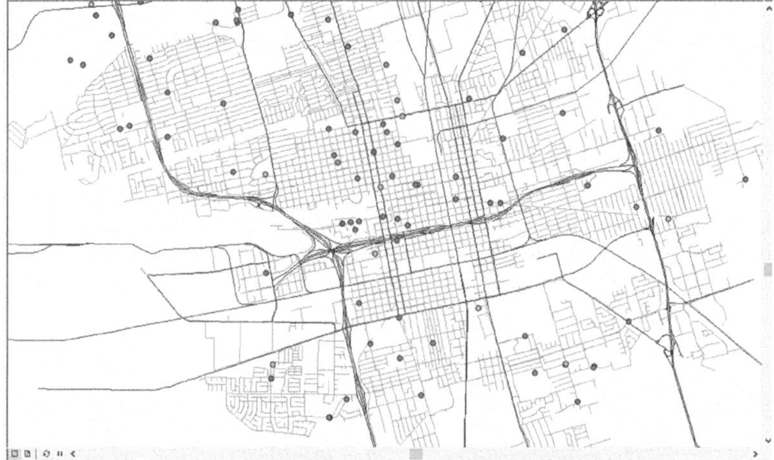

Figure 14.16 Critical facilities shapefiles. Dark lines represent highways and railroad tracks, dots represent critical infrastructure and light gray lines represent water mains.

Figure 14.17 Count of critical facilities per water main. Green 0–4, Yellow 5–8 and Red 8–10 (refer to online version of the book to visualize colors).

Once the spatial locations of these facilities were identified, the next step is to create a buffer around each water main segment. This step will help to count the number of critical facilities located in the zone of influence near each pipe segment. The size of the buffer can be determined based on the average lengths of the service lines connecting the water mains with the facilities or the distance between the longitudinal axis of the pipe and each facility's property line (Martínez García, 2019b). Consultation with a water utility (who are very well versed in the unique characteristics of the systems) will be very helpful. Figure 14.17 shows each water main with the count of critical facilities within the created buffer.

14.6.3 BRE matrix integrating LOF/COF

With the COF and LOF results, we could create a matrix to prioritize water main repairs and replacements. For example, Table 14.4 shows a categorization (Martínez García, 2019b) where it is proposed to categorize all water mains within a water district into six groups depending on the

Table 14.4 COF/LOF prioritization scheme (Martínez García *et al.*, 2018).

LOF				
	Consecutive Hot Spot – High z-score	Low priority, Re-assess individual cases 4	Close monitoring and maintenance 2	Immediate Risk Mitigation 1
	Consecutive Hot Spot – Lower Z-score	Low priority, Re-assess individual cases 4	Medium priority Re-assess indidivudal cases 3	Close monitoring and maintenance 2
	Other Hot Spot Categories	Minimum priority 5	Low priority, Re-assess individual cases 4	Medium priority Re-assess individual cases 3
	No pattern detected	Minimum Priority 6	Minimum priority 5	Low priority, Re-assess individual cases 4
	Category	***No critical facilities***	***0 to 10 critical facilities***	***11 or more critical facilities***
			COF	

Figure 14.18 Water mains classified based on the rating when combining COF/LOF.

combination of LOF/COF. Group 1 represents the water mains which should be prioritized because their probability of failure is high (failures occurring in nearby areas) and the number of affected facilities in case of water service stops is high. The categories in this table will be particular for each case depending on the water district and the available resources to perform the recommended water main replacements.

The final result would be to add the ratings into the water main pipelines as a new attribute for easier visualization (Figure 14.18). By incorporating this concept, water utilities can obtain significant new insights into enterprise asset management program design and management, deployment, implementation and execution. Implementing this type of programs to focus on maintaining water main integrity can save water utilities significant resources in avoided failures, reduced water loss, energy savings and customer service benefits. In particular, this methodology could be useful for water utilities that are expanding their water mains replacement programs for future budgets to allow for future replacements of pipelines identified as high risk.

14.7 PRACTICE PROBLEMS

14.7.1 R linear regression model

Let us develop a model that predicts pipe longevity using R given a dataset of water mains failure datasets. The attribute table for the water main set is included below. The variables include: diameter, material, reported date of failure, installation date of water main and length.

Step 1. Calculate the observed pipe longevity in Excel by subtracting the year of installation to the year of the failure: =YEAR(Failure)-YEAR(Installation).

Step 2. Sort all the data by Reported Date and add a column called 'Season', depending on the season add the corresponding name to each row in the database.

Step 3. Save the spreadsheet as a CSV (comma separated value) file (see table in Online Repository).

Step 4. Review the dataset. It is always recommended for any regression problems to review the data that will be used to see if it is complete, accurate and correct. Another factor to review, with categorical variables, is to make sure a certain variable takes more than a value. For example, a regression cannot be run if all water mains have the same diameter.

Step 5. In addition to step 4, we must make sure that there is enough sample size when carrying out the regression for all combinations. A recommendation could be to create a pivot table between material and diameter at least, to analyze how many elements are there for each combination

(Table 14.5). The pivot table will count the number of water mains for each combination of diameter and material. Some combinations are more extensive than others. In the table, remove the combinations with a small sample size (typically *n* less than 10). Results from this pivot table indicate that ductile iron elements must not be included in the dataset as well as transient pipes. It is also suggested to remove elements with diameters 10 and 20 in due to smaller datasets. Doing this prior to running the R model is highly recommended to avoid problems (i.e. unreliable modeling!) when running the regression.

Table 14.5 Summary of data.

Count of diameter	Column labels						
Row labels	2	4	6	8	10	20	Grand total
AC		13	48	17	2		80
CI	35	11	21	2	2		71
DI			2	1			3
STL	5	18	8	4		2	37
TRANS			6	1			7
Grand total	**40**	**42**	**85**	**25**	**4**	**2**	**198**

Step 6. Open R and create a new Directory to start a new project. It is recommended to save the backup data and the R project in the same folder.

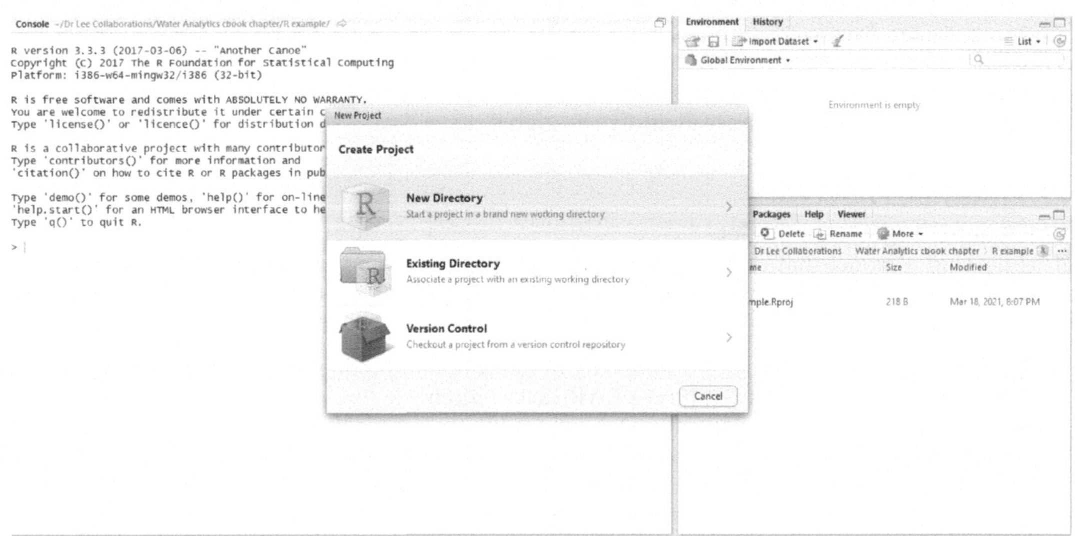

Figure 14.19 'R' welcome page.

Step 7. Create a new R Script and save it into the same folder.

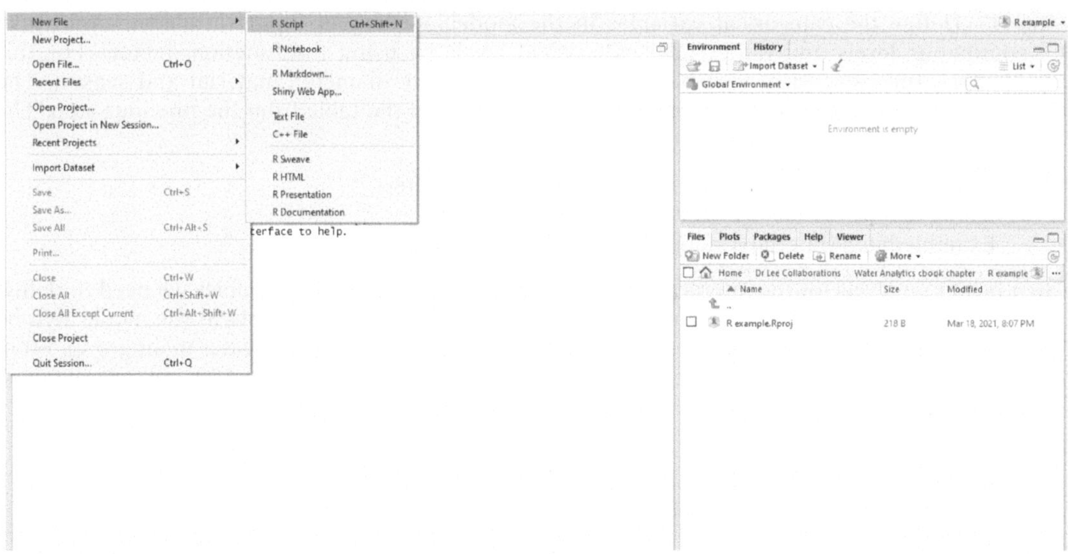

Figure 14.20 Opening a new file in 'R'.

Step 8. Let us start going through a simple R code to develop the model. The first step is to set a working directory. It is recommended to create a separate folder for the previously created. csv file and the R model. To call the directory type 'setwd("C:/…..)'. Each time you type an instruction, click on the 'Run' button.

Step 9. Install the library 'dplyr' by inserting the following command: **'install.packages**('dplyr')'. Once installed, this command can be deleted. After installation, the library will need to be called by typing 'library(dplyr)'.

Step 10. Then below insert 'rm(list = ls())'. This will erase any previous data.

Step 11. Name the model and all the file with the data to be used. In this case we are going to name the model 'Exampledata' and the.csv file where the data is stored is called 'rexampledataQAQC. csv'. These names will be set by the user:

Exampledata = data.frame(read.csv('rexampledataQAQC.csv'))

Step 12. Review the file data by typing 'head (…)'. Insert the name of your model, in this case 'head(Exampledata) '. The Console area below the code will show an extract of the water main failure data in the spreadsheet. Verify the data is accurate. In our example, the Console will show the following information:

```
> head(Exampledata)
  Age      ReportedDate        InstallDate Diameter Material Season   Length
1 59 05/01/2007 00:00 01/01/1948 00:00        2       CI  Winter 108.9022
2 62 22/09/2010 00:00 01/01/1948 00:00        2       CI    Fall 108.9022
3 59 05/01/2007 00:00 01/01/1948 00:00        2       CI  Winter 108.9022
4 49 06/12/2001 00:00 01/01/1952 00:00        2       CI  Winter 110.1174
5 62 28/10/2014 00:00 01/01/1952 00:00        2       CI    Fall 110.1174
6 55 21/12/2001 00:00 01/01/1946 00:00        2       CI  Winter 110.1796
> |
```

Figure 14.21 Extract of the data in 'R'.

Step 13. Define the categorical variables in the model. Remember these variables cannot take continuous levels and rather take categorical levels by using the function 'factor'. For this example, the categorical variables that we identified are diameter, material and season. The names should be similar to the names in the.csv file and the table from the previous step. The instruction will look like this:

> Exampledata$Diameter = factor(Exampledata$Diameter)
> Exampledata$Season = factor(Exampledata$Season)
> Exampledata$Material = factor(Exampledata$Material)

Step 14. Set up levels for the intercept. When working with categorical variables we need to define reference levels which will be taken by the intercept. All the coefficients in the model will be referenced to these levels. It is recommended to use the most common water main group. From the analysis on Step 5, we can see that AC pipes with a diameter of 6 in (150 mm) are the most common. To define the reference levels, we use the instruction 'relevel':

> Exampledata$Diameter = relevel(Exampledata$Diameter,ref = '6');
> Exampledata$Material = relevel(Exampledata$Material,ref = 'AC');
> Exampledata$Season = relevel(Exampledata$Season,ref = 'Spring')

Step 15. Set up the regression model using the function 'lm'. The regression model should look like this:

> Pipe longevity = α_i + $\beta 1$*material + $\beta 2$*diameter + $\beta 3$*season + $\beta 4$ * length + residual.

In R, the model is going to look like this:

> RegExampledata < lm(Exampledata$Age~Exampledata$Diameter + Exampledata$Material + Exampledata$Season + Exampledata$Length)

Step 16. Call the regression model results with the 'summary' function by typing summary (RegExamp ledata).

```
call:
lm(formula = Exampledata$Age ~ Exampledata$Diameter + Exampledata$Material +
    Exampledata$Season + Exampledata$Length)

Residuals:
    Min      1Q  Median      3Q     Max
-36.559  -6.144   1.161   8.518  23.441

Coefficients:
                          Estimate  Std. Error  t value     Pr(>|t|)
(Intercept)             40.21827868  4.83741163    8.314  0.000000000000027 ***
Exampledata$Diameter2   -0.66994340  3.01171795   -0.222              0.824
Exampledata$Diameter4   -2.49985208  2.63576691   -0.948              0.344
Exampledata$Diameter8   -0.19240419  3.08450507   -0.062              0.950
Exampledata$MaterialCI  17.35294206  2.68585703    6.461  0.000000001030988 ***
Exampledata$MaterialSTL -0.15298068  2.93443689   -0.052              0.958
Exampledata$SeasonFall  -0.06785262  2.86827320   -0.024              0.981
Exampledata$SeasonSummer -0.05799121 2.98571686   -0.019              0.985
Exampledata$SeasonWinter  3.76128040  3.15713977    1.191              0.235
Exampledata$Length      -0.00002768  0.02097316   -0.001              0.999
---
Signif. codes:  0 '***' 0.001 '**' 0.01 '*' 0.05 '.' 0.1 ' ' 1

Residual standard error: 12.77 on 172 degrees of freedom
Multiple R-squared:  0.3358,    Adjusted R-squared:  0.301
F-statistic: 9.662 on 9 and 172 DF,  p-value: 0.000000000006901
```

Figure 14.22 Regression results in 'R'.

14.7.2 Create grid to extract aggregated data

Step 1. Pipeline leak data in GIS is checked to see if it has a time attribute (date, time of leak) associated with each leak in the shape file's attribute table.

OBJECTID *	SHAPE *	Reported Date	Failureyear	WaterMainFailures.Count
1	Point	26/05/1992	1992	1
2	Point	15/06/1992	1992	1
3	Point	23/02/1990	1990	1
4	Point	17/01/1991	1991	1
5	Point	09/01/1992	1992	1
6	Point	20/05/1998	1998	1
7	Point	13/09/2003	2003	1
8	Point	19/01/2006	2006	1
9	Point	27/09/2016	2016	1
10	Point	10/01/2018	2018	1
11	Point	21/11/2020	2020	1
12	Point	22/11/2021	2021	1
13	Point	11/01/1999	1999	1
14	Point	09/08/2007	2007	1
15	Point	21/12/2021	2021	1
16	Point	15/09/2021	2021	1
17	Point	11/01/2005	2005	1
18	Point	19/11/1990	1990	1
19	Point	24/11/1999	1999	1
20	Point	13/06/2006	2006	1
21	Point	30/05/2011	2011	1
22	Point	05/10/2011	2011	1
23	Point	23/06/2014	2014	1
24	Point	05/01/2010	2010	1
25	Point	31/07/2009	2009	1
26	Point	18/10/2015	2015	1
27	Point	10/01/2017	2017	1
28	Point	26/10/2018	2018	1
29	Point	25/04/2012	2012	1
30	Point	06/08/2019	2019	1
31	Point	14/02/2020	2020	1
32	Point	15/02/2014	2014	1
33	Point	08/01/1990	1990	1
34	Point	04/05/2019	2019	1
35	Point	13/11/1990	1990	1
36	Point	27/05/1991	1991	1
37	Point	22/04/1992	1992	1
38	Point	22/04/1992	1992	1
39	Point	25/05/1992	1992	1

(0 out of 166 Selected)

WaterMainFailures

Figure 14.23 Water main failure attributes.

Step 2. The Measure tool in ArcMap is used to measure the distance between each pipeline leak/ break for a few selected leaks to get an estimate of how spatially separated the leaks are. More examination is to be done to estimate the average distance between the leaks to decide on the distance for creating buffers for identifying clusters of spatial data.

Step 3. Assigning leaks to a spatial cluster/group is done by grouping pipeline leaks within a certain specified distance of each other. To perform spatial clustering of pipeline leak data in GIS, we will create a grid. Separately, we could use the buffer tool in ArcMap 10.3.1, to create buffers of a certain specified distance say 300 ft (91 m) (assumed based on step 2) with the pipeline GIS data with the leak data information along with their spatial coordinate as the input feature. Then, a second step is to find clusters of leaks. This is to be done using the spatial join tool in ArcMap which can be used to find the number of leaks within each 300 ft (91 m) buffer.

Step 4. Use the Create Fishnet tool to create a grid. The template extent should be similar to the extent of the analysis area (see Figure 14.24). The size of each cell will depend on the findings of step 2. In this case we will use 100 m or 300 ft.

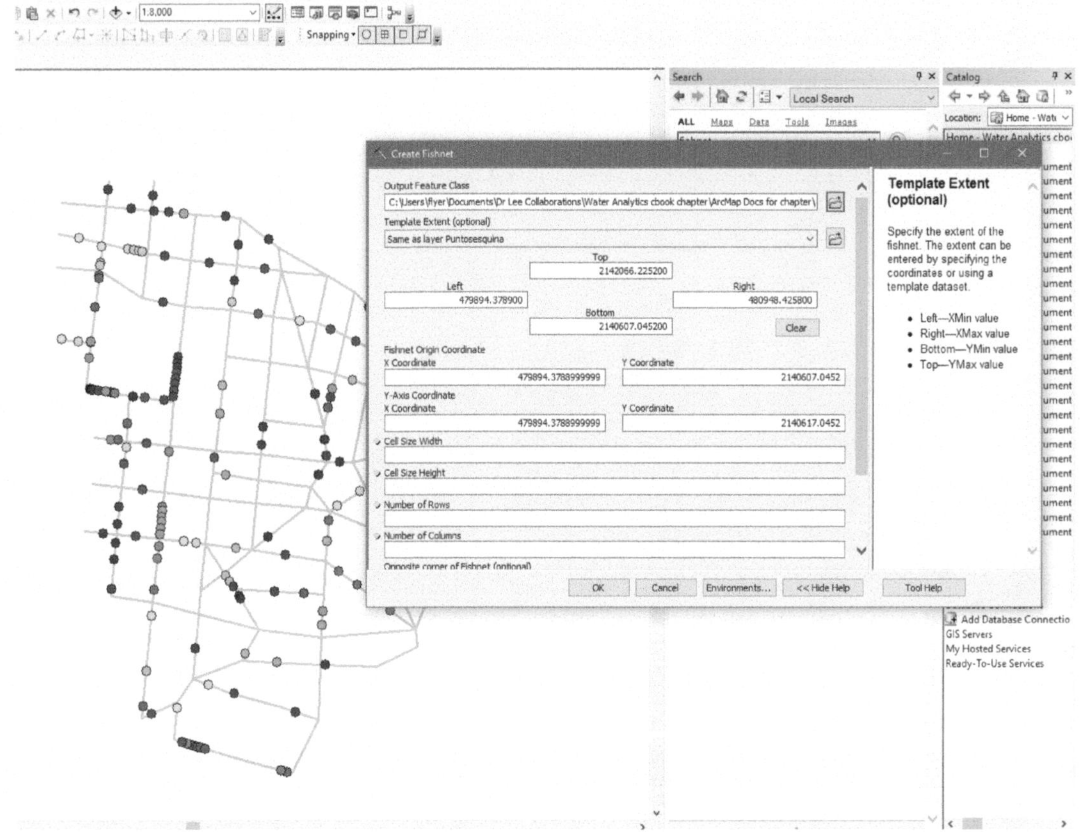

Figure 14.24 Creation of fishnet.

Step 5. Use the Create Fishnet tool to create a grid. In the Geometry Type choose polygon to generate a grid.

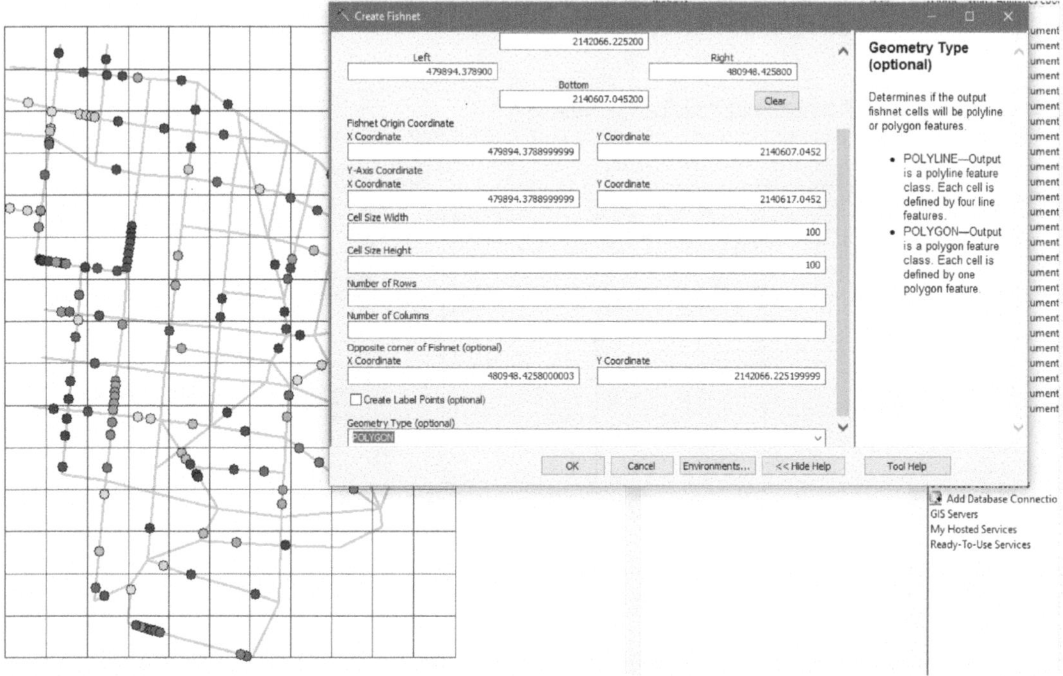

Figure 14.25 Creation of fishnet.

Step 6. To find clusters of leaks we will use the Spatial Join tool. The shapefile to be joined (water main failures) needs to have a count so the tool can add the failures. In this case, we should add an attribute and assign the number 1 to all the features. To do that, open the attribute table then click on 'table properties' and then 'Add Field'. Once the Field Calculator opens, add a new field. For example, 'WaterMainFailuresCount' and in the box below add number 1.

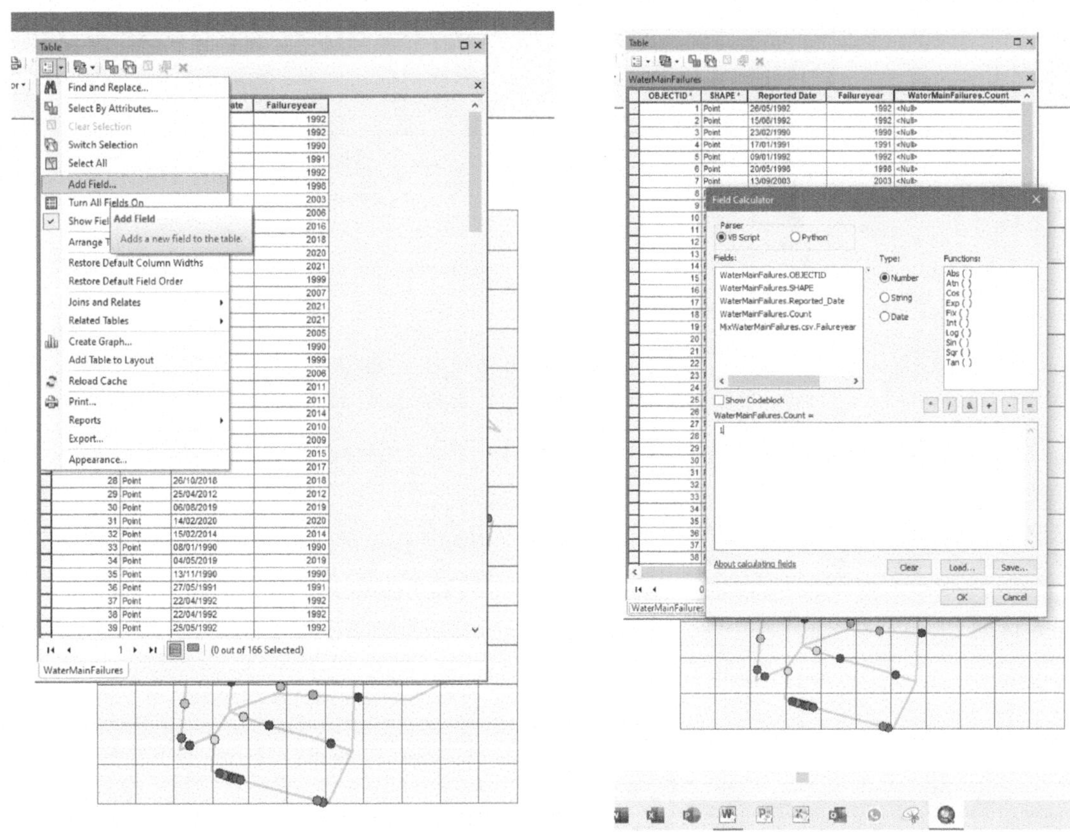

Figure 14.26 Adding a field with a value of 1 to count the number of failures.

Step 7. To find clusters of leaks we will use the Spatial Join tool. The target features are the bin that we just created, and we will join the water main failures. The target feature layer is the fishnet grid we created in last steps. The 'Join Features' will be the water main failure shapefile.

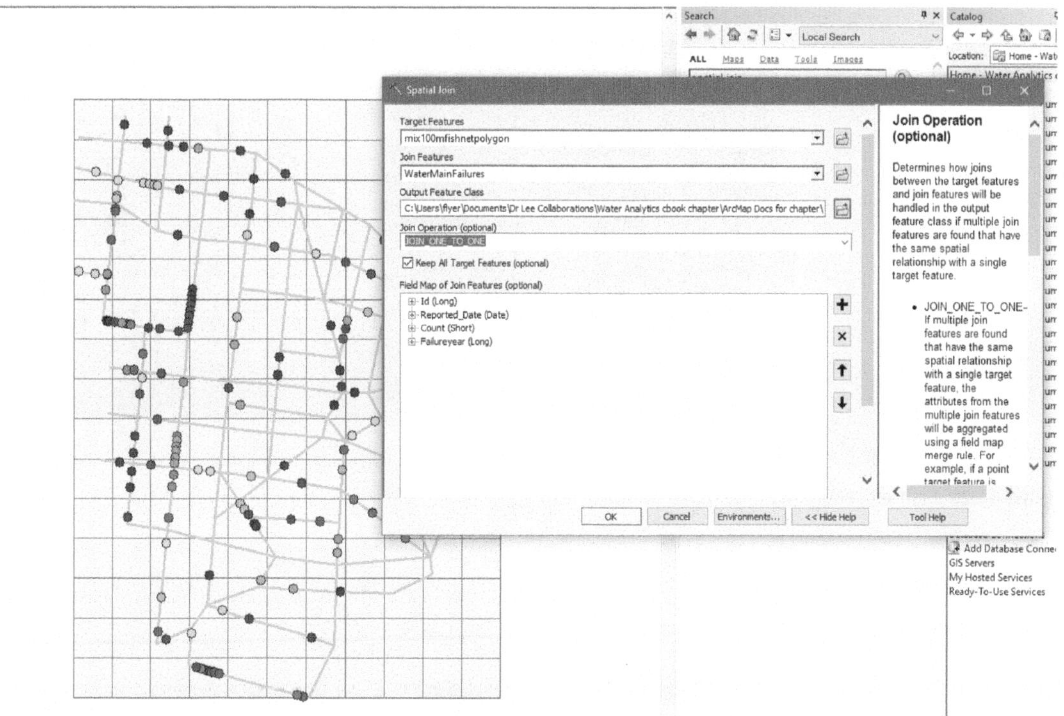

Figure 14.27 Spatial join tool.

Step 8. End Result using a 50, 100 and 200 m grid size. Note how the count of failures per increases as the bin size increase. The color scheme can be edited using the 'Layer Properties' menu when you right-click on the layer name.

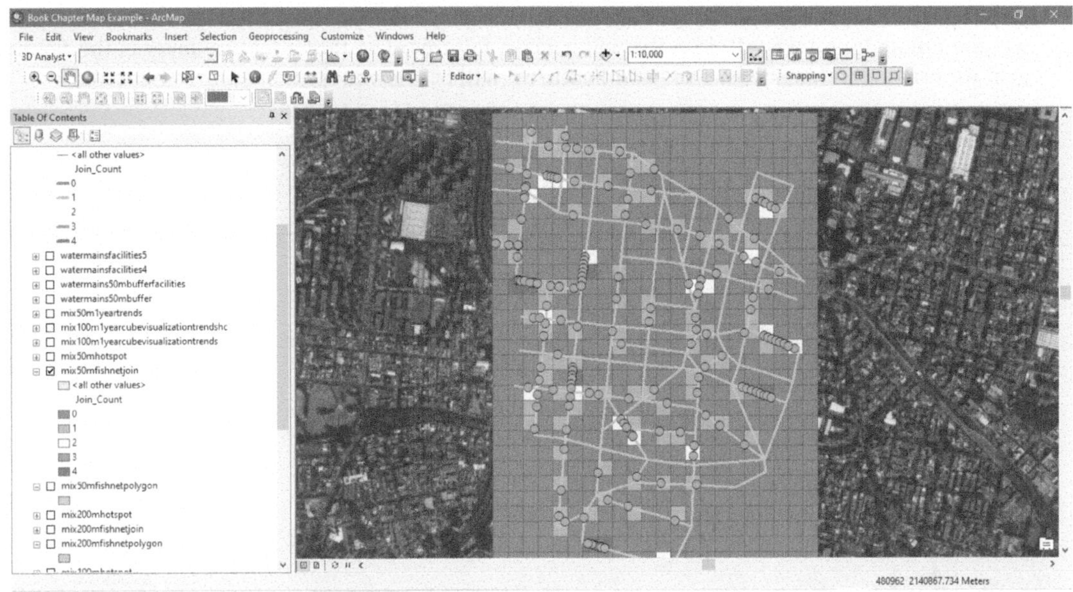

Figure 14.28 Fishnet results with a size of 20 m.

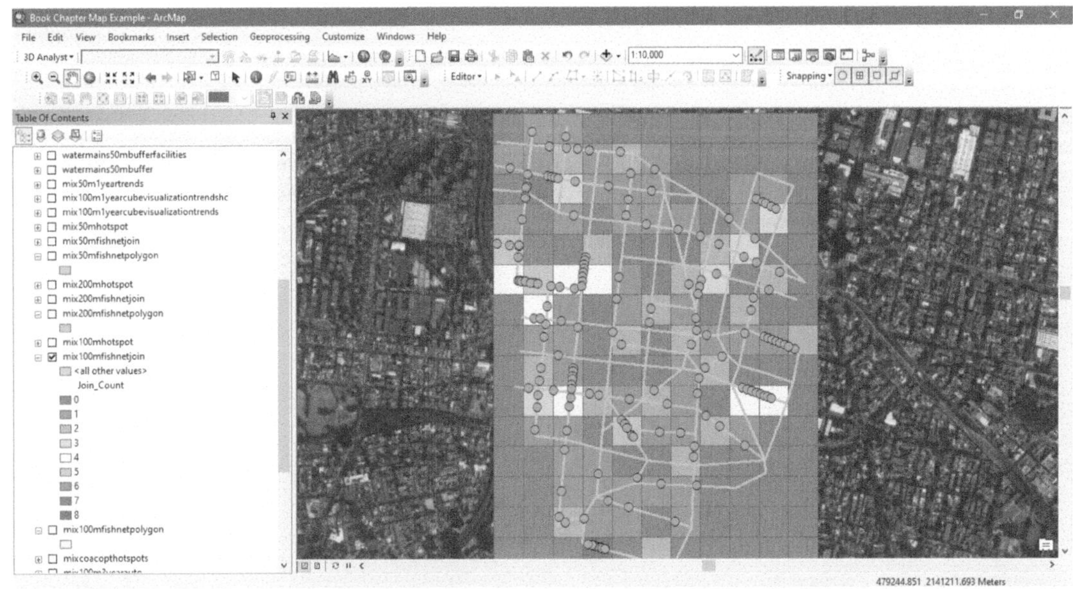

Figure 14.29 Fishnet results with a size of 50 m.

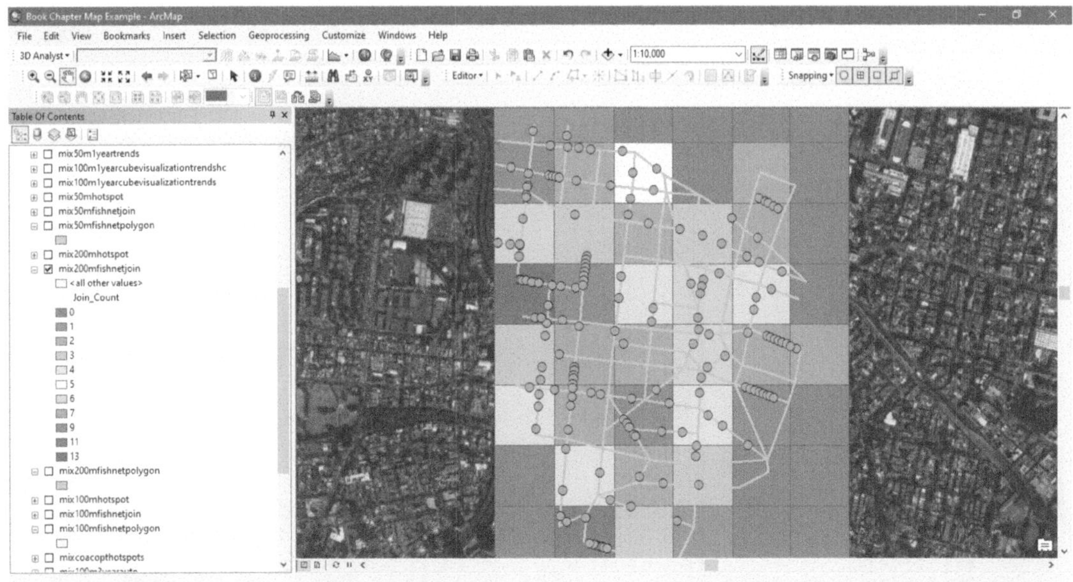

Figure 14.30 Fishnet results with a size of 100 m.

14.7.3 Hot spot analysis of water main failure

Step 1. Using the same dataset and the grid created in the last example, let us run the Hot Spot Analysis tool. Open the Hot Spot Analysis tool and use the shapefile created in the previous example as the 'Input Feature Class'.

Step 2. Select as 'Input Field' the count of failures created in the last example. Let us leave the other parameters as optional.

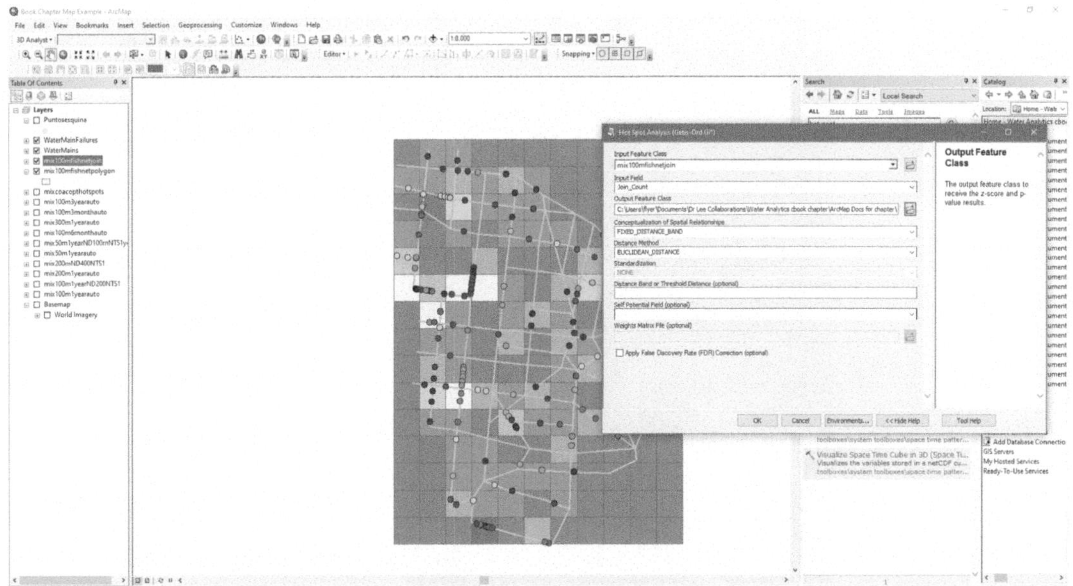

Figure 14.31 Hot spot analysis tool.

Step 3. Final results are shown below for the three bin sizes (50, 100 and 200 m).

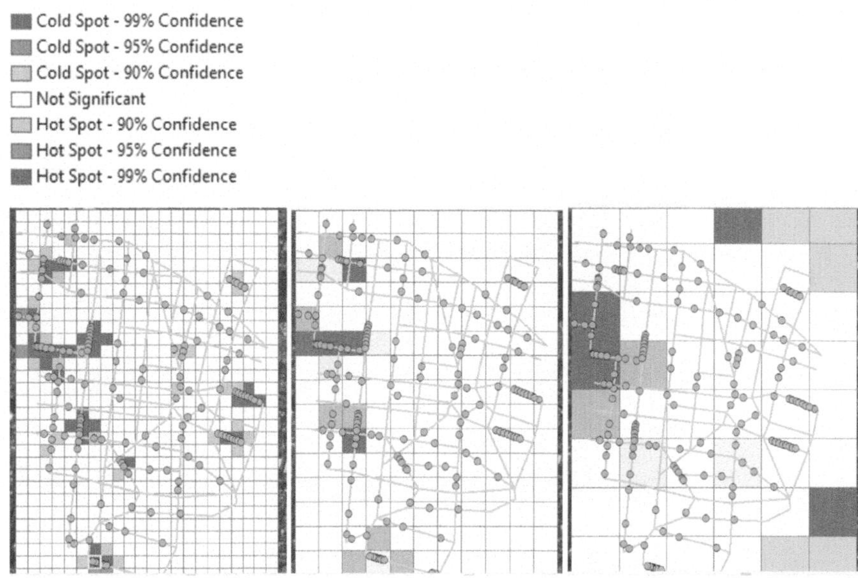

Figure 14.32 Hot spot analysis tool results with bin sizes of 50, 100 and 200 m (refer to online version of the book to visualize colors).

14.7.4 Create space time cube of water main failures

Step 1. Pipeline leak data in GIS is checked to see if it has a time attribute (date, time of leak) associated with each leak in the shape file's attribute table.

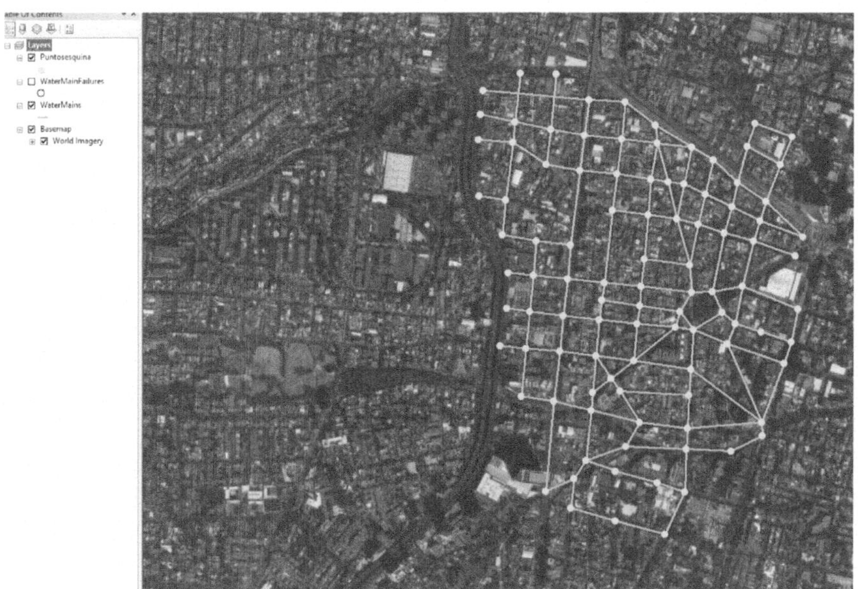

Figure 14.33 Water main grid.

Step 2. Each leak record is assigned/identified its lat/long coordinate in GIS as an attribute in the pipeline GIS shape file consisting of leaks/breaks information.

Figure 14.34 Water main grid with water main failures.

Step 3. The distance tools in ArcMap is used to measure the distance between each pipeline leak/ break using the spatial coordinate information. This will help determine distance parameters when creating the Space Time Cube.

Step 4. The Space Time Pattern Mining tool in ArcMap will be used to create spatio-temporal clusters. The Space Time Pattern Mining tools are used for analyzing data distributions and patterns in the context of both space and time. By specifying the time interval in the Create Space Time Cube, spatio-temporal clusters of pipeline leak/failure data.

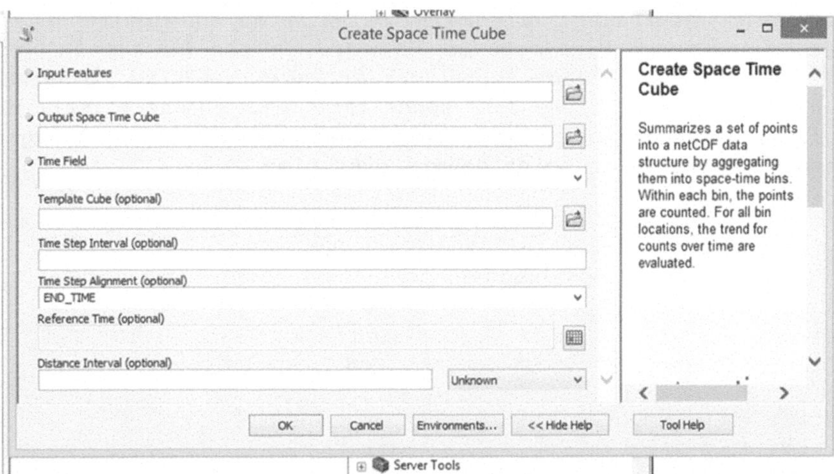

Figure 14.35 Create space time cube tool in ArcMap.

Step 5. Fill the requested fields as follows:
   Input features – Water main failure shapefile
   Output Space Time Cube – Create your own file
   Time field – Reported data in this case
   Time step – Six months
   Define Interval – 100 m

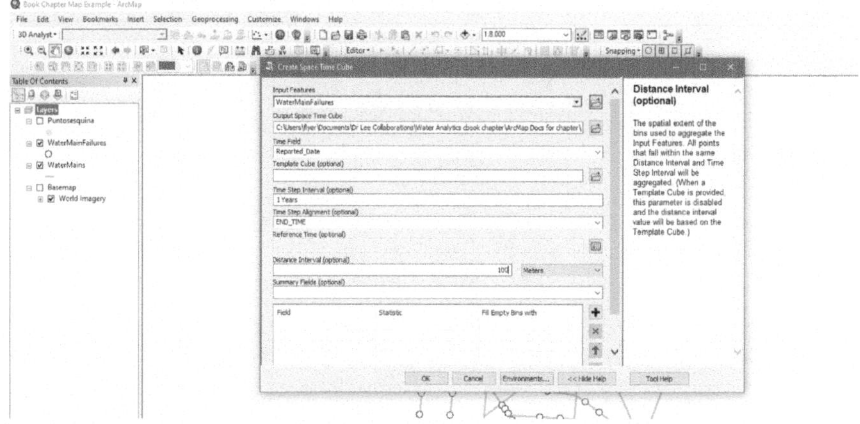

Figure 14.36 Create space time cube tool in ArcMap.

Step 6. Summary of the tool after running the Space Time Cube Tool. See bold text information that could be of interesting for our purposes. Confirm that the cube has the appropriate dimensions (see text in italics).

```
Executing: CreateSpaceTimeCube WaterMainFailures
Start Time: Sat Apr 17 09:40:12 2021
Running script CreateSpaceTimeCube…
The space time cube contains point counts for 165 locations over 64 time
step intervals. Each location is 100 meters by 100 meters spanning an
area 1100 meters west to east and 1500 meters north to south. Each of
the time step intervals is 1 year in duration so the entire time period
covered by the space time cube is 32 years. Of the 165 total locations, 81
(49.09%) contain at least one point for at least one time step interval.
These 81 locations comprise 2592 space time bins of which 142 (5.48%) have
point counts greater than zero. There is not a statistically significant
increase or decrease in point counts over time.

---------- Space Time Cube Characteristics ----------
Input feature time extent        1990-01-01 00:00:00
             to 2021-12-21 00:00:00
Number of time steps                     64
Time step interval                    6 months
Time step alignment                    End

First time step temporal bias          3.01%
First time step interval             after
              1989-12-21 00:00:00
              to on or before
              1990-12-21 00:00:00
```

```
Last time step temporal bias 0.00%
Last time step interval after
                    2020-12-21 00:00:00
                    to on or before
                    2021-12-21 00:00:00

Cube extent across space        (coordinates in meters)
Min X                   479856.4037
Min Y                   2140594.7895
Max X                   480956.4037
Max Y                   2142094.7895
Rows                    15
Columns                     11
Total bins                  5280


------------- Overall Data Trend - COUNT -------------
Trend direction         Not Significant
Trend statistic             0.4413
Trend p-value               0.6590
```

14.7.5 Emerging Hot spot analysis of water main failures

The outcome of the cluster analysis is to identify areas which are more prone for leaks and are at a high risk of pipeline failure warranting pipeline replacement in place of repair. The outcome of emerging hot spot analysis is the identification of areas which are in high risk of pipeline failure which may warrant replacement.

Step 1. Emerging hot spot analysis is to be done by using the existing spatio-temporal clusters created in ArcMap in the last example and using the Emerging Hot Spot Analysis tool in ArcMap. This tool makes use of user defined neighborhood distance, and neighborhood time steps for identifying new hotspots, intensifying existing hot spots, and also diminishing hot spots.

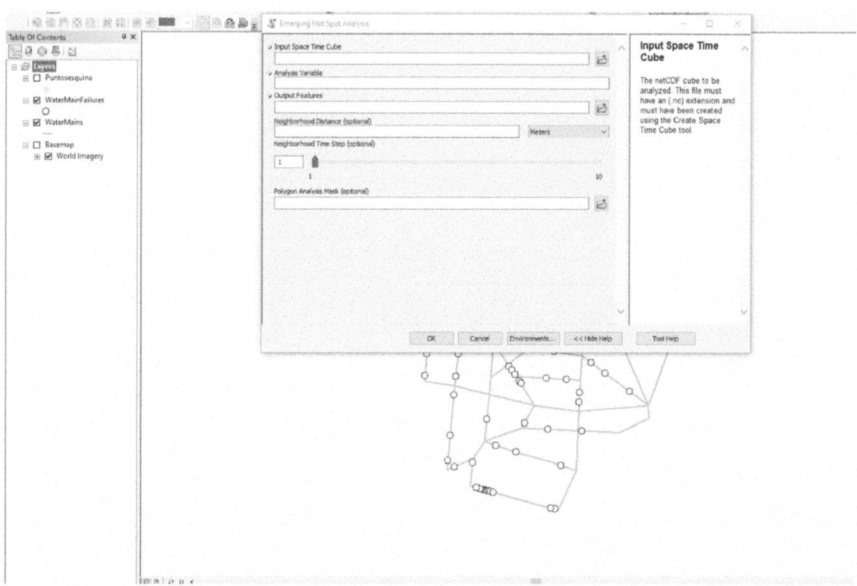

Figure 14.37 Emerging hot spot analysis tool in ArcMap.

Step 2. To start visualizing the behavior of this data, let us run the Emerging Hot Spot Analysis tool using automatic defined parameters by ArcGIS.

> Input Space Time Cube – Select the file created in the previous exercise with bin size 100 m and time step of six months.
> Analysis Variable – Count
> Output Feature – Insert a name of the newly created shapefile
> Neighborhood Distance and Neighborhood Time Step – leave blank for now. In next steps see how ArcGIS sets automatic parameters for you.

Step 3. Results
Start Time: Sat Apr 17 09:54:13 2021

```
Running script EmergingHotSpotAnalysis…
WARNING 110020: The default Neighborhood Distance is 174.557277 meters.
WARNING 110021: Setting the Neighborhood Time Step to: 1.
------------ Input Space Time Cube Details -------------
Distance interval 100 meters
Time step interval 6 months
Number of time steps 32
Number of locations analyzed 81
Number of space time bins analyzed 5184
% non-zero 0.00%
-------------------------------------------------------


---------------- Analysis Details ----------------
Neighborhood distance 174.557277 meters
Neighborhood time step intervals 1
(spanning 1 year)
-------------------------------------------------


--------- Summary of Results ---------
HOT COLD
New 3 0
Consecutive 0 0
Intensifying 0 0
Persistent 0 0
Diminishing 0 0
Sporadic 1 0
Oscillating 0 0
Historical 0 0
--------------------------------------
All locations with hot or cold spot trends: 4 of 81
```

Step 4. Let us visualize the trend z-scores (high z-scores reflect that there is an increasing number of failures in that location). Let us choose 'Layer Properties', then select 'Quantities' and 'Graduated Colors'. The map will show in red the areas with high z-scores.

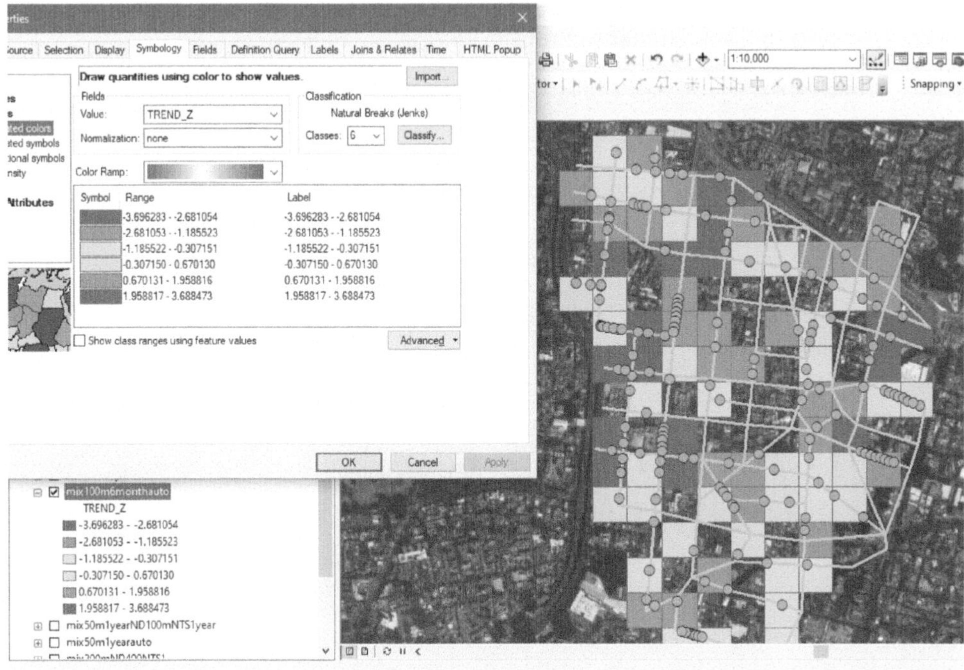

Figure 14.38 Visualization of emerging Hot spots *z*-scores (refer to online version of the book to visualize colors).

Step 5. ArcMap also allows the user to directly visualize trends. Let us choose 'Layer Properties', then select 'Quantities' and 'Graduated Colors'. However, now change the variable to 'Trends'. The map will show areas where the number of failures is increasing or decreasing and with which probability.

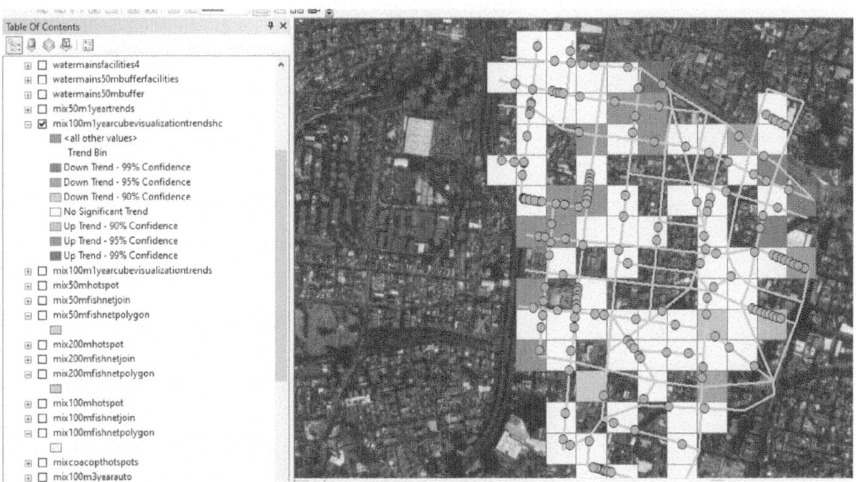

Figure 14.39 Visualization of emerging Hot spots trends (refer to online version of the book to visualize colors).

Step 6. Let us reduce the bin distance of the cube to 50 m to visualize patterns. Let us create a new cube with smaller bin size similar to Problem 14.7.

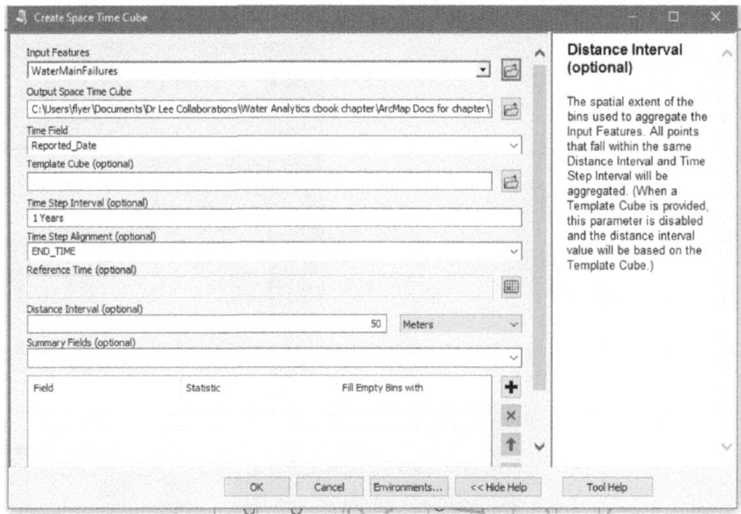

Figure 14.40 Create space time cube tool.

Step 7. Let us run the Emerging Hot Spot Analysis tool using automatic defined parameters by ArcGIS.

Input Space Time Cube – Select the file created in the previous exercise with bin size 50 m and time step of six months.
Analysis Variable – Count
Output Feature – Insert a name of the newly created shapefile
Neighborhood Distance and Neighborhood Time Step – Leave blank for now.

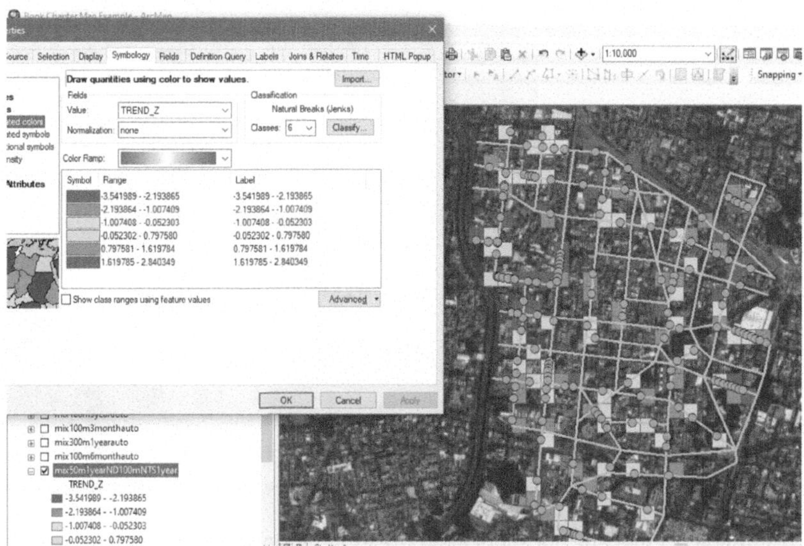

Figure 14.41 Visualization of emerging hot spots *z*-scores (refer to online version of the book to visualize colors).

Once emerging hot spots and densely clustered pipeline leaks areas are identified, it is proposed to analyze the repair cost versus replacement cost. If the replacement cost is 50% higher or more than the repair cost, then repair is a best option when compared to replacement.

14.7.6 Spatiotemporal based business risk exposure

Step 1. Using the same water main and water main failures shapefile, let us start by creating a buffer around each water main using the 'Buffer' tool. The parameters in the tool are:

Input Features – Water mains where buffers will be based on
Output Feature Class – Choose a name and folder to save the buffer shapefile
Distance – Size of the buffer starting from the centerline. As recommended, set it up as 150 ft or 50 m
Other parameters – Leave them as they appear in the tool

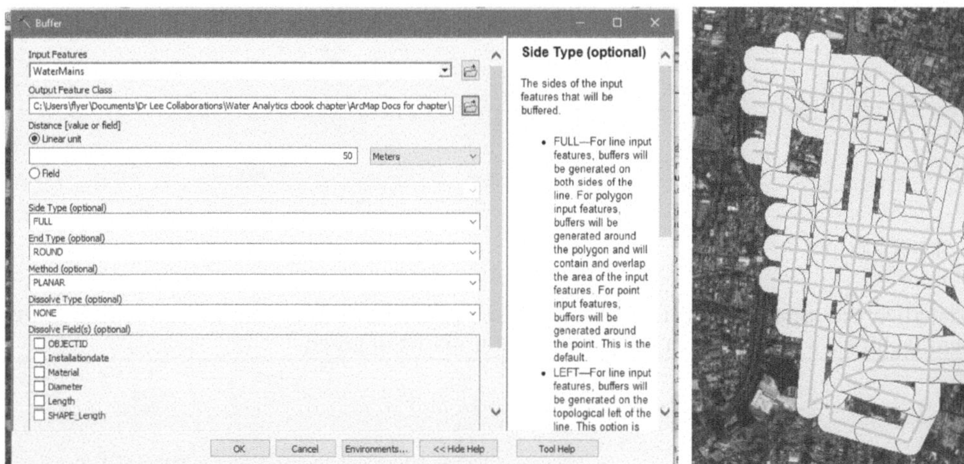

Figure 14.42 Buffer results around each water main.

Step 2. Find datasets for critical facilities. This step requires getting shapefiles with facilities where the consequence of failure is high, such as hospitals, schools, airports, and so on. For the purposes of this example, we created a shapefile with these facilities.

Step 3. Once the critical facilities shapefiles are loaded in ArcMap, run the 'Spatial Join' tool to obtain the count of these facilities within each water main's buffer. This will give us the number of critical facilities that could lose water supply in case of a water main failure. The parameters to run the tool are:

Input Features – Water mains buffer shapefile
Join Features – Shapefile that we want to join to the buffers, in this case the critical facilities
Output Feature Class – Choose a name and folder to save the spatial join shapefile
Join Operation – One to one

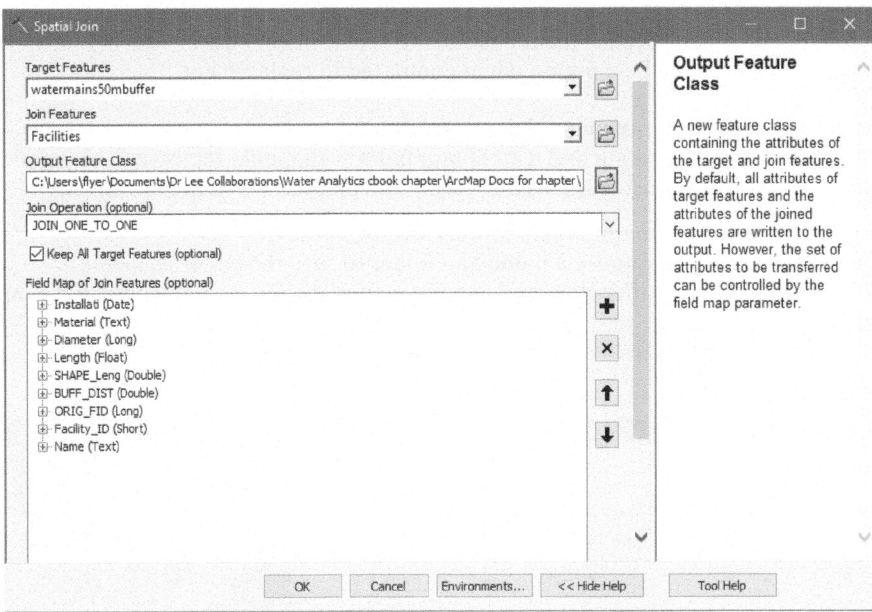

Figure 14.43 Spatial join tool.

Step 4. Visualize the 'Spatial Join' tool results. Red buffers have four or more critical facilities. Critical facilities are depicted in purple and water mains in green.

Figure 14.44 Buffer results depicting the number of critical facilities (refer to online version of the book to visualize colors).

Step 5. Run for a second time 'Spatial Join' tool to assign the count of critical facilities within the buffer distance to the water mains themselves. This will give us the number of critical facilities that could lose water supply in case of a water main failure. The parameters to run the tool are:

Input Features – Water mains
Join Features – Shapefile that we want to join to the buffers, in this case spatial join from the previous step
Output Feature Class – Choose a name and folder to save the spatial join shapefile
Join Operation – One to one
Match Option – Select Within

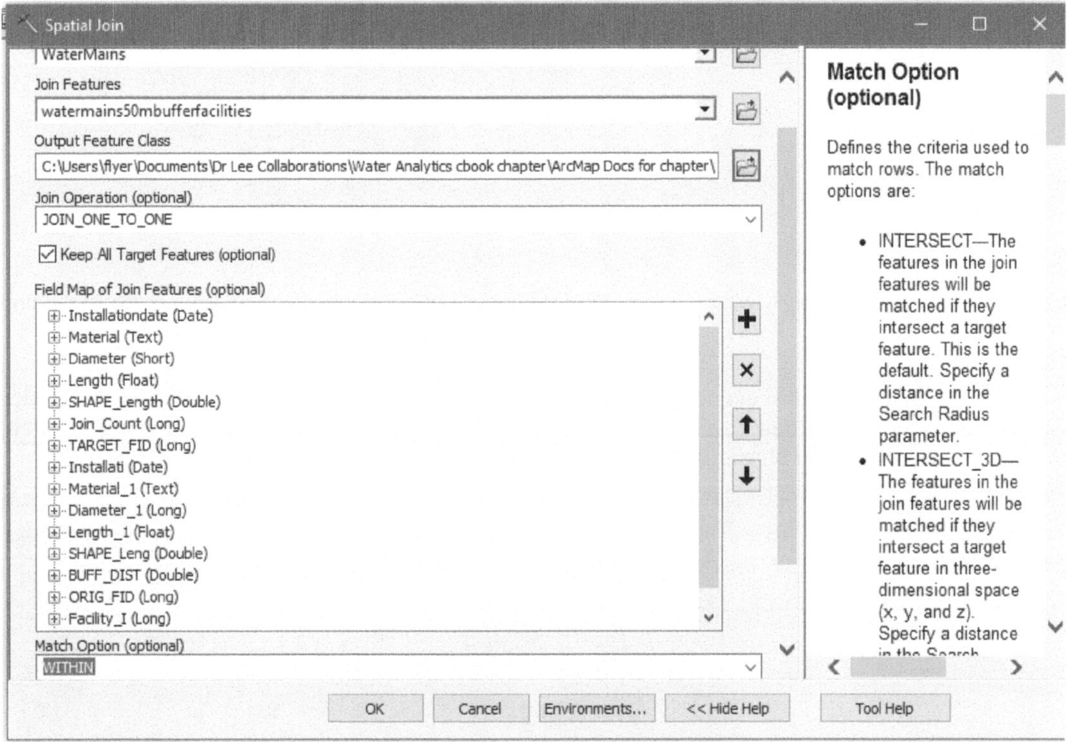

Figure 14.45 Spatial Join tool.

Step 6. Visualize the 'Spatial Join' tool results. Now water mains are assigned the critical facility count from their respective buffers. Critical facilities are depicted in purple and water mains in green.

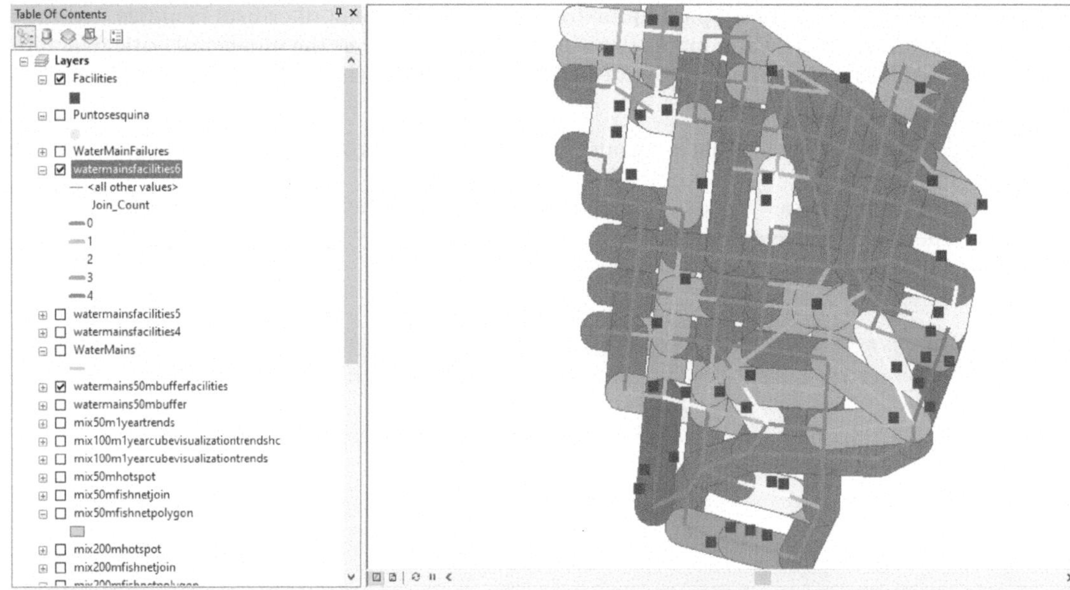

Figure 14.46 Spatial Join tool showing the number of failures per water main (refer to online version of the book to visualize colors).

> ***Step 7.*** Visualize the consequence of failure (COF) results. Water mains are assigned the critical facility count from their respective buffers. Critical facilities are depicted in dark blue, water main failures in light blue and water mains in green.

Figure 14.47 COF results (refer to online version of the book to visualize colors).

Step 8. Visualize the consequence of failure (COF) results and (LOF). LOF results are shown as the z-score from the Emerging Hot Spot Analysis

Figure 14.48 LOF results (refer to online version of the book to visualize colors).

Step 9. Visualization of water mains that need to be prioritized based on high COF and LOF.

Water Main A has eight failures and was categorized as a high z-score area which means there is a high likelihood of failure, in addition it has more than four critical facilities within the 150 ft (46 m) buffer. Therefore, this water main is a candidate for replacement.

Figure 14.49 Example of COF/LOF analysis for two water mains (refer to online version of the book to visualize colors).

Water Main B does not have any reported failure so its LOF is low, however it has more than four critical facilities within the 150 ft (46 m) buffer. Probably, is not a candidate for replacement but definitely has to be constantly monitored.

Figure 14.50 Example of COF/LOF analysis for two water mains (refer to online version of the book to visualize colors).

Water Mains C and D have 11 and eight failures respectively and were categorized as a high z-score area which means there is a high likelihood of failure Although they do not have critical facilities around, a failure in these water mains is probable and still would affect customers. Therefore, this water main is a candidate for replacement.

REFERENCES

ArcGIS (2016). ArcMap: Hot Spot Analysis (Getis-Ord Gi*). Available at: http://desktop.arcgis.com/en/arcmap/10.3/tools/spatial-statistics-toolbox/hot-spot-analysis.htm (accessed 24 May 2018).

AWWA (2017). Effective Utility Management: A Primer for Water and Wastewater Utilities. US Environmental Protection Agency, Washington, DC.

Baddeley A. (2010). Multivariate and Marked Point Processes. Handbook of Spatial Statistics. CRC Press, Boca Raton, FL.

Berardi L., Giustolisi O., Kapelan Z. and Savic D. A. (2008). Development of water main deterioration models for water distribution systems using EPR. *Journal of Hydroinformatics*, **10**(2), 113–126, https://doi.org/10.2166/hydro.2008.012

Bogárdi I. and Fülöp R. (2011). A spatial probabilistic model of water mainline failures. *Periodica Polytechnica Civil Engineering*, **55**(2), 161–168, https://doi.org/10.3311/pp.ci.2011-2.08

Christodoulou S., Charalambous C. and Adamou A. (2008). Rehabilitation and maintenance of water distribution network assets. *Water Science & Technology: Water Supply*, **8**(2), 231–239, https://doi.org/10.2166/ws.2008.066

de Oliveira D. P., Neill D. B., Garrett J. H. and Soibelman L. (2011). Detection of patterns in water distribution pipe breakage using spatial scan statistics for point events in a physical network. *Journal of Computing in Civil Engineering*, **25**(1), 21–30, https://doi.org/10.1061/ASCECP.1943-5487.0000079

Folkman S. (2018). Water Main Break Rates in the USA and Canada: A Comprehensive Study. Available at: https://digitalcommons.usu.edu/mae_facpub/174 (accessed 18 June 2020).

Ganesan S., Martínez García D., Lee J., Keck J. and Yang P. (2017). Spatio-temporal water mains integrity management program for California. Proceedings of the World Environmental and Water Resources Congress 2017, Sacramento, California, 21–25 May, https://doi.org/10.1061/9780784480625.049

Giustolisi O., Laucelli D. and Savic D. (2006). Development of rehabilitation plans for water mains replacement considering risk and cost–benefit assessment. *Civil Engineering and Environmental Systems*, **23**(3), 175–182, https://doi.org/10.1080/10286600600789375

Goulter I. C. and Kazemi A. (1988). Spatial and temporal groupings of water main pipe breakage in Winnipeg. *Canadian Journal of Civil Engineering*, **15**(1), 91–97, https://doi.org/10.1139/l88-010

Güngör-Demirci G., Lee J. and Keck J. (2018). Assessing the performance of a California water utility using two-stage data envelopment analysis. *Journal of Water Resources Planning and Management, ASCE*, **144**(4), 05018004, https://doi.org/10.1061/(ASCE)WR.1943-5452.0000921

Hot Spot Analysis (Getis-Ord Gi*) (Spatial Statistics). Hot Spot Analysis (Getis-Ord Gi*) (Spatial Statistics)-ArcGIS Pro | Documentation. (n.d.). Retrieved April 1, 2021, from https:///pro.arcgis.com/en/pro-app/2.8/tool-reference/spatial-statistics/hot-spot-analysis.html

Jacobs P. and Karney B. (1994). GIS development with application to cast iron water main breakage rates. In Proceedings of the 2nd International Conference on Water Pipeline Systems, Edinburgh. Mechanical Engineering Publication Ltd., London.

Kettler A. J. and Goulter I. 1985 An analysis of pipe breakage in urban water distribution networks. *Canadian Journal of Civil Engineering*, **12**(2), 286–293, https://doi.org/10.1139/l85-030

Lee J. and Tanverakul S. (2015). Price elasticity of residential water demand in California. *Journal of Water Supply: Research and Technology AQUA*, **64**(2), 211–218, https://doi.org/10.2166/aqua.2014.082

Lee J., Lohani V., Dietrich A. and Loganathan G. V. (2012). Hydraulic transients in plumbing systems. *Water Science and Technology: Water Supply*, **12**(5), 619–629, https://doi.org/10.2166/ws.2012.036

Martínez García D., Lee J., Keck J., Yang P. and Guzzetta R. (2018). Hot spot analysis of water main failures in California. *Journal of the American Water Works Association*, **110**(6), E39–E49, https://doi.org/1010.1002/awwa.1039

Martínez García D., Lee J., Keck J., Yang P. and Guzzetta R. (2019). Spatiotemporal and deterioration assessment of water main failures. *AWWA Water Science*, **1**(5), https://doi.org/10.1002/ aws2.1159

Martínez García D., Lee J., Keck J., Yang P. and Guzzetta R. (2019b). Utilizing spatiotemporal based business risk exposure to analyze cast iron water main failures in California. *Journal of Water Supply: Research and Technology AQUA*, **68**(2), 111–120, https://doi.org/10.2166/aqua.2019.120

Martínez García D., Lee J., Keck J., Kooy J., Yang P. and Wilfley B. (2020). Case study: pressure-based analysis of water main failures in California. *Journal of Water Resources Planning and Management*, **146**(9), 371–402, 05020016, https://doi.org/10.1061/%28ASCE%29WR.1943-5452.0001255

Rajani B. B. and Kleiner Y. (2001). Comprehensive review of structural deterioration of water mains: physically based models. *Urban Water*, **3**(3), 151–164, https://doi.org/10.1016/S1462-0758(01)00032-2

Shi W., Zhang A. and Ho O. (2013). Spatial analysis of water mains failure clusters and factors: a Hong Kong case study. *Annals of GIS*, **19**(2), 89–96, https://doi.org/10.1080/19475683.2013.782509

Tabesh M., Delavar M. R. and Delkhah A. (2010). Use of geospatial information system based tool for renovation and rehabilitation of water distribution systems. *International Journal of Environmental Science and Technology*, **7**(1), 47–58, https://doi.org/10.1007/BF03326116

The R Foundation (2016). Introduction to R & The R Environment. Available at: https://www.r-project.org/about.html (accessed 10 March 2020).

Wang Y., Moselhi O. and Zayed T. M. (2009). Study of the suitability of existing deterioration models for water mains. *Journal of Performance of Construction Facilities*, **23**(1), 40–49, https://doi.org/10.1061/(ASCE)0887-3828(2009)23:1(40)

doi: 10.2166/9781789062380_0381

Chapter 15
Decision Analysis

*Eftila Tanellari[1] and Juneseok Lee[2]**

[1]*Department of Economics, Radford University, Radford, VA, USA*
[2]*Department of Civil and Environmental Engineering, Manhattan College, Riverdale, NY, USA*
**Corresponding author: juneseok.lee@manhattan.edu*

LEARNING OBJECTIVES

At the end of this chapter, you will be able to:

(1) Explain nonmarket valuations.
(2) Perform contingent valuation and conjoint analysis in Excel.
(3) Explain pair-wise comparison to perform a rational decision making.
(4) Perform AHP in Excel.

15.1 NONMARKET VALUATION

Nonmarket valuation is a method that is used to estimate the total willingness to pay (WTP) for goods or a service that is not traded in the market. For goods that are traded in the market, the total willingness to pay can be easily estimated by the area under the demand curve (the demand curve represents the relationship between the price of a good or service and the quantity consumers are willing and able to purchase). However, this is a more challenging task in the case of nonmarket goods. Because these goods and services are not sold in the market, the demand curve does not exist. Instead, the willingness to pay is either revealed through consumers' choices or directly elicited through surveys.

There are two broad categories of valuation methods, revealed preference methods and stated preference methods. Revealed preference methods are based on actual choices that individuals make which in turn directly indicate the values that they may place on the good or service of interest. For example, by calculating how much households spend on bottled water, filters and water treatment devices in a given time period, a revealed preference method may infer the value that households place on clean water. The cost of such treatments and devices is directly incurred by households and is observable through the prices they pay. Stated preference methods elicit willingness to pay directly from consumers through surveys. Consumers are directly or indirectly asked to state their willingness to pay for a good or service. In this section, we will examine two widely used stated preference methods, contingent valuation and conjoint analysis.

15.2 CONTINGENT VALUATION

The contingent valuation method is a direct stated preference method. It uses a survey format to directly asks respondents about their willingness to pay *contingent* on a hypothetical scenario or market. The reliability of the contingent valuation results depends mainly on the design of the survey and the analysis of the data; thus, it is important to carefully consider these aspects.

There are five main components in a contingent valuation study. The first component is to identify the changes to be valued. This step includes defining the quality or quantity of the good before the policy change and after the policy change and the values to be estimated. This will be helpful when describing the good to respondents.

The second component is the selections of the data collection mode and the sample size. The data collection surveys can be conducted in person, via phone, via mail or web-based. While in person interviews are recommended by the NOAA panel (NOAA, 1993) and Mitchell and Carson (1989), they tend to be the most expensive mode for data collection. The survey should then be administered to the largest possible sample size given the resources available. According to Mitchell and Carson (2013), this is necessary given the large variance present in willingness to pay (WTP) responses.

The third component is choosing an elicitation method and designing the contingent valuation question. Generally, a detailed description of the good or market and the change that is taking place is provided before the elicitation question is asked. The information describes the good together with how it will be offered, how consumers would pay for it and for how long. Then the willingness to pay question is asked by choosing an elicitation format. The main formats that are used in the literature include the open-ended format, the payment card method and the dichotomous choice question. Below are some examples of the same question (adopted from Tanellari *et al.* 2015) asked in different formats.

> ***Open-ended***: the respondents are directly asked to state a dollar amount that represents their willingness to pay, for example What is the maximum amount you are willing to pay, through an increase in your quarterly water bill to upgrade the water distribution infrastructure in your utility service area?
>
> ***Payment card***: this method provides respondents with a series of dollar values using a card and asks them to choose the value that best represents their maximum willingness to pay. The values presented are preselected by the researchers, for example Which of the values below best represents the maximum amount you are willing to pay, through an increase in your quarterly water bill to upgrade the water distribution infrastructure in your utility service area?

$0	$10	$100
$1	$25	$150
$3	$50	$200
$5	$75	More than $200

> ***Dichotomous choice***: respondents are simply asked if they are willing to pay a given amount in a yes or no format. The dollar amount asked is varied randomly among respondents, for example Would you be willing to pay $10 (the bid amounts asked are not arbitrary; e.g., in Tanellari *et al.* (2015) they are based on the actual quarterly water bill values for residents in the study area), through an increase in your quarterly water bill to upgrade the water distribution infrastructure in your utility service area?

There are other variations of these formats such as double bounded or multiple bounded questions (see Boyle, 2003 for more detail). Each of the above response formats have advantages and disadvantages, however, the dichotomous choice format is generally the most widely used one. Regardless of which response format is used, respondents should have the choice of answering with $0. In addition, the survey may include follow up questions to the contingent valuation question to distinguish any protest answers.

The fourth component is the design of the survey. Besides the elicitation question, the survey should also include other relevant questions to collect information needed for the regression analyses of the data as well as socio-demographic questions that serve as explanatory variables.

The fifth and the last component is the collection and analysis of the data. Once the data are collected, the analysis of the willingness to pay will depend on the elicitation format used. Since dichotomous choice questions are the most commonly used method, we will focus on the analysis of data collected using this method. For more information on how to analyze data using other elicitation formats please see Boyle (2003).

Contingent valuation has proven valuable in estimating non-use values of a good or service and when revealed preference methods cannot be used. It is a very flexible valuation method and can be applied to a wide number of scenarios and range of goods by simply asking respondents about their willingness to pay. Although widely used, contingent valuation is not without faults as a method. One of the main weaknesses is the potential for biased responses. There are several potential biases that may affect the validity of the results. Respondents may have little incentive to answer the survey truthfully or consider their answers carefully, particularly since the choices presented are of a hypothetical nature. Others may intentionally answer in a way that may influence policy decisions towards a particular outcome. Biased results may also arise when respondents are not familiar with the good being evaluated or not enough information has been given at the beginning of the survey. The survey instrument used can also introduce bias. Consumers may be sensitive to the dollar values presented by the payment card and the dichotomous choice. In addition, the payment vehicle bias can result when respondents refuse to pay any amount, not because they do not value the good but because they do not like the means by which the WTP will be collected, generally increase in taxes or utility bills, donations, and so on. Careful survey design and data collection are important in reducing some of these biases and providing more reliable and valid results that can be useful in policy analysis. Several reports have provided guidelines and best practices in conducting a contingent valuation study (Boyle, 2003; Mitchell & Carson, 1989; NOAA, 1993) and should be carefully reviewed when considering such studies.

15.2.1 Analysis of CV data

In this section, we will focus on the analysis of a dichotomous choice WTP question. Once the data are collected, they need to be coded and prepared for the analysis. Questions with choices of 'yes' or 'no' produce discrete variables and are coded as '1' or '0'. Categorical variables are coded similarly by assigning dummy variables for each category while continuous variables are left as they are. Then the data needs to be checked for outliers and missing values. Imputation or mean substitution can be used in the case of missing data, whenever possible. Any outliers present in the data should be examined carefully and excluded only if they are in fact inaccurate.

Dichotomous choice data are generally analyzed using a logistic model. First we input the data in Excel. Respondents answered 'yes' (WTP = 1) or 'no' (WTP = 0) to the varying bid amounts offered. The data is arranged as presented in Table 15.1.

Before we run the regression analysis, we can conduct a descriptive analysis and cross-tabulation analysis of the data. The descriptive analysis can be conducted in Excel by using the multiple regression tool which is accessible when the Analysis Toolpak add-in is installed. To obtain the descriptive statistics, go to Data→Analyze→Data Analysis and choose Descriptive Statistics. Make sure the 'Summary Statistics' output option is checked. The summary statistics of the data are presented in Table 15.2.

Table 15.1 CV coded data in Excel.

ID	Bid	WTP Responses
1	10	1
2	25	0
3	50	1
4	100	0
5	1	1
6	100	0
7	50	1
8	1	1
9	25	1
10	150	0
11	50	0
12	100	1
13	10	1
14	5	1
15	1	1
16	5	1
17	200	0
18	25	1
19	150	0
20	10	1

Table 15.2 Summary statistics of CV data in Excel.

	Bid	WTP Responses
Mean	53.4	0.65
Median	25	1
Standard Deviation	59.69	0.49
Minimum	1	0
Maximum	200	1

The summary statistics indicate that 65% of the respondents answered 'yes' to the bid amount they were presented with and the mean bid amount asked is \$53.4. In addition, we can cross-tabulate the WTP responses by bid amount as presented in Table 15.3 to get a better understanding of the survey responses.

The data can then be used to run a regression analysis. The model we are trying to estimate is:

$$Y = \beta_0 + \beta_1(Bid) + \varepsilon \tag{15.1}$$

where Y is the WTP response, β_0 is the intercept, β_1 is the parameter we are trying to estimate and ε is the error term.

The regression analysis can be conducted in Excel by using the multiple regression tool which is accessible through the Analysis Toolpak. To do this, go to Data → Analyze → Data Analysis and choose Regression. Specify 'WTP Responses' as the dependent variable and the 'Bid' as the independent

Table 15.3 Cross-tabulation of CV data in Excel.

Bid	WTP		Number of Responses
	No = 0	Yes = 1	
1	0	3	3
5	0	2	2
10	0	3	3
25	1	2	3
50	1	2	3
100	2	1	3
150	2	0	2
200	1	0	1
			20

variable. In addition to the bid values, other explanatory variables can be added to the analysis that may help explain WTP responses, for example socioeconomic variables (gender, education, income, etc.). The regression results are presented in Table 15.4.

Based on the regression results the regression line is:

$$WTP = 0.97 - 0.006 * (bid) \tag{15.2}$$

The estimated coefficients indicate how the explanatory variables are related to the dependent variable. For example, a \$1 increase in the bid amount will decreases the probability that a respondent will answer 'yes' by 0.6%.

Table 15.4 CV regression analysis results in Excel.

Summary Output

Regression Statistics	
Multiple R	0.72289
R Square	0.52257
Adjusted R Square	0.49604
Standard Error	0.3474
Observations	20

ANOVA

	df	SS	MS	F	Significance F
Regression	1	2.377682	2.3777	19.702	0.00032
Residual	18	2.172318	0.1207		
Total	19	4.55			

	Coefficients	Standard Error	t Stat	P-value
Intercept	0.96647	0.105441	9.166	3×10^{-8}
Bid	−0.0059	0.001335	−4.4387	0.0003

The standard errors for the regression coefficients indicate the accuracy of the estimated coefficients and the R^2 indicates the goodness of fit for the model. In this case about 50% of the variation in WTP is explained by the bid amount included in the regression model.

The significance of the F-value of shows the significance of the model results. If this value is less than 0.05 (like in this case) than the results are significant. Similarly, the P-value indicates the significance of each coefficient estimate. These values should also be below 0.05 for most or all coefficients to indicate significance. If the F or P-values are not below 0.05, the explanatory variables should not be used and the regression should be rerun with a different set of variables.

The regression results can then be used to calculate the mean WTP per person.

$$Mean\,WTP = -(\beta_0/\beta_1) \tag{15.3}$$

Finally, the nonuse value for the good can be estimated by multiplying the mean WTP per person by the total number of people, as follows:

$$Nonuse\,Value = Mean\,WTP \times Total\,Population \tag{15.4}$$

15.3 CONJOINT ANALYSIS

Conjoint analysis, also referred to as an attribute-based method, is an indirect stated preference method used to elicit the economic value of a good through surveys. Conjoint analysis, unlike contingent valuation, elicits willingness to pay indirectly by asking respondents to choose among different bundles of goods. Each bundle includes different attributes with different levels where one of the attributes is the price. By choosing a bundle, respondents are indirectly indicating their willingness to pay for the good described.

There are several components in designing a conjoint analysis study. Similar to contingent valuation studies, the first component is to identify the changes to be valued. Any changes in the quality of a good and the values that are affected as a result of these changes should be identified. For example, valuation of the benefits from improving the drinking water quality in an area should include all the relevant changes associated with the water attributes, such as taste, clarity, odor, and so on.

The second component includes selecting the data collection mode and the sample size. The data can be collected by using in-person surveys, phone surveys, mail or web-based surveys. The survey should be conducted to the geographical site(s) that is/are affected by the proposed change while the mode of collection can depend on the type of survey as well as budget limitations.

The third component involves identifying the attributes and determining the experimental design. Researchers have to identify the most important attributes that may impact the use of the good and determine the number of attributes to include as well as the different levels of each attribute. The attribute bundles are then constructed using an experimental design. Because the number of possible combinations of attributes may be high, not all of them can be included in the bundles presented to respondents. An experimental design can be used to identify the optimal bundles to use in the survey. The main designs that can be used to create the optimal bundles are full factorial design, fractional factorial design or randomized design. For more details on strategies of experimental designs, please see Holmes and Adamowicz (2003).

The fourth component is the selection of the response format and the design of the survey. The three most popular response formats are ratings, rankings and choice (Holmes & Adamowicz, 2003). An example of a rating response format is provided in Table 15.5.

An example of a choice format questions is provided in Table 15.6. This represents only one choice set and respondents usually are presented with several different choice sets. In general, researchers recommend no more than eight choice sets be used (Holmes & Adamowicz, 2003).

The ranking response format is similar to the choice format but instead of picking one alternative, respondents are asked to rank the alternatives presented to them from the most preferred to the least preferred.

Table 15.5 A CA rating response format example.

How would you rate each of the following attributes of corrosion resistance for drinking water plumbing materials? (Please circle one number for each item)

	Very Undesirable		No Opinion		Very Desirable	Don't Know
May corrode under select conditions	1	2	3	4	5	6
Not susceptible to corrosion	1	2	3	4	5	6
Resists corrosion and oxidation	1	2	3	4	5	6

Table 15.6 A CA choice set example.

Assuming that the plumbing materials below were the only material available on the market, which of the plumbing materials would you purchase for your house? Please check only one box.

Plumbing Material A	Plumbing Material B	
• Not susceptible to corrosion • Compounds released from this material may give a bitter or metallic taste or odor to the water. • $10 for 2 cm diameter pipe	• May corrode under select conditions • Compounds released from this material may give a bitter or metallic taste or odor to the water. • $5 for 2 cm diameter pipe	I would stay with the product currently installed in my house

The fifth and the last component is the collection and analysis of the data. Once the data are collected, the analysis will depend on the response format that was used in the survey. Depending on the assumptions that are made about the distribution of the error term, a binary probit, a conditional logit or a multinomial logit model can be used to analyze the data. Other explanatory variables that may explain respondents' choices are also included in the analysis.

15.3.1 Analysis of CA data

In this section, we will focus on the analysis of attribute-based methods. Similar to the CV method when the data are collected, they need to be coded and prepared for the analysis.

Consider a conjoint analysis problem with three attributes and different levels as indicated in Table 15.7. We are trying to evaluate different attributes of drinking water plumbing materials.

A full factorial design will lead to a total combination of 18 different alternatives or profiles for the plumbing materials, as shown in Table 15.8.

Table 15.7 CA attributes and levels example.

Attributes/ Levels	Corrosion Properties	Taste and Odor Properties	Price for 2 cm Diameter Pipe
1	Not susceptible to corrosion	Compounds released from this material may give a bitter or metallic taste or odor to the water.	$5
2	May corrode under select conditions	Compounds released from this material may give a chemical or solvent taste and odor to the water.	$10
3		No adverse taste and odor to the water	$30

Table 15.8 CA full factorial experimental design example.

Alternatives	Corrosion Properties	Taste and Odor Properties	Price for 2 cm Diameter Pipe
1	Not susceptible to corrosion	Compounds released from this material may give a bitter or metallic taste or odor to the water	$5
2	Not susceptible to corrosion	Compounds released from this material may give a bitter or metallic taste or odor to the water	$10
3	Not susceptible to corrosion	Compounds released from this material may give a bitter or metallic taste or odor to the water	$30
4	Not susceptible to corrosion	Compounds released from this material may give a chemical or solvent taste and odor to the water	$5
5	Not susceptible to corrosion	Compounds released from this material may give a chemical or solvent taste and odor to the water	$10
6	Not susceptible to corrosion	Compounds released from this material may give a chemical or solvent taste and odor to the water	$30
7	Not susceptible to corrosion	No adverse taste and odor to the water	$5
8	Not susceptible to corrosion	No adverse taste and odor to the water	$10
9	Not susceptible to corrosion	No adverse taste and odor to the water	$30
10	May corrode under select conditions	Compounds released from this material may give a bitter or metallic taste or odor to the water	$5
11	May corrode under select conditions	Compounds released from this material may give a bitter or metallic taste or odor to the water	$10
12	May corrode under select conditions	Compounds released from this material may give a bitter or metallic taste or odor to the water	$30
13	May corrode under select conditions	Compounds released from this material may give a chemical or solvent taste and odor to the water	$5
14	May corrode under select conditions	Compounds released from this material may give a chemical or solvent taste and odor to the water	$10
15	May corrode under select conditions	Compounds released from this material may give a chemical or solvent taste and odor to the water	$30
16	May corrode under select conditions	No adverse taste and odor to the water	$5
17	May corrode under select conditions	No adverse taste and odor to the water	$10
18	May corrode under select conditions	No adverse taste and odor to the water	$30

$$2 \; corrosion \times 3 \; taste/odor \times 3 \; price = 18 \tag{15.5}$$

Let us assume that consumers are asked to rank the different alternatives by assigning a value from 0 to 10, from the least preferred to the most preferred.

We can use Excel to analyze the data from a conjoint analysis study. First, we input the data in an Excel spreadsheet as presented in Table 15.9.

Next, we have to code the data where each level of each attribute is listed as a variable. Then we code these as discrete variables by assigning a value of '1' if the characteristic is present in that particular alternative and '0' if it is not, as shown in Table 15.10.

Table 15.9 CA data in Excel.

Alternatives	Corrosion	Taste/Odor	Price	Ranking
1	1	1	1	7
2	1	1	2	5
3	1	1	3	4
4	1	2	1	7
5	1	2	2	5
6	1	2	3	5
7	1	3	1	10
8	1	3	2	9
9	1	3	3	8
10	2	1	1	6
11	2	1	2	5
12	2	1	3	2
13	2	2	1	6
14	2	2	2	5
15	2	2	3	2
16	2	3	1	8
17	2	3	2	6
18	2	3	3	4

Table 15.10 CA coded data in Excel.

Alternatives	No Corrosion	May Corrode	Bitter/ Metallic	Chemical/ Solvent	No Effect Taste/Odor	$5	$10	$30	Ranking
1	1	0	1	0	0	1	0	0	7
2	1	0	1	0	0	0	1	0	5
3	1	0	1	0	0	0	0	1	4
4	1	0	0	1	0	1	0	0	7
5	1	0	0	1	0	0	1	0	5
6	1	0	0	1	0	0	0	1	5
7	1	0	0	0	1	1	0	0	10
8	1	0	0	0	1	0	1	0	9
9	1	0	0	0	1	0	0	1	8
10	0	1	1	0	0	1	0	0	6
11	0	1	1	0	0	0	1	0	5
12	0	1	1	0	0	0	0	1	2
13	0	1	0	1	0	1	0	0	6
14	0	1	0	1	0	0	1	0	5
15	0	1	0	1	0	0	0	1	2
16	0	1	0	0	1	1	0	0	8
17	0	1	0	0	1	0	1	0	6
18	0	1	0	0	1	0	0	1	4

Table 15.11 CA regression data in Excel.

Alternatives	May Corrode	Bitter/Metallic	Chemical/Solvent	$10	$30	Ranking
1	0	1	0	0	0	7
2	0	1	0	1	0	5
3	0	1	0	0	1	4
4	0	0	1	0	0	7
5	0	0	1	1	0	5
6	0	0	1	0	1	5
7	0	0	0	0	0	10
8	0	0	0	1	0	9
9	0	0	0	0	1	8
10	1	1	0	0	0	6
11	1	1	0	1	0	5
12	1	1	0	0	1	2
13	1	0	1	0	0	6
14	1	0	1	1	0	5
15	1	0	1	0	1	2
16	1	0	0	0	0	8
17	1	0	0	1	0	6
18	1	0	0	0	1	4

In this setup, there is a linear dependence problem because there are sets of variables that be expressed as a linear combination of each other, for example by knowing the value of 'no corrosion' we can predict the value of 'may corrode'. In regression analysis, this creates problems due to singularity, so there is no unique solution to estimate the regression coefficients. To solve this issue, we omit one column from each group of attributes and it is not important which level is excluded. Doing so will make that attribute the reference category without excluding it from the regression analysis. For this example, we have omitted 'no corrosion', 'no effect taste/odor' and '$5' and the dataset is transformed as shown in Table 15.11.

The data is now ready to run a regression. The model we are trying to estimate is:

$$Y = \beta_0 + \beta_1(may\ corrode) + \beta_2(bitter/metallic) + \beta_3(chemical/solvent) + \beta_4(\$10) + \beta_5(\$30) + \varepsilon \quad (15.6)$$

where Y is the ranking by respondents, β_0 is the intercept, β_1 to β_5 are the parameters we are trying to estimate and ε is the error term. The coefficients for the reference categories ('no corrosion', 'no effect taste/odor' and '$5') are zero.

The regression analysis can be conducted in Excel by using the multiple regression tool which is accessible through the Analysis Toolpak. To run a regression analysis, go to *Data* → *Analyze* → *Data Analysis* and choose *Regression*. Specify '*Ranking*' as the dependent variable and the other five attribute columns as the independent variables. The regression results are presented in Table 15.12.

Based on the regression results, the regression line is:

$$Ranking = 9.94 - 1.78 * (may\ corrode) - 2.67 * (bitter/metallic) - 2.5 * (chemical/solvent)$$
$$- 1.5 * (\$10) - 3.17 * (\$30) \quad (15.7)$$

Table 15.12 CA regression analysis results in Excel.

Summary Output

Regression Statistics

Multiple R	0.94809
R Square	0.89888
Adjusted R Square	0.85674
Standard Error	0.8165
Observations	18

ANOVA

	df	SS	MS	F	Significance F
Regression	5	71.11111	14.2222	21.333	1.4×10^{-5}
Residual	12	8	0.66667		
Total	17	79.11111			

	Coefficients	Standard Error	t Stat	P-value
Intercept	9.94444	0.471405	21.0954	7×10^{-11}
May Corrode	−1.7778	0.3849	−4.6188	0.0006
Bitter/Metallic	−2.6667	0.471405	−5.6569	0.0001
Chemical/Solvent	−2.5	0.471405	−5.3033	0.0002
$10	−1.5	0.471405	−3.182	0.0079
$30	−3.1667	0.471405	−6.7175	2×10^{-5}

The estimated coefficients indicate how the explanatory variables are related to the dependent variable. For example, if the pipes may corrode under select conditions, ranking decreases by 1.78 points. These results are also useful in predicting the ranking based on different values of the explanatory variables.

The standard errors for the regression coefficients indicate the accuracy of the estimated coefficients and the R^2 indicates the goodness of fit for the model. In this case 90% of the variation in ranking is explained by the independent variables included in the regression model.

Similar to the CV regression analysis, the value of Significance F shows the significance of the model results, and the *P*-value indicates the significance of each coefficient estimate.

15.3.2 Analytical Hierarchical Process (AHP)

The Analytical Hierarchical Process (AHP) determines the preference for a decision-making unit using a pair-wise comparison of attributes. Assessing pair-wise preferences enables the decision maker to concentrate his/her judgment on two elements with regards to a single property. So, the decision maker does not need to think of other properties or elements while comparing and deriving the final decision. The formal process includes the following steps (Lee, 2008, 2015; Lee *et al.*, 2009, 2013):

Step 1 [Use the standard preference table]: A scale (1–9) of pair-wise preference weights are given in Table 15.13 (Saaty, 1980).

Step 2 [Develop the pair-wise preference matrix]: Instead of assessing the weight for attribute, i, directly, the relative weight $a_{ij} = w_i/w_j$ between attribute i and j is assessed, which is why we call it pair-wise comparison. Each participant is asked to fill in an $n \times n$ attribute matrix of pair-wise preferential weights using standard numerical scores (Tables 15.14 and 15.15).

Table 15.13 Standard numerical score.

Preference Level	Numerical Score, $a(i,j)$ 1–9 Scale
Equally preferred	1
Equally to moderately preferred	2
Moderately preferred	3
Moderately to strongly preferred	4
Strongly preferred	5
Strongly to very strongly preferred	6
Very strongly preferred	7
Very strongly to extremely preferred	8
Extremely preferred	9

Table 15.14 Pair-wise preference weight matrix [A].

	Attribute 1	Attribute 2	...	Attribute n
Attribute 1	w_1/w_1	w_1/w_2	...	w_1/w_n
Attribute 2	w_2/w_1	w_2/w_2	...	w_2/w_n
...	...	...	...	...
Attribute n	w_n/w_1	w_n/w_2	...	w_n/w_n
Sum	X/w_1	X/w_2	...	X/w_n

in which: $X = (w_1 + w_2 + ... + w_n)$.

Table 15.15 Rescaled pair-wise preference matrix [A_{norm}].

	Attribute 1	Attribute 2	...	Attribute n	Average
Attribute 1	w_1/X	w_1/X	...	w_1/X	$\text{Ave}(w_1/X)$
Attribute 2	w_2/X	w_2/X	...	w_2/X	$\text{Ave}(w_2/X)$
...	...	...	...	...	...
Attribute n	w_n/X	w_n/X	...	w_n/X	$\text{Ave}(w_n/X)$
Sum	1	1	1	1	1

in which: $X = (w_1 + w_2 + ... + w_n)$.

$$\begin{bmatrix} w_1/w_1 & w_1/w_2 & ... & w_1/w_n \\ w_2/w_1 & w_2/w_2 & ... & w_2/w_n \\ ... & ... & ... & ... \\ w_n/w_1 & w_n/w_2 & ... & w_n/w_n \end{bmatrix} \begin{Bmatrix} w_1 \\ w_2 \\ ... \\ w_n \end{Bmatrix} = n \begin{Bmatrix} w_1 \\ w_2 \\ ... \\ w_n \end{Bmatrix} \qquad (15.8)$$

in which: $\begin{Bmatrix} w_1 \\ w_2 \\ ... \\ w_n \end{Bmatrix} = \{w\} =$ required global preference vector (of weights). Note that Table 15.14 is a reciprocal matrix in that the off diagonal elements are reciprocals of each other, Numerical score,

$a(i,j) = [weight\,for\,criterion, i, w_i]/[weight\,for\,criterion, j, w_i] = 1/a(j,i)$

Step 3 [Evaluate the re-scaled pair-wise preference matrix]: A rescaled preference matrix is generated by dividing each column entry in Table 15.14 by that column's sum yielding Table 15.15. The last column of Table 15.15 (average column) shows the ranking of the attributes which is the relative preference vector between criteria 1, 2, and n obtained by averaging the columns of the rescaled pair-wise matrix.

Step 4 [Preference evaluation for piping materials]: Each participant is asked to complete an $m \times m$ material matrix of pair-wise preferential weights using standard numerical scores (Table 15.16). Results for $(n-1)$th attribute are shown in Table 15.16. Table 15.17 is the rescaled matrix in which the weights sum to 1. The procedure is iterated for n attributes. The results are shown in Table 15.18 as the $m \times n$ matrix. Multiplying the pipe material preference matrix (Table 15.18) and the attribute preference vector (Table 15.19) yields Table 15.20.

Based on the relative preference score in Table 15.20, final ranks are determined. It is critical that participant's comparison results be consistent enough to provide reliable estimates of his/her preferences. In Step 5, the consistency checks for both all matrices are performed. Participants should reassess the pair-wise weights if the consistency check failed.

Table 15.16 Pair-wise matrix for the $(n–1)$th attribute.

	Mat. 1	**Mat. 2**	**...**	**Mat. m**
Mat. 1	y_1/y_1	y_1/y_2	...	y_1/y_m
Mat. 2	y_2/y_1	y_2/y_2	...	y_2/y_m
...	...	...	...	...
Mat. M	y_m/y_1	y_m/y_2	...	y_m/y_m
Sum	Y/y_1	Y/y_2	...	Y/y_m

in which: $Y = (y_1 + y_2 + ... + y_m)$.

Table 15.17 Rescaled matrix for the $(n–1)$th attribute.

	Mat. 1	**Mat. 2**	**...**	**Mat. m**	**Average**
Mat. 1	y_1/Y	y_1/Y	...	y_1/Y	$Ave(y_1/Y)$
Mat. 2	y_2/Y	y_2/Y	...	y_2/Y	$Ave(y_2/Y)$
...	...	...	...	...	...
Mat. m	y_m/Y	y_m/Y	...	y_m/Y	$Ave(y_m/Y)$
Sum	1	1	1	1	1

in which: $Y = (y_1 + y_2 + ... + y_m)$.

Table 15.18 Average ranking of materials for each attribute (pipe material preference matrix).

	Attribute 1	**Attribute 2**	**...**	**Attribute (n–1)**	**Attribute n**
Mat. 1	$Ave(y_{(1,1)}/Y)$	$Ave(y_{(2,1)}/Y)$	...	$Ave(y_{((n-1),1)}/Y)$	$Ave(y_{(n,1)}/Y)$
Mat. 2	$Ave(y_{(1,2)}/Y)$	$Ave(y_{(2,2)}/Y)$	...	$Ave(y_{((n-1),2)}/Y)$	$Ave(y_{(n,2)}/Y)$
...	...	...	...	...	...
Mat. m	$Ave(y_{(1,m)}/Y)$	$Ave(y_{(2,m)}/Y)$	...	$Ave(y_{((n-1),m)}/Y)$	$Ave(y_{(n,m)}/Y)$

in which Ave $(y_{(n,m)}/Y)$ is the averaged value for material m regarding attribute n.

Table 15.19 Attribute preference vector.

	Average
Attribute 1	$\text{Ave}(w_1/X)$
Attribute 2	$\text{Ave}(w_2/X)$
...	...
Attribute n	$\text{Ave}(w_n/X)$

in which: $X = (w_1 + w_2 + \dots + w_n)$.

Table 15.20 Final preference matrix.

	Preference
Mat. 1	$\text{Ave}(y_{(1,1)}/Y)^* \text{Ave}(w_1/X) + \text{Ave}(y_{(2,1)}/Y)^* \text{Ave}(w_2/X) + \dots + \text{Ave}(y_{(n,1)}/Y)^* \text{Ave}(w_n/X)$
Mat. 2	$\text{Ave}(y_{(1,2)}/Y)^* \text{Ave}(w_1/X) + \text{Ave}(y_{(2,2)}/Y)^* \text{Ave}(w_2/X) + \dots + \text{Ave}(y_{(n,2)}/Y)^* \text{Ave}(w_n/X)$
...	...
Mat. m	$\text{Ave}(y_{(1,m)}/Y)^* \text{Ave}(w_1/X) + \text{Ave}(y_{(2,m)}/Y)^*\text{Ave}(w_2/X) + \dots + \text{Ave}(y_{(n,m)}/Y)^* \text{Ave}(w_n/X)$

Step 5 [Perform consistency check]: The maximum eigenvalue in Table 15.14 ($[\mathbf{A}]$) is:

$$\lambda_{\max} = n \tag{15.9}$$

where n is number of attributes. The eigenvalue of $\mathbf{A}$ can be found by solving:

$$\mathbf{Ax} = \lambda\mathbf{x} \tag{15.10}$$

where $\mathbf{x}$: $n \times 1$ matrix $\mathbf{x}$ (eigenvector). If the actual eigenvalue is different from n, there are inconsistencies in the weight assignments. Saaty (1980) defines a consistency index as:

$$\text{C.I.} = \frac{\lambda_{\max} - n}{n - 1} \tag{15.11}$$

Table 15.21 contains the Random Index (R.I.) values calculated from randomly generated weights as a function of the pair-wise matrix size (number of criteria). Based on many randomly simulated outcomes, Saaty (1980) suggests that if the ratio of C.I. to R.I.:

$$\frac{C.I.}{R.I.} < 0.1 \tag{15.12}$$

the preference assessments should be taken as consistent.

Table 15.21 Random Index (R.I.)

Matrix Size (*n*)	R.I.
3	0.58
4	0.9
5	1.12
6	1.24
7	1.32
8	1.41
9	1.45
10	1.49

	Price	Corrosion	Fire	Health	Longevity	Resale	Taste Odor
Price	1.00	0.50	1.00	0.14	0.20	1.00	0.14
Corrosion	2.00	1.00	2.00	0.33	1.00	2.00	0.25
Fire	1.00	0.50	1.00	0.14	0.20	0.25	0.14
Health	7.00	3.00	7.00	1.00	3.00	7.00	1.00
Longevity	5.00	1.00	5.00	0.33	1.00	5.00	0.33
Resale	1.00	0.50	4.00	0.14	0.20	1.00	0.14
Taste Odor	7.00	4.00	7.00	1.00	3.00	7.00	1.00
Sum	24.00	10.50	27.00	3.08	8.60	23.25	3.00

	P	C	F	H	L	R	T	Average
P	0.04	0.05	0.04	0.05	0.02	0.04	0.05	0.04
C	0.08	0.10	0.07	0.11	0.12	0.09	0.08	0.09
F	0.04	0.05	0.04	0.05	0.02	0.01	0.05	0.04
H	0.29	0.29	0.26	0.32	0.35	0.30	0.33	0.31
L	0.21	0.10	0.19	0.11	0.12	0.22	0.11	0.15
R	0.04	0.05	0.15	0.05	0.02	0.04	0.05	0.06
T	0.29	0.38	0.26	0.32	0.35	0.30	0.33	0.32
Sum	1.00	1.00	1.00	1.00	1.00	1.00	1.00	1.00

Figure 15.1 Attribute comparisons.

15.3.2.1 Example

Figure 15.1 shows the pair-wise preference weight matrix [A] for the criteria that we consider. In this problem, we consider price, corrosion, fire resistance, health, longevity, resale, and taste odor of different pipe materials. We obtain the rescaled [Anorm] matrix (Figure 15.1).

The pair-wise weight matrices for the three different pipe materials are also obtained (Figure 15.2).

We obtain the final ranking of the three different materials for the seven criteria. The average ranking of the criteria is shown in Figure 15.3. We also perform the consistency check as follows. CI/RI ratio is less than 0.1, so the decision maker's preference elicitation is consistent.

Price							
	Mat. A	Mat. B	Mat. C		Mat. A	Mat. B	Mat. C	Average
Mat. A	1.000	0.200	5.000	Mat. A	0.161	0.153	0.333	0.216
Mat. B	5.000	1.000	9.000	Mat. B	0.806	0.763	0.600	0.723
Mat. C	0.200	0.111	1.000	Mat. C	0.032	0.085	0.067	0.061
Sum	6.200	1.311	15.000	Sum	1.000	1.000	1.000	1.000

Corrosion Resistance							
	Mat. A	Mat. B	Mat. C		Material 1	Material 2	Material 3	Average
Mat. A	1.000	0.250	0.200	Material 1	0.100	0.024	0.153	0.092
Mat. B	4.000	1.000	0.111	Material 2	0.400	0.098	0.085	0.194
Mat. C	5.000	9.000	1.000	Material 3	0.500	0.878	0.763	0.714
Sum	10.000	10.250	1.311	Sum	1.000	1.000	1.000	1.000

Fire Retardant							
	Mat. A	Mat. B	Mat. C		Material 1	Material 2	Material 3	Average
Mat. A	1.000	5.000	1.000	Material 1	0.455	0.455	0.455	0.455
Mat. B	0.200	1.000	0.200	Material 2	0.091	0.091	0.091	0.091
Mat. C	1.000	5.000	1.000	Material 3	0.455	0.455	0.455	0.455
Sum	2.200	11.000	2.200	Sum	1.000	1.000	1.000	1.000

Health Effects							
	Mat. A	Mat. B	Mat. C		Material 1	Material 2	Material 3	Average
Mat. A	1.000	0.333	0.111	Material 1	0.077	0.053	0.085	0.071
Mat. B	3.000	1.000	0.200	Material 2	0.231	0.158	0.153	0.180
Mat. C	9.000	5.000	1.000	Material 3	0.692	0.789	0.763	0.748
Sum	13.000	6.333	1.311	Sum	1.000	1.000	1.000	1.000

Longevity/ Cold/Hot water							
	Mat. A	Mat. B	Mat. C		Material 1	Material 2	Material 3	Average
Mat. A	1.000	4.000	0.333	Material 1	0.235	0.400	0.217	0.284
Mat. B	0.250	1.000	0.200	Material 2	0.059	0.100	0.130	0.096
Mat. C	3.000	5.000	1.000	Material 3	0.706	0.500	0.652	0.619
Sum	4.250	10.000	1.533	Sum	1.000	1.000	1.000	1.000

Resale value of home							
	Mat. A	Mat. B	Mat. C		Material 1	Material 2	Material 3	Average
Mat. A	1.000	3.000	1.000	Material 1	0.429	0.429	0.429	0.429
Mat. B	0.333	1.000	0.333	Material 2	0.143	0.143	0.143	0.143
Mat. C	1.000	3.000	1.000	Material 3	0.429	0.429	0.429	0.429
Sum	2.333	7.000	2.333	Sum	1.000	1.000	1.000	1.000

Taste/ odor							
	Mat. A	Mat. B	Mat. C		Material 1	Material 2	Material 3	Average
Mat. A	1.000	0.111	0.111	Material 1	0.091	0.011	0.053	0.052
Mat. B	1.000	1.000	1.000	Material 2	0.091	0.099	0.474	0.221
Mat. C	9.000	9.000	1.000	Material 3	0.818	0.890	0.474	0.727
Sum	11.000	10.111	2.111	Sum	1.000	1.000	1.000	1.000

Figure 15.2 Materials comparisons for each attribute.

	Average
	0.041
	0.092

	P	C	F	H	L	R	T	Average
Mat. A	0.216	0.092	0.455	0.071	0.284	0.429	0.052	0.036
Mat. B	0.723	0.194	0.091	0.180	0.096	0.143	0.221	0.306
Mat. C	0.061	0.714	0.455	0.748	0.619	0.429	0.727	0.148

Average
0.057
0.320

	Final Preference
Mat. A	0.138
Mat. B	0.199
Mat. C	0.663

	Final Ranking
Mat. A	2
Mat. B	3
Mat. C	1

0.297	7.295
0.688	7.462
0.254	7.051
2.280	7.444
1.113	7.515
0.405	7.161
2.373	7.415
	7.335
CI	0.056
CI/RI	0.042

Figure 15.3 Final ranking and consistency check.

15.4 CONCLUSIONS

This chapter covered the three popular decision analysis methods – conjoint analysis, contingent valuation, and analytical hierarchical process in the context of water infrastructure. This type of decision modeling efforts can help make better informed decisions in water infrastructure systems.

REFERENCES

Boyle K. J. (2003). Contingent valuation in practice. In: A Primer on Nonmarket Valuation, P. A. Champ, K. J. Boyle and T. C. Brown (eds.), Kluwer Academic Publishers, Dordrecht, Netherlands, pp. 111–170.

Holmes T. P. and Adamowicz W. L. (2003). Attribute-based methods. In: A Primer on Nonmarket Valuation, P. A. Champ, K. J. Boyle and T. C. Brown (eds.), Kluwer Academic Publishers, Dordrecht, Netherlands, pp. 171–220.

Lee J. (2008). Two Issues in Premise Plumbing: Contamination Intrusion at Service Line and Choosing Alternative Plumbing Material. Doctoral dissertation, Virginia Tech, Blacksburg, VA.

Lee J. (2015). A holistic decision-making framework for selecting domestic piping materials. *Journal of Water Supply: Research and Technology – AQUA*, **64**(3), 326–332, https://doi.org/10.2166/aqua.2015.088

Lee J., Kleczyk E., Bosch D., Tanellari E., Dwyer S. and Dietrich A. (2009). Case study: preference trade-offs toward home plumbing attributes and materials. *Journal of Water Resources Planning and Management*, **135**(4), 237–243, https://doi.org/10.1061/(ASCE)0733-9496(2009)135:4(237)

Lee J., Kleczyk E., Bosch D. J., Dietrich A. M., Lohani V. K. and Loganathan G. V. (2013). Homeowners' decision-making in a premise plumbing failure–prone area. *Journal of the American Water Works Association*, **105**(5), E236–E241, https://doi.org/10.5942/jawwa.2013.105.0071

Mitchell R. C. and Carson R. T. (1989). Using Surveys to Value Public Goods: The Contingent Valuation Method. Resources for the Future, Washington, DC.

National Oceanic and Atmospheric Administration (NOAA). (1993). Natural resource damage assessments under the Oil pollution Act of 1990. *Federal Register*, **58**, 4601–4614.

Saaty T. L. (1980). The analytic hierarchy process (AHP). *Journal of the Operational Research Society*, **41**(11), 1073–1076.

Tanellari E., Bosch D., Boyle K. and Mykerezi E. (2015). On consumers' attitudes and willingness to pay for improved drinking water quality and infrastructure. *Water Resources Research*, **51**, 47–57, doi: 10.1002/2013WR014934

doi: 10.2166/9781789062380_0399

Chapter 16

Non-revenue water, what are their determinants?

*Gamze Güngör-Demirci[1] and Juneseok Lee[2],**

[1]*The Cadmus Group, Waltham, MA, USA*
[2]*Department of Civil and Environmental Engineering, Manhattan College, Riverdale, NY, USA*
**Corresponding author E-mail: Juneseok.Lee@manhattan.edu*

LEARNING OBJECTIVES

At the end of this chapter, you will be able to:

(1) Install and run R.
(2) Run a plm package for panel data.
(3) Interpret the fixed effects results and their interpretations in terms of explaining Non-Revenue-Water (NRW).
(4) Assess results based on model interpretability.

16.1 INTRODUCTION

Around the world, more than \$14 billion per year is lost due to water loss (Kingdom *et al.*, 2006), and these losses are primarily covered by paying customers. Water loss is a huge challenge for water utilities, which require fundamental understanding of the influencing factors (Güngör-Demirci *et al.*, 2018a). The Organization for Economic Co-operation and Development (OECD) found out that water loss can go up to 65% for developing countries (OECD, 2016). It is a challenging task to reduce the water loss, even in highly developed countries as well (Thornton *et al.*, 2008). For an effective water loss reduction program, it is critical to have a deep understanding of the causal factors as well as why its reduction is so challenging (van den Berg, 2015).

Many literatures cited environmental, managerial, physical, sociological, and technical factors. The examples include system age, pipe length/layouts of the systems, hydraulic conditions, external soil characteristics and topography, traffic loading, and service connection densities (Güngör-Demirci *et al.*, 2018a, 2018b, 2018c). A more realistic water loss target program can be developed once these factors are understood at each system level.

To accurately measure water loss in water distribution systems, the International Water Association (IWA) Task Forces on Water Losses and Performance Indicator recommended use of the term *non-revenue water* (NRW). NRW is defined as the difference between the input water volume (into water distribution systems) and the billed total volume of water (Alegre *et al.*, 2006). The IWA Water Balance Standard can be used to conduct water loss volume estimation in the network, and Figure 16.1 shows the overall framework for water balance. In the following is the mathematical formulation to understand

<table>
<tr><td rowspan="6">System Input Volume (SIV)</td><td rowspan="4">Authorized Consumption</td><td rowspan="2">Billed Authorized Consumption</td><td>Billed Metered Consumption</td><td rowspan="2">Revenue Water</td></tr>
<tr><td>Billed Unmetered Consumption</td></tr>
<tr><td rowspan="2">Unbilled Authorized Consumption</td><td>Unbilled Metered Consumption</td><td rowspan="6">Non-Revenue-Water (NRW)</td></tr>
<tr><td>Unbilled Unmetered Consumption</td></tr>
<tr><td rowspan="5">Water losses</td><td rowspan="2">Apparent Losses</td><td>Unauthorized Consumption</td></tr>
<tr><td>Customer Metering Inaccuracies and Data Handling Errors</td></tr>
<tr><td rowspan="3">Real Losses</td><td>Leakage on Transmission and/or Distribution Mains</td></tr>
<tr><td>Losses at Utility's Storage Tanks</td></tr>
<tr><td>Leakage on Service Connections up to Point of Customer Use</td></tr>
</table>

Figure 16.1 IWA Standard International Water Balance (adopted from Farley & Trow, 2003).

the factors in explaining the NRW for water systems. We believe that this type of study will help reduce NRW and set realistic targets for the water utilities that consider water loss reduction program.

16.2 REGRESSION MODEL SPECIFICATION

In this chapter, we will use the fixed effects panel regression model to understand the impacts of independent variables on the NRW. As covered in Chapter 2, the fixed effects model attempts to control for unobservable factors among each district by assigning unique time-invariant identifiers. A simplified form of a fixed effects model is as follows (Hsiao, 2003):

$$y_{it} = \alpha_i + \beta x_{it} + \mu_{it} \tag{16.1}$$

where y_{it} is the dependent variable observed for district i ($i=1....N$) at time t ($t=1...T$); α_i is the unknown intercept for each district (i.e., N entity-specific intercepts); x_{it} is the independent variable; β is the coefficient for that independent variable; and μ_{it}, is the error term. In this case, the intercept value, α_i, depends on omitted factors specific to each district i that are possibly correlated with the chosen independent variables, x_{it}. Any time-invariant variables that may have an effect on NRW are thus absorbed into the intercept term. The error term μ_{it} represents effects from unique district factors that were not accounted for or are uncorrelated with identified independent variables. District heterogeneity is assumed to have an influence on water loss, hence a fixed effects model is adopted. In addition, the Hausman test was performed to justify the adoption of fixed effects model over the random effects model (Tanverakul & Lee, 2015). The statistical program R was used to perform the analysis (Croissant *et al.*, 2016).

We set the model as follows:

$$(\text{NRW\_CON\_DAY})_{it} = \alpha_i + \beta_1 (\text{NETLEN}) + \beta_2 (\text{CON\_DENS})$$
$$+ \beta_3 (\text{LEAK}) + \beta_4 (\text{NET\_OPREV}) + \mu_{it} \tag{16.2}$$

where (NRW_CON_DAY) is expressed as m³ of water lost/connection/day, as a dependent variable. Four explanatory (i.e., independent) variables are included on the right-hand side of the panel regression equation (Equation 16.2). The calculated coefficients measure the elasticity of NRW, revealing how much NRW varies in response to a change in the various drivers.

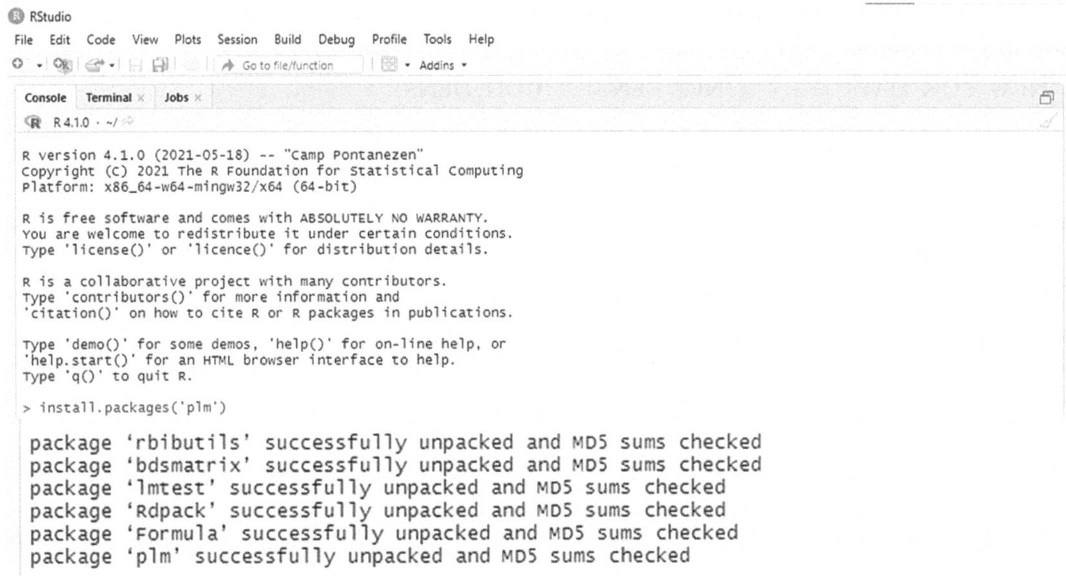

Figure 16.2 Installing plm package.

16.2.1 Example

plm package. 'lm: Linear Models for Panel Data' is an R package including a set of estimators and tests for panel data econometrics (Croissant *et al.*, 2021). In this chapter, we will see an example run for having fixed effects panel regression analysis.

Let us start with installing plm in RStudio by typing the following line of code (Figure 16.2):

```
install.packages('plm')
```

Package 'plm' Version 2.4-1 is installed on RStudio (RStudio Desktop 1.4.1717) as shown below. The RStudio is run on R 4.1.0 for Windows (86 megabytes, 32/64 bit).

16.2.2 Water utility example

Now that the plm package is installed, let us move on to our example. The data used for this example is considered to be panel data as it includes time series observations of a number of individual water utilities. In other words, the data structure involves two dimensions: a time series dimension and a cross-sectional dimension. For more information about the analysis of panel data, you can check Hsiao (2003). The example data belongs to five different utilities for the period from 1998 to 2014. The data set contains a total of 76 year-utility combinations since some utilities do not have data for some years.

In our fixed effects panel regression model, non-revenue water (NRW), expressed as m³ of water lost/connection/day (NRW_CON_DAY), is selected as the dependent variable. Our independent variables are: (1) network length (NET_LEN), expressed as km of network, (2) connection density (CON_DENS), expressed as number of connections per km of network, (3) the number of pipe failures per year (LEAKS), and (4) the difference between operating revenue and operation and maintenance cost per cubic meter of water sold (NET_OPREV). The final regression equation that we solve here is represented as follows:

$$\left(\text{NRW\_CON\_DAY}\right)_{it} = \alpha_i + \beta_1\left(\text{NET\_LEN}\right) + \beta_2\left(\text{CON\_DENS}\right)$$
$$+\beta_3\left(\text{LEAK}\right) + \beta_4\left(\text{NET\_OPREV}\right) + \mu_{it} \tag{16.3}$$

Here, i represents utility (i.e., 1 to 5) and t represents time (i.e., 1 to 17). α_i is the unknown intercept for each utility (i.e., five utility-specific intercepts); β is the coefficient for the independent variable; and μ_{it} is the error term. The calculated coefficients, β, will measure how much NRW varies in response to a change in these four different factors.

In this case, the intercept value, α_i, depends on omitted factors specific to each utility, i, that are possibly correlated with the chosen independent variables. Any time-invariant variables that may have an influence on NRW are thus incorporated into the intercept term. The error term, μ_{it}, represents impacts derived from unique utility factors that were not accounted for or are uncorrelated with the independent variables.

Our data file is called 'NRW_data.txt' and includes columns for year, ID_no (the ID number of the water utility; 1 to 5), NRW_CON_DAY, NET_LEN, CON_DENS, LEAKS and NET_OPREV.

Before we start, let us make sure that your plm package is active (Figure 16.3).

First, we need to read our txt file as a data table by typing the following line of code (Figure 16.4):

nrw <-read.table ('NRW_data.txt', header = TRUE, sep = '\t')

Remember, in this case, the data file is in the default working directory. If your data file is not in your default working directory, you can change it by typing:

setwd ('C:/(name of your data folder)')

Before starting our analysis, we should check multicollinearity. In other words, we have to make sure that there is no linear correlation among the independent variables we use. For this purpose, we

Files	Plots	Packages	Help	Viewer		
Install	Update					
	Name		Description		Version	
User Library						
	bdsmatrix		Routines for Block Diagonal Symmetric Matrices		1.3-4	
	digest		Create Compact Hash Digests of R Objects		0.6.27	
	Formula		Extended Model Formulas		1.2-4	
	lmtest		Testing Linear Regression Models		0.9-38	
	maxLik		Maximum Likelihood Estimation and Related Tools		1.4-8	
	miscTools		Miscellaneous Tools and Utilities		0.6-26	
✓	plm		Linear Models for Panel Data		2.4-1	
	rbibutils		Read 'Bibtex' Files and Convert Between Bibliography Formats		2.2.1	
	rDEA		Robust Data Envelopment Analysis (DEA) for R		1.2-6	
	Rdpack		Update and Manipulate Rd Documentation Objects		2.1.2	
	sandwich		Robust Covariance Matrix Estimators		3.0-1	
	slam		Sparse Lightweight Arrays and Matrices		0.1-48	
	truncnorm		Truncated Normal Distribution		1.0-8	
	truncreg		Truncated Gaussian Regression Models		0.2-5	
	zoo		S3 Infrastructure for Regular and Irregular Time Series (Z's Ordered Observations)		1.8-9	

Figure 16.3 Checking plm package.

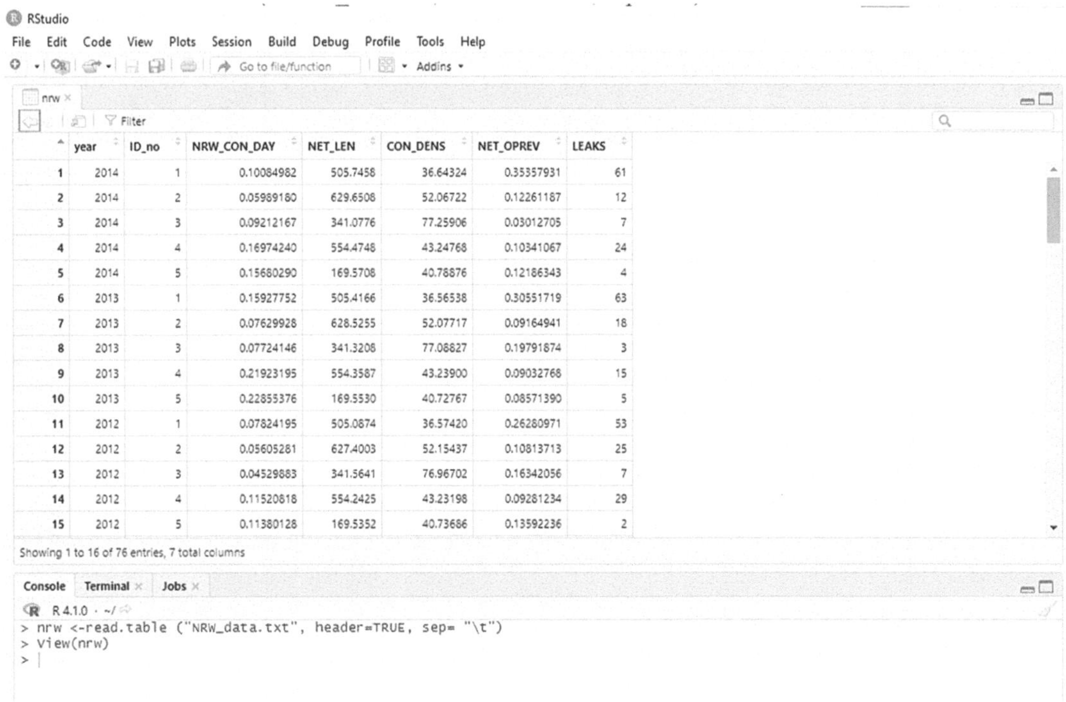

Figure 16.4 Data loading.

need to find the variance inflation factor (VIF) and it should be less than 5 to show an acceptable level of collinearity. For more information on VIF, please check Chapter 2.

To find VIF, we should first install the package 'car' by typing:

```
install.packages('car')
```

Then, we can write the following two lines of code (Figure 16.5):

```
nrw.lm <- lm(NRW_CON_DAY~NET_LEN+LEAKS+NET_OPREV+CON_DENS, data=nrw, x=TRUE)
  vif(nrw.lm)
```

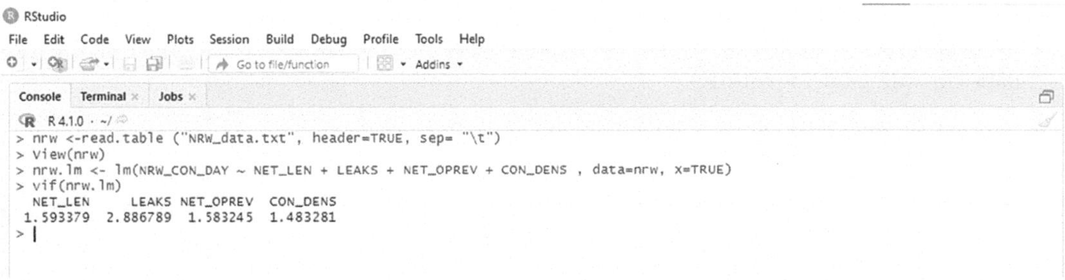

Figure 16.5 VIF calculation.

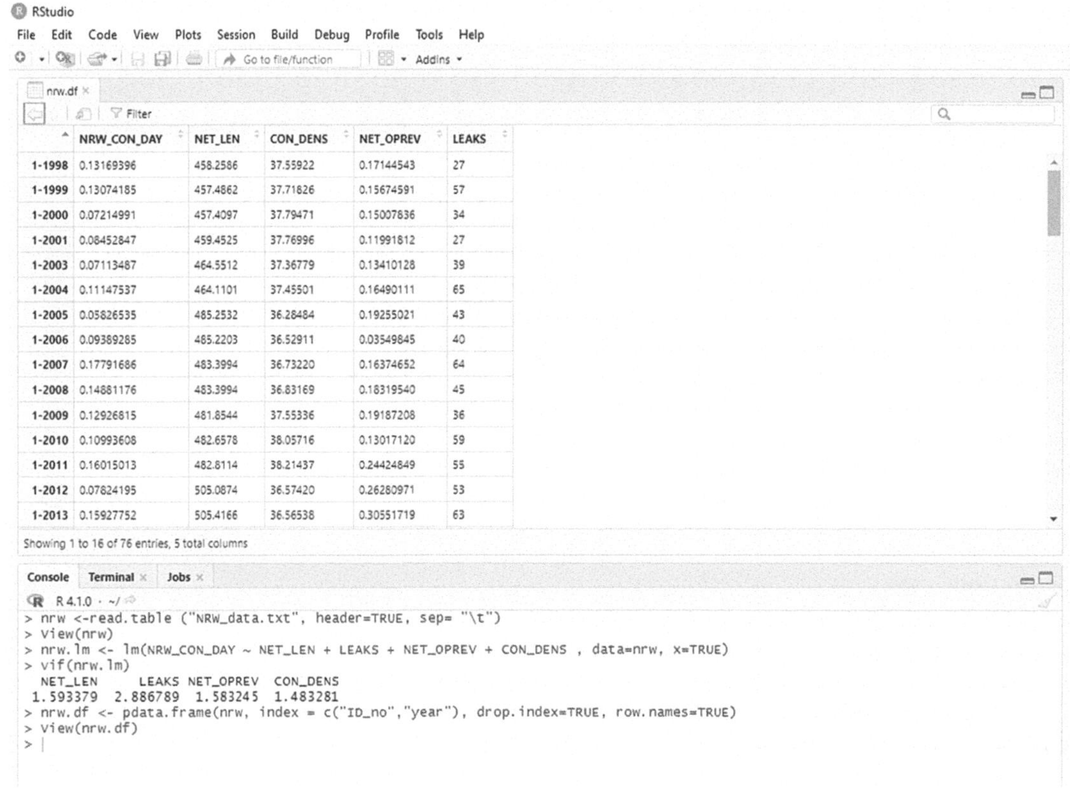

Figure 16.6 Conversion to panel data.

Since all VIFs are less than 5, we can use all these independent variables in our fixed effects model. Now, let us convert our data table (i.e., nrw) to panel data frame by typing (Figure 16.6):

nrw.df <- pdata.frame(nrw, index = c('ID_no', 'year'), drop.index = TRUE, row.names = TRUE)

'pdata.frame' is a data.frame with an index attribute that describes its individual and time dimensions. As you see in the below screenshot, utility ID numbers are now matched with the year to construct the panel data.

Now it is time to run our fixed effects model by typing the following line of code:

fixed_effects <- plm(NRW_CON_DAY~NET_LEN+CON_DENS+NET_OPREV+LEAKS, data = nrw.df, model = 'within')

We can see the summary of the results by typing (Figure 16.7):

summary(fixed_effects)

According to Endsley (2016), the R-squared value given by the plm does not represent the goodness-of-fit statistic for the full model. So, to find the R-squared value of our full model, let us write the following lines of code as follows (Figure 16.8):

```
RStudio
File   Edit   Code   View   Plots   Session   Build   Debug   Profile   Tools   Help
                                            Go to file/function          Addins

Console   Terminal    Jobs

R 4.1.0 · ~/
> nrw <-read.table ("NRW_data.txt", header=TRUE, sep= "\t")
> View(nrw)
> nrw.lm <- lm(NRW_CON_DAY ~ NET_LEN + LEAKS + NET_OPREV + CON_DENS , data=nrw, x=TRUE)
> vif(nrw.lm)
   NET_LEN      LEAKS NET_OPREV   CON_DENS
  1.593379   2.886789  1.583245   1.483281
> nrw.df <- pdata.frame(nrw, index = c("ID_no","year"), drop.index=TRUE, row.names=TRUE)
> View(nrw.df)
> fixed_effects <- plm(NRW_CON_DAY ~ NET_LEN + CON_DENS + NET_OPREV + LEAKS, data = nrw.df, model = "within")
> summary(fixed_effects)
Oneway (individual) effect Within Model

Call:
plm(formula = NRW_CON_DAY ~ NET_LEN + CON_DENS + NET_OPREV +
    LEAKS, data = nrw.df, model = "within")

Unbalanced Panel: n = 5, T = 11-17, N = 76

Residuals:
      Min.     1st Qu.     Median     3rd Qu.        Max.
 -0.1536478  -0.0350343  -0.0096304   0.0336560   0.1300530

Coefficients:
             Estimate   Std. Error  t-value  Pr(>|t|)
NET_LEN    -0.00113574   0.00079606  -1.4267   0.15831
CON_DENS   -0.01757078   0.00929184  -1.8910   0.06295 .
NET_OPREV  -0.18610270   0.12741329  -1.4606   0.14879
LEAKS       0.00199645   0.00105194   1.8979   0.06202 .
---
Signif. codes:  0 '***' 0.001 '**' 0.01 '*' 0.05 '.' 0.1 ' ' 1

Total Sum of Squares:      0.25221
Residual Sum of Squares:  0.2203
R-Squared:       0.12652
Adj. R-Squared: 0.022219
F-statistic: 2.42608 on 4 and 67 DF, p-value: 0.056394
>
```

Figure 16.7 Summary of fixed effect model.

```
sst <- with(nrw.df, sum((NRW_CON_DAY – mean(NRW_CON_DAY))^2))
fixed_effects.sse <- t(residuals(fixed_effects)) %*% residuals(fixed_effects)
(sst – fixed_effects.sse)/sst
```

16.2.3 Interpretations

Results obtained by the fixed effect model run above shows CON_DENS and LEAKS as statistically significant variables. The overall fit of the model (i.e., R^2 value) is found be 0.40. The negative correlation between NRW_CON_DAY and CON_DENS is explained as the loss of less water in more densely connected areas because of the lower network maintenance cost per connection (González-Gómez *et al.*, 2012). The positive sign of LEAK indicates that the number of pipe failures each year is correlated with NRW per connection per day as fewer pipe failures indicate a higher quality of maintenance and network integrity and hence a lower level of NRW.

It is important to note that any other utility using this model may get a different correlation result due to their unique physical/mechanical characteristics of water distribution systems as well as socioeconomic issues/status, utility's private–public structure, political situations, and so on. However, this approach/method can be applied to any water utilities to further understand the fundamental nature/determinants of NRW.

```
> nrw <-read.table ("NRW_data.txt", header=TRUE, sep= "\t")
> View(nrw)
> nrw.lm <- lm(NRW_CON_DAY ~ NET_LEN + LEAKS + NET_OPREV + CON_DENS , data=nrw, x=TRUE)
> vif(nrw.lm)
   NET_LEN     LEAKS NET_OPREV  CON_DENS
 1.593379  2.886789  1.583245  1.483281
> nrw.df <- pdata.frame(nrw, index = c("ID_no","year"), drop.index=TRUE, row.names=TRUE)
> View(nrw.df)
> fixed_effects <- plm(NRW_CON_DAY ~ NET_LEN + CON_DENS + NET_OPREV + LEAKS, data = nrw.df, model = "within")
> summary(fixed_effects)
Oneway (individual) effect Within Model

Call:
plm(formula = NRW_CON_DAY ~ NET_LEN + CON_DENS + NET_OPREV +
    LEAKS, data = nrw.df, model = "within")

Unbalanced Panel: n = 5, T = 11-17, N = 76

Residuals:
      Min.    1st Qu.     Median     3rd Qu.        Max.
-0.1536478 -0.0350343 -0.0096304  0.0336560  0.1300530

Coefficients:
             Estimate  Std. Error t-value Pr(>|t|)
NET_LEN    -0.00113574  0.00079606 -1.4267  0.15831
CON_DENS   -0.01757078  0.00929184 -1.8910  0.06295 .
NET_OPREV  -0.18610270  0.12741329 -1.4606  0.14879
LEAKS       0.00199645  0.00105194  1.8979  0.06202 .
---
Signif. codes:  0 '***' 0.001 '**' 0.01 '*' 0.05 '.' 0.1 ' ' 1

Total Sum of Squares:    0.25221
Residual Sum of Squares: 0.2203
R-Squared:      0.12652
Adj. R-Squared: 0.022219
F-statistic: 2.42608 on 4 and 67 DF, p-value: 0.056394
> sst <- with(nrw.df, sum((NRW_CON_DAY - mean(NRW_CON_DAY))^2))
> fixed_effects.sse <- t(residuals(fixed_effects)) %*% residuals(fixed_effects)
> (sst - fixed_effects.sse) / sst
          [,1]
[1,] 0.4064639
> |
```

Figure 16.8 r-squared results.

16.3 CONCLUSION

A utility is committed to managing water resources that is vital to the communities and customers they serve. Many water utilities are also enhancing the business processes that support their Water Audit and Loss Control Program. These improvements will provide a more accurate picture of the components that make up NRW, including unbilled authorized consumption, apparent losses, and real losses. As water infrastructure ages and continues to deteriorate, it is expected that NRW will continue in the field, so close monitoring and improvements to the utility's infrastructure renewal/ maintenance programs, business processes, as well as technology should be further developed. We believe that ongoing research efforts will be able to optimize all these efforts to efficiently and effectively reduce NRW.

REFERENCES

Alegre H., Baptista J. M., Cabrera E., Jr., Cubillo F., Duarte P., Hirner W., Merkel W. and Parena R. (2006). Performance Indicators for Water Supply Services, 2nd edn. IWA Publishing, London.

Croissant Y., Millo G., Tappe K., Toomet O., Kleiber C., Zeileis A., Henningsen A., Andronic L. and Schoenfelder N. (2016). *Linear Models for Panel Data*. Available at: https://cran.r-project.org/web/packages/plm/vignettes/plm.pdf (last accessed 3 March 2022).

Croissant Y., Millo G., Tappe K., Toomet O., Kleiber C., Zeileis A., Henningsen A., Andronic L. and Schoenfelder N. (2021). *Linear Models for Panel Data*. Available at: https://cran.r-project.org/web/packages/plm/index.html (last accessed 3 March 2022).

Endsley K. A. (2016). *Diagnostics for Fixed Effects Panel Models in R*. Available at: http://karthur.org/2016/fixed-effects-panel-models-in-r.html (last accessed 7 September 2021).

Farley M. and Trow S. (2003). Losses in Water Distribution Networks: A Practitioner's Guide to Assessment, Monitoring and Control. IWA Publishing, London.

González-Gómez F., Martínez-Espiñeira R., García-Valiñas M. A. and García-Rubio M. A. (2012). Explanatory factors of urban water leakage rates in southern Spain. *Utilities Policy*, **22**, 22–30, https://doi.org/10.1016/j.jup.2012.02.002

Güngör-Demirci G., Lee J., Keck J., Guzzetta R. and Yang P. (2018a). Determinants of non-revenue water for a water utility in California. *Journal of Water Supply: Research and Technology – AQUA*, **67**(3), 270–278, https://doi.org/10.2166/aqua.2018.152

Güngör-Demirci G., Lee J. and Keck J. (2018b). Assessing the performance of a California water utility using two-stage data envelopment analysis. *Journal of Water Resources Planning and Management*, **144**(4), 05018004, https://doi.org/10.1061/(ASCE)WR.1943-5452.0000921

Güngör-Demirci G., Lee J. and Keck J. (2018c). Measuring water utility performance using nonparametric linear programming. *Civil Engineering and Environmental Systems*, **34**(3–4), 206–220.

Hsiao C. (2003). Analysis of Panel Data. Cambridge University Press, New York.

Kingdom B., Liemberger R. and Marin P. (2006). The Challenge of Reducing Non-Revenue Water (NRW) in Developing Countries. How the Private Sector Can Help: A Look at Performance-Based Service Contracting. Water Supply and Sanitation Sector Board Discussion Paper Series, Paper No. 8. The World Bank, Washington, DC.

OECD. (2016). Factors shaping urban water governance. In: Water Governance in Cities. OECD Publishing, Paris, pp. 15–30. Available at: https://read.oecd-ilibrary.org/governance/water-governance-in-cities_9789264251090-en#page5 (last accessed 3 March 2022).

Tanverakul S. and Lee J. (2015). Impacts of metering on residential water use in California. *Journal American Water Works Association*, **107**(2), 78–79, https://doi.org/10.5942/jawwa.2015.107.0005

Thornton J., Sturm R. and Kunkel G. (2008). Water Loss Control. McGraw-Hill, New York.

van den Berg C. (2015). Drivers of non-revenue water: a cross-national analysis. *Utilities Policy*, **36**, 71–78, https://doi.org/10.1016/j.jup.2015.07.005

doi: 10.2166/9781789062380_0409

Chapter 17

Water utility performance measurements using data envelopment analysis

*Gamze Güngör-Demirci[1] and Juneseok Lee[2]**

[1]*The Cadmus Group, Waltham, MA, USA*
[2]*Department of Civil and Environmental Engineering, Manhattan College, Riverdale, NY, USA*
**Corresponding author. E-mail: Juneseok.Lee@manhattan.edu*

LEARNING OBJECTIVES

At the end of this chapter, you will be able to:

(1) Explain the water utility performance.
(2) Install and run the R and rDEA.
(3) Assess rDEA results based on interpretability and practical implications.

17.1 INTRODUCTION

Efficient water utility management practices have become more vital than ever because of the large gap between available water supply and rising customer demand, as well as unpredictable climate patterns due to changing climate. However, not all water utilities are functioning at the same level of efficiency in their operations. In this chapter, we will develop a useful performance measurement tool and apply it to individual water utility's operations. Measurement of performance assessments for each water utility will identify the opportunities to improve their management deficiencies/economic performances. Also, the performance measurements will provide in-depth insights toward a fully efficient water utility.

Data Envelopment Analysis (DEA) is an optimization tool for measuring efficiencies of organizational subunits, for example, production centers or departments. In addition to conventional DEA methods, we will explore two additional stages to examine the exogenous variables' impacts on the individual water utility's performance: double bootstrap truncated regression and Tobit regression. This chapter is based on previously published works (Güngör-Demirci *et al.*, 2017, 2018).

17.2 METHODS

17.2.1 Efficiency calculation by DEA

DEA is based on nonparametric linear programming (LP) and it measures the efficiencies of each unit/entity that we consider. This model was developed assuming constant return to scale (CRS) and input orientation. In other words, an increase in inputs will lead to a proportional increase in outputs. Alternatively, an output-oriented model minimizes inputs for a given or constant output. In the case of water utility's performances, it is more practically reasonable to use input-oriented models because water utilities must serve all customers that they are responsible for!

Input oriented CRS can be described as follows. For a sample of N Decision Making Units (DMUs) having K inputs and M outputs, the dataset consists of an input matrix of $X = K \times N$ and an output matrix of $Y = M \times N$. For the nth DMU, the inputs and outputs are represented by the column vectors x_n and y_n, respectively. The problem can be formulated as follows:

$$Minimize_{\theta,\lambda}\theta$$
$$subject\ to : -y_n + Y\lambda \geq 0, \quad \theta x_n - X\lambda \geq 0, \quad \lambda \geq 0 \tag{17.1}$$

where θ = the scalar measure of technical efficiency, λ = the $N \times 1$ constants (weights) with non-negativity, y_n = the $M \times 1$ vector of outputs produced by the nth DMU, x_n = the $K \times 1$ vector of inputs used by the nth DMU, Y = the $M \times N$ matrix of outputs of N DMUs in the sample, and X = the $K \times N$ matrix of inputs of the N DMUs (Coelli *et al.*, 2005).

Although the above CRS model may be valid under the condition of all DMUs operating at optimal conditions, this is usually not possible from a practical perspective. So, variable returns to scale (VRS) work better in terms of mimicking reality. VRS assumes that a given increase in inputs will result in a disproportionate output. The linear programming equation of VRS model is written as:

$$Minimize_{\theta,\lambda}\theta$$
$$subject\ to : -y_n + Y\lambda \geq 0, \quad \theta x_n - X\lambda \geq 0, \quad N1'\lambda = 1, \quad \lambda \geq 0 \tag{17.2}$$

where $N1$ = the $N \times 1$ vector of ones. The efficiency value calculated through VRS model (Equation 17.2) gives the pure technical efficiency, without taking scale efficiency into account. Therefore, in this study, the VRS based DEA was performed with an R package 'rDEA: Robust DEA for R' Version 1.2-4 (Simm & Besstremyannaya, 2020).

17.2.2 Input and output variables

A comprehensive literature review revealed that the major input/ouput variables for water water utility's performance measurements are as follows:

- *Input Variables*: operating expenses, capital expenses, network length, number of employees, energy expenses, staff expenses, and material expenses.
- *Output Variables*: operating revenue, the number of connections, the volume of water distributed, measures of water quality, and population served.

In this study, a stepwise procedure using a backward approach was used as the primary selection method (Wagner & Shimshak, 2007). All input and output variables were mean-normalized to remove any imbalance in the data magnitudes as well as a variety of processing of output/reporting problems (i.e., algorithmic/numerical convergence, and round-off errors; Ananda, 2014).

17.2.3 Bias correction for efficiency scores by bootstrapping

The overall deterministic nature of DEA can lead to limitations with respect to interpreting results. This shortcoming can be overcome by applying a bootstrap method in which empirical distributions of efficiencies are derived. The bootstrapping performs resampling with replacement from a given sample and then calculates the statistics from many iterative samples (Mirzaei *et al.*, 2015).

17.2.4 Exogenous variables

Exogenous variables can affect the technical efficiencies of DMUs but are not under control of the managers that are responsible for the operations of each utility. Exogenous variables are location specific (Marques *et al.*, 2014). In this chapter, five exogenous variables are considered: (1) number of connections, (2) customer density, (3) ratio of groundwater volume to total water production, (4) total number of leaks in the given year, and (5) total annual precipitation.

17.3 EXAMPLES

rDEA is an R package to perform a Data Envelopment Analysis and estimate robust DEA scores with or without environmental variables. In this chapter, we will see example runs for estimating DEA scores with conventional DEA, meaning without bias correction and without considering environmental variables, as well as with considering those.

Let us start with installing rDEA in RStudio.

'rDEA: Title Robust Data Envelopment Analysis (DEA) for R' Version 1.2-6 is installed on RStudio (RStudio Desktop 1.4.1717) as shown in Figure 17.1. The RStudio is run on R 4.1.0 for Windows (86 megabytes, 32/64 bit).

Next, let us move on to our water utility performance assessment example. However, before that, it is better to clean the environment that is still filled with data and values from our previous example. We can do this using the following line of code:

rm(list = ls())

For our water utility example, we first upload our data file 'water_utility_data.txt'. This is a tab-delimited text file. We have data for 22 different utilities in our dataset. Our actual data are all mean-normalized before creating this final data file since any imbalance in the data magnitudes can lead to a variety of processing and output/reporting problems (i.e., overall software execution, algorithmic/ numerical convergence, and round-off errors).

We use an input oriented, VRS model by having inputs (X) as energy expenditures (ENERGY) and other operating expenditures (OPEX) and output (Y) as operating revenue (OPREV). Our

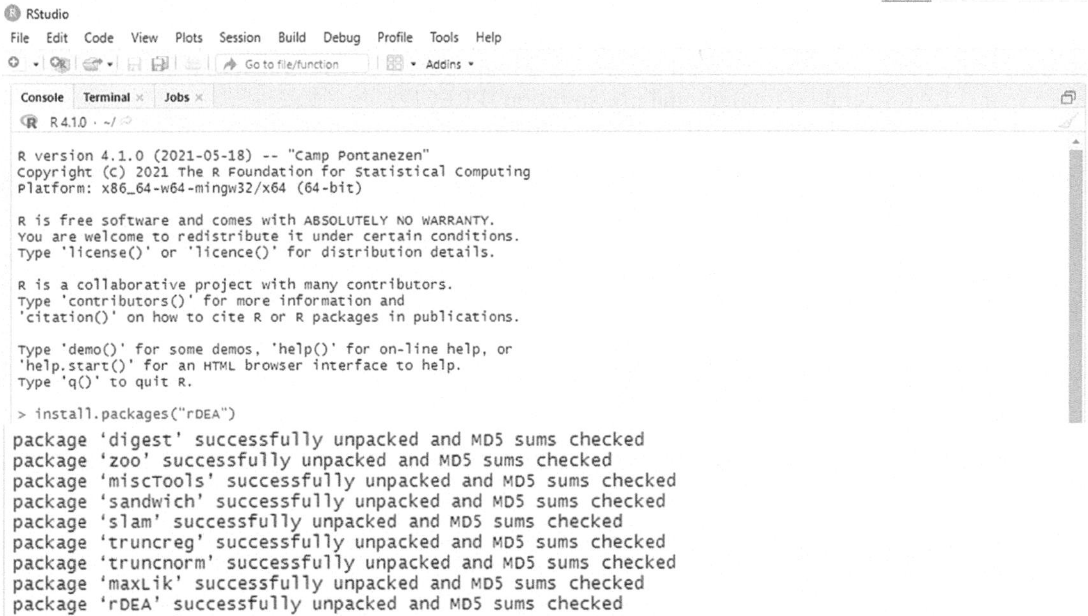

Figure 17.1 RStudio and rDEA installation.

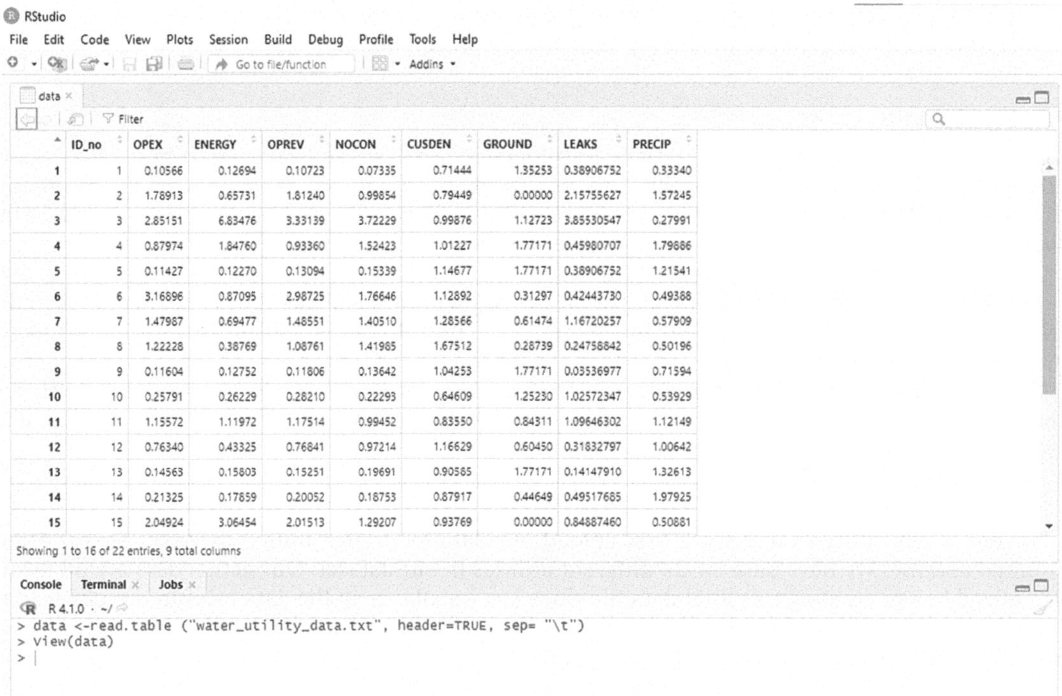

	ID_no	OPEX	ENERGY	OPREV	NOCON	CUSDEN	GROUND	LEAKS	PRECIP
1	1	0.10566	0.12694	0.10723	0.07335	0.71444	1.35253	0.38906752	0.33340
2	2	1.78913	0.65731	1.81240	0.99854	0.79449	0.00000	2.15755627	1.57245
3	3	2.85151	6.83476	3.33139	3.72229	0.99876	1.12723	3.85530547	0.27991
4	4	0.87974	1.84760	0.93360	1.52423	1.01227	1.77171	0.45980707	1.79886
5	5	0.11427	0.12270	0.13094	0.15339	1.14677	1.77171	0.38906752	1.21541
6	6	3.16896	0.87095	2.98725	1.76646	1.12892	0.31297	0.42443730	0.49388
7	7	1.47987	0.69477	1.48551	1.40510	1.28566	0.61474	1.16720257	0.57909
8	8	1.22228	0.38769	1.08761	1.41985	1.67512	0.28739	0.24758842	0.50196
9	9	0.11604	0.12752	0.11806	0.13642	1.04253	1.77171	0.03536977	0.71594
10	10	0.25791	0.26229	0.28210	0.22293	0.64609	1.25230	1.02572347	0.53929
11	11	1.15572	1.11972	1.17514	0.99452	0.83550	0.84311	1.09646302	1.12149
12	12	0.76340	0.43325	0.76841	0.97214	1.16629	0.60450	0.31832797	1.00642
13	13	0.14563	0.15803	0.15251	0.19691	0.90585	1.77171	0.14147910	1.32613
14	14	0.21325	0.17859	0.20052	0.18753	0.87917	0.44649	0.49517685	1.97925
15	15	2.04924	3.06454	2.01513	1.29207	0.93769	0.00000	0.84887460	0.50881

Showing 1 to 16 of 22 entries, 9 total columns

```
Console   Terminal ×   Jobs ×
R 4.1.0 · ~/
> data <-read.table ("water_utility_data.txt", header=TRUE, sep= "\t")
> View(data)
> |
```

Figure 17.2 RStudio and data upload.

environmental variables are: (1) number of connections (NOCON), (2) customer density (CUSDEN), (3) ratio of groundwater volume to total water production (GROUND), (4) total number of leaks in the given year (LEAKS), and (5) total annual precipitation (PRECIP).

First, we need to read our data by typing the following line of code (Figure 17.2):

data <-read.table ('water_utility_data.txt', header = TRUE, sep = '\t')

Remember that in this case the data file is in the default working directory. If your data file is not in your default working directory, you can change it by typing:

setwd ('C:/(name of the data folder')

Let us load our inputs and outputs as (Figure 17.3):

X = data[c('ENERGY', 'OPEX')]
Y = data[c('OPREV')]

When we run the 'dea' function in rDEA by typing below lines, we get the naïve (or in other words, conventional) DEA scores shown in Figure 17.4.

conv_dea_Score = dea(XREF = X, YREF = Y, X, Y, model = 'input', RTS = 'variable')
conv_dea_Score$thetaOpt

Now, with our input (X) and output (Y) matrices already loaded, let us load our environmental variables matrix (Z) by using variables NOCON, CUSDEN, GROUND, LEAKS, and PRECIP (Figure 17.5):

```
RStudio
File  Edit  Code  View  Plots  Session  Build  Debug  Profile  Tools  Help

Console   Terminal   Jobs

 R 4.1.0 · ~/
> data <-read.table ("water_utility_data.txt", header=TRUE, sep= "\t")
> view(data)
> X = data[c('ENERGY', 'OPEX')]
> Y = data[c('OPREV')]
> |
```

Figure 17.3 Loading input and output data.

Z = data[c('NOCON','CUSDEN','GROUND', 'LEAKS','PRECIP')]

We have data belonging to 22 utilities in this data set and we do our analysis for all of them. The total number of bootstrap iterations is taken as 2000 for both loops (L1 = 2000, L2 = 2000) while the confidence interval is 95% (i.e., alpha = 0.05). Let us input the lines of code below to calculate bias corrected DEA scores and its upper and lower bound:

second_stage_dea = dea.env.robust(X = X[utility,],   Y = Y[utility,],   Z = Z[utility,],   model = 'input',
    RTS = 'variable', L1 = 2000, L2 = 2000, alpha = 0.05)
second_stage_dea$delta_hat_hat
second_stage_dea$delta_ci_low
second_stage_dea$delta_ci_high

It may take a few seconds for rDEA to finish the run. Figure 17.6 shows the reciprocal of bias corrected DEA scores found by using the delta_hat_hat function along with the vector of the lower and upper bounds of confidence interval for delta_hat_hat (bias corrected reciprocal of DEA score). You need to take the reciprocals of these to find the bias corrected DEA scores.

Now it is time to see the effect of environmental variables by finding the regression coefficients and their upper and lower bounds. You will use beta_hat_hat and beta_ci functions for this purpose as shown in Figure 17.7. To find statistical significance of these environmental variables on utilities' efficiencies, you can refer to the article by Altman and Bland (2011).


```
RStudio
File  Edit  Code  View  Plots  Session  Build  Debug  Profile  Tools  Help

Console   Terminal   Jobs

 R 4.1.0 · ~/
> data <-read.table ("water_utility_data.txt", header=TRUE, sep= "\t")
> view(data)
> X = data[c('ENERGY', 'OPEX')]
> Y = data[c('OPREV')]
> conv_dea_score = dea(XREF=X, YREF=Y, X, Y, model="input", RTS="variable")
> conv_dea_score$thetaOpt
 [1] 1.0000000 1.0000000 1.0000000 0.9198125 1.0000000 1.0000000 0.9820593 0.9386396 0.9643813 0.9988634 0.9566218
[12] 0.9715443 0.9233074 0.8577494 0.8937163 1.0000000 0.9602528 0.9643328 0.8950190 0.8794448 1.0000000 0.9697775
>
```

Figure 17.4 Conventional DEA score.

```
RStudio
File  Edit  Code  View  Plots  Session  Build  Debug  Profile  Tools  Help
Go to file/function          Addins

Console   Terminal   Jobs
R 4.1.0 · ~/
> data <-read.table ("water_utility_data.txt", header=TRUE, sep= "\t")
> view(data)
> X = data[c('ENERGY', 'OPEX')]
> Y = data[c('OPREV')]
> conv_dea_score = dea(XREF=X, YREF=Y, X, Y, model="input", RTS="variable")
> conv_dea_score$thetaOpt
 [1] 1.0000000 1.0000000 0.9198125 1.0000000 1.0000000 0.9820593 0.9386396 0.9643813 0.9988634 0.9566218
[12] 0.9715443 0.9233074 0.8577494 0.8937163 1.0000000 0.9602528 0.9643328 0.8950190 0.8794448 1.0000000 0.9697775
> Z = data[c('NOCON','CUSDEN','GROUND', 'LEAKS','PRECIP')]
>
```

Figure 17.5 Loading environmental exogenous variables.

17.3.1 Results and interpretations

The seven utilities with a conventional DEA score of 1 are technically fully efficient compared to the remaining 15 utilities with a conventional DEA score of less than 1. These seven technically efficient districts constitute the best practice frontier. The average of conventional DEA scores is 0.958 with a standard deviation of 0.045. This means that, on average, the utilities can decrease their inputs by 4.2% (i.e., $(1.00-0.958) \times 100$) while keeping their output (their operating revenue in this case) constant. Among all the utilities, the most inefficient utility has a conventional DEA score of 0.858.

To deal with the uncertainty of the conventional DEA scores, bias-corrected DEA scores, as well as 95% confidence intervals, were calculated (i.e. results obtained by delta_hat_hat, delta_ci_low and delta_ci_high functions). The average efficiency decreased to 0.920, which can be translated as an 8% ($0.080 = 1.000 - 0.920$) input reduction requirement while holding output constant. The rankings of the utilities (in terms of efficiency scores) also change after bias correction. This is likely due to some degree of measurement noise in the initial conventional DEA, as later evidenced by the bootstrap.

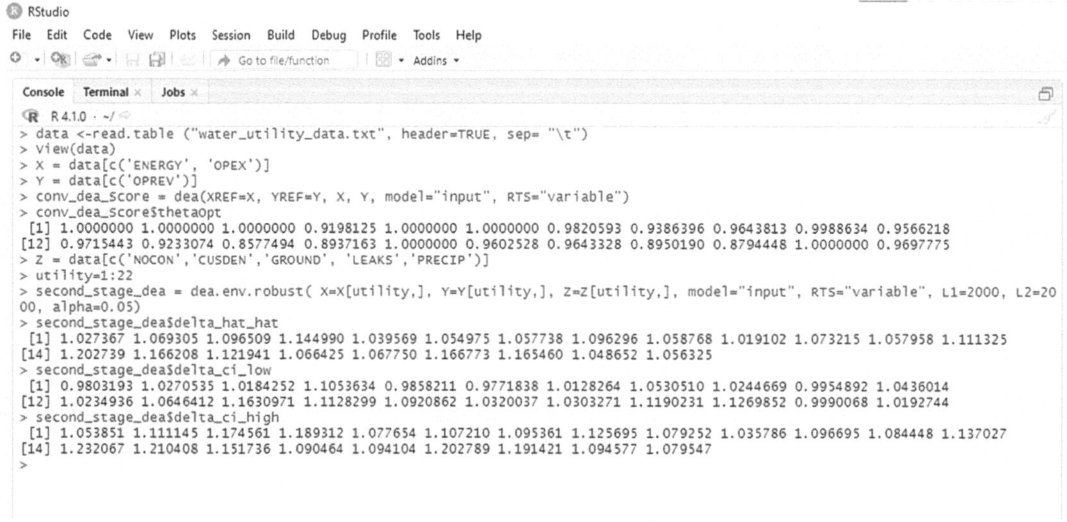

```
RStudio
File  Edit  Code  View  Plots  Session  Build  Debug  Profile  Tools  Help
Go to file/function          Addins

Console   Terminal   Jobs
R 4.1.0 · ~/
> data <-read.table ("water_utility_data.txt", header=TRUE, sep= "\t")
> view(data)
> X = data[c('ENERGY', 'OPEX')]
> Y = data[c('OPREV')]
> conv_dea_score = dea(XREF=X, YREF=Y, X, Y, model="input", RTS="variable")
> conv_dea_score$thetaOpt
 [1] 1.0000000 1.0000000 1.0000000 0.9198125 1.0000000 1.0000000 0.9820593 0.9386396 0.9643813 0.9988634 0.9566218
[12] 0.9715443 0.9233074 0.8577494 0.8937163 1.0000000 0.9602528 0.9643328 0.8950190 0.8794448 1.0000000 0.9697775
> Z = data[c('NOCON','CUSDEN','GROUND', 'LEAKS','PRECIP')]
> utility=1:22
> second_stage_dea = dea.env.robust( X=X[utility,], Y=Y[utility,], Z=Z[utility,], model="input", RTS="variable", L1=2000, L2=2000, alpha=0.05)
> second_stage_dea$delta_hat_hat
 [1] 1.027367 1.069305 1.096509 1.144990 1.039569 1.054975 1.057738 1.096296 1.058768 1.019102 1.073215 1.057958 1.111325
[14] 1.202739 1.166208 1.121941 1.066425 1.067750 1.166773 1.165460 1.048652 1.056325
> second_stage_dea$delta_ci_low
 [1] 0.9803193 1.0270535 1.0184252 1.1053634 0.9858211 0.9771838 1.0128264 1.0530510 1.0244669 0.9954892 1.0436014
[12] 1.0234936 1.0646412 1.1630971 1.1128299 1.0920862 1.0320037 1.0303271 1.1190231 1.1269852 0.9990068 1.0192744
> second_stage_dea$delta_ci_high
 [1] 1.053851 1.111145 1.174561 1.189312 1.077654 1.107210 1.095361 1.125695 1.079252 1.035786 1.096695 1.084448 1.137027
[14] 1.232067 1.210408 1.151736 1.090464 1.094104 1.202789 1.191421 1.094577 1.079547
>
```

Figure 17.6 Bias corrected DEA scores.

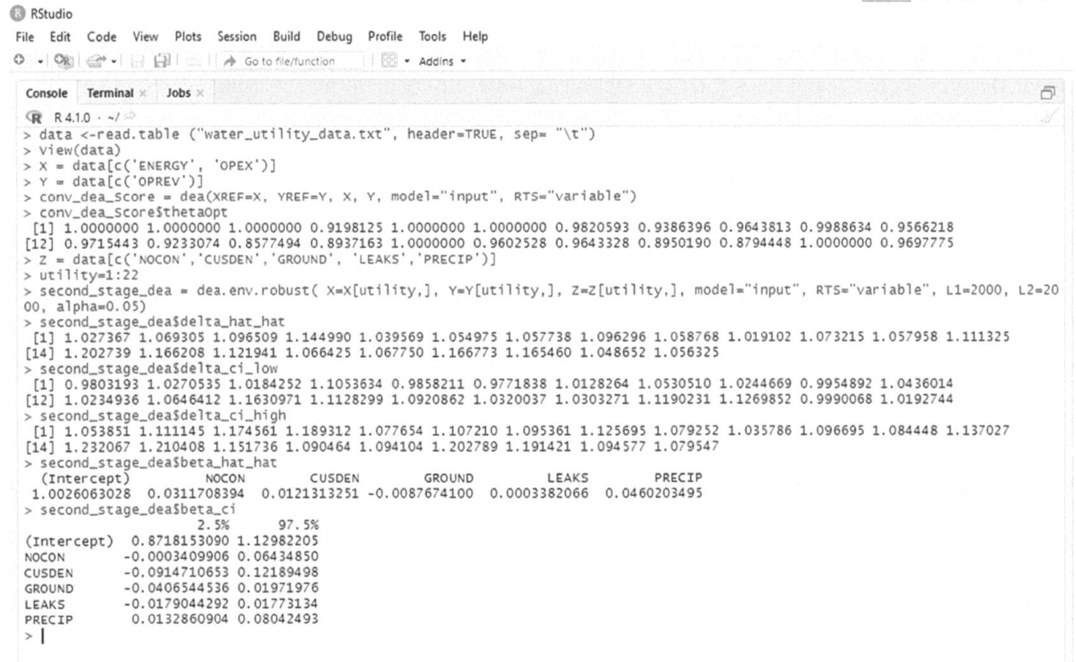

Figure 17.7 Effects of environmental exogenous variables.

Exogenous variables are selected to fully account for the variations in the DEA scores, with the goal being to explain factors that affect the utilities' efficiency that are beyond each water utility's control. The dependent variable used in the regression computations is greater than or equal to 1, so a positive sign for the estimated coefficient indicates a negative correlation. Similarly, a negative sign for the estimated coefficient indicates a positive correlation. The number of connections (NOCON) has a negative impact on efficiency, showing that an increase in the number of connections results in a decrease in efficiency. Precipitation (PRECIP) also has a negative correlation with efficiency. This is likely because a reduction of water used (sales) often accompanies patterns of increased rainfall (wet periods).

17.4 CONCLUSIONS

The findings of these type of studies are expected to be useful in guiding managerial improvement initiatives and actions at water utilities to help operational optimization efforts and provide better service to the utility customers. It is also worth noting that DEA can be very data-intensive and this requires not only analytical skills but also good communication with utility personnel to obtain the quality data which is useful for the analysis.

REFERENCES

Altman D. G. and Bland J. M. (2011). How to obtain the P value from a confidence interval. *British Medical Journal*, **343**, D2304, https://doi.org/10.1136/bmj.d2304

Ananda J. (2014). Evaluating the performance of urban water utilities: robust nonparametric approach. *Journal of Water Resources Planning and Management*, **140**, 04014021, https://doi.org/10.1061/(ASCE) WR.1943-5452.0000387

Coelli T. J., Rao D. S. P., O'Donnell C. J. and Battese G. E. (2005). An Introduction to Efficiency and Productivity Analysis. Springer Science & Business Media, Springer New York, NY. Available at: https://link.springer.com/book/10.1007/b136381 (last accessed 3 March 2022).

Güngör-Demirci G., Lee J. and Keck J. (2017). Measuring water utility performance using nonparametric linear programming. *Civil Engineering and Environmental Systems*, **34**(3–4), 206–220, https://doi.org/10.1080/10286608.2018.1425403

Güngör-Demirci G., Lee J. and Keck J. (2018). Assessing the performance of a California water utility using two-stage data envelopment analysis. *Journal of Water Resources Planning and Management*, **144**(4), 05018004, https://doi.org/10.1061/(ASCE)WR.1943-5452.0000921

Marques R. C., Berg S. and Yane S. (2014). Nonparametric benchmarking of Japanese water utilities: institutional and environmental factors affecting efficiency. *Journal Water Resources Planning & Management*, **140**(5), 562–571, https://doi.org/10.1061/(ASCE)WR.1943-5452.0000366

Mirzaei M., Huang Y. F., El Shafie A., Chimeh T., Lee J., Vaizadeh N. and Adamowski J. (2015). Uncertainty analysis for extreme flood events in a semi-arid region. *Natural Hazards*, **78**, 1947–1960, https://doi.org/10.1007/s11069-015-1812-9

Simm J. and Besstremyannaya G. (2020). *Package 'rDEA'*, Version 1.2-6. https://cran.r-project.org/web/packages/rDEA/index.html (last accessed 1 June 2021).

Wagner J. M. and Shimshak D. G. (2007). Stepwise selection of variables in data envelopment analysis: procedures and managerial perspectives. *European Journal of Operational Research*, **180**(1), 57–67, https://doi.org/10.1016/j.ejor.2006.02.048

Index

IWA Publishing's authorised EU representative for General Product Safety Regulations is Diane D'Arras, 15 rue Duret, 75116 Paris, France, e-mail: safety@iwap.co.uk.

Printed and bound by CPI Group (UK) Ltd, Croydon, CR0 4YY

12/05/2026

02108863-0012